DYNAMICS, EPHEMERIDES AND ASTROMETRY
OF THE SOLAR SYSTEM

INTERNATIONAL ASTRONOMICAL UNION

UNION ASTRONOMIQUE INTERNATIONALE

DYNAMICS, EPHEMERIDES AND ASTROMETRY OF THE SOLAR SYSTEM

PROCEEDINGS OF THE 172ND SYMPOSIUM OF THE
INTERNATIONAL ASTRONOMICAL UNION,
HELD IN PARIS, FRANCE, 3–8 JULY 1995

EDITED BY

S. FERRAZ-MELLO

*Instituto Astronómico e Geofísico,
Universidade de Sao Paulo, Brazil*

B. MORANDO and J.-E. ARLOT

Bureau des Longitudes, Paris, France

KLUWER ACADEMIC PUBLISHERS

DORDRECHT / BOSTON / LONDON

Library of Congress Cataloging-in-Publication Data

```
International Astronomical Union. Symposium (172nd : 1995 : Paris,
France)
  Dynamics, ephemerides, and astrometry of the solar system :
proceedings of the 172nd Symposium of the International Astronomical
Union, held in Paris, France, 3-8 July, 1995 / edited by S. Ferraz
-Mello, B. Morando and J.-E. Arlot.
     p.   cm.
  At head of title: International Astronomical Union.
  Includes index.
  ISBN 0-7923-4084-1
  1. Solar system--Congresses.    I. Ferraz-Mello, Sylvio.
II. Morando, Bruno.    III. Arlot, Jean-Eudes.    IV. Title.
QB500.5.I58   1995
523.2--dc20                                                96-2943
```
ISBN 0-7923-4084-1 (HB)

Published on behalf of
the International Astronomical Union
by
Kluwer Academic Publishers, P.O. Box 17, 3300 AA Dordrecht, The Netherlands.

Kluwer Academic Publishers incorporates
the publishing programmes of
D. Reidel, Martinus Nijhoff, Dr W. Junk and MTP Press.

Sold and distributed in the U.S.A. and Canada
by Kluwer Academic Publishers,
101 Philip Drive, Norwell, MA 02061, U.S.A.

In all other countries, sold and distributed
by Kluwer Academic Publishers Group,
P.O. Box 322, 3300 AH Dordrecht, The Netherlands.

Printed on acid-free paper

All Rights Reserved
©1996 International Astronomical Union

No part of the material protected by this copyright notice may be reproduced or utilized in any form or by any means, electronic or mechanical including photocopying, recording or by any information storage and retrieval system, without written permission from the publisher.

Printed in the Netherlands

TABLE OF CONTENTS

Introduction xi
List of Participants xiii

PART I - INTRODUCTION

Bruno Morando (1931-1995) 1
B. MORANDO – Deux cents ans de Mécanique Céleste sous les auspices du Bureau des Longitudes (*Lecture*) 3

PART II - PLANETS AND MOON: THEORY AND EPHEMERIDES

P. BRETAGNON – Analytical solution of the motion of the planets over several thousands of years. (*Lecture*) 17
E. M. STANDISH, Jr. and X X NEWHALL – New accuracy levels for solar system ephemerides. (*Lecture*) 29
X X NEWHALL, E. M. STANDISH, Jr. and J. G. WILLIAMS – Planetary and lunar ephemerides, lunar laser ranging and lunar physical librations. (*Lecture*) 37
E. V. PITJEVA – Using spacecraft range data and radar observations for the improvement of the orbital elements of planets and parameters of Mars rotation. 45
J. L. SIMON – New semi-analytic theory of the four outer planets 49
J. R. DONNISON and D. F. MIKULSKIS – The orbital stability of the Sun-Jupiter-Saturn system. 53
M. GHIL, F. VARADI and W. M. KAULA – On the secular motion of the jovian planets. 57
H. KINOSHITA and H. NAKAI – The motion of Pluto. (*Lecture*) 61
R. DVORAK and E. LOHINGER – Pluto's Lyapunov numbers in different dynamical models. 71
J. LASKAR – Marginal stability and chaos in the solar system (*Lecture*) 75
V. A. BRUMBERG – Theory compression with elliptic functions. (*Lecture*) 89
V. A. BRUMBERG and S. A. KLIONER – Numerical efficiency of the elliptic function expansions of first-order intermediary for general planetary theory. 101

J. F. CHANDLER – Pulsars and solar-system ephemerides. (*Lecture*) 105

K. D. PANG and K. K. YAU – The need for more accurate 4000–year ephemerides using lunar and spacecraft ranging, ancient eclipse and planetary data. 113

PART III - SATELLITES: THEORY AND EPHEMERIDES

L. DURIEZ – Reviewing the theories of motion of the satellites of Saturn. (*Lecture*) 117

P. J. MESSAGE – Foundations of a theory of the motion of the orbit plane of Hyperion. 127

N. O. KIRSANOV – The motion of Hyperion: On the accuracy of the observations. 137

K. X. SHEN and R. QIAO – Redetermination of the orbit of Iapetus. 141

A. VIENNE, L. DURIEZ and S. CHAMPENOIS – The Mimas–Tethys and Enceladus–Dione systems 143

R. VASUNDHARA, J.-E. ARLOT and P. DESCAMPS – New constants for Sampson–Lieske theory of the Galilean satellites of Jupiter from mutual occultation data. 145

K. G. HADJIFOTINOU and D. HARPER – The effect of step-size on the numerical integration of satellite orbits. 151

PART IV - ASTEROIDS: THEORY AND EPHEMERIDES

B. G. MARSDEN – From telescope to MPC: Organizing the minor planets. (*Lecture*) 153

A. LEMAITRE – Numerical mean elements for asteroid orbits. 165

B. ERDI – On the dynamics of Trojan asteroids. 171

S. FERRAZ-MELLO – On the Hecuba gap. 177

I. I. SHEVCHENKO and H. SCHOLL – Chaotic asteroidal trajectories exhibiting multiple bursts of eccentricity: A statistical analysis. 183

N. A. SOLOVAYA and E. M. PITTICH – Orbital stability of high inclination asteroids. 187

M. E. SANSATURIO, A. MILANI and L. CATTANEO – Nonlinear optimisation and the asteroid identification problem. 193

F. J. MARCO, J. A. LOPEZ and M. J. MARTINEZ – A time-dependent extension to Brouwer's method for orbital elements correction. ... 199

M. CARPINO and Z. KNEŽEVIĆ – Asteroid mass determination: (1) Ceres. 203
M. KUZMANOSKI – A method for asteroid mass determination. 207

PART V - COMETS AND METEORS

J. Q. ZHENG, M. J. VALTONEN, S. MIKKOLA and H. RICKMAN – Dynamics and orbital evolution of Oort cloud comets. 209
V. BATLLO – Hypothetical evolution of short-period comets. 213
E. M. PITTICH and J. KLAČKA – Taurid meteoroids and asteroid complex. 215
D. A. ANDRIENKO and I. I. MISHCHISHINA – Model of the comet outbursts based on a fragmentation of ice grains in its atmosphere. ... 221
R. S. GOMES and A. Y. MIGUELOTE – A numerical analysis of the process of capture into resonance. 223
B. J. GLADMAN, J. A. BURNS, H. LEVISON and M. J. DUNCAN – Ejecta transfer between terrestrial planets. 229

PART VI - EARTH AND DEFORMABLE CELESTIAL BODIES

J. GETINO and J. M. FERRANDIZ – Canonical treatment of dissipative forces between Earth mantle and core. 233
J. SOUCHAY – New series of rigid and non rigid Earth nutation. Comparison with observations. 239
Yu. V. BARKIN, J. M. FERRANDIZ and J. GETINO – About the application of action–angle variables to the rotation of deformable celestial bodies. 243
T. FUKUSHIMA – A numerical scheme to integrate the rotational motion of a rigid body. 245
R. MOLINA and A. VIGUERAS – Analytical integration of a generalized Euler-Poinsot problem: Applications. 249
V. SHKODROV and V. IVANOVA – On the tidal function between two real bodies. 251

PART VII - THE CALCULUS OF PERTURBATIONS

J. D. HADJIDEMETRIOU – Symplectic mappings. (*Lecture*) 255
A. DÉPRIT – L'algèbre symbolique en mécanique céleste. (*Lecture*) .. 267

T. V. IVANOVA – PSP: A new Poisson series processor. 283
E. A. GREBENIKOV – On the generators of new methods of perturbation theory. 285
Y. A. RYABOV – Analytic-numerical solutions of restricted non-resonance planar three-body problem. 289
C. FROESCHLÉ, A. GIORGILLI, E. LEGA and A. MORBIDELLI – On the measure of the structure around an invariant KAM torus. ... 293
L. FLORIA – On two–parameter linearizing transformations for uniform treatment of two–body motion. 299

PART VIII - GENERAL RELATIVITY. PHYSICS

M. H. SOFFEL – Relativistic equations of motion of celestial bodies. (*Lecture*) 303
S. A. KLIONER – Angular velocity of rotation of extended bodies in general relativity. (*Lecture*) 309
D. VOKROUHLICKÝ – Relativistic rotational effects: application to the Earth–Moon system. 321
J. M. GAMBI, P. ZAMORANO, P. ROMERO and M. L. GARCIA DEL PINO – On the orbital motion description with a proper reference frame in a complete Schwarzschild field. 325
A. V. VITYAZEV and A. G. BASHKIROV – Dynamical screening of interactions in gravitating systems and the ephemeris time. 327

PART IX - EPHEMERIDES REPRESENTATION

P. K. SEIDELMANN – Evolution of ephemerides representation and diffusion. (*Lecture*) 331
S. DÉBARBAT – Des éphémérides astronomiques annuelles en préliminaire à l'Annuaire du Bureau des Longitudes. 339
J. C. COMA, M. LARA and T. J. LÓPEZ MORATALLA – Fast evaluation of ephemerides by polynomial approximation in the Chebyshev norm. 345
A. A. TRUBITSINA – Experience of numerical integration and approximation with applying Chebyshev polynomials for constructing ephemerides of the solar system natural and artificial bodies. 347
Yu. CHERNETENKO, V. L'VOV, V. SHOR, R. SMEKHACHEVA and S. TSEKMEJSTER – Controlling the observational data on minor planets with the "CERES" software package. 353
N. V. EMELIANOV – Ephemerides software of natural satellites. 355

P. ROCHER and C. CAVELIER – Production d'éléments orbitaux de comètes sur PC. 357

M. TOULMONDE – Les diamètres du Soleil dans la Connaissance des temps depuis 1795. 361

PART X - SOLAR SYSTEM ASTROMETRY

S. J. OSTRO – Radar astrometry of asteroids, comets and planetary satellites. (*Lecture*) 365

D. PASCU – Long-focus CCD astrometry of planetary satellites. (*Lecture*) 373

J. V. SCOTTI and R. JEDICKE – Observation of small solar system objects with spacewatch. (*Lecture*) 389

L. V. MORRISON and M. E. BUONTEMPO – Carlsberg optical astrometry of the outer solar system. 399

A. S. KHARIN, Yu. B. KOLESNIK and O. E. CHUICHENKO – Accuracy estimation of new sets of the Sun and Planets observations. ... 407

P. J. SHELUS, R. L. RICKLEFS, J. G. RIES, A. L. WHIPPLE and J. R. WIANT – McDonald observatory lunar laser ranging: Beginning the second 25 years. 409

V. N. YERSHOV – Observation of the solar system bodies on board the future russian astrometric satellite. 415

R. VIEIRA MARTINS, C. H. VEIGA and M. ASSAFIN – Astrometric observations of faint satellites. 419

J.-E. ARLOT, W. THUILLOT, F. COLAS, P. DESCAMPS, J. BERTHIER, B. MORANDO, C. H. VEIGA, D. T. VU, Ch. RUATTI, J. LECACHEUX and P. LAQUES – The PHESAT95 campaign of observations of phenomena of the saturnian satellites. 423

T. P. KISSELEVA – The results of photographic observations of Galilean satellites with 26-inch refractor at Pulkovo. 427

D. H. P. JONES – Limitations on the accuracy possible in astrometric observations of the satellites of the major planets. 431

G. A. GONCHAROV – CCD astrometry oriented analysis of digitized multiple images (planets, satellites, etc.) obtained with the Pulkovo PVC. 435

D. HESTROFFER and B. MORANDO – Solar system objects observed by Hipparcos. 437

L. K. KRISTENSEN – Astrometry by asteroid occultations. 443

M. -L. BOUGEARD, J. -F. BANGE and A. BEC-BORSENBERGER – Hipparcos minor planets: Towards an improvement of the model analysis by detecting influence factors 447

O. P. BYKOV – CCD–observations of asteroids: Accuracy and the nearest perspectives for application of the Laplacian orbit determination method. 451

A. M. SVESHNIKOV and M. L. SVESHNIKOV – The "black drop" phenomenon and reduction of the Mercury transit observations. 453

PART XI - REFERENCE FRAMES. STELLAR ASTROMETRY

J. KOVALEVSKY and M. FEISSEL – Reference Frames (*Lecture*) ... 455

T. FUKUSHIMA – IAU standards - its future. (*Lecture*) 461

N. V. SHUYGINA and E. I. YAGUDINA – The FK5 equinox and equator from combined radar and optical data of the near–earth asteroids. 469

H. G. WALTER – Effects of precession uncertainties on planetary ephemerides. 475

Yu. B. KOLESNIK – Residual rotation of the FK5 system from optical observations of the Sun and planets 1960-1994. 477

S. RÖSER – An updated GSC as astrometric reference for minor planet observations. 481

A. LOPEZ GARCIA and L. I. YAGUDIN – Corrected GSC: Catalogue for CCD observations. 483

A. LOPEZ GARCIA, L. I. YAGUDIN and M. J. MARTINEZ – Application of GSC catalogue to minor planets plate reduction. 487

G. DAMLJANOVIĆ – The Hipparcos mission and the re-reduction of Belgrad zenith-telescope observations. 489

J. VONDRAK – Indirect linking of the Hipparcos catalog to extragalactic reference frame via Earth orientation parameters. 491

V. I. ALTUNIN, V. A. ALEXEEV, E. L. AKIM, T. M. EUBANKS, K. A. KINGHAM, R. N. TREUHAFT and K. G. SUKHANOV – Astrometry VLBI in space (AVS). 497

N. V. LEISTER, P. C. R. POPPE, M. EMILIO and F. LACLARE – Analysis of the Sun's observations with prismatic astrolabe. 501

Yu. B. KOLESNIK – Three campaigns of solar observations with an astrolabe at Simeiz. 505

Index of authors. 507

INTRODUCTION

This symposium is devoted to the oldest of the branches of Celestial Mechanics: the precise determination and prediction of the positions and movements of Solar System bodies. For short, the problem of the construction of ephemerides *in strictu sensu*. Since the antiquity, many human activities rely on ephemerides and the increasing request of precision has characterized this discipline throughout the time, with the use of the most advanced techniques. The past is so full of great achievements that we are always tempted to focus on them. But the scientific organizers of this Symposium decided to look rather forward and to emphasize those subjects which are expected to be at the center of the developments in the first decades of the next century.

The problem of ephemerides construction cannot be solved without a great deal of theory and observations. The existence of this scientific activity since the antiquity led to many popular misconceptions. Many people, even inside the astronomical community, still believe that Newton's laws with a pinch of relativistic correction in the perihelion motions is the current state of the art in Celestial Mechanics. In reality, however, it is no longer possible to satisfy all accuracy needs without a full consideration of Einstein's general relativity. Moreover, some particular observations are so accurate that investigations on the correctness of some of Einstein's basic assumptions are in progress. Another popular belief concerns the predictability of celestial motions. It is sometimes called the Laplace demon, since the science of Laplace's time was founded on the idea of full predictability of celestial motions. However, Poincaré's work on Celestial Mechanics has shown that chaotic behaviour is the rule in complex deterministic systems. Any deviation grows exponentially and ephemerides valid forever are not possible. This branch of Celestial Mechanics requests precise Physics, precise Mathematics and, above all, precise observations. The next decades are promising. The amount and accuracy of the available observational data is growing at an exponential pace. Modern ephemerides already rely on data obtained by radar and laser ranging, VLBI, VLA, space astrometry, millisecond pulsar timings and accurate CCD ground-based observations.

This Symposium was held under the invitation of the Bureau des Longitudes, Paris, to commemorate its bicentennial. Bureau des Longitudes is

COLAS François, Bureau des longitudes, France
COMA Juan C., Real Instituto y Observatorio de la Armada, Spain
DAMLJANOVIC Goran, Astronomical Observatory Belgrade, Yugoslavia
DEBARBAT Suzanne, Observatoire de Paris, France
DELMAS Christian, O.C.A./CERGA, France
DEPRIT André, National Institute of Standards and Technology, U.S.A.
DESCAMPS Pascal, Bureau des longitudes, France
DOGGETT Leroy, U.S. Naval Observatory, Washington, U.S.A.
DONNISON J.R., Goldsmiths College, University of London, U.K.
DOURNEAU Gérard, Observatoire de Bordeaux, France
DURIEZ Luc, Laboratoire d'Astronomie Université de Lille, France
DVORAK Rudolf, Institut für Astronomie, Austria
DYBCZYNSKI Piotr A., Astronomical Observatory, A. Mickiewicz University, Poland
EDELMAN Colette, Bureau des longitudes, France
EL BAKKALI Larbi, Dept of Physics, Faculty of Sciences, Morocco
ELIPE Antonio, Universidad de Zaragoza, Spain
ELMABSOUT Badaoui, Laboratoire des Systèmes Dynamiques U.F.R. de Mécanique, France
EMELIANOV Nicolai, Sternberg Astronomical Institute, Moscow, Russia
ERDI Balint, Department of Astronomy, Eotvos University, Hungary
EUBANKS Marshall, U.S. Naval Observatory, U.S.A.
FAIDIT Jean Michel, Planetarium, Université Montpellier III, France
FERNANDEZ Silvia, Observatorio Astronomico Córdoba, Argentina
FERRAZ-MELLO Sylvio, Instituto Astronomico e Geofisico, Universidade de São Paulo, Brasil
FLORIA Luis, Dept. de Mat. Aplicada a la Ingenieria, Universidad de Valladolid, Spain
FOLGUEIRA LOPEZ Marta, Instituto de Astronomia y Geodesia, Spain
FROESCHLÉ Claude, Observatoire de la Cote d'Azur, Nice, France
FUJISHITA Mitsumi, Kyushu Tokai University, Japan
FUKUSHIMA Toshio, Jet Propulsion Laboratory, MS 301-150, U.S.A.
GAMBI Jose M., Dep. Ingenieria, Universidad Carlos III, Spain
GETINO Juan, Dept. Matematica Applicada Fundamental, Universidad de Valladolid, Spain
GHIL Michael, Dept. of Atmospheric Sciences, University of California, U.S.A.
GLADMAN Brett, Cornell University, U.S.A.
GOFFIN Edwin, Hoboken, Belgium
GOMES Rodney S., Observatorio Nacional, Brasil
GONCHAROV George A., Pulkovo Observatory, Russia
GREBENIKOV Eugeniu, Institut des Super-Ordinateurs/R.A.S., Russia
GUSEVA Irina S., Pulkovo Observatory, Russia

HADJIDEMETRIOU John D., University of Thessaloniki, Greece
HADJIFOTINOU Katerina, Department of Mathematics, University of Thessaloniki, Greece
HARPER David, Astronomy Unit, Queen Mary and Westfield College, University of London, U.K.
HEMENWAY Paul D., Astronomy Department, University of Texas, U.S.A.
HENRARD Jacques, Departement de Mathematique, FUNDP, Belgium
HESTROFFER Daniel, ESTEC Keplerlaan 1, postbus 299, Netherlands
HILTON James, U.S. Naval Observatory, U.S.A.
HOHENKERK Catherine, Royal Greenwich Observatory, U.K.
ICHIKAWA Yuichi, Kyushu Tokai University, Japan
ICHTIAROGLOU Simos, University of Thessaloniki, Greece
INOUE Takeshi, Kyoto Sangyo University, Japan
IPATOV Sergei, Keldysh Institute of Applied Mathematics, R. A. S., Russia
IVANOVA Tamara V., Institute for Theoretical Astronomy, Russia
JANCART Sylvie, Fondation Biermans Lapotre, Ch 200 A, France
JEWITT David, IFA, Honolulu, U.S.A.
JONES Derek, Royal Greenwich Observatory, U.K.
KASPER Johannes N., Institut für Astronomie, Universität Wien, Austria
KHOLSCHEVNIKOV K. K., Astronomical Institute, St Petersburg University, Russia
KINOSHITA Hiroshi, National Astronomical Observatory, Japan
KIRSANOV Nikolai O., Institute for Theoretical Astronomy, Russia
KISSELEV, A. A., The Main Astronomical Observatory, R.A.S., Russia
KISSELEVA Tamara P.,The Main Astronomical Observatory, R.A.S., Russia
KLIONER Sergei A., Institute of Applied Astronomy, Russia
KLOKATCHEVA Ju., Institute for Theoretical Astronomy, Russia
KLOKOCNIK Jaroslav, Astronomical Institute, Č.A.S., Czech Republic
KNEŽEVIĆ Zoran, Astronomical Observatory, Belgrade, Yugoslavia
KOLESNIK Yuri, Institute of Astronomy, Moscow, Russia
KONACKI Maciej, Institute of Astronomy, Nicolaus Copernicus University, Poland
KOSEK Wieslaw, Space Research Centre, PAS, Poland
KOVALEVSKY Jean, Observatoire de la Cote d'Azur, CERGA, France
KOZAI Yoshihide, National Astronomical Observatory, Japan
KRISTENSEN Leif Kahl,Institute of Physics and Astronomy, University of Aarhus, Danemark
KUMAR Shiv S, University of Virginia, U.S.A.
KURZYNSKA Krystyna, Astronomical Observatory, A. Mickiewicz University, Poland
KUZMANOVSKI, Mathematical Faculty, Yugoslavia
LARA-COIRA Martine, Real Instituto y Observatorio de la Armada, Spain

LASKAR Jacques, Bureau des Longitudes, France
LE FLOCH J.C., Clichy, France
LE GUYADER Claude, Bureau des longitudes, France
LEISTER Nelson Vani, Instituto Astronomico e Geofisico Universidade de São Paulo, Brasil
LEMAITRE Anne, Departement de Mathématiques, FUNDP, Belgium
LIESKE Jay H., Jet Propulsion Laboratory, U.S.A.
LOHINGER Elke, Institut für Astronomie, Universität Wien, Austria
LOPEZ MORATALLA Teodoro, Real Instituto y Observatorio de la Armada, Spain
LOPEZ-GARCIA Alvaro, Astronomical Observatory, Valencia University, Spain
MAKHLOUF Amar, Université d'Annaba, Algeria
MANARA Alessandro, Osservatorio Astronomico di Brera, Italia
MARCO-CASTILLO Francisco Jose, Departament de Matematiques, Universitat Jaume I, Spain
MARSDEN Brian G., Harvard Smithsonian Center for Astrophysics, U.S.A.
MARTINEZ-USO Maria Jose, Departament de Matematiques, Universitat Jaume I, Spain
McHUGH Martin, Observatoire de Besançon, France
MESSAGE P. J., Dept. Applied Mathematics and Theoretical Physics, University of Liverpool, U.K.
MICHEL Patrick, Observatoire de la Cote d'Azur, France
MIGNARD François, O.C.A./CERGA, France
MILANI Andrea, Dipartimento di Matematica, Universita di Pisa, Italia
MIOC Vasile, Astronomical Institue, Astronomical Observatory Cluj-Napoca, Romania
MIZONY Michel, Université Claude Bernard Lyon I, France
MOISSON Xavier, Paris, France
MOLINA-LEGAZ Roque, Escuela Politecnica Superior, Spain
MOONS M., Department de Mathématique, FUNDP, Belgium
MORANDO B., Bureau des Longitudes, France
MORRISON L.V., Royal Greenwich Observatory, U.K.
NERON DE SURGY Olivier, A.S.D./Bureau des longitudes, France
NEWHALL X X, Jet Propulsion Laboratory, U.S.A.
OPRESCU Gabriela, Institutul Astronomic, Romania
PAKVOR Ivan, Astronomical Observatory of Belgrade, Yugoslavia
PANG Kevin D., La Canada, U.S.A.
PASCU Dan, U.S. Naval Observatory, U.S.A.
PINHEIRO Mery, Marseille, France
PITJEVA Elena V.,Institute of Applied Astronomy, R. A. S., Russia
PITTICH Eduard M., Astronomical Institute, S. A. S., Slovak Republic

LIST OF PARTICIPANTS

PRETKA Halina, Astronomical Observatory, A. Mickiewicz University,Poland
PROTITCH-BENISHEK V., Astronomical Observatory, Belgrade, Yugoslavia
QI Guan-Rong, Shaanxi Astronomical Observatory, China
QIAN Zhihan, Shanghai Observatory, China
RAFFERTY Theodore J., U. S. Naval Observatory, U.S.A.
RAPAPORT Michel, Observatoire de Bordeaux, France
REQUIÈME Yves, Observatoire de Bordeaux, France
ROCHER Patrick, Bureau des Longitudes, France
ROMERO PEREZ, Inst. de Astronomia y Geodesia, Facultad de Ciencias Matemáticas, Spain
RÖSER Siegfried, Astronomisches Rechen-Institut, Germany
RYABOV Yu.A., Moscow Auto and Road Construction Engineering University, Mathematics Dept, Russia
SAGNIER Jean-Louis, Bureau des longitudes, France
SANSATURIO Maria Eugenia, E.T.S. Ingenieros Industriales, University of Valladolid, Spain
SCHOLL H., Observatoire de Nice, France
SCOTTI James V., University of Arizona, U.S.A.
SEHNAL Ladislav, Astronomical Institut, Č.A.S., Czech Republic
SEIDELMANN P. Kenneth, U.S. Naval Observatory, U.S.A.
SEKIGUCHI Naosuke, National Astronomical Observatory, Japan
SHELUS Peter J., McDonald Observatory, U.S.A.
SHEN Kaixian, Shaanxi Observatory, China
SHEVCHENKO Ivan I., Institute for Theoretical Astronomy, Russia
SHKODROV Vladimir, Insitute of Astronomy, B.A.S., Bulgaria
SHOR V. A., Institute for Theoretical Astronomy, Russia
SHUYGINA Nadezhda, Institute of Applied Astronomy, Russia
SIMON Jean-Louis, Bureau des Longitudes, France
SMEKHACHEVA R.I., Institute for Theoretical Astronomy, Russia
SOFFEL Michael, Theor. Astrophys., Uni.Tubingen, Germany
SOKOLOV Leonid L., Astronomical Institute, St Petersburg University, Russia
SOKOLSKY Andrej, Institute for Theoretical Astronomy, Russia
SOLOVAYA Nina A., Sternberg State Astronomical Institute, Russia
SOMA Mitsuru, National Astronomical Observatory, Japan
SOUCHAY Jean, DANOF, Observatoire de Paris, France
STANDISH E. Myles, Jr., Jet Propulsion Laboratory, U.S.A.
STELLMACHER Irène, Bureau des Longitudes, France
STOEV Alexey, People's Astron. Observatory and Planetaria Yuri Gagarin, Bulgaria
SVESHNIKOV Michail, Institute for Theoretical Astronomy, Russia
TANGA Paolo, Instituto di Cosmogeofisica del CNR, Italia
TAYLOR D. B., Royal Greenwich Observatory, U.K.

THOLEN David, Instute for Astronomy, U.S.A.
THOMAS Fabrice, Observatoire de la Cote d'Azur, France
THUILLOT William, Bureau des longitudes, France
TICHA Jana, Klet Observatory, Czech Republic
TICHY Milos, Klet Observatory, Czech Republic
TOULMONDE Michel, DANOF, Observatoire de Paris, France
TRUBITSINA Anna, Institute for Theoretical Astronomy, Russia
TUPIKOVA Irina V., Institute for Theoretical Astronomy, Russia
URAS S., Cagliari Astronomical Observatory, CNR, Italia
VASILIEV Nikolai, Institute for Theoretical Astronomy, Russia
VASUNDHARA Raju, Indian Institute of Astrophysics, India
VEIGA Carlos, Observatorio Nacional, Brasil
VERNOTTE François, Observatoire de Besançon, France
VIATEAU B., Observatoire de Bordeaux, France
VIEIRA-MARTINS Roberto, Dep. Astronomia, Observatorio Nacional, Brasil
VIENNE Alain, Laboratoire d'Astronomie, Lille, France
VIGUERAS CAMPUZANO Antonio,, Depto. de Matematica Aplicada y Estadistica, Universidad de Murcia Spain
VOKROUHLICKY David, Astronomical Institute, Charles University Prague, Czech Republic
VONDRAK Jan, Astronomical Institute, Č. A. S., Czech Republic
VU D.T., Bureau des longitudes, France
WALTER Hans G., Astronomisches Rechen-Institut, Germany
WNUK Edwin, Astronomical Observatory, A. Mickiewicz University, Poland
WYTRZYSZCZAK Iwona, Astronomical Observatory A. Mickiewicz University, Poland
YAGUDIN Leonid, Astron. Obs., Valencia University, Spain
YAGUDINA Eleonora I., Astron. Obs., Valencia University, Spain
YALLOP Bernard D., Royal Greenwich Observatory, U.K.
YAO Zhen-Guo, Astrometry Department, U.S. Naval Observatory, U.S.A.
YERSHOV Vladimir, Pulkovo Observatory, Russia
YOSHIDA Haruo, National Astronomical Observatory, Japan
YOSHIKAWA Makoto, Communications Research Laboratory, Japan
ZALAMANSKY Gilles, Observatoire de Besançon, France
ZECH Gert, Astronomisches Rechen Institut, Germany
ZHENG Jia-Qing, Tuorla Observatory, University of Turku, Finland
ZHENG Xue-Tang,Nanjing University of Science and Technology, China

BRUNO MORANDO (1931-1995)

Bruno Morando est né en 1931 à Courbevoie, près de Paris. Après des études à la Faculté des Sciences de Paris, il est nommé assistant à la Faculté en 1960 et, en 1963, il entre au Bureau des longitudes, où il effectuera sa carrière d'astronome. Il soutient sa thèse en 1966 sur le thème:"Théorie planétaire semi-numérique sans introduction de termes séculaires; application à Vesta". Grâce à André Danjon, Jean Kovalevsky et Bruno Morando vont transformer le Service des calculs du Bureau des longitudes en un laboratoire de recherche en mécanique céleste mondialement réputé. Il en sera le directeur de 1971 à 1984.

Durant cette période, Bruno Morando s'est attaché à soutenir les initiatives des jeunes chercheurs, leur permettant d'introduire de nouveaux thèmes de recherche au Bureau des longitudes, thèmes auxquels il participera, telles les campagnes d'observation des phénomènes mutuels des satellites des planètes. Il s'est également investi en histoire des sciences et a notamment travaillé à la publication des oeuvres de D'Alembert. Depuis le début, il fit partie du projet Hipparcos et apporta une contribution importante à la réduction des observations d'astéroïdes grâce à la modélisation des courbes de modulation photométriques de ces objets non ponctuels. Enfin, il faut noter les grandes qualités d'enseignant de Bruno Morando qui donna des cours dès 1966, d'abord à la Faculté des Sciences de Paris, puis à l'Université Pierre et Marie Curie et enfin dans le cadre des enseignements du Diplôme d'Études Approfondies de l'Observatoire de Paris.

Mécanicien céleste de réputation internationale, Bruno Morando eut le plaisir de voir l'UAI donner son nom à l'astéroïde 1931 FC découvert à Heidelberg l'année de sa naissance. L'UAI doit beaucoup à Bruno Morando: membre de l'Union depuis 1964, membre des comités d'organisation des commissions 4 (éphémérides), 7 (mécanique céleste) et 20 (positions et mouvements des comètes, astéroïdes et satellites des planètes).

Il participa activement aux travaux de l'Union. Il présida la commission 4, de 1985 à 1988, et les working groups "orbites et éphémérides des satellites naturels" et "constantes astronomiques". Par ailleurs, il s'attacha toute sa vie à vulgariser l'astronomie, à donner de nombreuses conférences et à guider les jeunes dans l'observation du ciel. C'était un homme d'une grande culture, amateur de littérature, de musique et d'art. Tous ceux qui l'ont connu garderont le souvenir d'un homme exceptionnel, d'une gentillesse infinie et d'une grande ouverture d'esprit.

J.-E. ARLOT AND S. FERRAZ-MELLO

DEUX CENTS ANS DE MÉCANIQUE CÉLESTE SOUS LES AUSPICES DU BUREAU DES LONGITUDES

BRUNO MORANDO
Bureau des longitudes
Paris, France

Abstract. Two hundred years of Celestial Mechanics under the auspices of Bureau des Longitudes. Lagrange and Laplace were two of the first members of Bureau des longitudes which, among other tasks, were responsible for the improvement of astronomical tables and the progress of celestial mechanics. Between 1795 and 1850, many improved tables were published under the auspices of Bureau des longitudes: tables of the Sun by Delambre (1806), of the Moon by Burg (1806), Burckhardt (1812) and Damoiseau (1828), of Jupiter, Saturn and Uranus by Bouvard (1808, 1821), of Mercury by Le Verrier (1844), of the satellites of Jupiter by Delambre (1817) and Damoiseau (1836). In his tables, Bouvard showed there was a problem for Uranus. This led to the calculations of the elements of an unknown planet by Le Verrier and Adams and the discovery of Neptune in 1846. Le Verrier's calculations were published in *Connaissance des Temps* for 1849. In the second half of the XIXth century, two prominent members of Bureau des longitudes, Le Verrier and Delaunay, made major contributions to celestial mechanics by building elaborate theories for the motions of the Sun, the planets and the Moon. Other theories, which improved the above, appeared elsewhere at the end of the century, especially those of Newcomb, Hill and Brown. During the first half of the XXth century, there was a decline of the studies in celestial mechanics which seemed to have reached its limits owing to the difficulties of the computations involved. Yet Sampson's theory of the motion of the satellites of Jupiter and Chazy's first attempts to introduce general relativity into classical celestial mechanics should be quoted. In 1961, thanks to A. Danjon, Bureau des longitudes was reorganized so that its computation service became a research laboratory where, since then, important work in the theories of the planets, the Moon and the satellites has been made.

Après la mort de Newton, rien de concret n'a été fait pendant assez longtemps dans le domaine de la mécanique céleste. Les savants de l'époque dominaient encore trop mal le calcul intégral, mais un développement rapide se fera à partir de 1740. C'est un suisse, Euler, et deux français, Clairaut et d'Alembert, qui, les premiers, réussissent à mettre en équation le problème du mouvement d'un corps perturbé par un autre et à en trouver des solutions à l'aide des séries trigonométriques que les travaux d'Euler avaient rendues utilisables. Citons les travaux de Clairaut sur la figure de la Terre et le mouvement de la Lune (1752) ainsi que ses calculs de prédictions du retour de la comète de Halley pour 1759. En 1754, D'Alembert publie sa propre théorie de la Lune. Un autre groupe de savants français, Lagrange et Laplace vont, vingt ans plus tard, faire faire des progrès décisifs à la mécanique céleste. Ces deux astronomes furent tous deux membres fondateurs du Bureau des longitudes et commencent donc la période qui nous occupe.

Joseph-Louis Lagrange (1736-1813) est né à Turin d'une famille d'origine française. Il fonde en 1759 l'Académie de Turin. Il invente le calcul des variations et Euler le fait rentrer à l'Académie de Berlin où il est invité par Frédéric II en 1766. Après la mort du roi de Prusse en 1786, Lagrange est invité à Paris où il restera jusqu'à sa mort. Les travaux de Lagrange en mécanique céleste portent principalement sur son étude de certaines solutions du problème des 3 corps (découverte des *"points de Lagrange"*, sommets de triangles équilatéraux occupés par les 3 corps) et sur la méthode de la variation des constantes qui consiste à faire varier les paramètres caractérisant une orbite elliptique sous l'effet des perturbations extérieures. Il introduisit ainsi les célèbres équations de Lagrange et la notion de fonction perturbatrice. Ces travaux sont rassemblés dans la *"Mécanique analytique"* parue en 1788.

Pierre-Simon Laplace (1749-1827), né en Normandie, vint à Paris à l'âge de 18 ans où il fit une carrière glorieuse à l'École Normale et au Comité des Poids et Mesures. Il fut ministre de l'intérieur pendant quelques jours sous le consulat, puis sénateur, comte, marquis sous Louis XVIII. Ses deux grands ouvrages sont " *l'exposition du système du monde*", où il expose sa théorie de la formation du système solaire à partir d'une nébuleuse primitive et sa " *Mécanique céleste*" en cinq volumes publiés entre 1799 et 1825. L'œuvre de Laplace est considérable tant en physique qu'en mathématiques (calcul des probabilités). C'est surtout ses travaux de mécanique céleste qu'il faut rappeler ici: étude de l'accélération séculaire de la Lune, découverte de la grande inégalité de Jupiter et Saturne, développement de la fonction perturbatrice des planètes, stabilité du système solaire, théorie des satellites de Jupiter, etc.

Au début du XIXe siècle, la mécanique céleste est tournée presque exclusivement vers la résolution du problème des mouvements dans le système solaire, cas particulier du problème des n corps que la nature a mis sous les yeux des astronomes. Font exception les travaux d'Édouard Roche (1820-1883) sur la structure interne des corps célestes sous l'effet de l'attraction mutuelle des particules qui les composent et qui le conduiront à introduire la notion de *limite de Roche (Mémoires de l'Académie des Sciences de Montpellier*, tome 1, 1849) et ceux de Laplace et Poincaré sur les marées.

1. Développement de la fonction perturbatrice

On exprimera le hamiltonien du mouvement en fonction des variables si l'on a exprimé en fonction de ces variables une fonction des coordonnées des corps en présence appelée *fonction perturbatrice*. Ceci n'est plus possible qu'en représentant la fonction perturbatrice par une somme d'un plus ou moins grand nombre de termes où certaines des variables figurent sous des signes cosinus, les variables qui leur sont conjuguées figurant dans les coefficients de ces cosinus. Les contributeurs à ce problème du développement de la fonction perturbatrice ont été nombreux au XIXe siècle. Laplace avait introduit les fonctions dites *coefficients de Laplace* pour développer l'inverse de la distance de deux planètes. F. Bessel (1784-1846) fit usage pour la première fois de ses *fonctions de Bessel* dans un traité sur le développement de la fonction perturbatrice paru à Berlin en 1824. En 1831, Peter A. Hansen introduisit dans son " *Untersuchung über die gegenseitigen Störungen des Jupiters und Saturns*" des fonctions, dites *coefficients de Hansen*, qui permettent les développements des rayons vecteurs des planètes, ou des puissances quelconques de ceux-ci, en séries trigonométriques des anomalies moyennes, les coefficients étant des séries entières de l'excentricité. Notons le rôle important joué dans toutes ces questions par les *polynômes de Legendre* introduits par ce mathématicien en 1782 dans son traité sur la *Figure des planètes*.

Le premier développement de la fonction perturbatrice des planètes jusqu'au troisième ordre des excentricités et des inclinaisons avait été donné par Laplace. Philippe de Pontécoulant (1795-1874), dans son *Système du Monde*, l'étendit au sixième ordre. En 1849, Peirce publia également un développement au sixième ordre dans l'*Astronomical Journal*, vol. 1. Le Verrier poussa le développement au septième ordre dans le premier volume des *Annales de l'Observatoire de Paris*.

2. Résolution des équations du mouvement

Les équations canoniques étant formées et la fonction perturbatrice étant développée jusqu'à un certain ordre de précision, comment en obtenir une solution approchée ? S'il s'agit d'une théorie de planète on cherchera un

et lui demanda une précision que "la jugeant futile" celui-ci ne donnera pas. L'affaire en reste là.

Pendant ce temps, Le Verrier travaillait assidûment. Durant l'été de 1845, Arago l'avait persuadé de l'importance qu'il y aurait à résoudre le problème et l'avait convaincu d'abandonner momentanément les recherches sur les comètes qu'il avait entreprises. Contrairement à Adams, il publie, et ce au fur et à mesure de l'avancement de ses travaux: le 10 novembre 1845 présentation à l'Académie des Sciences d'un *premier Mémoire sur la Théorie d'Uranus* suivi, le 1er juin 1846, d'un mémoire intitulé *Recherches sur les mouvements d'Uranus* et, le 31 août, d'un troisième mémoire *Sur la planète qui produit les anomalies observées dans le mouvement d'Uranus, détermination de sa masse, de son orbite et de sa position actuelle*. Après la découverte il lira à l'Académie des Sciences, le 5 octobre 1846, un dernier mémoire intitulé *Sur la planète qui produit les anomalies observées dans le mouvement d'Uranus, cinquième et dernière partie, relative à la détermination de la position du plan de l'orbite*. Il regroupera tout ceci dans un mémoire unique, publié ensuite dans la *Connaissance des Temps* pour 1849, sous le titre: *Recherches sur les mouvements de la planète Herschel* (dite Uranus).

Finalement il présenta à l'Académie des Sciences le 31 août 1846 les résultats suivants: masse, 1/9322 fois la masse du Soleil; demi-grand axe, 36.1539 unités astronomiques; excentricité, 0.10761; longitude du périhélie au 1er janvier 1847, 284°45'8''; longitude héliocentrique de la planète à la même date, 326°32'. Restait à trouver la planète. Ce ne fut pas si simple. Il semble que la plupart des observateurs, en particulier en France, se soient montrés sceptiques et réticents. On lit souvent que Le Verrier s'est finalement adressé à Galle parce qu'il savait que l'observatoire de Berlin disposait des bonnes cartes du ciel de Bremiker mais rien ne le prouve, Le Verrier n'en fait pas mention dans la lettre qu'il adresse à Galle et il semble même, comme nous le verrons, que Galle lui-même n'a pensé à cette carte qu'au dernier moment. Quoi qu'il en soit, le 18 septembre, Le Verrier écrit à Johan Gottfried Galle (1812-1910), astronome à l'observatoire de Berlin, dont le célèbre Encke était alors directeur. Galle était connu de Le Verrier auquel il avait envoyé peu de temps auparavant sa thèse de doctorat. Il reçoit la lettre le 23 septembre et, le soir même, aidé du jeune astronome Henri d'Arrest, il pointe l'excellente lunette de 9 pouces d'ouverture (23 cm) de Fraunhofer vers la région indiquée et ne trouve pas le disque de 3'' de diamètre que, selon Le Verrier, la planète devait présenter. Cela n'a rien d'étonnant car le pouvoir de résolution théorique de son instrument était de 0''5 et il aurait fallu un ciel particulièrement pur pour voir un disque de 3'' de diamètre. Grâce aux récentes bonnes cartes de Bremiker on trouve enfin la planète à 52' de la position prévue le 23 septembre 1846.

4. Les théories du mouvement de la Lune

La lune étant proche de la Terre et ayant un assez faible éclat, son déplacement parmi les étoiles est rapide et facilement observable. Ceci explique que de nombreuses particularités de son mouvement, comme la rotation de la ligne des nœuds, l'équation du centre, l'évection, etc., étaient connues d'Hipparque, de Ptolémée et des astronomes arabes. Tycho Brahé observa de nouvelles inégalités du mouvement mais il fallut évidemment attendre Newton et les mécaniciens célestes du XVIIIe siècle pour que l'on vît apparaître des théories mathématiques du mouvement s'appuyant sur la loi de la gravitation universelle. Où en était-on à la mort de Laplace ? Rappelons que Clairaut, en 1752, avait publié une théorie dont la précision sur la position de la Lune était de l'ordre de $1'5$. Il faut ensuite citer les théories de D'Alembert, Tobias Mayer et les deux théories d'Euler avant d'arriver à celle de Laplace publiée en 1802, qui a une précision de $0'5$ et qui a le mérite d'essayer d'expliquer l'accélération séculaire à laquelle nous consacrerons plus loin un paragraphe. Des tables publiées par Bury en 1806 et par Burckardt en 1812 exploitent directement la théorie de Laplace.

Dans la *Connaissance des Temps* pour 1824, puis pour 1827, fut publiée la théorie de la Lune de Damoiseau (1768-1846), analogue à celle de Laplace. Le franco-italien G. Plana publia en 1832 une théorie de la Lune entièrement littérale.

La théorie que Poisson a publiée en 1833, n'est pas une théorie complète du mouvement de la Lune car il n'a pas pu pousser assez loin les calculs auxquels elle le conduisait. Il suppose, en première approximation que le demi-grand axe, l'excentricité et l'inclinaison de l'orbite sont constants et que la longitude moyenne de la Lune, celle du périgée et celle du nœud sont des fonctions linéaires du temps. Ces quantités sont substituées dans les seconds membres des équations qui, développés, sont intégrés terme à terme. On obtient une meilleure approximation pour le demi-grand axe, l'excentricité, etc. que l'on substitue dans les seconds membres et l'on réitère ensuite le processus. Évidemment, les calculs, surtout avec les moyens de l'époque, deviennent vite inabordables, mais la méthode sera utilisée avec profit plus tard dans certains problèmes de mécanique céleste quand les ordinateurs auront apparu.

Philippe, comte de Pontécoulant (1795-1874), savant de grand mérite et dont on reparlera à propos de Jupiter et de Saturne, a commencé à publier ses travaux sur la Lune en 1837. Sa théorie date de 1846; elle donna lieu plus tard à des publications en 1860 dans les *Monthly Notices of the Royal Astronomical Society*, puis en 1862 dans les *Comptes-Rendus de l'Académie des Sciences de Paris*. Les variables qu'utilise Pontécoulant sont la longitude, la latitude et le rayon vecteur de la Lune et il introduit dès l'orbite perturbatrice dans le tome 1 des *Annales de l'observatoire de Paris*. Cette publication fut d'ailleurs fondée à l'initiative de Le Verrier qui y fit figurer par la suite tous les résultats de ses travaux sur les planètes. Un peu plus tard les *Astronomical Papers of the American Ephemeris* joueront le même rôle auprès de Newcomb.

La construction de théories planétaires passe par différentes étapes:

• Obtenir une solution des équations du mouvement qu'il faut mettre en table faute, à l'époque de Le Verrier, de pouvoir dans la pratique calculer directement les positions.

• Calculer les positions de la planète considérée pour toutes les dates des observations disponibles et modifier les éléments de l'orbite de façon à réduire l'écart entre ces positions et celles que les observations ont fournies.

• Enfin, s'il s'avère que c'est impossible par cette méthode, modifier les valeurs de certaines des constantes qui interviennent dans le problème, en particulier les masses des planètes.

• Il se peut que même ce dernier recours ne soit pas satisfaisant et il faut alors faire des hypothèses nouvelles sur les causes de tels phénomènes. Si l'on se représente les moyens de calcul de l'époque on s'imagine l'ampleur de la tâche.

Dans le tome 2 des *Annales de l'observatoire de Paris*, Le Verrier fit paraître en 1852 les résultats de la réduction des 9000 observations méridiennes des planètes faites par Bradley entre 1750 et 1762. La théorie du mouvement du Soleil figure dans le tome 4. Le Verrier y introduit des perturbations du 2e ordre par rapport aux masses perturbatrices qui modifient de $5''$ la longitude moyenne du Soleil pour 1850. Cette théorie l'a conduit à diminuer la masse adoptée alors pour Mars de 10% et à augmenter celle de la Terre de 10%. On sait maintenant, à la lumière des théories modernes basées sur des intégrations numériques, que les erreurs de la théorie de Le Verrier peuvent atteindre $1'',5$. La théorie du mouvement de Mercure fait l'objet du tome 5 (1859) des *Annales*. Le Verrier y rencontra l'avance inexpliquée du périhélie qui fera l'objet d'un paragraphe spécial un peu plus loin. Dans le tome 6 paraît en 1861 la théorie du mouvement de Vénus. On y voit une inégalité nouvelle découverte par Airy et une inégalité du 2e ordre due à la Terre et à Mars. La théorie du mouvement de Vénus amène Le Verrier à diminuer la masse alors adoptée pour Mercure et à augmenter celle de la Terre. Il a utilisé pour déterminer les constantes d'intégration, outre des observations méridiennes faites de 1751 à 1857, deux passages de Vénus sur le disque du Soleil observés en 1761 et 1769. On sait maintenant que les erreurs sur la position de Vénus calculée par la théorie de Le Verrier peuvent atteindre $19''$.

Dans ce même tome 6 figure la théorie du mouvement de Mars qui s'appuie sur des observations méridiennes faites entre 1751 et 1858 et sur une conjonction de la planète avec les étoiles ψ_1, ψ_2, ψ_3 du Verseau observée en 1672, en France par Picard et à Cayenne par Richer, dans le but de calculer la parallaxe du Soleil et donc sa distance à la Terre. Le Verrier est amené à introduire dans la longitude de Mars un terme d'amplitude $1'',5$ et de période 40 ans inconnu jusque là ainsi que deux inégalités du deuxième ordre, l'une due à la Terre et à Vénus et l'autre à la Terre et à Jupiter. Les travaux de Le Verrier sur le Soleil, Mercure, Vénus et Mars lui valurent en 1868 la médaille d'or de la Royal Astronomical Society.

Le problème des mouvements de Jupiter, Saturne, Uranus et Neptune est un problème beaucoup plus difficile, que Le Verrier n'aura pas le temps de mener à terme avant sa mort. En novembre 1872 paraît un premier mémoire sur les théories des quatre planètes supérieures qui traite des termes séculaires de ces planètes. En mars 1873 paraît la théorie du mouvement de Jupiter, en juillet 1873 celle de Saturne, en novembre 1874 celle d'Uranus et en décembre celle de Neptune.

Mais à l'époque les moyens de calcul dont on dispose font qu'il n'est pas question d'établir des éphémérides en sommant des séries trigonométriques, comme on peut le faire de nos jours avec des ordinateurs. Il faut dresser des tables, en général à double entrée, qu'on interpole pour calculer les diverses inégalités à un instant donné. Les tables de Jupiter paraissent en juin 1874, celles de Saturne en août 1875, les tables d'Uranus en novembre 1876. Les tables de Neptunes achevées par Gaillot sont présentées par celui-ci en octobre 1877, un mois après la mort de Le Verrier. Tous ces travaux sur les grosses planètes figurent dans les tomes 10, 11, 12,13 et 14 des *Annales de l'observatoire de Paris*.

Les théories des mouvements de Jupiter et de Saturne sont difficiles d'une part à cause de l'importance des forces perturbatrices, ces deux planètes étant massives et loin du Soleil, et d'autre part parce que le rapport du moyen mouvement de Jupiter à celui de Saturne est presque exactement égal à 5/2 ce qui entraîne la *grande inégalité* des longitudes des deux planètes dont la période est proche de 850 ans et l'amplitude environ $20'$ pour Jupiter et $1°$ pour Saturne. Jacobi avait trouvé une méthode pour isoler dans la fonction perturbatrice les termes de résonance correspondants et publié ce résultat dans son mémoire intitulé *"Versuch einer Berechnung der grossen Ungleichheit des Saturns nach einer strengen Entwickelung"* (*Astronomische Nachrichten*, tome 28, 653-654). La théorie de Le Verrier permet de représenter les observations de Jupiter faites entre 1750 et 1869 avec une erreur en longitude inférieure à $1''$ et, après que des corrections furent apportées par Gaillot, les observations de Saturne faites de 1750 à 1890, avec une erreur en longitude inférieure à $3''$ en valeur absolue. Les théories modernes montrent que la situation n'a pas empiré par la suite et que ces erreurs étaient toujours à peu près les mêmes vers les années 1970.

8. Autres travaux - Comètes

La découverte de Cérès, le 1er janvier 1801, fut suivie de nombreuses autres découvertes de petites planètes; mais les moyens mouvements de Jupiter et de certaines petites planètes sont souvent presque commensurables, ce qui introduit des résonances, phénomène dont nous avons déjà eu l'occasion de parler. Aussi les méthodes classiques, même si elles ont pu être utilisées avec succès par Perrotin pour Vesta (*Annales de l'observatoire de Toulouse*, tome 1, 1880) ou essayées par Damoiseau pour Cérès et Junon (*Addition à la Connaissance des Temps* pour 1846), sont peu efficaces.

Le Verrier avait trouvé qu'un terme de résonance de période 800 ans dans le mouvement de Pallas avait une amplitude de 895″. A. Cauchy (1789-1857), bien que mathématicien et non astronome, fut chargé par l'Académie des Sciences de vérifier ce résultat et inventa en six semaines une méthode ingénieuse qu'il publia dans les *Comptes Rendus de l'Académie des Sciences* (tome 20, 1845). V. Puiseux a exposé la méthode dans les *Annales de l'Observatoire de Paris* (tome 7, 1863) ainsi que Tisserand dans le tome 4 de son *Traité de Mécanique Céleste*.

En juin 1770, Messier avait découvert une comète dont A. Lexell (1740-1784), membre de l'Académie des Sciences de Saint-Petersbourg, étudia l'orbite et à laquelle il donna son nom. Il trouva que c'était une orbite elliptique dont le demi-grand axe était égal à 3 unités astronomiques, ce qui donnait à la comète une période de 5 ans et 7 mois. Une période aussi courte aurait dû permettre la découverte de la comète bien longtemps auparavant; par ailleurs, la comète avait disparu quelques années après sa découverte. On montra alors, en particulier Burckardt à l'instigation de Laplace, que des passages de la comète très près de Jupiter avaient profondément modifié l'orbite à deux reprises. Ceci posait un problème particulier que les techniques de calcul numérique de l'époque ne permettaient pas de résoudre d'une façon satisfaisante. Laplace, suivant en cela une idée de D'Alembert, avait alors introduit la notion de sphère d'activité: sur une certaine partie de la trajectoire d'une comète perturbée par Jupiter on a intérêt à considérer le mouvement héliocentrique et à déterminer les perturbations par Jupiter. Près de Jupiter au contraire, à l'intérieur de sa *sphère d'activité*, il est plus avantageux de considérer le mouvement jovicentrique de la comète et de calculer les perturbations par le Soleil. Le Verrier en 1857 reprendra cette méthode et l'appliquera à la comète de Lexell dans son mémoire *"Sur la théorie de la comète périodique de 1770"* (*Annales de l'observatoire de Paris*, tome 3). Un problème intéressant a été posé et résolu par Tisserand dans le *Bulletin Astronomique* (*"Sur la théorie de la capture des orbites périodiques"*, vol. 6, 1889) et reproduit dans le tome 4 de son *Traité de Mécanique Céleste* (1896). Il a donné lieu à la règle connue sous le nom de *critère de Tisserand*. Le problème est de savoir si deux comètes périodiques apparues à deux époques différentes sont ou non le même objet dont les éléments de l'orbite ont été modifiés par un passage proche de Jupiter. Utilisant une intégrale première du problème restreint des trois corps trouvée par Jacobi, Tisserand montre qu'il s'agit bien du même objet si l'on a:

$$\frac{1}{2a} + \sqrt{\frac{a(1-e^2)}{a'^3}} \cos i = cte$$

où a est le demi-grand axe de l'orbite de la comète, a' celui de l'orbite de Jupiter, e l'excentricité de l'orbite de la comète et i l'inclinaison du plan de l'orbite de la comète sur le plan de l'orbite de Jupiter. L'excentricité de l'orbite de Jupiter est négligée mais Callandreau dans son mémoire *"Étude sur la théorie des comètes périodiques"* (*Annales de l'observatoire de Paris*, tome 20, 1892) a montré comment tenir compte de la première puissance de cette excentricité.

Il faut bien sûr rappeler que la période qui nous occupe a vu deux passages de la célèbre comète de Halley, en 1835 et en 1910. Pour les calculs de prédiction du passage de 1835 nous retrouvons les deux compétiteurs au concours sur le mouvement de la Lune: Damoiseau et Pontécoulant. Damoiseau en 1820 prédit le retour de la comète pour le 17 novembre 1835 (*Mémoires de l'Académie Royale des Sciences de Turin*, 24,1). Il avait tenu compte des perturbations par Jupiter, Saturne et Uranus (planète inconnue au passage précédent) pendant la période s'étendant de 1682 à 1835. Il ajouta ensuite les perturbations par la Terre dans un second mémoire paru en 1829 (publié comme *supplément à la Connaissance des Temps* pour 1832), ce qui ramena la date du passage au soir du 4 novembre. Pontécoulant, avec les mêmes hypothèses que Damoiseau, publia trois mémoires en 1830, 1834 et 1835 (respectivement *suppléments à la Connaissance des Temps* pour 1833 et 1837 et *Mémoires présentés à divers savants de l'Académie des Sciences*, publié en 1835). Sa prédiction finale pour le passage est le 12 novembre à 22h. Le passage au périhélie eut lieu le 16 novembre à 10h. Pontécoulant entreprit aussi la prédiction du passage de 1910 dès 1864.

9. Deux "petites difficultés"

C'est ainsi que Tisserand appelle le problème de l'avance du périhélie de Mercure et l'accélération séculaire de la Lune qui semblaient mettre en défaut la loi de la gravitation universelle. La première fut découverte par Le Verrier qui trouva une avance inexpliquée de $38''$ par siècle. La seconde fut étudiée par Laplace qui pensa expliquer l'accélération de la Lune de $20''$/siècle/siècle par les perturbations planétaires de l'excentricité de l'orbite terrestre. En fait, la première difficulté fut résolue grâce à la Relativité Générale, la seconde par le ralentissement de la rotation de la Terre due aux marées.

Le début du siècle est marqué par l'œuvre majeure d'Henri Poincaré (1854-1912). Indépendamment de ses travaux mathématiques et philosophiques, il est connu pour sa théorie des orbites périodiques et asymptotiques, ses travaux sur la convergence des séries du problème des n corps, sur les intégrales premières, les exposants caractéristiques, les invariants intégraux, les marées. Son ouvrage le plus célèbre est *"Les méthodes nouvelles de la mécanique céleste"* paru entre 1892 et 1899.

La mort de Poincaré marque la fin d'une époque triomphale de l'astronomie classique désormais débordée par l'astrophysique.

La mécanique céleste semblait bloquée par l'ampleur des calculs qu'il aurait fallu pouvoir faire, calculs qui paraissaient d'ailleurs inutiles compte-tenu de la précision des observations. Il faudra attendre 1960, les ordinateurs, les premiers satellites artificiels pour voir un renouveau de la mécanique céleste en France, et en particulier au Bureau des longitudes.

ANALYTICAL SOLUTION OF THE MOTION OF THE PLANETS OVER SEVERAL THOUSANDS OF YEARS

P. BRETAGNON
Bureau des Longitudes
77, avenue Denfert-Rochereau, 75014 Paris, France

1. Introduction

The results of a planetary theory built by an iterative method are given here in order to show the relation with the secular variation theories and the meaning of the mean elements in these latter theories. The general theories have a validity span of several millions years but a weak precision; on the contrary, the secular variation theories reach a great precision over several thousand years. Two applications of the analytical planetary theories are presented : the relation between the barycentric coordinates and the geocentric ones; the determination of the terms of precession and nutation for the rigid Earth.

2. General theories

2.1. FORM OF THE SOLUTION

The general theories represent the motion of the planets in Fourier series :

$$a_i = a_{i0} + \sum_{\Phi^*} A_{i,\Phi^*} \cos \Phi^* + \sum_{\Phi} A_{i,\Phi} \cos \Phi$$

$$\lambda_i = \lambda_{i0} + n_{i0}t + \delta n_i t + \sum_{\Phi^*} B_{i,\Phi^*} \sin \Phi^* + \sum_{\Phi} B_{i,\Phi} \sin \Phi$$

$$k_i = \sum_{k=1}^{8} \lambda_{ik} M_k \cos \psi_k + \sum_{k=1}^{8} M_{i,\psi_k} \cos \psi_k$$
$$+ \sum_{\Phi^*} \epsilon_{\Phi^*} M_{i,\Phi^*} \cos \Phi^* + \sum_{\Phi} \epsilon_\Phi M_{i,\Phi} \cos \Phi$$

TABLE 1. Amplitude of the long period terms in the longitudes of Jupiter and Saturn. The unit is the arcsecond.

Argument	Period	B_{J,Φ^*}	B_{S,Φ^*}
$\psi_5 - \psi_6$	54 017	−4 653	11 492
$\psi_5 - \psi_7$	1 124 076	1 606	−2 681
$\psi_5 - \psi_6 - \theta_6 + \theta_7$	1 996 218	986	−2 430
$\psi_6 - \psi_7$	51 540	345	−832
$2\psi_5 - 2\psi_6$	27 009	−272	668
$2\psi_5 - \psi_6 - \psi_7$	56 744	62	−154
$2\psi_5 - 2\psi_6 - \theta_6 + \theta_7$	52 594	40	−98
$\psi_5 - 2\psi_6 + \psi_7$	26 375	−39	95

TABLE 2. Amplitude of the long period terms in the longitudes of Uranus and Neptune. The unit is the arcsecond.

Argument	Period	B_{U,Φ^*}	B_{N,Φ^*}
$\psi_5 - \psi_7$	1 124 076	−20 583	9 996
$\psi_7 - \psi_8$	535 721	−431	263
$\psi_5 - \psi_6$	54 017	−248	13
$\psi_5 - \psi_8$	362 810	220	−250
$\theta_7 - \theta_8$	562 640	159	−43
$\psi_6 - \psi_7$	51 540	−97	22
$2\psi_5 - 2\psi_7$	562 038	−13	−34

2.3. LONG PERIOD VARIATIONS OF THE PLANETARY ORBITS

The general theories are developed at the Bureau des Longitudes for all the planets since 1970. They take into account the mutual perturbations of all the planets one another, the relativistic effects, the lunar perturbations. The long period terms of the variables k, h, q, p give the variations of the planetary orbits over several millions years. These solutions bear a fundamental part in paleoclimatology because they allow to date with a great accuracy the glaciations of the Earth of the quaternary period (Berger, 1973), the climatic variations of Mars (Ward, 1979), (Borderies, 1980).

3. Secular variation theories

The general theories represent the motion of the planets over very large time spans but with a weak precision. The secular variation theories attempt to reach a great precision over time spans restricted to a few thousands of years. Therefore it is useless to keep the long period terms in a periodic form. By construction, we directly determine the time polynomial corresponding to the long period part of the formulas (1) and the Poisson series corresponding to the short period terms. Thus, to the long period terms of the formula (3) corresponds the time polynomial of the semi major axis of Saturn :

$$a_{Saturn} = 9.554\,909\,1915 - 21.3896 \times 10^{-6} t + 444 \times 10^{-10} t^2 \\ + 670 \times 10^{-10} t^3 + 110 \times 10^{-10} t^4 \qquad (5)$$

where t is reckoned in thousands Julian years from J2000.

The first determinations of these secular terms of the semi major axes were obtained by (Simon and Chapront, 1974). Duriez (1978) has given a proof of the Poisson theorem and established that these terms are of order equal or greater than 3 with respect to the planetary masses.

The expansion of the long period terms with respect to time reduces considerably the size of the series and allows to reach a great precision. Thus, the longitude of Saturn of the formula (4) becomes :

$$\lambda_{Saturn} = 0.874\,016\,284 + 213.299\,104\,960 t + 0.000\,366\,597 t^2 \\ - 806 \times 10^{-9} t^3 - 557 \times 10^{-9} t^4 \\ + 0.013\,944\,575 \sin(2\lambda_5 - 5\lambda_6) + 0.002\,196\,781 \cos(2\lambda_5 - 5\lambda_6) \\ - 0.001\,590\,423 t \sin(2\lambda_5 - 5\lambda_6) + 0.005\,404\,368 t \cos(2\lambda_5 - 5\lambda_6) \\ - 0.001\,061\,109 t^2 \sin(2\lambda_5 - 5\lambda_6) - 0.000\,474\,470 t^2 \cos(2\lambda_5 - 5\lambda_6) \\ + \cdots \qquad (6)$$

In the formula (6), the coefficients are in radians.

The polynomial part of a variable is the mean element of this variable. It represents the development with respect to time of the long period part of the general theories. The mean element contains the most important variations but it does not represent a good approximation of the solution, particularly for the outer planets which include very large short period perturbations as we see in the expression (6).

The present analytical theories VSOP82 (Bretagnon, 1982), TOP82 (Simon, 1983), VSOP87 (Bretagnon and Francou, 1988) have, over one century around J2000, an accuracy of about 10^{-7} for Jupiter and Saturn and of about 10^{-8} for the other planets. They include the mutual perturbations of all the planets one another up to an order with respect to the masses equal or greater than 3, the relativistic contributions of the Schwarzschild problem, the perturbations by the Moon and some asteroids. They are computed using the planetary mass values of IAU (Grenoble, 1976); the integration constants are determined by comparison to DE200 numerical integration of JPL (Standish, 1982).

The precisions of these solutions seem good enough for the two following applications : the determination of the difference between the barycentric time TCB and the geocentric time TCG and the computation of the terms of precession and of nutation for the rigid Earth.

4. Relation between TCB and TCG

The analytical solutions of the motion of the planets and of the Moon ELP2000/82 (Chapront-Touzé and Chapront, 1983) were used in the computation of the relation between the Barycentric Coordinate Time (TCB) and the Geocentric Coordinate Time (TCG) par (Hiramaya et al 1987), (Fairhead and Bretagnon, 1990).

Restricted to the terms proportional to c^{-2}, the relation between TCB and TCG is written :

$$\begin{aligned}TCB &= TCG + c^{-2} \int_{TCB_0}^{TCB} (U_E + \tfrac{1}{2} v_E^2) \, \mathrm{d}(TCB') \\ &= TCG + L_C \, TCB \\ &\quad + 1\,656.674\,564 \, \mu s \, \sin(6\,283.075\,850 \, TCB + 6.240\,054) + \cdots \end{aligned}$$

with $L_C = 1.480\,826\,8475 \times 10^{-8}$ and TCB in thousands Julian years.

U_E represents the external mass force function evaluated at the geocentre and has been computed taking into account the Sun, the Moon and the planets from Mercury to Neptune.

By comparison to numerical integrations, T. Fukushima and A. Irwin have shown that this solution has an accuracy of 1.8 ns over (1980–2000). At the beginning, the solution of (Fairhead and Bretagnon, 1990) retained only the periodic terms greater than 0.1 ns. To obtain the precision of 1.8 ns, we have taken into account the 971 periodic terms greater than 0.01 ns.

To improve the relation between TCB and TCG, we have :
- to compute U_E and v_E with planetary motion solutions using recent values of the planetary masses, for instance the ones of the IERS Standards 1992 (McCarthy, 1992);
- to take into account Pluto : $\Delta L_C^{Pluto} \sim 2 \times 10^{-18}$;
- to take into account asteroids (Fukushima, 1995) : $\Delta L_C^A \sim 4.5 \times 10^{-18}$;
- to determine the terms proportional to c^{-4}. For these terms, (Moisson, 1995) finds :

$$\begin{aligned}\Delta(TCB - TCG) &= c^{-4} \int_{TCB_0}^{TCB} (\tfrac{1}{8} v_E^4 + \tfrac{3}{2} U_E v_E^2 - \tfrac{1}{2} U_E^2) \, \mathrm{d}(TCB') \\ &= 1.0965 \times 10^{-16} \, TCB - 0.10 \times 10^{-20} \, TCB^2 \\ &\quad - 7^s.3 \times 10^{-12} \sin(\lambda_3) - 31^s.9 \times 10^{-12} \cos(\lambda_3) + \cdots\end{aligned}$$

5. Precession and nutation for the rigid Earth

5.1. EQUATIONS OF THE MOTION

We have established the motion equations for the rigid Earth with the Euler angles ψ, ω, φ reckoned in the positive direction and ω being the rotation angle from ecliptic J2000 to the equator of date. We have therefore :

$$\psi = -\psi_A$$
$$\omega = -\omega_A$$

where ψ_A and ω_A represent the luni-solar precession and the obliquity with the notations of (Lieske et al 1977). The sidereal time φ is given by :

$$\varphi = \varphi_0 + \varphi_1 t + \Delta\varphi$$

with :

$$\varphi_0 = 4.903\,562\,579\,35$$
$$\varphi_1 = 2\,301\,216.753\,1542 \text{ rd/thousand Julian years}$$

We also defined :

$$\tilde{\varphi} = \varphi + \alpha$$

with :

$$\alpha = -14°.95$$

longitude of major axis of equatorial ellipse (Bursa, 1992).
Then, the equations are written :

$$\ddot{\omega} + \frac{C}{A}\sin\omega_0 \varphi_1 \dot{\psi} = \frac{L}{A} + F_2 + \frac{B-A}{A}F_1$$
$$\sin\omega_0 \ddot{\psi} - \frac{C}{A}\varphi_1 \dot{\omega} = \frac{M}{A} + G_2 + \frac{B-A}{A}G_1$$
$$\ddot{\varphi} = \frac{N}{C} + H_2 + \frac{B-A}{C}H_1 \qquad (7)$$

with :

$$F_2 = -\frac{C}{A}\dot{\psi}\varphi_1(\sin\omega - \sin\omega_0) - \frac{C}{A}\dot{\psi}\Delta\dot{\varphi}\sin\omega - \frac{C-A}{A}\dot{\psi}^2 \sin\omega\cos\omega$$
$$G_2 = -\ddot{\psi}(\sin\omega - \sin\omega_0) + \frac{C}{A}\dot{\omega}\Delta\dot{\varphi} - \frac{A+B-C}{A}\dot{\psi}\dot{\omega}\cos\omega$$
$$H_2 = -\ddot{\psi}\cos\omega + \dot{\psi}\dot{\omega}\sin\omega \qquad (8)$$

and :

$$\begin{aligned}
F_1 &= \ddot{\psi}\sin\tilde{\varphi}\cos\tilde{\varphi}\sin\omega + \dot{\psi}\dot{\varphi}(\cos^2\tilde{\varphi} - \sin^2\tilde{\varphi})\sin\omega - 2\dot{\omega}\dot{\varphi}\sin\tilde{\varphi}\cos\tilde{\varphi} \\
&\quad -\ddot{\omega}\sin^2\tilde{\varphi} + \dot{\psi}^2\cos^2\tilde{\varphi}\sin\omega\cos\omega \\
G_1 &= 2\dot{\psi}\dot{\varphi}\sin\tilde{\varphi}\cos\tilde{\varphi}\sin\omega + \ddot{\omega}\sin\tilde{\varphi}\cos\tilde{\varphi} + \dot{\omega}\dot{\varphi}(\cos^2\tilde{\varphi} - \sin^2\tilde{\varphi}) \\
&\quad +\dot{\psi}^2\sin\tilde{\varphi}\cos\tilde{\varphi}\sin\omega\cos\omega - \ddot{\psi}\cos^2\tilde{\varphi}\sin\omega \\
H_1 &= \dot{\omega}^2\sin\tilde{\varphi}\cos\tilde{\varphi} - \dot{\psi}\dot{\omega}(\cos^2\tilde{\varphi} - \sin^2\tilde{\varphi})\sin\omega - \dot{\psi}^2\sin\tilde{\varphi}\cos\tilde{\varphi}\sin^2\omega
\end{aligned}$$
(9)

Let $Oxyz$ be the reference frame ecliptic J2000 and $OXYZ$ the non-rotating equator of date. From $Oxyz$, we define $OXYZ$ with 2 rotations:
- a rotation with ψ angle around z axis;
- a rotation with ω angle around X axis.

The quantities L, M, N of the formula (7) represent thus the components, in $OXYZ$, of the torque of the external forces with respect to the geocenter O. A, B, C are the moments of inertia.

5.2. USED MODEL

In the computation of the quantities L, M, N, we take into account, for the influence by the Moon, the terms of the terrestrial potential depending on $C_{n,0}$ for n from 2 to 5, on $C_{2,2}$, $S_{2,2}$ and $C_{3,k}$, $S_{3,k}$ for k from 1 to 3; for the influence by the Sun : $C_{2,0}$, $C_{3,0}$, $C_{2,2}$, $S_{2,2}$, $C_{3,1}$, $S_{3,1}$; for the influence by the planets from Mercury to Neptune : $C_{2,0}$. The lunar theory used is ELP2000/82 (Chapront-Touzé and Chapront, 1983); the one of the planets and the Sun is VSOP87A (Bretagnon and Francou, 1988). In this study, we take the following choices :
a) We study the variations of the rigid Earth equator with respect to the ecliptic and the equinox J2000. So, we have to take into account the perturbations of the equator by the Moon, the Sun and the planets, the motion of which is expressed in rectangular coordinates with respect to the ecliptic and the equator J2000.
b) The solution is expanded in Fourier and Poisson series, the angles of which are linear combinations of the planet longitudes λ_i reckoned from the equinox J2000 and of the Delaunay angles which do not depend on the origin. In consequence, the 18.6 year period perturbation is represented by the angle $\lambda_3 + D - F$ which differs from the longitude Ω of the node of the Moon referred to the equinox of date :

$$\lambda_3 + D - F = \Omega + 180° - p \times t$$

where p is the constant of the precession in longitude and t the time reckoned from J2000.
c) In the Fourier and Poisson series, we keep only the linear part of the mean longitudes of the planets and of the Delaunay angles D, F, l_M; the poly-

nomial parts, the degree of which is equal or greater than 2 are expanded in Poisson series.

d) The perturbations due to the Moon are computed as a whole. We do use a representation of the lunar motion in rectangular coordinates, containing the perturbations of the main problem, the direct and indirect planetary perturbations, the perturbations due to the terrestrial potential, the tidal effects. In the same way, we use a solution of the Sun in rectangular coordinates reckoned with respect to the Earth but not to the Earth-Moon barycenter.

e) The computation was performed with the value of the precession constant given by (Williams et al 1991) and used by (Simon et al 1994) :

$$p = 50\,288''.200/\text{thousand Julian years.}$$

This value corresponds to :

$$\left(\frac{d\psi_A}{dt}\right)_{t=0} = 50\,385''.0672.$$

The value of the geodesic precession p_g determined by (Brumberg et al 1991) is :

$$p_g = 19''.1988.$$

We have therefore solved equations (7) fixing the value of the moment of inertia C :

$$C = 1.805\,465\,872 \times 10^{-15} (\text{m}_S \text{ au}^2) \tag{10}$$

in order to obtain :

$$-\left(\frac{d\psi}{dt}\right)_{t=0} = 50\,385''.0672 + 19''.1988 = 50\,404''.2660. \tag{11}$$

The value of the obliquity is :

$$\varepsilon_0 = 23°26'21''.412.$$

In the relation (10), m_S is the mass of the Sun :

$$m_S = 332\,946.045 m_E$$

At this value of C corresponds the dynamical ellipticity H_d :

$$H_d(50\,288''.2) = 0.003\,273\,800\,45$$

For $p = 50\,287''.7$, one obtains :

$$H_d(50\,287''.7) = 0.003\,273\,767\,98$$

TABLE 3. Most important perturbations of the nutation. Amplitudes are in $10^{-6}{''}$, periods in days.

Origin	Argument	Amplitude in ψ	Amplitude in ω	Period
Moon $C_{2,0}$	$\lambda_3 + D - F$	17 292 345.65	9 227 970.05	6 793.48
$C_{3,0}$	$\lambda_3 + D - l$	104.05	88.95	3 232.61
$C_{4,0}$	$\lambda_3 + D - F$	0.73	6.84	6 793.48
$C_{5,0}$	$\lambda_3 + D - l$	0.01	0.00	3 232.61
$C_{2,2} - S_{2,2}$	$2\lambda_3 + 2D - 2\varphi$	29.44	11.71	0.52
$C_{3,1} - S_{3,1}$	$\lambda_3 + D + \varphi$	38.44	15.25	0.96
$C_{3,2} - S_{3,2}$	$\lambda_3 + D - 2\varphi$	0.39	0.14	0.51
$C_{3,3} - S_{3,3}$	$\lambda_3 + D - 3\varphi$	0.41	0.20	0.34
Sun $C_{2,0}$	$2\lambda_3$	1 276 723.69	552 395.17	182.63
$C_{3,0}$	λ_3	0.26	0.22	365.26
$C_{2,2} - S_{2,2}$	$2\lambda_3 - 2\varphi$	12.32	4.90	0.50
$C_{3,1} - S_{3,1}$	$\lambda_3 + \varphi$	2.79	1.11	1.00
Mercury $C_{2,0}$	$\lambda_1 - 4\lambda_3$	1.03	0.43	2 432.11
Venus $C_{2,0}$	$3\lambda_2 - 5\lambda_3$	216.71	90.76	2 959.21
Mars $C_{2,0}$	$\lambda_3 - 2\lambda_4$	11.55	0.95	5 764.01
Jupiter $C_{2,0}$	$2\lambda_5$	104.41	45.69	2 166.29
Saturn $C_{2,0}$	$2\lambda_6$	12.15	5.16	5 379.61
Uranus $C_{2,0}$	$2\lambda_7$	0.65	0.29	15 344.24
Neptune $C_{2,0}$	$2\lambda_8$	0.40	0.16	30 091.15
Complements	$\lambda_3 + D - F$	15 361.43	0.04	6 793.48

5.3. RESULTS

We give in table 3 the most important perturbation of each component. For the planets, it is the direct influence which is concerned. Complements correspond to the quantities F_1, G_1, H_1, F_2, G_2, H_2 of the equations (7). Table 4 gives the different components to the secular terms of ψ_A and of ω_A. In table (5) we compare the polynomial parts of ψ_A and ω_A to the results of (Simon et al 1994) and of (Williams, 1994). The difference with Williams et al in $\psi_A(t)$ results from a different choice of precession constant and an insufficient model in Simon et al explains the difference in $\omega_A(t)$. Besides, Williams takes into account the secular variation of J_2 that explains the discrepancy in $\psi_A(t^2)$. With the value of $\dot{J}_2$ (Bursa, 1992) :

$$\dot{J}_2 = (-2.8 \pm 0.3) \times 10^{-9}/\text{century}$$

we compute the following perturbation :

$$\Delta\psi_A(\dot{J}_2) = -0{''}.651\,804 t^2 + 0{''}.001\,849 t^3 + 0{''}.000\,022 t^4 + \cdots \\ -0{''}.000\,447 t \sin(\lambda_3 + D - F) + \cdots$$

TABLE 4. Secular term of ψ_A and of ω_A in arcseconds per thousand years.

Origin		ψ_A	ω_A
Moon	$C_{2,0}$	34 455.298 798	−0.254 417
	$C_{3,0}$	−0.000 057	−0.000 011
	$C_{4,0}$	0.025 192	
Sun	$C_{2,0}$	15 948.860 274	0.002 923
	$C_{3,0}$	−0.000 026	−0.000 005
Mercury	$C_{2,0}$	0.003 698	−0.000 088
Venus	$C_{2,0}$	0.181 582	−0.016 814
Mars	$C_{2,0}$	0.005 999	0.000 357
Jupiter	$C_{2,0}$	0.117 060	0.002 804
Saturn	$C_{2,0}$	0.005 208	0.000 220
Uranus	$C_{2,0}$	0.000 100	0.000 001
Neptune	$C_{2,0}$	0.000 029	0.000 001
Complements		−0.231 857	
$-p_g$		−19.198 800	
		50 385.067 200	−0.265 029

$$\Delta\omega_A(\dot{J}_2) = \quad 0''.000\,003 t^2 - 0''.000\,088 t^3 + 0''.000\,150 t^4 + \cdots \\ + 0''.000\,239 t \cos(\lambda_3 + D - F) + \cdots$$

TABLE 5. Secular variations of ψ_A and of ω_A.

	t	t^2	t^3	t^4
ψ_A	50 385.067 200	−107.246 837	−1.144 309	1.329 708
Simon et al	50 385.067 200	−107.237 4	−1.142 4	1.327 9
Williams	50 384.565 010	−107.897 7	−1.141	1.33
ω_A	−0.265 029	5.129 643	−7.732 154	−0.004 852
Simon et al		5.129 4	−7.727 6	−0.004 8
Williams	−0.244 00	5.126 8	−7.727	

From ψ_A and ω_A, we have computed the variables p_A and ε_A and compared to the KS solution of (Kinoshita and Souchay, 1990). Kinoshita and Souchay use the following value of the precession constant :

$$p = 50\,290''.966.$$

So, we have multiplied their solution by :

$$\text{coef} = \frac{50\,404.266}{50\,407.032}$$

TABLE 6. Difference p_A−coef KS. Unit is $10^{-6''}$.

Argument	sin	cos	Period
$2\lambda_5 - 5\lambda_6$	−499	−660	883 y
$3\lambda_3$	216	888	121.75 d
$\lambda_3 + D - F$	−25	−384	6 793.48 d
$2\lambda_3 + 2D - F - l_M$	20	281	6 167.21 d
$4\lambda_3 - 8\lambda_4 + 3\lambda_5$	−52	166	1 783 y
$\lambda_3 + D - l_M$	−106	96	3 232.61 d
$8\lambda_2 - 13\lambda_3$	79	31	239 y
$2\lambda_3 + 2D$	81	−3	13.66 d
$F - l_M$	−33	33	2 190.35 d

in order to make the solutions comparables. We give in table 6 the most important discrepancies p_A − coef KS.

References

Berger, A. (1973) Théorie astronomique des paléoclimats, *Thèse, Louvain*.
Borderies, N. (1980) La rotation de Mars. *Thèse, Toulouse*.
Bretagnon, P. (1982) *Astron. Astrophys.*, **114**, p. 278-288.
Bretagnon, P., Francou, G. (1988) *Astron. Astrophys.*, **202**, p. 309-315.
Bretagnon, P., Simon, J.L. (1990) *Astron. Astrophys.*, **239**, p. 387-398.
Bretagnon, P., Francou, G. (1992) *IAU Symposium no. 152*, p. 37-42.
Brumberg, V.A., Bretagnon, P., Francou, G. (1991) in : Capitaine N. (ed.) *Systèmes de références spatio-temporels. Journées 1991, Obs. de Paris*, p. 141.
Bursa, M. (1992) *Bull. Geod.*, **66-2**, p. 193.
Chapront-Touzé, M., Chapront, J. (1983) *Astron. Astrophys.*, **124**, p. 50.
Duriez, L. (1978) *Astron. Astrophys.*, **68**, p. 199.
Fairhead, L., Bretagnon, P. (1990) *Astron. Astrophys.*, **229**, p. 240.
Fukushima, T. (1995) *Astron. Astrophys.*, **294**, p. 895.
Hirayama, Th., Fujimoto, M.-K., Kinoshita, H., Fukushima, T. (1987) *IAG Symposia at IUGG*, **Tome I**, p. 91
Kinoshita, H., Souchay, J. (1990) *Celest. Mech.*, **48**, p. 187.
Lieske, J.H., Lederle, T., Fricke, W., Morando, B. (1977) *Astron. Astrophys.*, **58**, p. 1-16.
McCarthy, D.D. (1992), IERS Standards (1992) *IERS Technical Note* **13,** Observatoire de Paris.
Moisson, X. (1995), Rapport de stage de DEA *Observatoire de Paris*.
Simon, J.-L., Chapront, J. (1974) *Astron. Astrophys.*, **32**, p. 51.
Simon, J.L. (1983) *Astron. Astrophys.*, **120**, p. 197-202.
Simon, J.L., Bretagnon, P., Chapront, J., Chapront-Touzé, M., Francou, G., Laskar, J. (1994) *Astron. Astrophys.*, **282**, p. 663-683.
Standish, E.M. (1982) *Astron. Astrophys.*, **114**, p. 297-302.
Ward, W.R. (1979) Present Obliquity Oscillations of Mars : Fourth-Order Accuracy on Orbital e and I. *J. Geophys. Res.*, **84** p. 237.
Williams, J. G., Newhall X X, Dickey, J. O. (1991) *Astron. Astrophys.*, **241**, L9.
Williams, J. G. (1994) *Astron. J.*, **108** (2), p. 711.

NEW ACCURACY LEVELS FOR SOLAR SYSTEM EPHEMERIDES

E. M. STANDISH AND X X NEWHALL
JPL / Caltech
301-150; Pasadena, CA 91109; USA

Abstract. DE403/LE403 is the latest JPL Planetary and Lunar Ephemeris. It represents a number of changes and improvements to previous JPL ephemerides: the reference frame is now that of the IERS, newer and more accurate observations are used in the adjustment process, some of the data reduction techniques have been refined, and improved dynamical modeling has been incorporated into the equations of motion. As a result, the internal accuracy of the inner four planets has been improved. Further, various measurements accurately tie Jupiter onto the IERS Reference Frame. In the future, use of CCD measurements and the Hipparcos Catalogue should improve the ephemerides of the outermost four planets.

DE403/LE403 has been integrated over 6000 years, from 3000 BC to 3000 AD. A more condensed representation has been made from this, named DE404/LE404. It replaces DE102 as the new JPL "Long Ephemeris".

1. Introduction

The latest JPL Planetary and Lunar Ephemeris, DE403/LE403, was created in May 1995 and described by Standish *et al.* (1995). The JPL ephemerides are now numbered in the 400's, signifying that they are based upon the reference frame of the IERS Radio Catalogue. This has recently been made possible by including two new sets of data into the ephemeris adjustments: 1) ΔVLBI observations of the Magellan Spacecraft orbiting Venus and of the Phobos Spacecraft during its approach to Mars and 2) a frame-tie determination between previous JPL ephemerides and IERS Radio Catalogue. The internal accuracy of the ephemerides has also been improved. Recent radar-ranging observations to Mercury, Venus, and Mars have been acquired; the complete set of Viking Lander ranges is now used; VLA thermal emission measurements of the Jovian planets and ΔVLBI measurements of the Magellan, Phobos, and Ulysses spacecraft have been added to the data set; the tracking files of the Voyager 1 encounter of Jupiter have been re-reduced; the set of transit observations has been extended, including photoelectric transit measurements of Pluto; and Lunar Laser Ranging observations up to the present are included.

The data reduction processes and the dynamical modeling of the forces have also been refined: the surfaces of Mercury and Venus are modeled more accurately, the perturbations of 300 asteroids are now accounted for, and the formulation for integrating the lunar librations has been improved.

The IERS Reference Frame is discussed in Section II; the data sets to which the ephemerides are fit, in Section III; the improvements to the data reductions and to the dynamical modeling are discussed in Section IV; DE403 is compared with DE200 and DE102 in Section V; and the creation of DE404 is described in Section VI.

2. The IERS Reference Frame

The choice of the IERS Reference Frame for the ephemerides is advantageous for the following reasons:

- the ephemerides are now adjusted to a number of accurate ($\pm 0\rlap{.}''001$ to $\pm 0\rlap{.}''005$) observations which are referenced to the IERS Frame,
- the frame-tie between the JPL ephemerides and the IERS Frame has been accurately ($\pm 0\rlap{.}''003$) measured,
- the IERS Frame itself is accurate: the source positions are now determined at the sub-milliarcsecond level,
- the IERS Frame is accessible and well-defined: the catalogue of source positions is published on an annual basis and is available worldwide,
- the IERS Frame is stable: an IAU directive to the IERS states that any further adjustments to the coordinates of the catalogue must be done in such a fashion that there is no net rotation introduced into the system as a whole, and
- the timing and polar motion information used for the orientation of the earth is referred to the IERS Frame.

The observations mentioned in the first two items above are discussed in the next section. They allow the ephemerides to be oriented automatically onto the IERS Frame. Of course, for the sake of consistency, any observations referred to any other external reference frame (e.g., optical FK5 Frame) must be appropriately adjusted.

3. The Observational Data

The observational data to which the ephemerides have been fit are discussed in more detail by Standish *et al.* (1995). The sets include

- Optical observations: The transit observations are visual, photoelectric, and now, CCD. There are also astrometric plate observations of Pluto, astrolabe observations of Mars through Uranus, timings of occultations of stars by the rings of Uranus and the disk of Neptune.
- Radar ranging: The ranges to the terrestrial planets have been extended through 1991 for Mercury, 1990 for Venus, and 1993 for Mars. The modeling of these is discussed briefly in the next section.
- Spacecraft ranging: The ranging to Mars (Mariner 9 and Viking), Jupiter (Voyagers 1 and 2), and Mercury and Venus (Mariner 10) have been previously discussed by Standish *et al.* (1976, 1990, and 1995).
- Thermo-electric emission: These measurements of the Galilean satellites, Titan, Uranus and Neptune, taken at the VLA, are discussed by Muhleman *et al.* (1985) and by Berge *et al.* (1988). They refer the center of the thermally-emitting disk to the IERS reference radio sources with an accuracy of about ($\pm 0''\!.03$).
- Spacecraft tracking files. These are the standard doppler and range data which are used by the JPL Navigation Team for determining spacecraft trajectories. If the spacecraft is close enough to the planet to be influenced by its gravity, the ephemeris of the target planet is adjusted as part of the orbit determination process.
- Laser-ranging to the moon. The twenty-five years of data show a general increase in accuracy from 30 cm in the early years to less than 3 cm at present.
- IERS-based ΔVLBI of Phobos approaching Mars and Magellan orbiting Venus: These observations have been reported by Folkner *et al.* (994a,b). In conjunction with the planet-centered spacecraft ephemeris, these observations provide a position of the planet wrt the IERS Frame which is generally accurate to about ($\pm 0''\!.003$).
- Frame-tie measurement. (Folkner *et al.*., 1993a) have measured the frame-tie between JPL ephemerides and the IERS Reference Frame: the IERS Catalogue is connected to the VLBI antennae through the VLBI measurements; the antennae are connected to the LLR telescopes via geodetic ground surveys; the LLR telescopes are connected to the Moon by the LLR observations. Since the LLR data is sensitive to the positions of the sun, the whole inner planet system is connected to the Moon. Thus, the planetary positions are effectively linked to the IERS Catalogue.

4. Data Reduction and Dynamical Modeling

Most of the standard data reduction and dynamical modeling in the JPL ephemeris improvement process have been discussed previously (see Standish, 1990 for the former; Newhall et al., 1983 for the latter). There have been some more recent refinements, however.

4.1. OPTICAL DATA

The observations are initially transformed onto the FK5 system. They are then corrected for more recent improvements to the value of precession. Finally, they are transformed onto the IERS Frame using a rotation matrix determined in the ephemeris solution itself.

4.2. RADAR RANGES

The radar-ranging to the surfaces of the terrestrial planets is directly affected by uncertainties in the topographical surface of the planet. Refinements come from improved modeling of these surfaces. For Mercury, an elliptically shaped equator has been fit and removed from the residuals, using the form, $-c\Delta\tau = R_{\text{Mercury}} + a \cos(2\lambda) + b \sin(2\lambda)$, where λ is the longitude of the echo point on the surface of Mercury. For Venus, a topographical model of the surface, derived from Pioneer-Venus Orbiter data (Pettengill et al., 1980), was used.

4.3. ASTEROID PERTURBATIONS

There are quite a few asteroids whose perturbations produce a non-negligible effect upon the orbits of Mars and the Earth. For the JPL ephemerides, these are handled in two different ways.

The "Big 3": Ceres, Pallas, and Vesta, are handled individually: they are assumed to follow periodic Keplerian orbits: osculating elements are first transformed into mean elements (Williams, 1989); from the mean elements, positions for each asteroid are computed throughout one of its periods; these positions are fit with Chebychev polynomials, to be interpolated later during the integration; the interpolated positions are combined with a pre-assigned mass to compute the force which is applied to every planet. Corrections to the mass of each of the three asteroids may be solved for.

The 297 next most important asteroids are handled in a different manner: these asteroids are grouped according to approximate taxonomic class: C, S, or M; osculating elements are transformed into mean elements; the volume of each asteroid is computed from its estimated diameter, and a density is assigned according to its taxonomic class; vectors of the forces upon

Mars, the earth and the moon are computed for each asteroid, summed over all of the asteroids in the class, and stored in a file to be interpolated during the integration of the equations of motion. Corrections to the density of each of the three classes may be solved for.

It should be noted in both cases, that the asteroids' contributions to the location of the center of mass of the solar system are accounted for.

The masses of the asteroids, Ceres, Pallas and Vesta, determined for DE403 are 4.64×10^{-10}, 1.05×10^{-10}, and 1.34×10^{-10} solar masses, respectively. These are 7% , 25% and 11% lower than those estimated by Standish and Hellings (1989). Interestingly, these newer results for Pallas and Vesta are quite close to the original values obtained by Schubart (1975); however, the uncertainties involved, approximately the sizes of the changes themselves, show this to be probably coincidental. On the other hand, the new value for the mass of Ceres agrees quite closely with results reported at this conference by Carpino and Knežević (1995) and by Viateau (1995).

For the densities of the three asteroid classes, the determinations for the M-class seem quite untrustworthy; consequently, that value was adopted as 5.0 gm/cm^3 and not solved-for. The determined values for the C- and S-classes, 1.80 and 2.40, must be regarded as quite tentative.

4.4. LUNAR MODELING

The set of solution parameters includes orbits of moon and earth, physical librations, observatory and reflector coordinates, lunar gravity harmonics and moment of inertia combinations, lunar Love numbers and rotational dissipation, GM of earth+moon, precession and nutation of the earth, earth rotation corrections (mainly to the earlier data span) and drifts, and tidal acceleration parameters. Plate motion is imposed upon the stations.

Since the IERS celestial coordinates are not aligned with the J2000 equator it is necessary to solve for the offset of the equator from the zero-declination plane of the IERS system (or equivalently, to solve for the offset of the rotation pole from the IERS north pole). The precession and obliquity rates have been adopted, the ecliptic plane is constrained, and the offsets are solved for. The offsets were treated as constant corrections to the two nutation components, which is an approximation. With T in centuries from J2000,

$\Delta\epsilon = -0\rlap{.}''00399 - 0\rlap{.}''024 \; T +$ periodic nutation corrections, and

$\sin \epsilon \Delta\psi = -0\rlap{.}''01536 - 0\rlap{.}''1193 \; T +$ periodic nutation corrections.

The periodic nutation corrections are adopted improvements in the nutations plus a solution for 18.6 yr terms using the constraints of Williams *et al.* (1991). The LLR data also determine the difference between the orbital longitudes of the geocentric moon and sun.

References

Berge, G. L. , Muhleman, D. O. , and Linfield, R. P. : 1988, "Very Large Array Observations of Uranus at 2.0 cm", *Astron. J.* **96**, #1, 388–395.
Folkner, W. M. , Kroger, P. M. , and Iijima, B. A. , and Hildebrand, C. E. : 1992a, "VLBI measurement of Mars on 17 February 1989", JPL IOM 335.1-92-26.
Folkner, W. M. , Kroger, P. M. , and Iijima, B. A. , and Hildebrand, C. E. : 1992b, "VLBI measurement of Mars on 25 March 1989", JPL IOM 335.1-92-27.
Folkner, W. M. , Kroger, P. M. , and Hildebrand, C. E. : 1992c, "Preliminary results from VLBI measurement of Venus on September 12, 1990", , JPL IOM 335.1-92-25
Folkner, W. M. , Charlot, P. , Finger, M. H. , Williams, J. G. , Sovers, O. J. , Newhall, X X, and Standish, E. M. : 1993a, "Determination of the extragalactic frame tie from joint analysis of radio interferometric and lunar laser ranging measurements", *Astron. Astrophys.* **287**, 279–289.
Folkner, W. M. , Kroger, P. M. , and Iijima, B. A. : 1993b, "Results from VLBI measurement of Venus on March 29, 1992", JPL IOM 335.1-93-22
Folkner, W. M. , Kroger, P. M. , Iijima, B. A. , and Border, J. S. : 1994a, "Results from VLBI measurements of Venus on December 22 and December 23, 1991", JPL IOM 335.1-94-006
Folkner, W. M. , Border, J. S. , and Iijima, B. A. : 1994b, "Results from VLBI measurement of Venus on April 1, 1994", JPL IOM 335.1-94-014
Fricke, W. : 1982, "Determination of the Equinox and Equator of the FK5", *Astron. Astrophys.* **107**, L13-L16
Carpino, M. and Knežević, Z.: 1995, this volume.
Lieske, J. H. : 1979, "Precession Matrix Based on IAU (1976) System of Astronomical Constants", *Astron. Astrophys.* **73**, 282-284.
Muhleman, D. O. , Berge, G. L. , Rudy, D. J. , Niell, A. E. , Linfield, R. P. and Standish, E. M. : 1985, "Precise Position Measurements of Jupiter, Saturn and Uranus Systems with the Very Large Array", *Cel. Mech.* **37**, 329–337.
Newhall, X X, Standish, E. M. and Williams, J. G. : 1983, "DE102: a numerically integrated ephemeris of the Moon and planets spanning forty-four centuries", *Astron. Astrophys.* **125**, 150–167.
Pettengill, G. H. , Eliason, E. , Ford, P. G. , Loriot, G. B. , Masursky, H. , and McGill, G. E. : 1980, "Pioneer Venus Radar Results: Altimetry and Surface Properties", *J. Geophys. Res.* **85**, A13, 8261–8270; table of values transmitted to the authors via W L Sjogren.
Schubart, J. : 1975, "The mass of Pallas", *Astron. Astrophys.* **39**, 147–148.
Standish, E. M. , Keesey, M. S. W. and Newhall, X X : 1976, "JPL Development Ephemeris Number 96", JPL Tech. Rep. , 32-1603, Pasadena.
Standish, E. M. and Hellings, R. W. : 1989, "A Determination of the Masses of Ceres, Pallas and Vesta from their Perturbations upon the Orbit of Mars", *Icarus* **80**, 326-333.
Standish, E. M. : 1990, "The observational basis for JPL's DE200, the planetary ephemerides of the Astronomical Almanac", *Astron. Astrophys.* **233**, 252-271.
Standish, E. M. , Newhall, X X, Williams, J. G. and Folkner, W. F. : 1995, "JPL Planetary and Lunar Ephemerides, DE403/LE403", JPL IOM 314.10-127.
Viateau, B. : 1995, private communication.
Williams, J. G. : 1989, "Harmonic Analysis", *BAAS* **21**, # 3, 1009.
Williams, J. G. , Newhall, X X, and Dickey, J. O. ; 1991, "Luni-solar Precession: Determination from Lunar Laser Ranges", *Astron. Astrophys.* **241**, L9-L12

PLANETARY AND LUNAR EPHEMERIDES, LUNAR LASER RANGING, AND LUNAR PHYSICAL LIBRATIONS

X X NEWHALL, J. G. WILLIAMS, AND E. M. STANDISH, JR.
Jet Propulsion Laboratory
California Institute of Technology
Pasadena, California 91109-8099 USA

1. The Planetary and Lunar Ephemerides

The Jet Propulsion Laboratory (JPL) has recently produced a new integrated planetary and lunar ephemeris DE403/LE403. This ephemeris spans the interval JED 624912.5 (December 2, −3002, Julian) − JED 2817104.5 (November 14, 3000, Gregorian) and is an improvement on DE102 (Newhall *et al.*, 1983) and DE200 (Standish, 1990). This integration carries the Cartesian states of the Sun, Moon, and planets, along with the three Euler angles describing the lunar physical librations.

1.1. DYNAMICAL MODEL

The dynamical model of the solar system consists of the following point-mass effects:

(1) The relativisitic n-body equations of motion (see Newhall *et al.*, 1983) for the point-mass Sun, Moon, and planets;

(2) The three major asteroids Ceres, Pallas, and Vesta perturbing the Moon and planets; and

(3) The resultant accelerations on the Earth, Moon, and Mars due to 297 additional asteroids. (These accelerations are not computed simultaneously with the integration of the planets but are produced separately and read from a file.)

In addition, perturbations on the orbit of the Earth-Moon barycenter are included from the interaction of the point-mass Sun with the figure and solid-body tides of both the Earth and Moon; perturbations on the geocentric lunar orbit include effects of each of the Earth and Moon acting on the figure and solid tides of the other.

1.2. NUMERICAL INTEGRATION

The numerical integration of the solar-system bodies was done using the subroutine package DIVA, a variable-order, variable-step Adams method (Krogh, 1974). The planets, Moon, and lunar librations were integrated; the state of the Sun was computed from the relativistic conservation of center of mass. The step size was approximately 0.3 days; for DE403 there were 7.4 million steps.

The integration was done in nine successive sections. To preserve continuity and avoid noise and integration errors introduced by restarts at the beginning of each section, the integrator's difference tables were saved at the end of each integration section and then reloaded later to allow seamless continuation.

1.3. EPHEMERIS REPRESENTATION

The distributed ephemerides are in the form of Chebyshev polynomials. For the Sun, planets, and Moon the polynomials represent Cartesian position coordinates; they are solar-system barycentric for the Sun and planets and geocentric for the Moon. The lunar physical librations are expressed as three Euler angles defining a rotation between the Earth mean equator and equinox of J2000 and the lunar selenographic system. For a detailed description of Chebyshev representation of these ephemerides, see Newhall (1989).

Polynomial interpolation error is the difference between an interpolated value and the corresponding value obtained from the original integrator difference tables. For DE403, the maximum interpolation error is less than 0.5 mm for the Sun, Moon, and planets; for the libration angles the maximum interpolation error is less than 20 μas. A significantly compressed form of DE403, designated DE404, in which the interpolation-error criteria are relaxed, is available (Standish and Newhall, 1995). On that ephemeris, the maximum interpolation error is less than 25 meters for the Sun and planets and less than 1 meter for the Moon. The lunar librations are not included on DE404.

2. Lunar Laser Ranging

Since 1969, laser ranges (round-trip light travel times) to retroreflectors on the Moon have been acquired by terrestrial observatories. Beginning in early 1970, ranges were sufficiently accurate for use in the determination of astronomical quantities. As more data were accumulated, estimates were made of the lunar ephemeris, the lunar physical librations, and, ultimately, parameters associated with General Relativity. This section contains a brief summary of the present status of lunar laser ranging (LLR); for a thorough review article, see Dickey *et al.* (1994).

2.1. THE LLR DATA SET

The present set of LLR data consists of more than 9900 normal points. From 1970 through 1983 the data are from only the McDonald Observatory of the University of Texas. Beginning in 1984, ranges were acquired at Mt. Haleakala on the island of Maui and then at CERGA in Grasse, France.

Uncertainties on the data were initially 20–30 cm. In 1976, instrumentation improvements yielded uncertainties of 10–15 cm; subsequent continuing improvements have provided uncertainties in the 2–3 cm range since 1991.

2.2. THE OBSERVATION MODEL

Computed ranges between each station and reflector are required for parameter estimation. Errors in the modeled values must be substantially less than the uncertainties of the ranges. The present model includes contributions from:

(1) Planetary and lunar ephemerides;
(2) UT1 and polar motion, including sub-daily effects in UT1;
(3) Geocentric station locations;
(4) Selenocentric reflector locations;
(5) Solid tides on the Earth and Moon, produced on each body by the gravity field of the other and by that of the Sun;
(6) Pole tides on the Earth, arising from the elastic deformation of the Earth caused by the displacement of the instantaneous pole from the mean pole;
(7) Precession and nutation of the Earth;
(8) Physical librations of the Moon;
(9) Atmospheric delay; and
(10) Relativistic effects, including the Lorentz contraction of the Earth and Moon, the time delay in the ranges caused by the combined potential of the Sun and Earth, and the transformation between TAI (atomic time) and coordinate time.

2.3. ESTIMATED PARAMETERS

The data set permits the estimation of several parameters to which the observations are sensitive. In the estimation process, LLR data are combined with planetary data to allow the estimation of additional solar-system parameters. The solution set includes:
- Earth orbit initial conditions
- Lunar orbit initial conditions
- Lunar physical libration initial conditions
- The length of the astronomical unit
- $GM_{\text{Earth+Moon}}$
- $M_{\text{Earth}}/M_{\text{Moon}}$
- Station locations
- Reflector locations
- Lunar moments of inertia
- Lunar harmonics to third degree
- Lunar Love numbers k_2, h_2, and l_2
- Lunar dissipation time delay
- Earth dissipation time delays (and the consequent secular acceleration of the lunar longitude)
- Precession correction
- Long-period nutation terms
- Fast (sub-daily) UT1 terms
- Relativity parameters: β, γ, $\dot{G}/G$, $M_{\text{Gravitational}}/M_{\text{Inertial}}$ for the Earth, and geodetic precession
- UT1 and polar motion

The estimation of UT and polar motion treats those variables as stochastic parameters. A Kalman-filter approach is used to estimate a value of UT and the two components of polar motion at every data point. This process is done simultaneously with the estimation of a single (global) value of each of the remaining parameters in the set. For the single-station (McDonald) data before 1983, only the two parameters UT0 and variation of latitude are estimated. For data after that time, *a priori* covariances are sufficiently accurate to allow estimation of UT1 and both components of polar motion.

3. The Lunar Physical Librations

3.1. DEFINITION OF THE LIBRATIONS

Lunar laser ranging (LLR) has enabled the determination of the lunar ephemeris and rotation history (physical librations) to a high degree of accuracy. The librations consist of a set of three Euler angles that define a rotation matrix between the Earth mean equator and equinox of J2000 and the selenographic system. The angles are ϕ, the angle along the Earth's equator from the vernal equinox to the ascending node of the lunar equator; θ, the inclination of the lunar equator to the Earth's equator; and ψ, the angle along the lunar equator from its node on the Earth's equator to the lunar prime meridian.

3.2. DESCRIPTION OF THE LIBRATIONS

There are two basic types of librations: longitude and latitude. The libration in longitude refers to the behavior of the combination $\Delta\psi + \Delta\phi \cos\theta$, the total variation of the angular position of the Moon about its axis. It is analogous to a variation in UT1 on Earth, except that the lunar libration has no known stochastic component.

The lunar libration in latitude refers to the variation of the position of the lunar pole in space (the combined effect of the two variations $\Delta\phi$ and $\Delta\theta$) and is the equivalent of a combination of precession, nutation, and polar motion on Earth. The lunar north pole precesses one full revolution about the ecliptic pole every 18.6 years and is displaced from the ecliptic pole by 1.54°. By comparison, the Earth's north pole precesses one revolution about the ecliptic pole every 26,000 years and is displaced from the ecliptic pole by 23.4°. (For a thorough description of the lunar librations, see Eckhardt (1981).)

3.3. FORCED VS. FREE LIBRATIONS

Forced librations are those variations in lunar rotation and orientation caused by external time-varying torques originating from the gravitational effects of the Earth, Sun, and planets. *Free librations* are variations in rotation and orientation due to three natural libration modes and frequencies. They can be stimulated by impacts and internal processes. The free libration period is also a function of the lunar orbit period, which is increasing due to the effect of the Earth's tides. Passage of the libration period through a resonance with the external forcing terms can stimulate a free libration mode (Eckhardt, 1993).

The amplitude and phase of the free librations cannot be predicted; they must be measured. The free libration in longitude has a predicted period of 1056 days. The free libration in latitude has two predicted modes: a Chandler-like wobble with a period of about 74.6 years (when expressed in the rotating lunar frame), and a precession-like motion with a 79-year period (when expressed in the inertial frame).

3.4. THE PHYSICAL LIBRATION MODEL

The librations are the result of external torques on the Moon. Sources for torques include the point-mass Sun, Earth, Venus, and Jupiter interacting with the lunar figure; the effect of the point-mass Sun and Earth acting on the lunar solid tides; and the mutual interaction between the Earth and Moon figures.

Lunar elasticity and dissipation are modeled by assuming a time-varying moment of inertia $\mathbf{I}(t;\tau)$ that is a function of the Earth-Moon vector $\mathbf{r}_{EM}$ and lunar angular velocity ω at a retarded time $t - \tau$:

$$\mathbf{I}(t;\tau) = f[\mathbf{r}_{EM}(t-\tau), \omega(t-\tau)]$$

The value of τ is typically 4 hours.

When the model is integrated backwards in time, the dissipative terms will contribute to an exponential growth in libration amplitude. Further, in a backwards integration the retarded-time formulation functions mathematically as an advanced-time equation and is inherently unstable, augmenting the exponential growth in amplitude.

The fundamental differential equation for the lunar rotation in terms of the lunar angular velocity ω is:

$$\dot{\omega}(t) = \mathbf{I}(t;\tau)^{-1} \{-\omega(t) \times \mathbf{I}(t;\tau)\omega(t) - \dot{\mathbf{I}}(t;\tau)\omega(t) + \mathbf{N}_{\text{pt-mass}} + \mathbf{N}_{\text{fig-fig}}\}$$

where $\mathbf{I}(t;\tau)$ is the inertia tensor defined above, $\mathbf{N}_{\text{pt-mass}}$ is the net torque resulting from point-mass effects, and $\mathbf{N}_{\text{fig-fig}}$ is the torque from the mutual interaction from the Earth and Moon figures.

When the above equation is integrated, the three libration angles ϕ, θ, and ψ are obtained by integrating their corresponding differential equations:

$$\dot{\phi} = (\omega_x \sin \psi + \omega_y \cos \psi)/\sin \theta$$
$$\dot{\theta} = \omega_x \cos \psi - \omega_y \sin \psi$$
$$\dot{\psi} = \omega_z - \dot{\phi} \cos \theta$$

where ω_x, ω_y, and ω_z are the components of the angular velocity vector ω in the selenographic system.

3.5. THE LIBRATION STUDY

A large rotational dissipation was discovered from LLR analyses and is a clue about the lunar interior. Its origin is uncertain. Solid-body friction is an unlikely source; however, observations are consistent with viscous damping at the surface of a small fluid core interacting with the mantle (Yoder, 1981; Dickey et al., 1994).

3.5.1. *Motivation*

Free librations were reported by Calame (1977). The damping time should be geologically short, due to the small lunar Q (about 26), and impacts appear statistically too infrequent to serve as the only source of excitation. Yoder (1981) shows that turbulence in a core-mantle boundary could function as a continuing, more frequent internal source of excitation and could account for the existence of free librations in the presence of strong dissipation. If free librations can be established, their presence would support the existence of a fluid core.

3.5.2. *Separation of Free and Forced Librations*

Previous researchers have compared observations with both an analytical theory and a numerical integration of forced librations. In this study a Fourier analysis approach is taken. First, initial conditions for the physical librations were derived from a fit to LLR data. Next, the librations were integrated simultaneously with the solar system bodies over the time span of interest. Then an iterative procedure was used for each of the three libration angles: (1) A Fourier analysis is done over the selected span (718 years [262,144 days] in this study) to identify the largest amplitudes; (2) these amplitudes are least-squares estimated; (3) the corresponding frequencies are then estimated; and (4) the resulting trigonometric series is evaluated and subtracted from the values of the input angles to provide the modified array of angle values to be Fourier analyzed. Iteration was continued until all components down to a selected amplitude level had been determined.

3.6. FREE LIBRATION RESULTS

In the case of longitude, an amplitude of $1.8''$ near the free libration period is seen; however, the free libration is blended with two forced amplitudes due to Venus (Eckhardt, 1982) having nearly identical periods of 1056.3 and 1056.4 days and resists separation. The free period cannot be determined precisely; if the two forced terms are subtracted out, the resulting free amplitude is strongly dependent on what free period is assumed. To illustrate this sensitivity, an assumed 1056.00-day period yields a free amplitude of $1.4''$; a 1056.20-day period yields an amplitude of $0.9''$; and a 1056.28-day period gives an amplitude of $0.2''$.

The free libration in latitude exhibits a $3'' \times 8''$ elliptical component in the pole oscillation with a 74.6-year period. There are no known nearby theoretical forced terms.

The libration in latitude and the blend in longitude correspond to Calame's (1977) results.

It should be noted that the span of LLR data is 25 years. One might wonder how a 718-year analysis can be done on a support of only 25 years of data. The underlying assumption is that the physical model, data, and integration software are precise enough to permit extrapolation to the desired resolution. It is encouraging that the Fourier estimation-iteration procedure has yielded amplitudes and frequencies predicted by theory and that nothing unduly dramatic or spurious has appeared.

4. Acknowledgement

The research described in this paper was performed at the Jet Propulsion Laboratory, California Institute of Technology, under a contract with the National Aeronautics and Space Administration.

References

Calame, O. (1977) Free librations of the Moon from Lunar Laser Ranging, in *Scientific Applications of Lunar Laser Ranging*, J. D. Mulholland, ed., D. Reidel, Dordrecht, pp. 53–63

Dickey, J. O., Bender, P. L., Faller, J. E., Newhall, X X, Ricklefs, R. L., Ries, J. G., Shelus, P. J., Veillet, C., Whipple, A. L., Wiant, J. R., Williams, J. G., and Yoder, C. F. (1994) Lunar Laser Ranging: A Continuing Legacy of the Apollo Program, *Science*, **265**, pp. 482–490

Eckhardt, D. H. (1981) Theory of Libration of the Moon, *The Moon and the Planets*, **25**, pp. 3–49

Eckhardt, D. H. (1982) Planetary and Earth Figure Perturbations in the Librations of the Moon, in *High-Precision Earth Rotation and Earth-Moon Dynamics*, O. Calame, ed., D. Reidel, Dordrecht, pp. 193–198

Eckhardt, D. H. (1993) Passing Through Resonance: The Excitation and Dissipation of the Lunar Free Libration in Longitude, *Celest. Mech. Dyn. Astron.*, **57**, pp. 307–324

Krogh, F. T. (1974) Changing Stepsize in the Integration of Differential Equations Using Modified Divided Differences, in *Proceedings of the Conference of the Numerical Solution of Ordinary Differential Equations*, Oct. 1972, Lecture Notes in Mathematics, **362**, Springer-Verlag, New York, pp. 22–71

Newhall, X X, Standish, E. M., Jr., and Williams, J. G. (1983) DE102: a numerically integrated ephemeris of the Moon and planets spanning forty-four centuries, *Astron. Astrophys.*, **125**, pp. 150–167

Newhall, X X (1989) Numerical Representation of Planetary Ephemerides, *Celest. Mech.*, **45**, pp. 305–310

Standish, E. M., Jr. (1990) The observational basis for JPL's DE200, the planetary ephemerides of the Astronomical Almanac, *Astron. Astrophys.*, **233**, pp. 252–271

Standish, E. M., Jr. and Newhall, X X (1995) New Accuracy Levels for Solar System Ephemerides, in *Proceedings of IAU Symposium 172: Dynamics, Ephemerides, and Astrometry of Solar System Bodies*, Paris (this volume)

Yoder, C. F. (1981) The free librations of a dissipative Moon, *Phil. Trans. R. Soc. Lond.* Series A **303**, pp. 327–338

USING SPACECRAFT RANGE DATA AND RADAR OBSERVATIONS FOR THE IMPROVEMENT OF THE ORBITAL ELEMENTS OF PLANETS AND PARAMETERS OF MARS ROTATION

E.V. PITJEVA

Institute of Applied Astronomy, Russian Academy of Sciences
8 Zhdanov st., St.Petersburg, 197042 Russia
E-mail evp@ipa.rssi.ru

Abstract. The extremely precise Viking (1972–1982) and Mariner data (1971–1972) were processed simultaneously with the radar-ranging observations of Mars made in Goldstone, Haystack and Arecibo in 1971–1973 for the improvement of the orbital elements of Mars and Earth and parameters of Mars rotation. Reduction of measurements included relativistic corrections, effects of propagation of electromagnetic signals in the Earth troposphere and in the solar corona, corrections for topography of the Mars surface. The precision of the least squares estimates is rather high, for example formal standard deviations of semi-major axis of Mars and Earth and the Astronomical Unit were 1–2 m.

A set of range measurements of the two martian landers Viking-1 and Viking-2, obtained by the Jet Propulsion Laboratory of the USA during the years 1976–1982 surpass in accuracy any planetary radar-ranging observations. With Mariner-9 tracking data they have been one of the major contributors to the accuracy of the JPL planetary ephemerides.

It may be thought that the direct radar-ranging to planet surfaces becomes useless because the observations of this type are seriously corrupted by peculiarities of the planet relief. Nevertheless we believe that these observations are valuable for two reasons. Firstly, the relief errors may be taken into account with the help of the modern planet hypsometric maps of high accuracy. Secondly, the radar range observations now cover almost

30 years whereas the Mariner-9 (1971–72) and the Viking (1976–82) data are restricted to two short intervals of time.

In this paper we present results of the first stage of our investigation when the Viking and Mariner data, obtained from Standish (1994), were processed simultaneously with the radar-ranging observations of Mars made in Goldstone, Haystack and Arecibo in 1971–1973. Later, we intend to combine them with the all set of Russian and American radar observations, obtained during 1961–1995. The set of the data used now is given in Table 1

Table 1. Observations used in the ephemeris solutions.

Type of observations	Date	Number	A priori accuracy
Mariner-9 normal points	10.1971–09.72	645	40–400m
Viking Lander range points	07.1976–09.82	2462	7–12m
Haystack Mars radar ranging	04.1971–11.73	3490	70–7000m
Goldstone Mars radar ranging	06.1971–11.73	31274	70–500m
Arecibo Mars radar ranging	10.1973–11.73	1644	70–300m

The ranging observations near conjunctions with the Sun were corrected for the solar corona effect with simultaneous determinations of parameters of a corona model.

Great attention was given to the reduction of radar observations for the topography of Mars. This correction may be carried out by two methods. The first method makes use of hypsometric maps of Mars (Sherman, 1978). The second method uses a representation of the global topography of Mars by an expansion of spherical functions of 16–18 degrees (Bills and Ferrari, 1978). We have combined both these approaches. Usually the topographic reduction was made using spherical harmonics, but for some areas (for example, Olympus Mons) the heights were computed making use of the hypsometric map of this areas. The further details of the reduction and residuals of these observations can be founded in the paper (1995).

A least-squares solution for 23 unknowns was produced, where the unknowns were the elements of Mars and Earth (only those independent of the three-dimensional rotations), the Astronomical Unit, the scale correction to the reference surface of Mars, the parameter B of the solar corona, the lander locations, the rate of Mars rotation ($\dot{V}$), the angles of the Mars orientation (Ω_q and I_q) and their variations.

The formal standard deviations of the adjustment of the orbital parameters from the range data of Viking-1,-2, Mariner-9 and radar observations of Mars are given in Table 2.

Table 2. The formal deviations of the adjustment of orbital parameters.

Mars

$\Delta a/a$	$\Delta(\sin I \cos \Omega)$	$\Delta(\sin I \sin \Omega)$	$\Delta(e \cos \pi)$	$\Delta(e \sin \pi)$	$\Delta \lambda$
0."000003	0."000159	0."000217	0."000014	0."000014	0."000050

Earth-moon barycenter

$\Delta a/a$	$\Delta(e \cos \pi)$	$\Delta(e \sin \pi)$
0."000001	0."000009	0."000009

$\Delta AU = 2.135$ m

As seen from Table 2, the accuracy of the least-squares estimations of the parameters is rather high. It should be remembered however, that such formal statistics tend to be highly optimistic when there are unmodeled forces; in this case, the perturbations of many small asteroids cannot be modeled precisely, since their masses are unknown.

The positions of the landers must be computed taking into account the precession and nutation of Mars. Therefore, the tracking data allow one not only to improve the orbital elements of planets, but also to study the precession and nutation of the spin axis in order to provide a better understanding of planet geophysics.

The values of Mars's orientation parameters and the Viking lander coordinates from our solution are given below:

$$V = (350°.89198649 \pm 0°.00000038)/\text{day}$$
$$\Omega_q = 35°.34746 \pm 0°.00037$$
$$\dot{\Omega}_q = (0°.230 \pm 0°.014)/\text{century}$$
$$I_q = 25°.18231 \pm 0°.00018$$
$$\dot{I}_q = (-0°.0292 \pm 0°.0080)/\text{century}$$

These formal standard deviations are in agreement with those of Standish (1990), notwithstanding that somewhat different observational data were used in this work.

We have an estimation of the precession constant of Mars: $(-750'' \pm 36'')$/century, which is in a good agreement with the result derived from the analyses of Viking Lander range and doppler data covering a time span of nearly four years (1976–1980) (Michael and Kelly, 1981) and also with that of $-708''$/century, predicted by Lowell (1914).

It is seen from post-fit residuals that some systematic component is present, although post-fit residuals for the measurements obtained during the first 14 months are consistent with the values from the paper by Reasenberg et al.(1979). Different numerical experiments were carried out in attempts to eliminate this systematic component. We tried to explain it by a polar motion of Mars or by corrections to the principal nutation terms, but all the attempts failed. Possibly, this systematic component is due to insufficiency of our dynamical modeling of the perturbations from asteroids. At present there are only the perturbations from five most massive asteroids in our model, but we are planing to take into account perturbations of asteroids in more full scale.

It would be desirable to continue such experiments and obtain new data from landers on the surface of other planets, especially on the surface of Mercury. Such observations could give an opportunity to improve not only Mercury's elements and its rotation but to estimate the variability of the gravitational constant with the standard error $10^{-12} - 10^{-13}$ per year. But only Mars landers are planned: American project MESUR and the Russian project MARS-96 (this missions will place in July 1997). We expect that including new data from these missions to Mars into our processing would improve essentially the estimate of the Mars precession constant and make it possible to determine the amplitudes of the dominant short periodic nutations. As shown by our computations the inclusion of data within a year from the MESUR lander yields the estimate of the Mars precession constant with a standard error of $10'' - -15''$/century. But if the duration of the MESUR mission is only one month the improvement of this estimate will be insignificant.

The author would like to thank both French colleagues and Soros Foundation for financial support which enabled her participation in the Symposium 172. It is pleasure to acknowledge Dr. Standish for useful comments on the manuscript.

References

Bills, B.G. and Ferrari, A.J. (1978) Mars topography harmonics and geophysical implications, *J.Geophys.Res.*, **83**, **B7**, pp. 3497-3508.

Lowell, P., (1914) *Astron.J.*, **28**, pp. 169-171.

Michael, W.H.Jr. and Kelly, G.M. (1981) Dynamical constants and reference parameters for Mars, *Union Colloquium N 56 "Reference coordinate systems for Earth Dynamics"*, Warsaw, pp. 325-328.

Pitjeva, E.V. (1995) Using range observations of spacecrafts VIKING-1, VIKING-2, Mariner-9 for improvement of orbital elements of planets and parameters of Mars rotation, *Proceeding of the Third Workshop on Positional Astronomy and Celestial Mechanics, in Cuenca, October, 17-21, 1994*, Spain, Editor A. Lopez et al. (in print).

Reasenberg, R.D. et al. (1979) Viking relativity experiment: verification of signal retardation by solar gravity, *Astroph.J.*, **234**, L219-L221.

Sherman, S.C. (1978) Mars synthetic topographic mapping, *Icarus*, **33 no. 3**, pp. 417-440.

Standish, E.M.Jr. (1990) The observation basis for JPL's DE200, planetary ephemerides of the Astronomical Almanac, *Astron.Astroph.*, **233**, pp. 252-271.

Standish, E.M.Jr. (1994) (private communication).

NEW SEMI-ANALYTIC THEORY OF THE FOUR OUTER PLANETS

J.L. SIMON
Bureau des longitudes
77, avenue Denfert-Rochereau
75014, Paris FRANCE

1. Introduction

A planetary theory of the planets Jupiter, Saturn, Uranus and Neptune is presented here. It is a classical planetary theory where the perturbations are computed in the form of Poisson series of only one angular variable. It is built with modern values of the planetary masses and fitted to the numerical integration DE245 of the Jet Propulsion Laboratory (Standish, 1994). Its validity time span is of several thousand of years.

2. Form of the solution

2.1. NOTATIONS

The classical elliptic variables are used: $a, \lambda, e, \varpi, \gamma = \sin \frac{i}{2}, \Omega, k = e \cos \varpi, h = e \sin \varpi, q = \gamma \cos \Omega, p = \gamma \sin \Omega$. The mean mean longitude $\bar{\lambda}$ is given by: $\bar{\lambda} = \lambda^0 + Nt$ where λ^0 is the integration constant for the mean longitude λ, N is the mean mean motion and t is the time. The planets Jupiter, Saturn, Uranus and Neptune are numbered from 5 to 8.

2.2. THE REPRESENTATION

In a classical planetary theory the variables x are developed under the form of Poisson series:

$$x = x_0 + x_1 t + \ldots + x_j t^j + S_0 + tS_1 + \ldots + t^j S_j \tag{1}$$

where: x_0 is the integration constant for x, t is the time, x_q are numerical coefficients, S_q are Fourier series in the mean mean longitudes.

Here, the solutions are also Poisson series of the form (1), but S_q are Fourier series in only one argument μ:

$$S_q = \sum_r \{A_r \cos r\mu + B_r \sin r\mu\} \qquad (2)$$

with: $\mu = (N_5 - N_6)t/880 = 0.3595t$, where t is measured in thousands of years from J2000. The period of μ is 17 485 years.

μ is related to the mean mean longitudes $\bar{\lambda}_i$ by: $\bar{\lambda}_i = q_i\mu + \varepsilon_i t$, where q_i are integers and where ε_i are small quantities compared with N_i.

This representation was choosen because perturbations are more convergent under this form than under the classical form of Poisson series of the mean mean longitudes, for Jupiter and Saturn (Simon et al., 1992). For the other planets the convergence of perturbations is the same than under the classical form.

3. Construction of the solution

3.1. THE METHOD

The solution is computed by the iterative method of (Simon and Joutel, 1988). The equations are Lagrange equations under the form: $\dfrac{dx}{dt} = F(\mu, t)$

$F(\mu, t_i)$ are computed, by harmonical analysis, for 11 values of the time $t_0, (J2000), t_1, \ldots, t_{10}$. After interpolation and integration the solution has the form (1) with $j = 10$. For each iteration, the truncation error is $0''.0001$ over $-1000, +1000$ (0=J2000) or $0''.01$ over $-6000, +6000$.

3.2. PERTURBATIONS BY THE INNER PLANETS

Perturbations by the inner planets developed up to the third order of the masses are issued from VSOP82 (Bretagnon, 1982). They are transformed in series in μ and added to the solution. Then, they are introduced in the iterations. So, are computed the perturbations by the inner planets to the order 4 of the masses and a part of the perturbations of superior orders.

3.3. DETERMINATION OF THE INTEGRATION CONSTANTS

This theory is built using the IERS Standards 1992 masses of the outer planets (McCarthy, 1992) and it is fitted to DE245. Note that the masses of the outer planets used in DE245 are the masses of IERS 1992, except a very small difference for the mass of Saturn.

4. Results

4.1. DIFFERENCES WITH DE245

Table 1 gives the differences, after fit, between our theory and DE245 for the heliocentric variables, longitude, latitude and radius vector, over 1891–2000. These differences are small: some $0''.0001$ for the longitudes and the latitudes and some 10^{-8} au for the radius vector. Note that a large part of these differences is probably due to the truncation error. So, a few iterations with a truncation error of $0''.00001$ over $-1000, +1000$ are intended.

TABLE 1. Differences between our theory and DE245 for the heliocentric coordinates, over 1891–2000.

	Longitude $(0''.001)$	Latitude $(0''.001)$	Radius Vector $(10^{-8}\,\text{au})$
Jupiter	5	2	7
Saturn	8	3	19
Uranus	8	5	50
Neptune	4	2	59

4.2. PRECISION OF THE THEORY

The precision of our theory is computed by comparison to internal numerical integration of the eight major planets run over J2000−6000, J2000+6000. Figure 1 shows the differences between the theory and the numerical integration for the mean longitudes. For Jupiter, Uranus and Neptune the differences are smaller than $0''.01$ over $-1000, +1000$ and smaller than $0''.8$ over $-6000, +6000$. For Saturn, the precision is about $0''.02$ over $-1000, +1000$ and about $2''$ over $-6000, +6000$. For the other elliptic variables, the differences theory−numerical integration are of some $0''.001$ over $-1000, +1000$ and smaller than $0''.5$ over $-6000, +6000$.

The gain in precision, in connection with the best theories actually available, as VSOP87 (Bretagnon and Francou, 1987) and JASON84 (Simon and Bretagnon, 1984), is about 10.

5. Conclusion

We are going to obtain a semi-analytic theory of the outer planets built with modern values of the planetary masses, fitted to DE245 and showing real progress compared to the theories actually available.

Besides, due to the best convergence of the mutual perturbations Jupiter-Saturn, compact solutions, useful for historians, could be extracted from our theory.

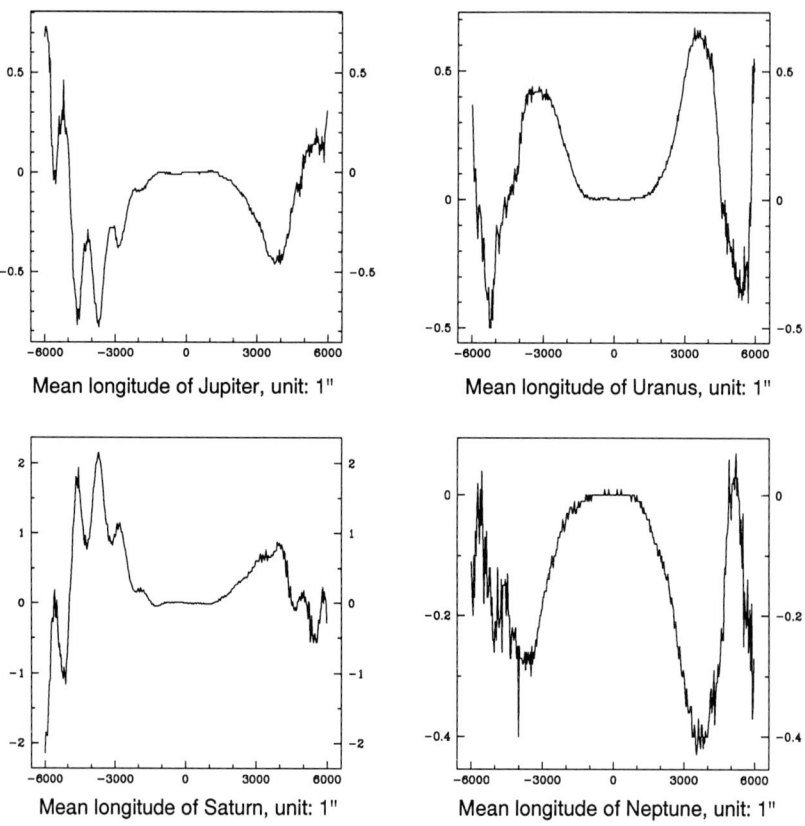

Figure 1. Mean longitudes of the outer planets. Comparison between our theory and an internal numerical integration of the eight major planete over −6000, +6000. 0 is J2000.

References

Bretagnon, P. (1982) *Astron. Astrophys.*, Vol. no. 114, p. 278–288.
Bretagnon, P., Francou, G. (1987) *Astron. Astrophys.*, Vol. no. 202, p. 309–315.
McCarthy, D.D. (1992), IERS Standards (1992) *IERS Technical Note . 13*, Observatoire de Paris
Simon, J.L. (1983) *Astron. Astrophys.*, Vol. no. 120, p. 197–202.
Simon, J.L., Bretagnon, P. (1984) *Astron. Astrophys.*, Vol. no. 138, p. 169–178.
Simon, J.L., Joutel, F. (1988) *Astron. Astrophys.*, Vol. no. 205, p. 328–334.
Simon, J.L., Joutel, F., Bretagnon, P. (1992) *Astron. Astrophys.*, Vol. no. 265, p. 308–323.
Standish, E.M. (1982) *Astron. Astrophys.*, Vol. no. 114, p. 297–302.
Standish, E.M. (1994), *Private Communication*

THE ORBITAL STABILITY OF THE SUN-JUPITER-SATURN SYSTEM

INVESTIGATION OF THE TRIPLE SYSTEM BY MASS ENHANCEMENT AND VARIATION OF ECCENTRICITY

J. R. DONNISON AND D. F. MIKULSKIS
Department of Mathematical Sciences
Goldsmiths College, University of London
New Cross, London, SE14 6NW

Abstract. Kuiper(1973) suggested that the stability of the Solar System may be meaningfully investigated by studying the stability of the Sun-Jupiter-Saturn system. Numerical investigations by Nacozy(1976) showed that mass enhancement of the two planets beyond a factor of 29.25 led to instabilities in the system. In this new investigation similar mass enhancements were studied in detail numerically and compared with the analytical values derived from the c^2H method. In addition, the eccentricities of the two planets were varied as well as their masses. It was found that the system soon showed signs of instability for the increased eccentricities when the masses of the planets were enhanced by fairly small factors.

1. Introduction

Complex systems such as star clusters can be broken down into a series of hierarchical systems where successive separations of the component masses increase by large factors and can be described by a series of two-body motions. In particular, hierarchical three-body systems composed of a binary system with a third body moving on an approximately Keplerian orbit relative to the centre of mass of the binary has been investigated in some detail by number of authors (see Donnison & Mikulskis (1995)). It was suggested by Kuiper(1973) that the stability of such triple systems might have a strong bearing on the stability of the solar system, with the stability of the Sun-Jupiter-Saturn system being the dominant criterion for the continued existence of the planetary system. The stability of this system was investigated analytically by Szebehely and McKenzie (1977), and by

Nacozy(1976) by short-time numerical integration of the system by increasing the masses of Jupiter and Saturn. This enhancement of masses allows any instabilities to emerge more rapidly and thus evolve the system more quickly.

For comparison, a restricted three-body Hill model with the mass of Saturn taken to be infinitesimal and both the planetary orbits assumed to be circular gives a critical planetary mass enhancement factor γ for Jupiter of 18.45. A full three-body planar model using the c^2H parameter (c angular momentum and H the energy of the system) to control the topology of the zero-velocity surfaces, gave, using a two-body approximation for the energy, a value of 13.65 (Szebehely and McKenzie (1977). An improved model by Walker and Roy (1981) with the exact expression for the energy and a range of initial configurations reflected by a range of true anomalies gave $18.34 \leq \gamma \leq 24.16$, where the upper limit represents a straight-line configuration with Jupiter at aphelion and Saturn at perihelion. These models maintain hierarchical or Hill stability with exchange of the component masses or crossing of the orbits ruled out, though the escape of Saturn from the system would not violate such criteria. To establish more general orbital stability, Nacozy(1976) integrated the system numerically for a wide range of γ. No secular trends were found for $\gamma \leq 29.0$, though as γ was increased beyond $\gamma = 29.25$ such trends became immediately apparent in all the orbital elements, with Saturn eventually being ejected in a few thousand years. In the current investigation the numerical approach was extended to cover variations in the eccentricities of both Jupiter and Saturn as well as γ.

2. Dynamical model

In Jacobian coordinates the equations governing the motion are

$$\ddot{\underline{r}}_J = -G(M_\odot + M_J)\left[\frac{\underline{r}_J}{r_J^3} + \frac{M_S}{M_\odot + M_J}\left(\frac{\underline{r}_S}{r_S^3} - \frac{\underline{r}_{JS}}{r_{JS}^3}\right)\right]$$

and

$$\ddot{\underline{r}}_{CS} = -G\frac{M_\odot + M_J + M_S}{M_\odot + M_J}\left(\frac{M_\odot}{r_S^3}\underline{r}_S + \frac{M_J}{r_{JS}^3}\underline{r}_{JS}\right)$$

where $\underline{r}_J$ and $\underline{r}_S$ are the position vectors of Jupiter and Saturn with masses M_J and M_S relative to the Sun. $\underline{r}_{JS} = \underline{r}_S - \underline{r}_J$ and $\underline{r}_{CS}$ is the position of the vector of Saturn relative to the barycentre of the Sun-Jupiter system.

These equations were integrated numerically using a 12(10) embedded Runge-Kutta-Nystrom procedure (Dormand et al(1987)), with variable step length adjusted to a tolerance of 10^{-14}. The mass, position and velocity of each component mass was specified for a straight line configuration simulating maximum perturbations. The stability criterion was that of Laplace where the orbital elements show no secular or large periodic variations over the time scale of the integration, in practice this meant that instability was deemed to have set in if the semi-major axis increased by 10% or the eccentricity by 0.1. An additional check on the performance of the numerical code was the maintenance to high degree of accuracy of the constancy of the energy integral. The various systems considered were evolved numerically for about 1000 orbits of the Sun-Jupiter subsystem and in some cases for substantially more. In most cases any instability present tended to become apparent within the first 100 orbits and was marked by rapid secular changes in the orbital elements.

3. Results

In the first set of experiments the system was integrated for a range of γ between 20 and 100, with the orbital elements taken to have their present values. This is similar to the set of calculations to those carried out by Nacozy(1976). It was found that for $\gamma \leq 28$, the system obeyed our strict definition of stability. In the range $28.0 \leq \gamma \leq 29.2$, it was found that although the semi-major axes and eccentricities showed changes exceeding the 10% level, the changes appeared periodic in nature. For $\gamma > 29.2$, the orbits were found to not only violate the basic criteria but were clearly secular, increasing rapidly with Saturn eventually escaping from the system altogether. These results agree with those of Nacozy(1976) who found clear indications of instability for $\gamma > 29.25$; the value $\gamma = 29.2$ was not integrated by Nacozy. It seems clear that Nacozy(1976) used a criteria relating to large changes in eccentricity to determine instability. Graziani and Black(1981), when testing their numerical code against Nacozy's results, found that the system was stable when γ was 20 or less and strongly unstable when γ of 30 or more was used. Nothing was explicitly stated about the intervening values.

In the next series of numerical experiments the eccentricity of Jupiter's orbit, e_J, was increased for the same range of γ as before. Jupiter was thus initially released closer to Saturn producing greater perturbations in Saturn's orbit. The conditions under which varying the orbit of Jupiter will disrupt the orbit of Saturn and the nature of the disruption were determined. The results are displayed in table I. It was found that for an increase in e_J to 0.1 only $\gamma = 1$ was completely stable, though periodic variations were apparent for small changes in e_J or small γ values. These are likely to be unstable over longer time scales. The instabilities were largely, as expected, linked to the disturbance of Saturn's orbit which increases with

γ and e_J.

A similar series of numerical experiments were carried out by increasing the eccentricity of Saturn, e_S, while that of Jupiter remains at its current value. The initial perturbations on Saturn's orbit were thus increased compared to the present situation. The results are displayed in table II. The general mode of instability was the ejection of Saturn from the system with e_J increased.

From the series of calculations performed it was clearly found that, as the mass ratio γ was increased, instability in the present system became apparent as value around 29 were reached. Similar calculations varying e_J or e_S, as well as γ, showed that instability sets in rapidly for all but the smallest changes in the eccentricities.

e_J \ γ	1	5	10	15	18	20	25	29	35
0.48			STABLE						
0.1									
0.15	PERIODIC/UNSTABLE								
0.2									
0.25						UNSTABLE			
0.3									
0.5									

e_S \ γ	15	18	20	35
0.1	STABLE	PERIODIC/STABLE		
0.2	PERIODIC/STABLE			
0.3	PERIODIC/STABLE		UNSTABLE	
0.5				
0.8				

Table I variations in γ and e_J. Table II variations in γ and e_S

References

Donnison J.R.& Mikulskis, D.F.(1995) *MNRAS*, **272**, 1.
Dormand J.R., El-Mikkawy M.E.A. & Prince P.J.(1987) *IMA J. Num. Analy.*, **7**, 235.
Kuiper G.(1973) *Celest. Mech.*,**9**, 321.
Nacozy P.E.(1976) *Astron. J.*,**81**, 787.
Szebehely V.& McKenzie R.(1977) *Astron.J.*,**82**, 79.
Walker I.W.& Roy A.E.(1981) *Celest. Mech.*,**24**, 195.

ON THE SECULAR MOTION OF THE JOVIAN PLANETS

M. GHIL, F. VARADI AND W. M. KAULA
IGPP, University of California, Los Angeles, CA 90095-1567

Abstract. The motion of the Jovian planets is investigated using Hamiltonian perturbation theory and numerical integrations. Experiments varying the mass of Neptune exhibit 1:1 secular resonance between the perihelion motions of Jupiter and Uranus.

The current view of Solar system dynamics on the time-scale of millions of years is based on secular perturbation theory and various features of the dynamics are explained in terms of resonances and associated separatrices (Laskar, 1990; Sussman and Wisdom, 1992; Duncan and Quinn, 1993). The road to chaos goes through bifurcations: to understand "islands" of chaotic and regular motion in the present Solar system we need to explore the parameter space around it. We carried out numerical simulations of the motion of the Jovian planets and analyzed the results using Hamiltonian perturbation theory in order to see how well we can explain the dynamics and to find possible sources of chaos.

While developing a Hamiltonian secular perturbation theory for the Jovian planets, one has to deal with the 2:5 mean-motion near-resonance between Jupiter and Saturn, the so-called Great Inequality (GI). We found that straightforward averaging is not likely to provide a convergent perturbation expansion: the contribution from the GI at third order in the masses is comparable to its contribution at second order (Varadi *et al.*, 1995).

To better understand the results of our numerical integrations, we derived a Lie series, retaining only 9000 terms out of about one hundred thousand, to remove all short-periodic perturbations, except for the GI. The GI prevents us at present from deriving a sufficiently accurate secular system. It does not contribute to the quadratic part of the secular Hamiltonian but it does affect higher-degree terms. The diagonalization of this quadratic Hamiltonian decouples the linear part of the secular motion for

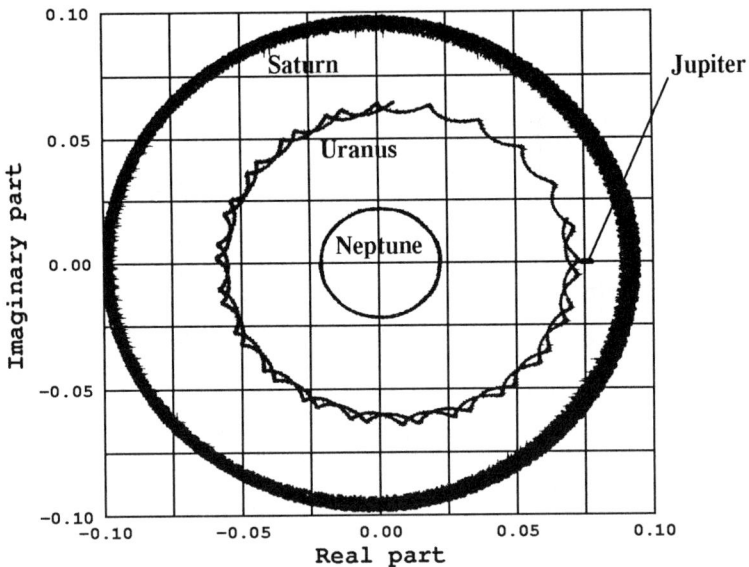

Figure 1. The diagonalized secular motion of the Jovian planets relative to Jupiter's, in ζ variables (see text for details).

each planet from that of the others, further simplifying the actual orbits. We use complexified canonical versions of the usual h and k variables, i.e., $\zeta = \sqrt{2\sqrt{a}(1 - \sqrt{1-e^2})} \exp(-i\varpi)$, where a is the semi-major axis, e the eccentricity, and ϖ the longitude of perihelion. The diagonalizing transformation introduces linear combinations of the original ζ variables and it is customary to associate these new variables with particular planets.

After the above transformations, the motion still exhibits large, supposedly nonlinear, variations: these are mostly dependent on the phase of the "diagonalized Jupiter's" ζ. One can plot the same data relative to this phase, i.e., rotated (h, k) variables, to obtain Fig. 1. Comparing this plot to a typical one for the original variables (e.g., Quinn *et al.*, 1991), we obtain a much simpler picture. One can observe nearly circular "orbits" but the centers of these are shifted from the origin: Saturn's to the left, Uranus' and Neptune's to the right. The low- and high-variability phases of Saturn on the left and on the right of the plot, as it turns out, are caused by the GI. Furthermore, Saturn goes around the center in about 50 thousand years, disturbing the motion of Uranus and causing its oscillations in (h, k).

How far is the system from a possible 1:1 secular resonance between Jupiter and Uranus? The quadratic part of the secular Hamiltonian predicts the location of secular resonances in parameter space. The simplest and least intrusive way to bring Jupiter and Uranus close to this resonance

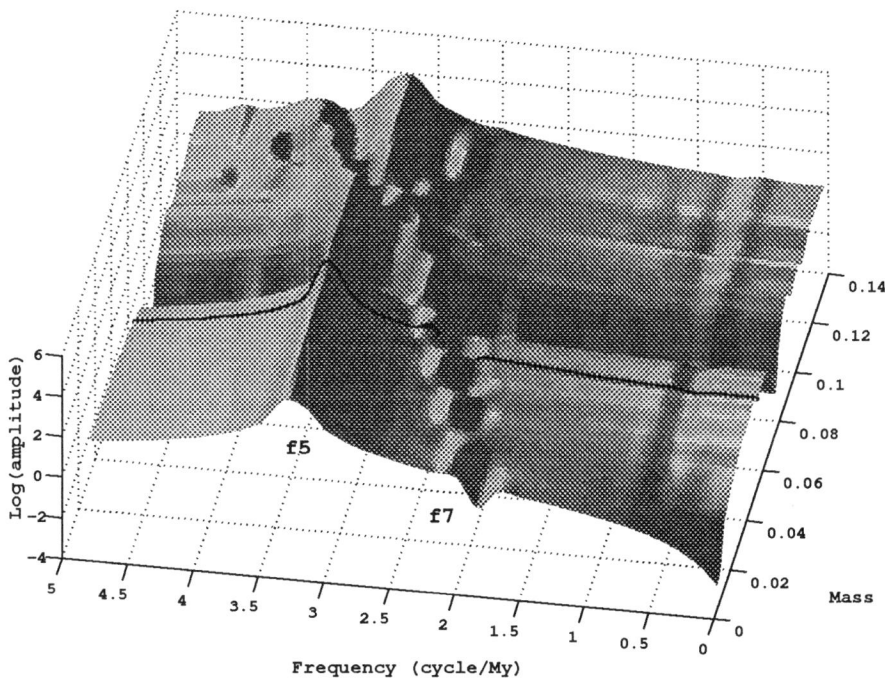

Figure 2. Composite raw Fourier spectra of Jupiter's canonical, *non-diagonalized*, $k = e\cos\varpi$ variable for 40 numerical integrations. The "mass" coordinate is m_N/m_J. The actual Solar System is indicated by a heavy line on the surface. The crossing by f_7 of the f_5 ridge indicates passage through the 1:1 resonance.

is to change the mass of Neptune. This does not alter the location of the mean-motion resonances nor the dynamics of the GI appreciably. Neptune's actual mass is 0.054 times that of Jupiter, $m_N = 0.054 m_J$, and the predicted secular near-resonance is around $m_N = 0.11 m_J$.

Near a 1:1 secular resonance, the identification of the eigenvalues and the eigenvectors of the quadratic part of the Hamiltonian breaks down. There are two nearly equal eigenvalues, and it is no longer clear how to match these to particular planets. In order to locate and understand this resonance, we carried out about 40 numerical integrations, each 6 million years long, varying m_N. The symplectic mapping technique of Wisdom and Holman (1991) was used with a 100-day stepsize. Additional integrations were carried out with a 16th order Cowell-Störmer integrator developed by W. I. Newman. The case of $m_N = 0.11 m_J$ shows large changes not only in the motion of Uranus and Neptune, but also in the motion of Jupiter. As opposed to the familiar annulus picture (e.g., Quinn *et al.*, 1991), the plot of Jupiter's non-diagonalized ζ fills a whole disk.

Each of the 40 integrations was analyzed separately, computing raw Fourier spectra for the non-diagonalized variables. We compiled the spectra of Jupiter's $Re(\zeta)$ into a function which also depends on the mass of Neptune. This enables us to see how the main lines associated with Jupiter (f_5) and Uranus (f_7) behave. We obtain a surface which shows that the system does go through the 1:1 secular resonance (Fig. 2). As the mass of Neptune increases the frequency f_7 increases and, around $m_N = 0.11 m_J$, f_7 and f_5 cannot be easily associated with individual planets. For larger Neptune masses the identity of the frequencies is clear again since f_5's amplitude is larger than f_7's. We find that the f_7 frequency is larger than f_5, i.e., the system does go through the 1:1 secular resonance.

There are some intriguing features in the composite Fourier spectra on Fig. 2. We see a large valley between $m_N = 0.06 m_J$ and $m_N = 0.08 m_J$ which might be a strong resonance, probably the actual 1:1 secular resonance. Also, f_7's ramp leading up to the main ridge of f_5 has steps which are associated with small valleys; perhaps these are other resonances. It is too early to draw definite conclusions, as the length and the accuracy of the integrations should be improved further. Also, the Wisdom-Holman mapping might have artificial bifurcations; these are being investigated by W. I. Newman and M. Haberkorn. The main feature, however, i.e., the passage through the 1:1 resonance, was confirmed by more accurate integrations (not shown).

Acknowledgements

We thank Dr. W. I. Newman and Mr. K. R. Grazier for numerous discussions on the topic and for providing us the Cowell-Störmer integrator. The financial support of NSF Grant ATM90–13217 (MG and FV), NASA Grant NAGW–2269 (WMK & FV) and the Condorcet Chair of the Ecole Normale Supérieure, Paris (MG) is gratefully acknowledged.

References

Duncan, M. J. and Quinn, T.: 1993, 'The long-term dynamical evolution of the Solar System', *Annu. Rev. Astron. Astrophys.*, **31**, 265

Laskar, J.: 1990, 'The chaotic motion of the Solar System: a numerical estimate of the size of the chaotic zones', *Icarus*, **88**, 266-291

Quinn, T. Q., Tremaine, S. and Duncan, M.: 1991, 'A three million year integration of the Earth's orbit', *Astron. J.*, **101**, 2287-2305

Sussman, G. J. and Wisdom, J.: 1992, 'Chaotic evolution of the Solar System', *Science*, 256-257

Varadi, F., Ghil, M. and Kaula, W. M.: 1995, 'The Great Inequality in a Hamiltonian planetary theory', in *From Newton to Chaos*, A. E. Roy and B. A. Steves (eds.), Plenum Publ. Co., 103-108

Wisdom, J. and Holman, M.: 1991, 'Symplectic maps for the N-body problem', *Astron. J.*, **102**, 1528-1538

THE MOTION OF PLUTO OVER THE AGE OF THE SOLAR SYSTEM

HIROSIII KINOSHITA AND HIROSHI NAKAI

National Astronomical Observatory
2-21-1 Osawa, Mitaka, Tokyo, Japan
E-mail(internet):Kinoshita@c1.mtk.nao.ac.jp

Abstract. Pluto's motion is chaotic in the sense that the maximum Lyapunov exponent is positive and the Lyapunov time (the inverse of the Lyapunov exponent) is about 20 million years (Myr). We have carried out the numerical integration of Pluto over the age of the solar system (5.7 billion years towards the past and 5.5 billion years towards the future), which is about 280 times of the Lyapunov time. Our integration does not show any indication of gross instability in the motion of Pluto. The time evolution of Keplerian elements of a nearby trajectory of Pluto at first grow linearly with the time and then start to increase exponentially. These exponential divergences stop at about 420 Myr and saturate. The exponential divergences are suppressed by the following three resonances that Pluto has:
(1) Pluto is in the 3:2 mean motion resonance with Neptune and the libration period of the critical argument is about 20000 years.
(2) The argument of perihelion librates around 90 degrees and its period is 3.8 Myr.
(3) The motion of the Pluto's orbital plane referred to the Neptune's orbital plane is synchronized with the libration of the argument of perihelion (a secondary resonance). The libration period associated with the second resonance is 34.5 Myr.
We briefly discuss the motions of Kuiper belt objects in a 3:2 mean motion resonance with Neptune and several possible scenarios how Pluto evolves to the present stable state.

1. Numerical Exploration of Pluto's motion

Pluto was found on January 21,1930 by Clyde W. Tombaugh of Lowell Observatory. Since the discovery Pluto moves only one third of her orbit. From this limited observation we discuss the orbital motion over the age of the solar system.

Cohen and Hubbard 1965 carried out numerical integration of outer planets over 120000 years and found Pluto is locked in the 3:2 mean motion resonance with Neptune. The critical argument of this resonance ($\theta_1 = 3\lambda_p - 2\lambda_n - \varpi_p$) librates around 180 degrees and its period is about 20000 years and its amplitude is 76 degrees. The libration means that Neptune passes Pluto near at her aphelion, it prevents a close approach of Pluto to Neptune and keeps Pluto at least 16 AU. from Neptune. Their results show that secular like change in the eccentricity and inclination of Pluto.

Next important investigation on long-period behavior of Pluto was made by William and Benson 1971. They integrated over 4.5 Myr the averaged equations of Pluto's motion disturbed by Jupiter, Saturn, Uranus, and Neptune whose motions are taken from the secular perturbation theory by Brouwer and van Woerkom 1950. Their treatment is equivalent to the first-order secular perturbation.

They found the argument of Pluto's perihelion ($\theta_2 = \varpi_P - \Omega_P$) librates about 90 degrees with an amplitude of 24 degrees and a period of about 4 Myr. The secular like changes in the eccentricity and the inclination found in Cohen and Hubbard 1965 are parts of this long-period motions. The libration of θ_2 prevents Pluto's perihelion from getting close to the plane of the disturbing planets, especially to Neptune's orbital plane. The dominant periodic variations of the eccentricity and the inclination are synchronized with the libration of the argument of perihelion, which is qualitatively explained by the secular perturbation theory (Kozai 1962) by assuming an exact 3:2 mean motion resonance ($\theta_1 = 180$ degrees).

Nacozy and Diehl 1972 and 1978 analytically investigated Pluto's motion with use of canonical perturbation method and confirmed the libration of the argument of pericenter discovered by Williams and Benson 1971.

Kinoshita and Nakai 1984 confirmed the libration of the argument of perihelion directly from the numerical integration over 5 Myr of outer planets.

Based on the 4.5 Myr integration, William and Benson 1971 made the following two conjectures on Pluto's motion:
1) The circulation period of $\theta_3 = \Omega_P - \Omega_N$ is equal to the libration period of the argument of perihelion.
2) $\theta_4 = \varpi_P - \varpi_N + 3(\Omega_P - \Omega_N)$ librates.
If these conjectures are correct, these effects increase the minimum distance

between Pluto and Neptune and stabilize Pluto-Neptune system.

The first conjecture was confirmed by Milani et al. 1989 using Longstop, 100 Myr numerical integrations of outer planets (Nobili et al. 1989). This type of resonance is called a 1:1 secondary resonance. The libration period associated with this secondary resonance is 34.5 Myr in both Longstop and our 5.5 billion years integration of outer planets (Kinoshita and Nakai 1995), which is discussed in the next section.

The second conjecture was confirmed by Kinoshita and Nakai 1995 with use of the 5.5 billion years. The argument θ_4 librates about 180 degrees with a period of 570 Myr.

The fact that the conjectures based on the first-order secular perturbation are confirmed may indicates that the planetary system in the age of the solar system could be investigated with use of averaged equations of motion. Laskar 1988 and 1989 discussed the long-term behavior of the planetary system (excluding Pluto) with use of the secular perturbation theory of second order with respect to the planetary masses and five degrees in eccentricity and inclination, which is higher by one order than Brouwer and van Woerkom's theory 1950. Sussman and Wisdom 1992 carried out 100 Myr integration of the whole planetary system including Pluto and the relativistic effect and their results in the timescale of 100 Myr are in excellent agreement with the results by Laskar et al. 1992, which are obtained from the integration of the averaged equations of motion. However it is not clear that whether a chaotic behavior of the system could be discussed with use of averaged equations of motion.

2. Pluto's motion is chaotic

In 1988 a shocking news "Pluto's motion is chaotic" came from Sussman and Wisdom. Their orbital computation was carried out Digital Orrery which is a special purpose computer. The maximum Lyapunov characteristic exponent is positive. The Lyapunov time(the inverse of Lyapunov exponent) is about 20 Myr. During their 850 Myr integration Pluto does not show any global change in the orbital elements. Then Sussman and Wisdom 1992 carried out an integration of the whole solar system including Pluto over 100 Myr. The evolution of Pluto in this integration is similar to that of Pluto found in 845 Myr integration and this integration also does not give any indication of a gross instability in the motion of Pluto. According to Wisdom 1992, " Since the system is apparently chaotic, we cannot rule out the possibility of gross instability. Recall some chaotic asteroid trajectories have been seen to evolve chaotically for 100 Lyapunov times at low eccentricity and then suddenly jump to large eccentricity. It will be very interesting to see a number of integrations of the whole solar system for the age of the solar system and longer."

Then we carried out orbital computation of outer planets over the age of the solar system forward to 5.5 billion years (Kinoshita and Nakai 1995) and backward to 5.7 billion years. As an integrator we used a linear symmetric multistep integrator developed by Quinlan and Tremaine 1990, whose truncation errors do not produce secular errors in the energy and angular momentum. In order to reduce the effect due to the accumulation of round-off errors, we adopted mixed precision arithmetic: the mutual gravitational forces are computed in double precision and the evaluations of the integrator are done in quadruple precision.

Figure 1 shows the orbital motions of five outer planets over 11.2 billion years. The x-axis in this figure is the direction of the perihelion of Pluto. Each planets in this figure are plotted at every 4×10^8 days$\simeq$ 1 Myr. All outer planets move regularly over the age of the solar system. Three resonances in Pluto's motion (1) the 3:2 mean motion resonance, 2) the libration of the argument of perihelion, and 3) the secondary resonance between θ_2 and θ_3, which are discovered from the previous shorter integrations, are well kept during 11.2 billion years.

Then the question is what Pluto's chaotic motion is in the time scale of the age of the solar system. We investigate the time evolution of the behavior of a nearby object of Pluto. At first the differences of the orbital elements between Pluto and its nearby orbit grow linearly with time and then grows exponentially and saturate. The exponential growth corresponds with the fact that the Lyapunov exponent is positive. The saturation in the semi-major axis (see Figure 2, A1) is related to the 3:2 mean motion resonance (see Figure 2, A2 and B). The saturated value (0.31 AU.) in the semi-major axis is just twice of the amplitude of the 3:2 resonance libration. The saturations in the eccentricity, the inclination, the argument of perihelion, and the longitude of node are related to both the libration of the argument of perihelion and the secondary resonance between θ_2 and θ_3 (see Kinoshita and Nakai 1995 for more detailed discussion).

We can say even though Pluto's motion is chaotic in the sense that Pluto's Lyapunov exponent is positive, Pluto does stay in very stable configuration, which are protected at least in the age of the solar system by the three resonances.

In relation with the positive Lyapunov exponent of Pluto, we computed a Lyapunov exponent of Pluto in the following five simplified models.

- A) a circular restricted three body problem in three dimensions.
  - A1) Neptune's motion is circular.
  - A2) Neptune's motion is elliptic.

THE MOTION OF PLUTO

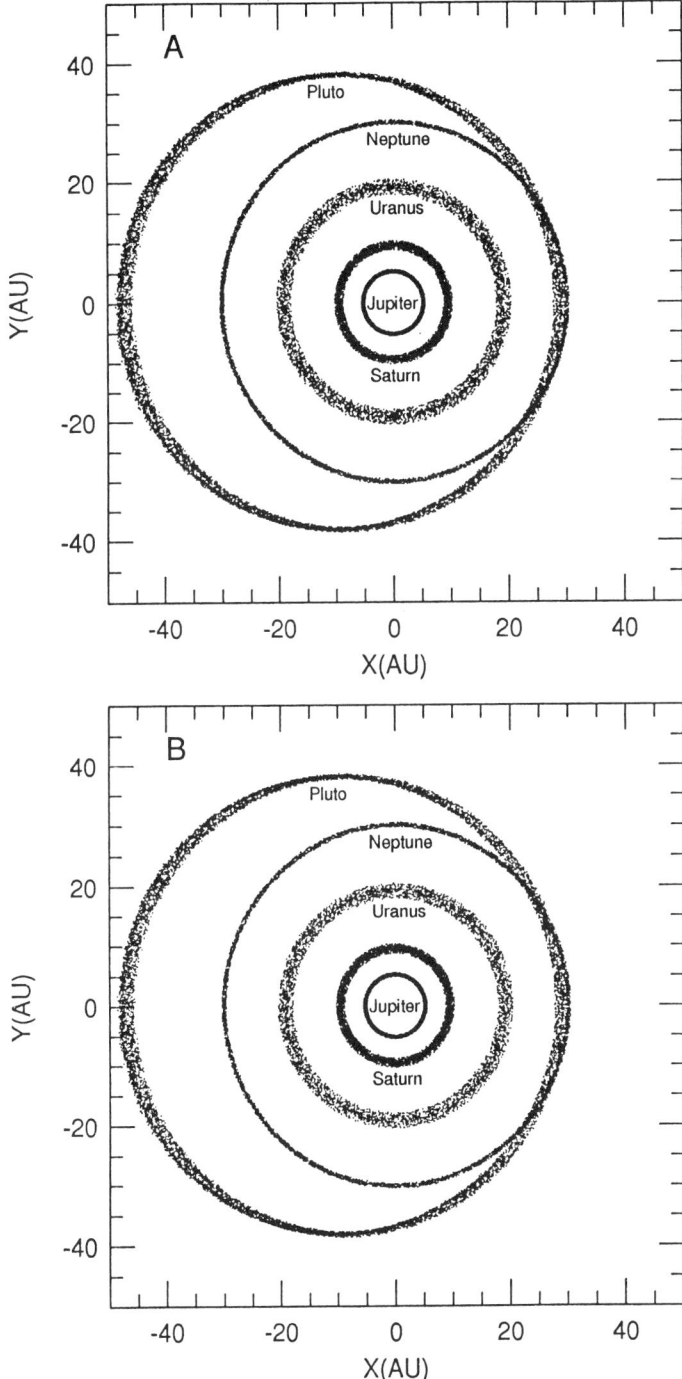

Figure 1. Orbital motions of outer planets over 11.2 billion years. The x-axis is the direction of Pluto's perihelion. A) past 5.7 billion years. B) future 5.5 billion years.

lar and the period of the critical argument of 3:2 mean motion resonance is about 17 thousands years. After about 3 Myr the mean motion resonance is destroyed and then the orbit becomes chaotic. We integrated a nearby orbit whose initial distance in the phase space is different by 10^{-8} relatively from 1994TB. The 3:2 mean motion resonance in this nearby orbit is conserved over about 15 Myr but then is destroyed. So far we could not identify the origin of the onset of this irregular motion. At least we could say this is not caused by the close approach to Uranus, even the minimum distance to Uranus is shorter than the distance to Neptune. After the destruction of the resonance, 1994TB comes to close to Uranus and then the orbit 1994TB begins to change drastically. The osculating elements of 1994TB adopted here are determined by assuming the 3:2 mean motion resonance (Minor Planet Electronic Circular (MPEC) 1994-E11). According to MPEC 1995-M07, by adding new observations a 4:3 mean motion resonance for 1994TB is possible but the possibility of a 3:2 libration is not excluded and further observations are needed in order to distinguish the 4:3 and 3:2 solutions.

Other than the 3:2 resonance objects, the candidates in a 4:3, 5:3, and 2:1 resonances with Neptune have been detected. It is necessary to investigate the resonance structure in the Kuiper belt and the dynamical similarity or difference from the resonance structure of the main asteroidal belt, which are studied by Yoshikawa 1989, 1990, and 1992, Morbidelli and Moons 1993, and Moon and Morbidelli 1995.

4. Origin of Pluto

As we have seen in the previous numerical simulations, Pluto is protected from global chaos by the three resonances, especially the 3:2 mean motion resonance and the libration of the argument of perihelion. So how Pluto evolves into the present state with three resonance lockings is a very interesting and challenging problem. There are many scenarios on the origin of Pluto, which may be divided in two categories, one type assumes a catastrophic event such as a close encounter or encounters and one type assumes a resonance sweeping. The scenarios of an encounter model are proposed by Lyttleton 1936 (an encounter between Pluto as a satellite and Triton), Harrington and van Flandern 1979 (a single encounter of Neptune with a massive solar system body), Dormand and Wolfson 1980 (an extension of Lyttleton hypothesis), Olsson-Steel 1988 (close encounters between Pluto as a planet and Neptune), and Levison and Stern 1995 (encounters between a planet Pluto and primordial Kuiper belt objects).

Resonance sweeping scenarios are proposed by Colombo and Franklin 1970 (evolution in the presence of a resisting medium whose density decreases with time) and recently Malhotra 1993 and 1995 (Pluto is trapped into the 3:2 mean motion resonance by Neptune whose semi-major axis is increasing).

Levison and Stern 1995 showed that with use of numerical simulations Pluto with a nearly circular and low inclined orbit placed at near the 3:2 mean motion resonance with Neptune could evolve to the present highly eccentric and inclined orbit due to purely gravitational interactions of the giant planets in the present configuration in a timescale about 10^7 years or longer. This amplification of the eccentricity and inclination could push Pluto into the state of the 3:2 mean motion resonance and the libration of the argument of perihelion. However the amplitude of the critical argument θ_1 is too large. In order to reduce the amplitude for the stabilization of Pluto's motion Levison and Stern propose two different scenarios: 1) collisional and gravitational scattering between planet Pluto and small neighboring bodies in the Kuiper belt or 2) a single, inelastic, giant collision.

According to Malhotra's scenario 1993 and 1995, at the early stage of the solar system, when planets are forming from planetesimals, Neptune's orbit gradually expands by getting angular momentum from planetesimals. Pluto at this stage was near circular and low inclined. Neptune's orbit gradually expands and when Neptune reaches at the place where the 3:2 mean motion is possible, and Pluto is captured in a 3:2 mean motion resonance and is kept in the resonance after this capture. The argument of perihelion, however, is not locked and hops between 90 and 270 degrees over 10 Myr timescales.

Her scenario is very attractive but her scenario bases on the assumption that Neptune's orbit expands at planetary formation period. At present time there is no concrete and definite scenario on the evolution of a semi-major axis of outer planets at their formation period.

References

Brouwer, D. and van Woerkom, A. J. J. (1950) The Secular Perturbations of the Orbital Elements of the Principal Planets, *Astron. Papers Am. Ephemeris.* **13**(Pt. 2), pp. 85-107.

Cohen, C. J. and Hubbard, E. C. (1965) Libration of the Close Approaches of Pluto to Neptune, *Astron. J.* **70**, pp. 10-13.

Colombo, G. and Franklin, F. A. (1970) On the Evolution of the Solar System and the Pluto-Neptune Case, in *Periodic Orbits, Stability and Resonances* Giacaglia (ed.), Reidel Pub. Comp. , pp. 328-331.

Dormand, J. R. and Woolfson, M. M. (1980) The Origin of Pluto, *Mon. Not. R. astr. Soc.* **193**, pp. 171-174.

Harrington, R. S. and van Flandern, T. C. (1979) The Satellites of Neptune and the Origin of Pluto, *Icarus.* **39**, pp. 131-136.

Kinoshita, H. and Nakai, H. (1984) Motions of the Perihelion of Neptune and Pluto, *Celestial Mechanics.* **34**, pp. 203-217.

Kinoshita, H. , Yoshida, H. , and Nakai, H. (1991) Symplectic Integrators and Applications to Dynamical Astronomy, *Celestial Mechanics.* **50**, pp. (59-71).

Kinoshita, H. and Nakai, H. (1995) Long-Term Behavior of the Motion of Pluto over 5. 5 Billion Years, in *The Small Bodies in the Solar System and their Interactions with Planets* Rickman (ed.), Kluwer, in press.

Kozai, Y. (1962) Secular perturbation of asteroids with high inclination and eccentricity, *Astron. J.* **67**, pp. 591–598.

Laskar, J. (1988) Secular Evolution of the Solar System over 10 Million Years, *Astron. Astrophys.* **198**, pp. 341-362.

Laskar, J. (1989) A numerical experiment on the chaotic behavior of the solar system, *Nature.* **338**, pp. 237–238.

Laskar, J. , Quinn, T. R. , and Tremaine, s. (1992) Confirmation of Resonant Structure in the Solar System, *Icarus.* **95**, pp. 148-152.

Levison, H. F. and Stern, S. A. (1995) Possible Origin and Early Dynamical Evolution of the Pluto-Charon Binary, *Icarus.* **116**, pp. 315-339.

Lyttleton, R. A. (1936) On the Possible Results of an Encounter of Pluto with the Neptunian System, *Mon. Not. Roy. Astron. Soc.* **97**, pp. 108-115.

Malhotra, R. (1993) The Origin of Pluto's Peculiar Orbit, *Nature.* . **365**, pp. 819-821.

Malhotra, R. (1995) The Origin of Pluto's Orbit:Implications for the Solar System beyond Neptune, *Astron. J.* **110**, pp. 420-429.

Milani, A. , Nobili, A. M. and Carpino, M. (1989) Dynamics of Pluto, *Icarus.* **82**, pp. 200-217.

Morbidelli, A. and Moons, M. (1993) Secular Resonances in Mean Motion Commensurabilities: The 2/1 and 3/2 Cases, *Icarus.* **102**, pp. 316-332.

Moons, M. and Morbidelli, A. (1995) Secular Resonances in Mean Motion Commensurabilities: The 4/1, 3/1, 5/2, and 7/3 Cases, *Icarus.* **114**, pp. 33-50.

Nacozy, P. E. and Diehl, R. D. (1972) On the Long-Term Motion of Pluto, *Celestial Mechanics.* **8**, pp. 445-454.

Nacozy, P. E. and Diehl, R. D. (1978) A Discussion of the Solution for the Motion of Pluto, *Celestial Mechanics.* **17**, pp. 405-421.

Nakai, H. , Kinoshita, H. , and Yoshida, H. (1992) Dependence on computer's arithmetic precision in calculation of Lyapunov characteristic exponent, in *Proceedings of 25 the Symposium on Celestial Mechanics* Kinoshita, H. and Nakai, H (eds.), pp. 1–10.

Nobili, A. M. , Milani, A. and Carpino, M. (1989) Fundamental frequencies and small divisors in the orbits of the outer planets, *Astron. Astrophys.* **210**, pp. 313-336.

Olsson-Steel, D. I. (1988) Results of Close Encounters between Pluto and Neptune, *Astron. Astrophys.* **195**, pp. 327-330.

Quinlan, D. and Tremaine, S. (1990) Symmetric multistep methods for the numerical integration of planetary orbits, *Astron. J.* **100**, pp. 1694-1700.

Sussman, G. J. and Wisdom, J. (1988) Numerical evidence that the motion of Pluto is chaotic, *Science.* **241**, pp. 433-437.

Sussman, G. J. and Wisdom, J. (1992), Chaotic evolution of the solar system, *Science.* **257**, pp. 56-62.

Williams, J. G. and Benson, G. S. (1971) Resonances in the Neptune-Pluto system, *Astron. J.* **76**, pp. 167-177.

Wisdom, J. (1992) Long term evolution of the solar system, in *Chaos, Resonance and Collective Dynamical Phenomena in the Solar System* S. Ferraz-Mello (ed.), Kluwer, pp. 17-24.

Yoshikawa, M. (1989) A Survey on the Motion of Asteroids in Commensurabilities with Jupiter, *Astron. Astrophys.* **213**, pp. 436-458.

Yoshikawa, M. (1990) Motions of Asteroids at the Kirkwood Gaps. I. On the 3:1 resonance with Jupiter, *Icarus.* **87**, pp. 78-102.

Yoshikawa, M. (1990) Motions of Asteroids at the Kirkwood Gaps. II. On the 5:2, 7:3, and 2:1 resonance with Jupiter, *Icarus.* **92**, pp. 94-117.

PLUTO'S LYAPUNOV NUMBERS IN DIFFERENT DYNAMICAL MODELS

R. DVORAK AND E. LOHINGER
Institut für Astronomie, Universität Wien,
Türkenschanzstraße 17, A-1180 Vienna, Austria.

Abstract. We present the results of numerical integrations of Pluto and some fictitious Plutos in three different models (the circular and the elliptic restricted three body problem and the outer solar system). We determined the "extension" of the stable region in these models by means of the Lyapunov Characteristic Numbers and by an analysis of the orbital elements.

1. Introduction

Since the study of Sussman & Wisdom (1988, 1992), Pluto's orbit is known to be chaotic in the sense that its Lyapunov exponent is positive. But, until now, no investigation was carried out which could show that this indication of a positive Lyapunov exponent is connected with an unstable motion. Even integrations up to half of the life-time of our solar system do not at all show significant changes in the orbital elements of Pluto (cf. Dvorak et al., 1995). The aim of this investigation is to study Pluto's motion in simple models which will be compared with more realistic ones, moreover we computed the largeness of the region in which Pluto can move without showing the above mentioned changes of the orbital elements.

2. Three dynamical models and three different methods

We studied Pluto's motion in the three dimensional circular restricted three body problem (Sun + Neptune), in the elliptic restricted problem and finally in a model of the outer solar system (=OSS) where the perturbations of the planets Jupiter to Neptune were taken into account. In all cases, the masses of the neglected planets were added to that of the Sun. The orbits of Pluto and 11 fictitious bodies — with semimajor axes $a = a_{Pluto} - 0.1 \cdot k$ AU, $k = 1, \ldots, 8$ AU and $a = a_{Pluto} + 0.1 \cdot l$ AU, $l = 1, 2, 3;$ — were ex-

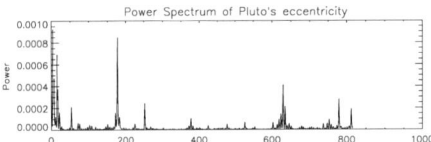 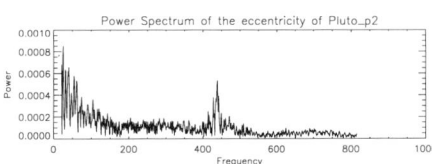

Figure 1. Fourier spectrum of the eccentricity of (a) Pluto in the model of the outer solar system and (b) of a fictitious Pluto outside the "stable" region in the restricted problem.

amined using three different approaches: 1st by means of a FFT applied to the orbital elements, 2nd by the calculation of the Lyapunov Characteristic Numbers (=LCNs) and 3rd we computed the distribution of the largest Local Lyapunov Number (LLCN) of Pluto's orbit. We have to point out that many of the studied orbits are not in 2:3 resonance with Neptune. The numerical calculations were carried out by means of two methods: the Lie integrator and the Bulirsch-Stoer integrator. The latter was used for the computation of the Lyapunov numbers — these are the local variations of the tangent vectors to the flow (Benettin et al., 1980; C. Froeschlé, 1984). In order to replace the evolved tangent vectors by new orthonormalized ones, the Gram-Schmidt procedure was employed. This program was provided by R. Gonczi (of the OCA Nice).

3. The results

The application of the FFT to the time evolution of the orbital elements showed that in all models Pluto's orbit has a signal which corresponds to a more or less quasiperiodic orbit over the time interval of 40 million years (see e.g. fig. 1.a). A former study of Pluto's orbital elements showed that the width of the "stability zone" of Pluto — stable in the sense that the orbital changes are not significant — turned out to be approximately 0.7 AU in semimajor axis (cf. Dvorak et al., 1995). The present study of the Fourier spectra of the different fictitious Plutos confirmed these former results. In fig. (1.b) the signal of a chaotic orbit in the vicinity of Pluto is shown by way of comparison. The calculations of the LCNs in the two models of the restricted problem led to similar results, that is, in case of the inner bodies, the largest LCN (LCN_1) is positive only for pluto_m8 (with a semimajor

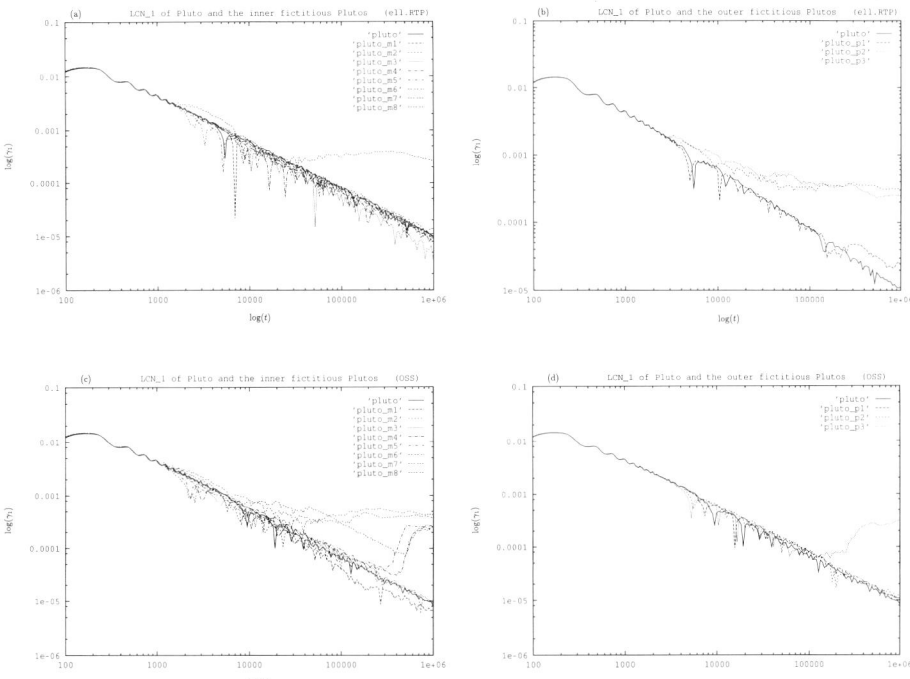

Figure 2. The largest LCNs of (a) Pluto and the "inner" Plutos, (b) Pluto and the "outer" Plutos in the spatial elliptic restricted problem and the same in the OSS (c and d). The integration time was 10^6 years into the future.

axis $a = a_{Pluto} - 0.8$ AU) (see fig. 2.a) — which was even thrown out of the system. For the region outside (see fig.2.b) — outside in the sense that the initial semimajor axis is larger than Pluto's — only LCN_1 of the orbit next to Pluto tends to zero within the considered time; this is consequently a quasi-periodic orbit. From the body pluto_p2 on (with a semimajor axis $a = a_{Pluto} + 0.2$ AU) the largest LCNs indicate chaotic motion. In the more realistic model (OSS) the region of stable motion seems to be shifted outwards, and became also narrower: From pluto_m4 (with a semimajor axis $a = a_{Pluto} - 0.4$ AU) on the LCN_1 of all inner orbits turned out to be positive. Outside Pluto's orbit, only pluto_p3 (with a semimajor axis $a = a_{Pluto} + 0.3$ AU) has a positive LCN_1. The extensions of the stable region in the restricted three body problem and in the outer solar system are presented in fig. 3 where the diminution and the shift outwards in case of the OSS is clearly seen. How does this fit to the result that Pluto's Lyapunov exponent is positive as mentioned in the introduction? We used relatively short time intervals for the determination of the LCNs: therefore

million years, using a vectorized computer specially designed for the task (Applegate et al., 1986; Sussman and Wisdom, 1988). This latter integration showed that the motion of Pluto is chaotic, with a Lyapunov exponent of 1/20 million years. But since the mass of Pluto is very small (1/130000000 the mass of the Sun), this does not induce macroscopic instabilities in the rest of the solar system, which appeared relatively stable in these numerical studies.

2. Chaos in the Solar System

My approach was different and more in the spirit of the analytical works of Laplace and Le Verrier. Indeed, since these pioneer works, the Bureau des Longitudes has traditionally been the place for development of analytical planetary theories (Brumberg and Chapront, 1973; Bretagnon, 1974; Duriez, 1979). All these studies are based on classical perturbation series; thus, implicitly, they assume that the motion of the celestial bodies are regular and quasiperiodic. The methods used are essentially the same which were used by Le Verrier, with the additional help of the computers. Indeed, such methods can provide very satisfactory approximations of the solutions of the planets over a few thousand years, but they will not be able to give answers to the question of the stability of the solar system over time span comparable to its age. This difficulty which is known since Poincaré is one of the reasons which motivated the previously quoted long time numerical integrations. Nevertheless, it should be stressed that, until 1991, the only numerical integration of a realistic model of the full solar system was the ephemeris DE102 of JPL (Newhall et al., 1983) which spanned only 44 centuries.

A first attempt consisted to extend as far as possible the classical analytical planetary theories, but it was realized quite rapidly that this was hopeless when considering the whole solar system, because of severe convergence problems encountered in the secular system of the inner planets (Laskar, 1984). I thus decided to proceed in two very distinct steps: a first one, purely analytical, consisted in the averaging of the equations of motion over the rapid angles, that is the motion of the planets along their orbits. This process was conducted in a very extensive way, without neglecting any term, up to second order with respect to the masses, and through degree 5 in eccentricity and inclination. The system of equations thus obtained comprises some 150000 terms, but it can be considered as a simplified system, as its main frequencies are now the precessing frequencies of the orbits of the planets, and no longer comprise their orbital periods. The full system can thus be numerically integrated with a very large stepsize of about 500 years. Contributions due to the Moon and to the general relativity are added without difficulty.

This second step, i.e. the numerical integration, is very efficient because of the symmetric shape of the secular system, and was conducted over 200 millions years in just a few hours on a super computer. The main results of this integration was to reveal that the whole solar system, and more particularly the inner solar system (Mercury, Venus, Earth, and Mars), is chaotic, with a Lyapunov exponent of 1/5 million years (Laskar, 1989). An error of 15 meters in the Earth's initial position gives rise to an error of about 150 meters after 10 million years; but this same error grows to 150 million km after 100 million years. It is thus possible to construct ephemerides over a 10 million year period, but it becomes practically impossible to predict the motion of the planets beyond 100 million years.

This chaotic behavior essentially originates in the presence of two secular resonances among the planets: $\theta = 2(g_4 - g_3) - (s_4 - s_3)$, which is related to Mars and the Earth, and $\sigma = (g_1 - g_5) - (s_1 - s_2)$, related to Mercury, Venus, and Jupiter (the g_i are the secular frequencies related to the perihelions of the planets, while the s_i are the secular frequencies of the nodes) (Laskar, 1990). The two corresponding arguments change several times from libration to circulation over 200 million years, which is also a characteristic of chaotic behavior. When these results were published, the only possible comparison was the comparison with the 44 centuries ephemeris DE102, which already allowed to be confident on the results (Laskar, 1986, 1990). At the time, there was no possibility to obtain similar results with direct numerical integration. In fact, partly due to the very rapid advances in computer technology, and in particular to the development of workstations, only two years later, Quinn *et al.* (1991) were able to publish a numerical integration of the full solar system, including the effects of general relativity and the Moon, which spanned 3 million years in the past (completed later on by an integration from -3Myrs to +3Myrs). Comparison with the secular solution of (Laskar, 1990) shows very good quantitative agreement and confirms the existence of secular resonances in the inner solar system (Laskar *et al.*, 1992a). Later on, using a symplectic integrator directly adapted towards planetary computations which allowed them to use a larger stepsize of 7.2 days, Sussman and Wisdom (1992) made an integration of the solar system over 100 million years which confirmed the existence of the secular resonances as well as the value of the Lyapunov exponent of about 1/5 Myrs for the solar system.

3. Planetary evolution over Myr

The planetary eccentricities and inclinations present variations which are clearly visible over a few million of years (Fig .1). Indeed, this was known since Laplace and LeVerrier (for a detailed account see Laskar 1992b), us-

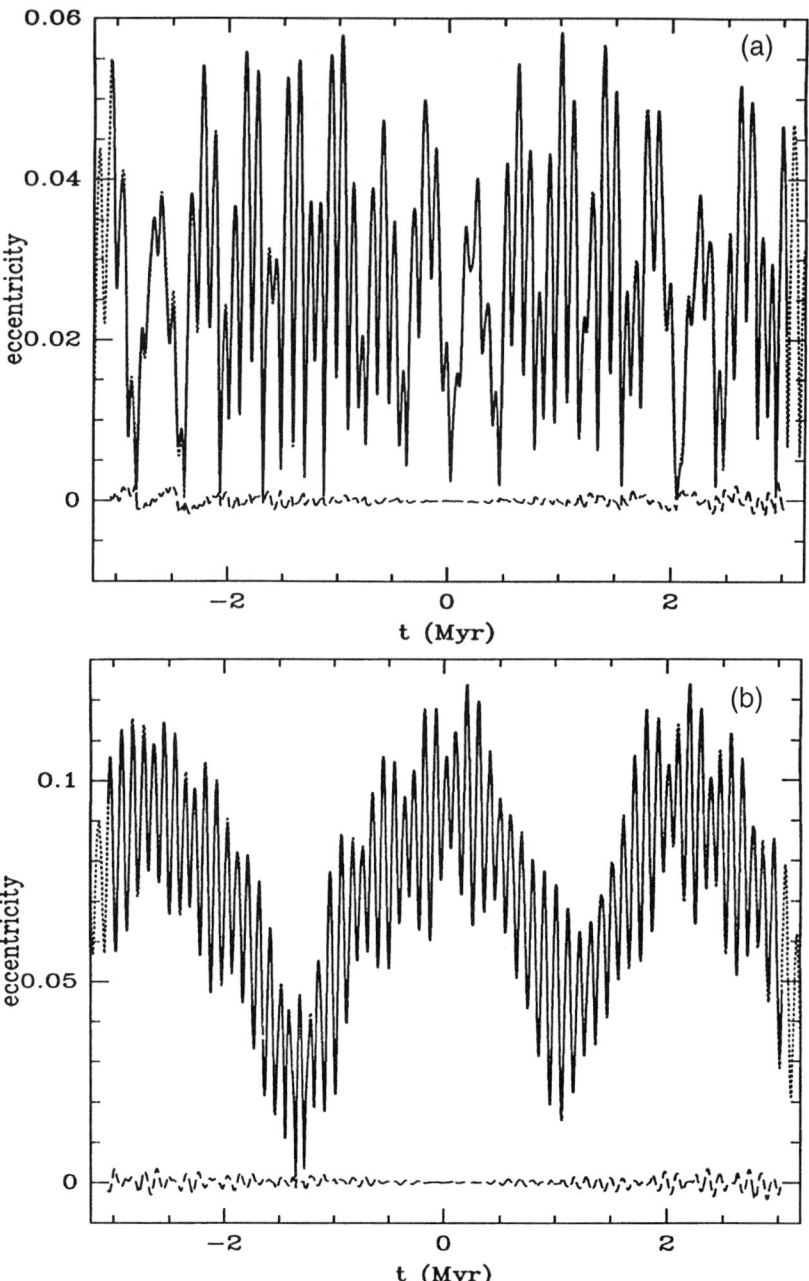

Figure 1. The eccentricity of the Earth (a) and Mars (b) during a 6 Myr timespan centered at the present. The solid line is the numerical solution from Quinn *et al.* (1991) and the dotted line the integration La90 of the secular equations (Laskar, 1990). For clarity, the difference between the two solutions is also plotted (from Laskar, *et al.*, 1992).

ing the secular equations. Over 1 million years, perturbation methods will give a good account of these variations which are mostly due to the linear coupling present in the secular equations. These secular variations involve the precessional periods of the orbits, ranging from 40 000 years to a few million of years. From -200 Myr to +200 Myr, the behavior of the solutions for the outer planets (Jupiter, Saturn, Uranus and Neptune) are very similar to the behavior over the first million years and the motion of these planets appears to be very regular, which was also shown very precisely by mean of frequency analysis (Laskar, 1990).

For the Earth, over such time span, the chaotic effect will induce a lost of predictability for the orbit. The additional change of eccentricity resulting from the chaotic diffusion is moderate and may be estimated to about 0.01 for the Earth (Laskar, 1992a,b). The most perturbed planet is Mercury, the effects of its chaotic dynamics being clearly visible over 400 million years (Laskar, 1992a,b).

It should be stressed that the exponential divergence of the orbits revealed by the computation of the Lyapunov exponent result mostly from the change from libration to circulation of the resonant precession angles, which induce after some time a total indeterminacy of the precessional angles of the orbit, that is its orientation in space. The eccentricity and inclination (which are action-like variables) variations due to the chaotic diffusion is much less rapid, and an important question is to estimate their wandering over the time of the life of the solar system.

4. Planetary evolution on Gyr time scales

If the motion of the solar system were close to quasiperiodic, that is close to a KAM tori, then it could be expected that some bound on the possible diffusion of the orbit over 5 Gyr would result from a Nekhoroshev-like theorem. In fact, as it was shown in (Laskar, 1990), although the system reduced to the outer planets may be considered as close to a KAM tori, the full solar system evolves far from a KAM tori of maximal dimension and diffusion of the action-like variables (eccentricity and inclination) occurs. The natural question is thus to estimate this diffusion. Let us remind that contrarily to two degrees of freedom systems, where the diffusion may be bounded, in such many degrees of freedom system (15 independent degrees of freedom for the secular system), there exist no results on the existence on invariant set which will bound the evolution of the system on infinite time span.

One may be tempted to try to integrate the motion of the solar system over 5 Gyr, that is over its expecting time life. For direct numerical integrations, this can be considered as an interesting challenge as it is still out of

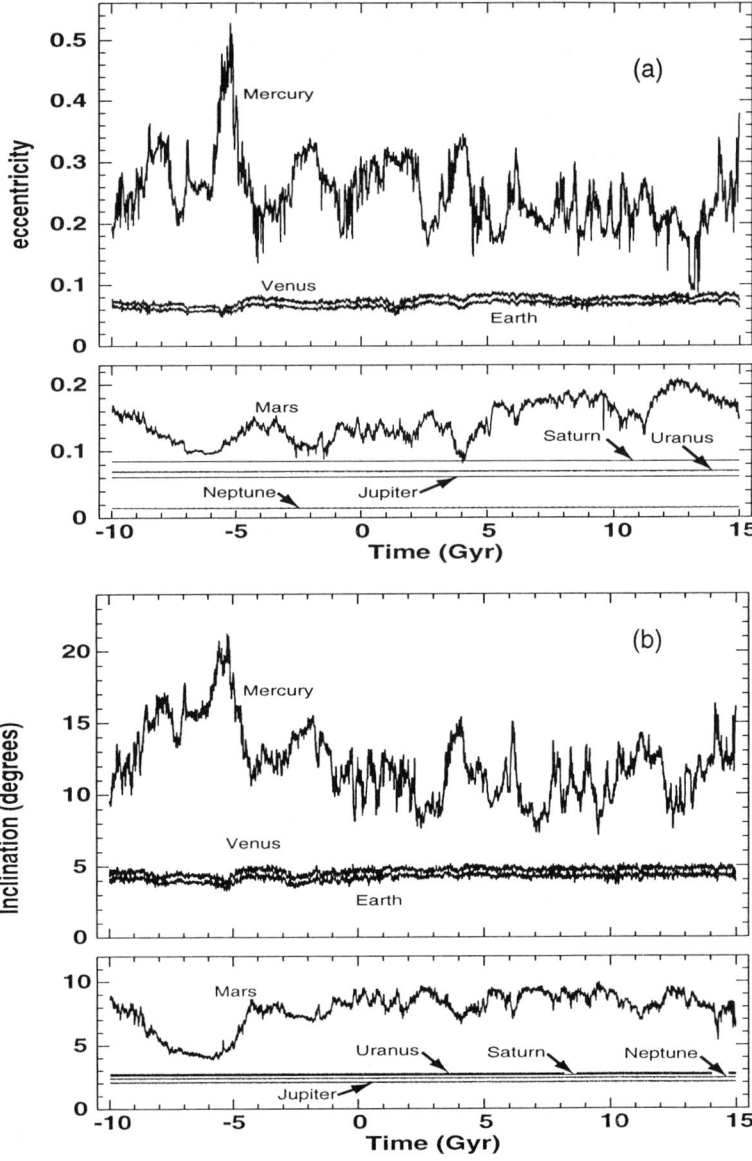

Figure 2. Numerical integration of the averaged equations of motion of the solar system 10 Gyr backward and 15 Gyr forward. For each planet, the maximum value obtained over intervals of 10 Myr for the eccentricity (a) and inclination (in degrees) from the fixed ecliptic J2000 (b) are plotted versus time. For clarity of the figures, Mercury, Venus and the Earth are plotted separately from Mars, Jupiter, Saturn, Uranus and Neptune. The large planets behavior is so regular that all the curves of maximum eccentricity and inclination appear as straight lines. On the contrary, the corresponding curves of the inner planets show very large and irregular variations, which attest to their diffusion in the chaotic zone.(Laskar, 1994)

reach of present computer technology, but it should be stressed, that by no means it can be considered as the description of the evolution of the solar system over 5 Gyr. Indeed, because of the exponential divergence with a Lyapunov time of 5Myr, after about 100 Myr the computed solution will be very different from the real solution followed by the actual solar system. Such a solution still present some interest, as it gives one of the possible behavior of the solar system, but it is much more important to obtain some description of the chaotic zone where the solar system evolves. In particular, it is more interesting to estimate the speed of diffusion in this chaotic zone. For such a goal, an integration of the solar system over 5 Gyr can be used, but will not be sufficient. Quite surprisingly, we can use integrations over even longer time span, which will act as scout exploring this chaotic zone. We can also send multiple of these explorers with very close initial conditions, in order to reach a larger portion of the phase space which can be attained by the solar system in 5 Gyr.

In order to achieve this task, it becomes quite obvious that we need to be able to integrate very rapidly the motion of the solar system, and the secular system of equations was even more simplified (Laskar, 1994), retaining only 50 000 terms and conserving the symmetries of the equations. Doing that, only about 6000 terms really need to be computed during the evaluation of the second hand member of the equations and the computations could be achieved on an IBM RS6000/370 workstation at a rate of about 1 day of CPU time per Gyr, without any loss in the precision.

As we want to understand the dynamics of this secular system, it is actually necessary to make the integration with great accuracy. The secular system is an approximation of the real equations of motion, but by understanding completely the global dynamical behavior of this system, we will obtain a lot of information on the original system.

Some first integrations were conduced over 25 Gyr (-10Gyr to + 15 Gyr) (Fig.2). It may seem strange to try to track the orbit of the solar system over such an extended time, longer than the age of the solar system, but one should understand that it is done in order to explore the chaotic zone where the solar system evolves and, after 100 Myr, can give only an indication of what can happen. On the other hand, if there is a sudden increase of eccentricity for one planet after 10 Gyr, this still tells us that such an event could probably also occur over a much shorter time, for example in less than 5 Gyr. In the same way, what happens in negative time can happen as well in positive time.

In order to follow the diffusion of the orbits in the chaotic zone, one needs quantities which behave like action variables, that is quantities which will be constant for a regular (quasiperiodic) solution of the system. Such quantities are given here by the maximum eccentricity and inclination attained by each planet during intervals of 10 Myr (Fig. 2).

The behavior of the large planets is so regular that all the corresponding curves appear as straight lines (Fig. 2). On the contrary, the maxima of eccentricity and inclination of the inner planets show very large and irregular variations, which attest to their diffusion in the chaotic zone. The diffusion of the eccentricity of the Earth and Venus is moderate, but still amounts to about 0.02 for both planets. The diffusion of the eccentricity of Mars is large and reaches more than 0.12, leading to values higher than 0.2 for the eccentricity of Mars. For Mercury, the chaotic zone is so large (more than 0.4) that it reaches values larger than 0.5 at some time. The behavior of the inclination is very similar.

Strong correlations between the different curves appear in figure 2. Indeed, as the solar system wanders in the chaotic zone, it is dominated by the linear coupling among the proper modes of the averaged equations (Laskar, 1990), which induces a very similar behavior for the maximum eccentricity and inclination of Venus and the Earth. This coupling is also noticeable in the solution of Mars. On the other hand, an angular momentum integral exists in the averaged equations and explains why when Mercury's eccentricity and inclination increase, the similar quantities for Venus, the Earth and Mars decrease. Thus it appears that, despite the small values of the inner planets' masses, the conservation of angular momentum plays a decisive role in limiting their excursions in the chaotic zone.

5. Escaping planets

At some time, Mercury suffered a large increase in eccentricity (Fig. 2) rising up to 0.5. But this is not sufficient to cross the orbit of Venus. The question then arises whether it is possible for Mercury to escape from the solar system in a time comparable to its age. A first attempt to answer this was made by slightly changing the initial conditions for the planets. Indeed, because of the chaotic behavior, very small changes in the initial conditions lead to completely different solutions after 100 Myr. Using this, I decided to change only one coordinate in the position of the Earth, amounting to a physical change of about 150 meters (10^{-9} in eccentricity). The full system was integrated with several of these modified solutions, but this led to similar (although different) solutions. In fact, it should not be too easy to get rid of Mercury, otherwise it would be difficult to explain its presence in the solar system.

I thus decided to guide Mercury somewhat towards the exit. A first experiment was done for negative time: for 2 Gyr, the solution is left unchanged, then, 4 different solutions are computed for 500 Myr, in each of which the position of the Earth is shifted by 150 meters, in a different direction (due to the exponential divergence, this corresponds to a change smaller than Planck's length in the original initial conditions).

The solution which leads to the maximum value of Mercury's eccentricity is retained up to the nearest entire Myr, and is started again. In 18 of such steps, Mercury attains eccentricity values close to 1 at about -6 Gyr (Fig 2) when the solution enters a zone of greater chaos, with Lyapunov time $\approx$ 1 Myr, giving rise to much stronger variations of the orbital elements of the inner planets. A second solution was also computed in positive time, with changes in initial condition of only 15 meters instead of 150 meters. As anticipated, this led to a similar increase in Mercury's eccentricity, this time in only 13 steps and about 3.5 Gyr (Fig 2).

While the eccentricity increases, the inclination of Mercury can change very much but the computation of the relative positions of the intersection of the orbits of Mercury and Venus with their line of nodes demonstrated that the orbits effectively intersect at about 3.5 Gyr. At this time, the two planets can experience a close encounter which can lead to the escape of Mercury or to collision.

For very high eccentricity of Mercury, the model used here no longer gives a very good approximation to the motion of Mercury, but it is very important to know that in this approximation, the chaotic zone allows the escape of a planet from the solar system in a time smaller than the expected life of the solar system, due to diffusion in the chaotic zone. Even more, in this averaged system, the degrees of freedom corresponding to semi major axes and mean longitudes are removed, but in the real system the addition of these extra degrees of freedom will probably lead to even stronger chaotic behavior, as in general, addition of degrees of freedom increases the stochasticity of the motion.

Similar computations were made for Mars and the Earth, but did not lead up to now to an escaping solution. For the Earth, the maximum eccentricity reached after 5 Gyr is about 0.1, while for Mars, the eccentricity attained 0.25 after 5 Gyr. With such a high eccentricity, Mars comes very close to the Earth, and it may be possible to find some escaping solution for Mars when considering the complete equations, but this probably needs the next generations of computers.

6. Marginal stability of the solar system

The existence of an escaping orbit for Mercury does not mean that this escape is very likely to occur. In fact, the solution computed here which lead to an escape was very carefully tailored, by selecting at each step one solution among 4 equivalent ones. The result is the existence for an escaping orbit, but does not tell us the probability for this escape to occur.

The computation of an estimate of this probability would require to take into account the full equations in order to be accurate. From the present computation, it can be thought that this probability is small, but not null, which is compatible with the present existence of Mercury.

Even without speaking of escaping orbits, the very large diffusion of the inner planets orbits is very striking. Even after the discovery of the chaotic behavior of the solar system, and despite the results of (Laskar, 1990), many assumed that the chaotic diffusion in the solar system was very small. Here is clearly demonstrated that for the inner planets, it is not the case. More, for the inner planets, the excursion of the eccentricity and inclination variables seems to be essentially constrained by the angular momentum conservation. This is quite surprising, when considering that the essential part of the angular momentum comes from the outer planets. In fact, the outer planets system is very regular, and practically no diffusion will take place among the degrees of freedom related to the outer planets. Thus, exchanges of angular momentum among the proper modes related to the inner planets result from the chaotic diffusion. This explains that when the maximum eccentricity of Mercury increases, the maximum eccentricity of Venus, the Earth and Mars decreases. One can also notice that the eccentricity curves of Venus and the Earth are very similar. This is due to the strong linear coupling between the proper modes of these two planets.

On figure 2, it is evident that the less massive planets are subject to the largest variation of eccentricity. This becomes obvious when considering that these variations are essentially bounded by the angular momentum conservation, which for each planets is proportional to $m\sqrt{a}$, where m is the mass of the planet, and a its semi major axis.

If, for each planet, we consider the maximum diffusion of the eccentricity over 5 Gyr (Fig. 3) we find that Mercury's eccentricity can go sufficiently high to allow Mercury's orbit to cross the orbit of Venus, Venus and the Earth's eccentricity can go up to 0.1, and Mars as high as 0.25. Apart from some small place in between Venus and the Earth, or the Earth and Mars, all the inner solar system is swept by the planetary orbits, and the small planets (Mercury and Mars) are the planets which present the largest excursions. Practically, we can conclude that the inner solar system is full. That is there is no room for any extra planet. Indeed, even if there are some place which seems not to be possibly reached in 5 Gyr, the additional planet orbit will present some eccentricity variations, and thus most probably will intersect with one of the already existing orbits. If we add a large planet, of the size of the Earth or Venus, its orbital elements will not vary much but it will induce strong short periods perturbations. On the contrary, a small object will suffer large orbital variations, as it will not be much constrained

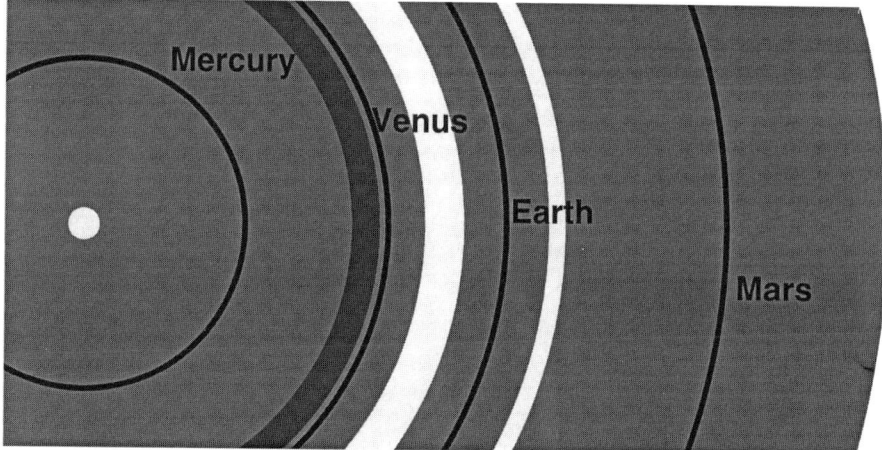

Figure 3. Estimates of the zones possibly occupied by the inner planets of the solar system over 5 Gyr. The circular orbits correspond to the bold lines, and the zones visited by each planet resulting from the possible increase of eccentricity are the shaded zones. In the case of Mercury and Venus, these shaded zones overlap. Mars can go as far as 1.9 AU, which roughly corresponds to the inner limit of the asteroid belt (Laskar, 1995).

by the angular momentum conservation. In this case, encounters with the already existing planets is very probable. It could be said that the variations which are plotted in fig 3 are the maximum variations possible over 5 Gyr, and not the most probable variations. This is true, but the addition of an extra planet will most probably increase very much the diffusion by increasing the numbers of degrees of freedom, and these maximum possible variations can probably be considered as the probable variations over 5 Gyr in the eventuality of the addition of an extra planet in the inner solar system. It becomes thus interesting to speak of marginal stability when considering the solar system. Maybe there was some extra planet at the early stage of formation of the solar system, and in particular in the inner solar system, but this lead to so much instability that one of the planets (probably among the smallest ones, of the size of Mercury or Mars) suffers a close encounter, or a collision with the other ones. This lead eventually to the escape of this planet and the remaining system gets more stable. In this case, at each stage, the system should have a time of stability comparable with its age, which is roughly what is achieved now, when ones finds that escape of one of the planets (Mercury) can occur within 5Gyr.

7. Conclusions

The analysis of the possible diffusion of the planetary orbits over 5 Gyr gives new insight on several question on the formation and evolution of the solar system. First of all, the existence of an escaping orbit for Mercury demonstrates that the the solar system in not stable, even when considering the strongest meaning of this word. On the contrary, although the solar system is not stable, it can be considered as marginally stable, that is. strong instabilities (collision or escape) can only occur on a time scale comparable to its age, that is about 5 Gyr. Some extra inner planets may have existed, but their existence gave rise to a much more unstable system, leading to the escape or collision of one of the planets. The organization of the inner planetary system is thus most probably due to its long run orbital evolution, and not uniquely to its rapid (less than 100 million years) formation process. This result is important for the understanding of the formation of the solar system, as it tells us that the solar system at the end of its formation process may have been significantly different from the present one, and has then evolved towards the present configuration because of the gravitational instabilities. It should be said that the outer system is very stable, but the long time recent numerical integrations (Gladman and Duncan, 1990; Holman and Wisdom, 1993; Levison and Duncan, 1993) also demonstrate that the outer solar system is full, that is most of the objects introduced in this system will escape on time scale much shorter than 5 Gyr. Apart from some special locations, stable zones only begins at about 40 AU, where some objects were recently founded. The inner solar system is also full, from the 0 AU to about 2 AU, which coincide with the beginning of the asteroidal belt. It should be interesting to investigate this point further using simulations with the addition of an extra planets, but many features have already been deduced here from the present computations. In particular, in the repartition of the inner planets, which can be thought as the so called Titius-Bode law, it is very striking that the spacing between the planets does not seem to be related to their masses. Indeed, when considering the most direct perturbation, that is the short-period perturbations, the zone depleted by a planet should increase with its mass, due to the overlap of the mean motion resonances. This seems to be primordial for the outer planets system (although more complicated combination of resonance may be present involving also secular resonances, as for the asteroid dynamics), but does not work for the inner planets system, where the short-period perturbation are not very important. In this case, as was presented above, the smallest a planet is, the largest will be its diffusion due to the chaotic behavior of the secular system. There will thus be an equilibrium between the short term perturbations which increases with the mass of the planet, and the long time diffusion of the orbital eccentricity and inclination, which is larger for the small planets. These competing effects could end up with an apparent rapartition which does not depend any longer on the masses. In any case, the marginal stability of the solar system revealed by the analysis of its long time behavior over 5 Gyr is an indication that its present organization results from its dynamical evolution.

In particular, one may be now tempted to answer to the question of what will be a generic planetary system ?

Considering the present results on our solar system, I would think that a generic planetary system will always be in a state of marginal stability, resulting from its gravitational interactions. If the formation process is such that there exists some large outer planets and some small inner planets, after 5 Gyr, the inner planets will therefore be subject to some instabilities similar to the present ones, and thus so will be their obliquities (Laskar and Robutel, 1993), with all the climate implications (Laskar, 1993; Laskar et al.1993). In particular, a planetary system with only one or two planets should be excluded, or, if it does exist, would be crowded with asteroids everywhere which would be the original remaining planetesimals, not ejected by planetary perturbations.

Most of the results presented here rely on the analysis of the secular equations of the solar system and not on the complete equations. This was the price to pay for allowing a more global approach on the problem of the stability and long time evolution of the solar system. It is quite obvious that some integrations of the full equations are still needed, but it is doubtful that these future integrations will change much the global landscape of the dynamics of the solar system portrayed here.

References

Applegate, J.H., Douglas, M.R., Gursel, Y., Sussman, G.J. and Wisdom, J.: 1986, 'The solar system for 200 million years,' *Astron. J.* **92**, 176–194

Bretagnon, P.: 1974, Termes à longue périodes dans le système solaire, *Astron. Astrophys* **30** 341–362

Brumberg, V.A., Chapront, J.: 1973, Construction of a general planetary theory of the first order, *Cel. Mech.* **8** 335–355

Carpino, M., Milani, A. and Nobili, A.M.: 1987, Long-term numerical integrations and synthetic theories for the motion of the outer planets, *Astron. Astrophys* **181** 182–194

Cohen, C.J., Hubbard, E.C., Oesterwinter, C.: 1973, , *Astron. Papers Am. Ephemeris* **XXII** 1

Duriez, L.: 1979, 'Approche d'une théorie générale planétaire en variable elliptiques héliocentriques, *thèse* Lille

Gladman, B., Duncan, M.: 1990, On the fates of minor bodies in the outer solar system *Astron. J.*, **100**(5)

Holman, M.J., Wisdom, J.: 1993, Dynamical stability in the outer solar system and the delivery of short period comets *Astron. J.*, **105**(5)

Kinoshita, H., Nakai, H.: 1984, Motions of the perihelion of Neptune and Pluto, *Cel. Mech.* **34** 203

Laskar, J.: 1984, *Thesis*, Observatoire de Paris

Laskar, J.: 1986, Secular terms of classical planetary theories using the results of general theory,, *Astron. Astrophys.* **157** 59–70

Laskar, J.: 1989, A numerical experiment on the chaotic behaviour of the Solar System *Nature*, **338**, 237–238

Laskar, J.: 1990, The chaotic motion of the solar system. A numerical estimate of the size of the chaotic zones, *Icarus*, **88**, 266–291

Laskar, J.: 1992a, A few points on the stability of the solar system, in Symposium IAU 152, S. Ferraz-Mello ed., 1–16, Kluwer, Dordrecht

Laskar, J.: 1992b, La stabilité du Système Solaire, in *Chaos et Déteminisme*, A. Dahan et al., eds., Seuil, Paris

Laskar, J.: 1993, La Lune et l'origine de l'homme, *Pour La Science*, **186**, *avril 1993*

Laskar, J.: 1994, Large scale chaos in the solar system, *Astron. Astrophys.* **287** L9-L12

Laskar, J.: 1995, Large scale chaos and Marginal stability of the solar system, XIème Colloque ICMP, *Paris july, 1994*, International Press, p. 75–120

Laskar, J., Quinn, T., Tremaine, S.: 1992a, Confirmation of Resonant Structure in the Solar System, *Icarus*, *95*,148–152

Laskar, J. Robutel, P.: 1993, The chaotic obliquity of the planets, *Nature*, **361**, 608–612

Laskar, J., Joutel, F., Robutel, P.: 1993, Stabilization of the Earth's obliquity by the Moon, *Nature*, **361**, 615–617

Levison, H.F., Duncan, M.J.: 1993, The gravitational sculpting of the Kuiper belt, *Astrophys. J. Lett.*, **406**, L35-L38

Newhall, X. X., Standish, E. M., Williams, J. G.: 1983, DE102: a numerically integrated ephemeris of the Moon and planets spanning forty-four centuries, *Astron. Astrophys.* **125** 150–167

Nobili, A.M., Milani, A. and Carpino, M.: 1989, Fundamental frequencies and small divisors in the orbits of the outer planets, *Astron. Astrophys.* **210** 313–336

Quinn, T.R., Tremaine, S., Duncan, M.: 1991, 'A three million year integration of the Earth's orbit,' *Astron. J.* **101**, 2287–2305

Sussman, G.J., and Wisdom, J.: 1988, 'Numerical evidence that the motion of Pluto is chaotic.' *Science* **241**, 433–437

Sussman, G.J., and Wisdom, J.: 1992, 'Chaotic evolution of the solar system', *Science* **257**, 56–62

THEORY COMPRESSION WITH ELLIPTIC FUNCTIONS

VICTOR A. BRUMBERG

Bureau des Longitudes
77, av. Denfert-Rochereau, Paris 75014, France

(On leave from Institute of Applied Astronomy
8, Zhdanovskaya st., St.-Petersburg 197042, Russia)

Abstract. Introduction of Jacobi elliptic functions in planetary, satellite and cometary problems of celestial mechanics is a transformation of variables to present the analytical theories of motion in the more compact form as compared with the traditional series in multiples of mean longitudes or mean anomalies.

1. Introduction

Analytical techniques of celestial mechanics survive now not the best period in their history. In relation to the two-century anniversary of Bureau des Longitudes, the world-recognized center of analytical celestial mechanics, it may be reasonable to discuss once again the role of analytical techniques in celestial mechanics. It is true that numerical integration represents now the best tool to investigate the empirical evolution of dynamical systems and to produce the high accuracy ephemerides of specific bodies. But analytical theories enable one to get a more profound insight into physical and mathematical laws of motion. Indeed, analytical techniques of celestial mechanics are aimed

- to obtain a general solution of the equations of motion as explicit function of initial values and parameters;
- to present a solution in the physically adequate form;
- to be used as a framework to investigate small effects due to extra additive forces in the right-hand members of the equations of motion.

It should be added that analytical techniques of celestial mechanics contributed a lot into mathematical and natural sciences in such domains as special functions, perturbation theory, stability and resonance theory, periodic solutions, algebraic systems, etc.

It is of interest to compare the role of numerical integration and analytical approach in typical problems of celestial mechanics.

1. Two-body problem
   It is hardly possible to imagine it without the analytical solution in spite of the fact that the closed-form time-explicit solution of this problem has been actually obtained only quite recently by Osácar and Palacián (1994) with the use of dilogarithmic function. This is an example where analytical treatment due to its simplicity and compactness is far beyond numerical integration approach.
2. Lunar theory
   The most recent semi-analytical theory of the motion of the Moon is ELP by Chapront and Chapront-Touzé (1995). This theory is used in Connaissance des Temps but as compared with numerical integration its accuracy is not completely sufficient to analyze the high-precision LLR observations. Nevertheless, by its domain of applicability beyond the scope of LLR this theory significantly surpasses numerical lunar ephemerides.
3. Planetary theories
   The semi-analytical theories VSOP of the motion of the major planets by Bretagnon (1982) are used in Connaissance des Temps and in space research as well as numerical planetary ephemerides. Along with this they are of great benefit for astrometry, IERS activity and Earth's sciences.
4. Satellite theory
   It is evident that for the analysis of high-precision SLR observations one needs numerical ephemerides. But for a large class of research problems a simple first-order analytical theory may be quite adequate. Kaula's theory is just an example of such wide-purpose analytical theory.
5. Cometary motion
   In spite of the efforts of Hansen, Gylden and their followers (see below) this problem remains to be a challenge for analytical techniques and the cometary orbits are computed mostly by numerical integration.
6. Rotation of celestial bodies
   Analytical solution of this problem like recent theory of precession and nutation by Kinoshita and Souchay (1990) is quite competent with numerical integration by its accuracy but is much more informative

in respect of physical interpretation of different components of the solution.

7. GTM (general theory of translatory-rotational motion)
At present, most analytical theories representing translatory or rotational motion of celestial bodies are constructed for the sake of compactness with fictitious secular terms. The physically adequate form of the solution implies that the angular variables are expressed as linear functions of time and the action variables are represented by slowly changing quasi-periodic functions of time. Such theory for the solar system bodies may be called GTM. Its compactness may be achieved by introducing an adequate intermediary (Hill-like type for the translatory motion), separating fast and slow variables in the sense of Zeipel and performing a normalizing (Birkhoff-like) transformation to a secular system. Such a problem is still awaiting its practical solution.

Twenty or thirty years ago it seemed that the desired accuracy of analytical theories could be always achieved by using very long Poisson series constructed by means of Poisson series processors. Now it became evident that the increase of the number of terms in Poisson series cannot be too efficient tool for this aim. The main recipe nowadays is to develop compact-form analytical theories with the aid of more sophisticated specialized software based on some universal computer algebra system (*Maple, Mathematica,* etc.). A compact-form analytical theory may be understood as a theory with the large value of the ratio of the achieved accuracy to the needed number of terms (like the commercial quality to price ratio). It is evident that the vast arsenal of compression transformations of classical celestial mechanics may be used now in combination with present computer software facilities.

2. Compact-Form Series in Classical Celestial Mechanics

Many analytical techniques of classical perturbation theory are based on the two-body problem Fourier expansions in multiples of mean anomaly M, true anomaly v or eccentric anomaly g

$$\left(\frac{r}{a}\right)^n \exp imv = \sum_{s=-\infty}^{\infty} X_s^{n,m}(e) \exp isM \qquad (2.1)$$

$$= \sum_{s=-\infty}^{\infty} Y_s^{n,m}(e) \exp isv \qquad (2.2)$$

$$= \sum_{s=-\infty}^{\infty} Z_s^{n,m}(e) \exp isg . \qquad (2.3)$$

Expansion (2.1) enables one to represent the coordinates as explicit functions of time. But Hansen coefficients $X_s^{n,m}(e)$ for moderate and large values of eccentricity e decrease not so fast with the increase of $|s|$ so that this expansion may be too long. Expansions (2.2) and (2.3) are more compact in this respect (moreover, they reduce to finite trigonometric polynomials for $n < 0$ and $n > m > 0$, respectively). In classical celestial mechanics the eccentric anomaly series based on (2.3) were used in particular by Hansen, Newcomb and Hill. Later on the true anomaly series based on (2.2) were used by Brown and Shook (1933). From time to time expansions (2.2) and (2.3) may be met in contemporary papers. But it seems that most people share the negative opinion by Zeipel. In his encyclopaedic paper Zeipel (1912) comes to conclusion that the advantages of the compact representation of the disturbing function based on the eccentric anomaly series might be lost during the integration by means of the transformation to the mean anomalies with the aid of Bessel functions. We shall return below to the key problem of integration with different trigonometric arguments.

The idea to use elliptic functions to find more efficient expansions than (2.1)–(2.3) for compression of analytical theories of celestial mechanics was suggested by Gylden more than a century ago. His idea is that 'one views the mean (true, eccentric) anomaly ... as the elliptic amplitude of a new (independent) variable' (see Nacozy, 1977). Unfortunately, this idea was related to the Hansen method of partial anomalies for the analytical representation of cometary motion. This method involves the division of a cometary orbit into several (at least two) parts with its own independent argument for each part. In spite of all efforts of Gylden and his followers the partial anomaly technique remained and still remains rather cumbersome for wide application (Nacozy, 1969; Skripnichenko, 1972). Since then the idea of Gylden was regarded only as an attempt to improve the partial anomaly technique and numerical integration became the most widespread tool for investigation of the cometary motion.

3. Elliptic Function Expansions as Convergence Accelerators

In realizing the Gylden's idea one has not to deal with problems admitting a solution in terms of elliptic functions (as the problem of two fixed centres, for example). Moreover, one has to do not with elliptic functions themselves but rather with their Fourier expansions. The main idea is to find, if possible, a transformation of variables

$$(x, y) \to (k, u) \tag{3.1}$$

reducing a function $f(x, y)$ of one power ($0 \le x \le 1$) and one trigonometric variable ($0 \le y \le 2\pi$) to Jacobi elliptic function $g(k, u)$ with modulus k and argument u. If such transformation (3.1) exists then the ordinary Fourier expansion

$$f(x,y) = \sum f_m(x) \exp imy \qquad (3.2)$$

of function $f(x,y)$ will be replaced by the Fourier expansion

$$g(k,u) = \sum g_m(q) \exp imw, \qquad w = \frac{\pi u}{2K(k)} \qquad (3.3)$$

of function $g(k,u)$. Here q is Jacobi nome remaining comparatively small even for the values of k close to 1. $K(k)$ is the complete elliptic integral of the first kind. For large $|m|$ coefficients g_m decrease generally much faster than coefficients f_m and one may expect that series (3.3) will be much more compact than series (3.2).

One should underline two points frequently overlooked, i.e.

- transformation (3.1) is made in function $f(x,y)$ but not in series (3.2) which is not needed at all;
- compactness of series (3.3) with respect to (3.2) is due mainly to the k-dependent angular variable w and in lesser extent due to the compactness of coefficients g_m themselves. Coefficients f_m may be computed without power series expansions (Laplace coefficients in planetary problems or Hansen coefficients in satellite problems) or may be represented by closed form expressions (coefficients of (2.2) and (2.3) series) but this has nothing to do with the slow convergence of series (3.2) itself.

The standard Fourier series for simple combinations of Jacobi elliptic functions may be found in many textbooks. They are collected as (2.5.66)–(2.5.78) in (Brumberg, 1995). The Fourier series for any rational function of Jacobi elliptic functions may be derived by recurrence relations starting with these standard expansions.

As it is well known even more fast converging expansions are provided by theta functions. But in using these expansions one meets two difficulties, i.e.

- the absence of such expansion for periodic part of elliptic amplitude needed for the inversion and interrelation problem;
- the necessity to perform operations on rational functions of Poisson series.

The latter difficulty occurs also in applying the Landen transformation $(k,u) \to (k_1,v)$

$$k = \frac{2k_1^{1/2}}{1+k_1}, \qquad u = (1+k_1)v \qquad (3.4)$$

enabling one to operate with the Jacobi nome $q_1 = q^2$.

Application of the standard expansions for Jacobi elliptic functions demands a simpler software to perform symbolic operations on trigonometric series with rational coefficients.

4. Elliptic Functions in Planetary Problems

The starting point of many analytical investigations in planetary problems is the expansion of a generating function

$$\gamma(n, x, y, \nu, \alpha, \zeta) = \alpha^n (1 - \alpha \zeta^{-1})^x (1 - \alpha \zeta)^y (-\zeta)^\nu \tag{4.1}$$

occurring in the right-hand members of the equations of motion. Here α is a real parameter (the ratio of the semi-major axes), ζ is an exponential function of the mean longitudes λ and λ'

$$\zeta = \exp i(\lambda - \lambda'), \tag{4.2}$$

n and ν are integers, x and y are real numbers or more precisely

$$x = -\frac{K}{2}, \quad y = -\frac{L}{2} \tag{4.3}$$

where K and L are positive odd integers. Traditional Fourier expansion of function (4.1) has the form

$$\gamma(n, x, y, \nu, \alpha, \zeta) = \sum_{\sigma=-\infty}^{\infty} \tilde{\gamma}_\sigma(n, x, y, \nu, \alpha) \zeta^\sigma. \tag{4.4}$$

Functions of type (4.1) and their expansions (4.4) were considered by Newcomb, Cauchy and Gylden (see Zeipel, 1912). Later on this function was studied by Brown and Shook (1933) and was intensively used in constructing GPT, general planetary theory (Brumberg, 1995). Quite recently it was applied by Laskar and Robutel (1995) to derive a new expansion of the planetary disturbing function. The main difficulty in applying (4.4) is the slow convergence of this expansion especially for large values of α. The possible remedy is to find transformation (3.1) of variables $(\alpha, \lambda - \lambda')$ to elliptic variables (k, u). Indeed, the transformation

$$2 \operatorname{am} u = M, \quad M = \pi - (\lambda - \lambda') \tag{4.5}$$

and

$$k^2 = \frac{4\alpha}{(1+\alpha)^2} \tag{4.6}$$

reduces function (4.1) to the form

$$\gamma(n,x,y,\nu,\alpha,\zeta) = \frac{\alpha^n(-\zeta)^\nu}{[(1+\alpha)\,\mathrm{dn}\,u]^{\max\{K,L\}}}\left(1 - \alpha\zeta^{\mathrm{sgn}(K-L)}\right)^{\frac{|K-L|}{2}} \quad (4.7)$$

with

$$\zeta = -\exp(-\mathrm{i}2\,\mathrm{am}\,u). \quad (4.8)$$

It is easy to see that function (4.7) may be expanded in Fourier series in multiples of

$$w = \frac{\pi u}{K(k)} \quad (4.9)$$

or in the exponential form

$$\gamma(n,x,y,\nu,\alpha,\zeta) = \sum_{\sigma=-\infty}^{\infty} \gamma_\sigma(n,x,y,\nu,\alpha)\tau^\sigma, \quad \tau = \exp iw. \quad (4.10)$$

The difference in the arguments (3.3) and (4.9) is due to the fact that the real periods of function (4.7) and of general elliptic function (3.3) are $2K(k)$ and $4K(k)$, respectively. Coefficients γ_σ of (4.10) may be easily found in the closed form with respect to Jacobi nome q. Indeed, these coefficients are expressed in terms of auxiliary functions $I_{2m,n}$ defined by

$$I_{p,n}(u) = \frac{\exp(ip\,\mathrm{am}\,u)}{(\mathrm{dn}\,u)^n} \quad (4.11)$$

and for $p = 2m$ (integer m) these functions are easily expanded in τ-series by means of the recurrence relations (see, for example, Howland, 1988). The most difficult case in the planetary problems is provided by the pair Venus–Earth. For this pair one has $\alpha = 0.723$, $k = 0.987$ and $q = 0.215$. Even for this pair the τ-expansion (4.10) of generating function (4.1) is much more compact than traditional ζ-series (4.4).

Relationship with time is realized with the aid of (4.5). This relation represents the Lagrange implicit function equation of the form

$$w + \sum_{m=1}^{\infty} d_m(q)\sin mw = M \quad (4.12)$$

with

$$d_m(q) = \frac{4}{m}\frac{q^m}{1+q^{2m}}. \quad (4.13)$$

The inversion of this equation has the form

$$w = M + \sum_{m=1}^{\infty} c_m(q)\sin mM. \quad (4.14)$$

Moreover, for any real s (E.Brumberg et al., 1995)

$$\exp isw = \sum_{r=-\infty}^{\infty} E_r^{(s)}(q)\exp i(s+r)M \qquad (4.15)$$

and

$$\exp isM = \sum_{r=-\infty}^{\infty} F_r^{(s)}(q)\exp i(s+r)w. \qquad (4.16)$$

Recent results by E.Brumberg (1995) for the analogous relations for high eccentricity orbits show that coefficients $F_r^{(s)}$ decrease with large $|r|$ much faster than coefficients $E_r^{(s)}$.

Transformation (4.5) and (4.6) known already in classical celestial mechanics was applied by Richardson (1982) in his research on planetary intermediate orbits constructed by Lie transforms. Later on the same transformation has been used by Williams et al. (1987) in the attempt to represent the first-order classical planetary theories in k, u variables. At the same time Chapront and Simon (1988) have developed a first-order CPT (compact planetary theory). Replacing mean longitudes λ and λ' by linear function of time l and elliptic argument u

$$\lambda = l - \operatorname{am} u + \frac{\pi}{2}, \qquad \lambda' = l + \operatorname{am} u - \frac{\pi}{2} \qquad (4.17)$$

they have compared the traditional perturbation theory series

$$x = \sum_{k=0}^{m} [\tilde{x}_k + \tilde{S}_k(\lambda, \lambda')] t^k \qquad (4.18)$$

with the elliptic argument series

$$x = \sum_{k=0}^{m} [x_k + S_k(l, w)] w^k. \qquad (4.19)$$

It turned out that series $S_k(l, w)$ are much shorter than series $\tilde{S}_k(\lambda, \lambda')$ for all couples of the major planets. Series (4.18) and (4.19) were computed in the semi-analytical form by application of FFT (fast Fourier transform) to the equations of motion. This work is still awaiting its completion.

Transformation (4.5) and (4.6) has been intensively used in elaborating GPT, a planetary theory without fictitious secular terms (Brumberg, 1995). GPT is constructed by the series in powers of the eccentricity and inclination variables with the coefficients depending on the differences of the mean longitudes of the planets. These series reduce the original equations of motion to an autonomous secular system. In dependence on the representation of the GPT series one may distinguish three forms of GPT, i.e.

(a) ζ-series form as (4.4) extended for the N-planet case

$$\zeta_{ij} = \exp i(\lambda_i - \lambda_j), \qquad i,j = 1,2,\ldots,N. \tag{4.20}$$

(b) first-order closed form in terms of Jacobi elliptic functions with arguments u_{ij} and modulii k_{ij}

$$2\,\mathrm{am}\,u_{ij} = M_{ij}, \qquad M_{ij} = \pi - (\lambda_i - \lambda_j), \qquad k_{ij}^2 = \frac{4a_i a_j}{(a_i + a_j)^2}, \tag{4.21}$$

a_i, a_j being the semi-major axes. This form involves also the elliptic quadratures, for example,

$$\Phi_n(\mathrm{am}\,u, k, s, \alpha) = \exp(i\alpha\,\mathrm{am}\,u)\int I_{s-\alpha,n-1}(u)du \tag{4.22}$$

with integer n and s and real α. These quadratures admit a closed form representation only for integer α. For real α one has to use different approximations as integration by parts (with respect to the fractional part of $s - \alpha$), series expansion, etc.

(c) τ-series form as (4.10) with

$$\tau_{ij} = \exp i w_{ij}, \qquad w_{ij} = \frac{\pi u_{ij}}{K(k_{ij})}. \tag{4.23}$$

Actual computation of the intermediate orbits for all major planets (Brumberg and Klioner, 1995) demonstrates the compactness of τ-series with respect to ζ-series.

Construction of the second-order theory by means of τ-series involves the quadratures of the form

$$I = \int f(u_{ij}, u_{ik})dt, \tag{4.24}$$

f being a trigonometric series in multiples of w_{ij} and w_{ik}. One may use three ways to take these quadratures, i.e.

- to express integrand f in terms of M_{ij} and M_{ik}, to perform the integration and to return to variables w_{ij} and w_{ik};
- using the three-anomaly relation

$$2(\mathrm{am}\,u_{ij} + \mathrm{am}\,u_{jk} + \mathrm{am}\,u_{ki}) = 3\pi \tag{4.25}$$

to reduce integral (4.24) to the form

$$I = \int g(u, u')du \tag{4.26}$$

with the relation

$$2\,\mathrm{am}\,u' = 2p\,\mathrm{am}\,u + c, \qquad c = \mathrm{const}, \qquad |p| < \frac{1}{2} \qquad (4.27)$$

and to apply the Hansen's device expressing w' in terms of w and $w^* = pw + c$ (Brumberg, 1995);
- to integrate (4.26) by parts resulting at step k in the integral of the same type with factor p^k.

Analytical integration of (4.24) by any of these ways may be rather cumbersome. In constructing a semi-analytical theory with numerical coefficients one may combine these tools with the FFT technique. Anyway, the use of elliptic function expansions permits to reduce the number of terms of the first-order theories and it remains to find the most efficient way for the higher-order theory.

5. Elliptic Functions in Satellite Problems

The main difficulty in analytical treatment of highly eccentric orbits is due to the slow convergence of traditional M-series like (2.1) for large values of eccentricity e. It turns out (E.Brumberg, 1992) that transformation of variables $(e, M) \to (k, u)$

$$k = e, \qquad \mathrm{am}\,u = g + \frac{\pi}{2}, \qquad w = \frac{\pi u}{2K(k)} - \frac{\pi}{2} \qquad (5.1)$$

might be a possible remedy to overcome this difficulty. Classical expansions (2.1)–(2.3) are replaced therewith by rather compact expansion in multiples of the elliptic anomaly w

$$\left(\frac{r}{a}\right)^n \exp i m v = \sum_{s=-\infty}^{\infty} B_s^{n,m}(q) \exp i s w. \qquad (5.2)$$

Most simply the coefficients of (5.2) may be computed from the recurrence relations (E.Brumberg and Fukushima, 1994).

This technique was first applied to extend for highly eccentric orbits the first-order Kaula's theory of the satellite motion in the field of the non-spherical primary (E.Brumberg et al., 1995). Later on (E.Brumberg, 1995) this technique was extended to include the third-body perturbations and was improved by using expansions (4.16) instead of (4.15).

It should be added that similar elliptic anomalies based on the transformation of the true anomaly v instead of the eccentric anomaly g were introduced earlier by Bond and Janin (1981) and Nacozy (1977) to improve the efficiency of the numerical integration of highly eccentric orbits. In such a way all suggestions by Gylden mentioned above are now realized.

6. Elliptic Functions for Nearly Intersecting Orbits

For analytical investigation of nearly intersecting orbits (of comets, some asteroids and space probes) one may apply the elliptic function expansions in combination with the development of the disturbing function elaborated by Boda (1931), Brown and Shook (1933), Petrovskaya (1970, 1972) and Yuasa and Hori (1979). As suggested in the last paper one may introduce as the initial approximation for the mutual distance Δ between two bodies

$$\Delta_0^2 = r^2 + r'^2 - 2rr'\mu\cos(W - W') \tag{6.1}$$

with the orbital longitude

$$W = \Omega + \omega + v \tag{6.2}$$

and the inclination factor

$$\mu = (cc' - ss')^2, \qquad s = \sin\frac{i}{2}, \qquad c = \cos\frac{i}{2}. \tag{6.3}$$

All designations are evident. Primed quantities are referred to the disturbing body. For any integer n the power expansion

$$\Delta^{-n} = \Delta_0^{-n} \sum_{k=0}^{\infty} \frac{(-1)^k \left(\frac{n}{2}\right)_k}{(1)_k} \left(\frac{\Delta^2 - \Delta_0^2}{\Delta_0^2}\right)^k \tag{6.4}$$

converges everywhere excepting the points of the actual collision. The negative powers of Δ_0 may be expanded in trigonometric series

$$\Delta_0^{-n} = \sum_{j=0}^{\infty} \left(\Delta_0^{-n}\right)_j \cos j(W - W') \tag{6.5}$$

with coefficients depending on r and r' by means of hypergeometric polynomials of different arguments (Brumberg, 1995). Any function $f(r, r')$ occurring in (6.4) may be expanded in the symbolic form

$$f(r, r') = \left(\frac{r}{a}\right)^D \left(\frac{r'}{a'}\right)^{D'} f(a, a'), \qquad D = a\frac{\partial}{\partial a}, \qquad D' = a'\frac{\partial}{\partial a'}. \tag{6.6}$$

Further expansions in terms of Jacobi nome q of the disturbed body and eccentricity e' of the disturbing major planet are performed with the aid of (5.2) and (2.1) with replacing integer index n by symbol D or D', respectively. One returns to the classical problem of calculating Newcomb operators but in combination with elliptic anomaly expansion (5.2).

References

Boda, K.: 1931, 'Entwicklung der Störungsfunction und ihrer Ableitungen in Reihen, welche für beliebige Exzentrizitäten und Neigungen konvergieren', *Astron. Nachr.*, **243**, 17

Bond, V.R. and Janin, G.: 1981, 'Canonical Orbital Elements in Terms of an Arbitrary Independent Variable', *Celes. Mech.* **23**, 159

Bretagnon, P.: 1982, 'Théorie du mouvement de l'ensemble des planètes. Solution VSOP82', *Astron. Astrophys.* **114**, 278

Brown, E.W. and Shook, C.A.: 1933, *Planetary Theory*, Cambridge Univ. Press

Brumberg, E. : 1992, 'Perturbed Two-Body Motion with Elliptic Functions', *Proc. 25th Symposium on Celestial Mechanics* (eds. H.Kinoshita and N.Nakai), 139, NAO, Tokyo

Brumberg, E.: 1995, 'Elliptic Anomaly Expansions to Construct High-Eccentricity Satellite Theory', Abstract 6a3, *IAU Symposium No. 172*, Paris

Brumberg, E. and Fukushima, T.: 1994, 'Expansions of Elliptic Motion Based on Elliptic Function Theory', *Celes. Mech.* **60**, 69

Brumberg, E., Brumberg, V.A., Konrad, Th. and Soffel, M.:1995, 'Analytical Linear Perturbation Theory for Highly Eccentric Satellite Orbits', *Celes. Mech.* **61**, 369

Brumberg, V.A.: 1995, *Analytical Techniques of Celestial Mechanics*, Springer

Brumberg, V.A. and Klioner, S.A.: 1995, 'Intermediate Orbit for General Planetary Theory in Elliptic Functions', Abstract A16, *IAU Symposium No. 172*, Paris

Chapront, J. and Chapront-Touzé, M.: 1995, 'Comparaison de la théorie du mouvement de la Lune ELP aux observations: la boite á outils', *Notes sci. et techn. du BDL* **S050**, 105

Chapront, J. and Simon, J. L.: 1988, 'Perturbations du premier ordre pour des couples de planètes', Bureau des Longitudes (unpublished)

Howland, R.A.: 1988, 'A New Approach to the Librational Solution in the Ideal Resonance Problem', *Celes. Mech.* **44**, 209

Kinoshita, H. and Souchay, J.: 1990, 'The Theory of the Nutation for the Rigid Earth Model at the Second Order', *Celes. Mech.* **48**, 187

Laskar, J. and Robutel, Ph.: 1995, 'Stability of the Planetary Three-Body Problem. I. Expansion of the Planetary Hamiltonian. *Celes. Mech.* (in press)

Nacozy, P.: 1969, 'Hansen's Method of Partial Anomalies: An Application', *Astron. J.* **74**, 544

Nacozy, P.: 1977, 'The Intermediate Anomaly', *Celes. Mech.* **16**, 309

Osácar, C. and Palacián, J.: 1994, 'Decomposition of Functions for Elliptic Orbits', *Celes. Mech.* **60**, 207

Petrovskaya, M.S.: 1970, 'Expansions of the Negative Powers of Mutual Distance Between Bodies', *Celes. Mech.* **3**, 121

Petrovskaya, M.S.: 1972, 'Expansions of the Derivatives of the Disturbing Function in Planetary Problems', *Celes. Mech.* **6**, 328

Richardson, D.L.: 1982, 'A Third-Order Intermediate Orbit for Planetary Theory', *Celes. Mech.* **26**, 187

Skripnichenko, V.I.: 1972, 'On the Application of Hansen's Method of Partial Anomalies to the Calculation of Perturbations in Cometary Motions', in G.A.Chebotarev, E.I.Kazimirchak-Polonskaya, and B.G.Marsden (eds.), *The Motion, Evolution of Orbits, and Origin of Comets*, p. 52, Reidel, Dordrecht

Williams, C.A., Van Flandern, T., and Wright, E.A.:1987, 'First Order Planetary Perturbations with Elliptic Functions', *Celes. Mech.* **40**, 367

Yuasa, M. and Hori, G.: 1979, 'New Approach to the Planetary Theory', in R.L.Duncombe (ed.), *Dynamics of the Solar System*, p. 69, Reidel, Dordrecht

Zeipel, H.: 1912, 'Entwicklung der Störungsfunktion', *Encyklopädie der math. Wiss.* **6** (2), 557

NUMERICAL EFFICIENCY OF THE ELLIPTIC FUNCTION EXPANSIONS OF THE FIRST-ORDER INTERMEDIARY FOR GENERAL PLANETARY THEORY

VICTOR A. BRUMBERG AND SERGEI A. KLIONER

Institute of Applied Astronomy,
197042, St.Petersburg, Russia

Abstract. We compare numerical efficiency of the two kinds of series for the first-order intermediate orbit for general planetary theory: (1) the classical expansion involving mean longitudes of the planets; (2) an expansion resulting from the theory of elliptic functions. We conclude that mutual perturbations of close couples of planets (the ratio of major semi-axes ~ 1) can be represented in more compact form with the aid of the second kind of series.

1. Two kinds of series for the first-order intermediary

In spite of significant progress of numerical approaches to construct the theories of motion of the major planets, analytical theory of planetary motion remains to be a challenging and interesting scientific task. Recently the interest in constructing analytical planetary theories has been revived by the idea to use an elliptic function of time (instead of time itself) as independent variable in order to make the resulting series more compact (see, Brumberg (1996) and references therein). Our aim is to compare the numerical efficiency of the classical series involving mean longitudes of planets and the series resulting from the application of elliptic functions for the particular case of the first-order intermediate orbit for general planetary theory (Brumberg, 1994).

We consider two kind of series for the first order intermediary $\mu \underset{1}{T}_{j}^{(i)}+$

defined by Eq. (5.19) of Brumberg (1994). The ρ-series are

$$\mu T_{1j}^{(i)+} = \sum_{k=-\infty}^{\infty} A_k^{ij} \rho_{ij}^k, \tag{1}$$

$$\rho_{ij} = \exp\left[\overset{\circ}{i} l_{ij}\right], \quad l_{ij} = 2\varphi_{ij}, \quad \varphi_{ij} = \frac{1}{2}(\pi - (\lambda_i - \lambda_j)), \tag{2}$$

λ_i is the mean longitude of the ith planet. The τ-series read

$$\mu T_{1j}^{(i)+} = \sum_{k=-\infty}^{\infty} B_k^{ij} \tau_{ij}^k, \tag{3}$$

$$\tau_{ij} = \exp\left[\overset{\circ}{i} w_{ij}\right], \quad w_{ij} = \frac{\pi}{K(k_{ij})} F(\varphi_{ij}, k_{ij}), \quad k_{ij}^2 = \frac{4a_i a_j}{(a_i + a_j)^2}, \tag{4}$$

where a_i is major semi-axis of the ith planet, $F(\varphi, k)$ and $K(k)$ are elliptic integral and complete elliptic integral of the first kind. Our aim is to compute numerically the coefficients A_k^{ij} and B_k^{ij}.

2. Numerical techniques used to compute A_k^{ij} and B_k^{ij}

In order to evaluate numerically Eq.(5.19) of Brumberg (1994) defining $\mu T_{1j}^{(i)+}$, we have to be able to evaluate numerically functions $\Phi_3(\varphi, k, s, \alpha)$ for $s = -2, 0, 2$ and $\alpha = \pm 2m_{ij}$ defined as

$$\Phi_n(\varphi, k, s, \alpha) = \exp(\overset{\circ}{i} \alpha\varphi) \left(C + \int_0^\varphi \frac{1}{\delta^n(\varphi')} \exp\left(\overset{\circ}{i}(s - \alpha)\varphi'\right) d\varphi' \right), \tag{5}$$

where $\delta(\varphi) = (1 - k^2 \sin^2 \varphi)^{1/2}$, and C is a constant of integration. Due to the symmetry (see Eq.(4.62) of Brumberg (1994)) $\Phi_n(\varphi, k, -s, -\alpha) = \overline{\Phi}_n(\varphi, k, s, \alpha)$, it is sufficient to compute $\Phi_3(\varphi, \ldots)$ only for $\alpha = 2m_{ij}$. For any even s, $\Phi_3(\varphi, \ldots)$ has a period π. Therefore, both $\Phi_3(l, \ldots)$ and $\Phi_3(w, \ldots)$ are periodic functions with a period 2π. Since the real part of $\Phi_3(\varphi, \ldots)$ is an even function and the imaginary part of $\Phi_3(\varphi, \ldots)$ is an odd function, it is sufficient to know $\Phi_3(\varphi, \ldots)$ for $\varphi \in [0, \pi/2]$.

It is easy to see that the condition $\Phi_n(0, k, s, \alpha) = \Phi_n(\pi, k, s, \alpha)$ allows one to determine C uniquely:

$$\Phi_n(0, k, s, \alpha) = C = \frac{1}{\exp[-\overset{\circ}{i} \pi\alpha] - 1} \int_0^\pi \frac{\exp \overset{\circ}{i}(s - \alpha)\varphi}{\delta^n} d\varphi. \tag{6}$$

The integral in (6) can be evaluated numerically for any given k, s and α. One can check that it is the initial value (6) which results in Φ_3 having the secular part defined by Eq. (4.59) of Brumberg (1994). Numerical integration of (5) and (6) for $\varphi \in [0, \pi/2]$ allows us to compute $\Phi_n(l, \ldots)$ for any l. $\Phi_3(w, \ldots)$ can be computed as $\Phi_3(\varphi(w), \ldots)$ where $\varphi(w)$ can be computed from (4).

For each couple of planets, we compute both $\mu T_{1\,j}^{(i)+}(l_{ij})$ and $\mu T_{1\,j}^{(i)+}(w_{ij})$ numerically in a sufficiently large number of points uniformly distributed on the interval $[0, 2\pi[$ and then apply Fast Fourier Transform to get numerical values of the coefficients A_k^{ij} and B_k^{ij} respectively. We checked that the errors of numerical integration of (5) and (6), the errors of discretization, and the round-off errors do not influence the values shown in the Tables below.

3. Results

In Table I, for the couple Venus-Earth, we give the number of harmonics in the ρ-series (N_ρ) and τ-series (N_τ) whose magnitude is higher than $\varepsilon/2$ for several ε. This couple of planets is the most difficult one since $a_2/a_3 \sim 0.723$. Δ_ρ and Δ_τ are actual maximal errors of the ρ-series and the τ-series, respectively, provided that the specified number of harmonics are retained.

Table II shows the number of harmonics of the ρ-series and the τ-series with magnitudes higher than $5 \cdot 10^{-14}$. This cut-off level is sufficient to represent the first-order intermediate motion of the major planets with errors negligible in comparison with contemporary accuracy of observations.

Table III gives the total number of harmonics whose magnitudes are higher than $\varepsilon/2$ for the whole system of 8 planets. Δ_ρ and Δ_ε are maximal absolute errors of the corresponding series representing $\mu T_{1\,j}^{(i)+}$ among all the couples.

The Tables show that the τ-series are more efficient as compared to the ρ-series when we need to represent mutual perturbations of close couples of planets (when the ratio of the major semi-axes $a_i/a_j \sim 1$) with relatively high accuracy. Since it is mutual perturbations of close couples (primarily Venus-Earth and Jupiter-Saturn) which present the most difficult task, we hope that the τ-series facilitate the problem. However, it is easy to see that for the couples with $\min(a_i/a_j, a_j/a_i) \ll 1$ or if relatively low accuracy is required, the ρ-series may be more preferable.

The research described in this publication was made possible in part by Grant No. NSC000 from the International Science Foundation and by Grant No. NSC300 from the International Science Foundation and Russian Government.

TABLE 1. Number of harmonics for the couple Venus-Earth.

ε	10^{-6}	10^{-7}	10^{-8}	10^{-9}	10^{-10}	10^{-11}	10^{-12}	10^{-13}	10^{-14}
N_ρ	11	17	24	33	41	52	61	72	86
N_τ	17	23	26	29	33	37	41	44	47
N_ρ/N_τ	0.65	0.74	0.92	1.14	1.24	1.41	1.49	1.64	1.83
Δ_ρ/ε	0.66	0.90	1.3	1.2	1.7	1.1	0.99	2.7	2.9
Δ_τ/ε	0.87	0.31	0.42	0.81	0.65	0.48	0.34	0.35	0.60

TABLE 2. Number of harmonics of the ρ-series (left values) and the τ-series (right values) with magnitudes higher than $5 \cdot 10^{-14}$.

$i \setminus j$	Mercury	Venus	Earth	Mars	Jupiter	Saturn	Uranus	Neptune
Mercury		38/29	25/22	15/15	13/14	10/11	7/ 9	7/ 8
Venus	29/20		72/44	28/23	17/17	12/13	8/10	8/ 9
Earth	18/17	74/37		48/35	21/19	14/15	10/11	8/10
Mars	11/13	29/22	56/31		28/23	17/17	12/12	10/11
Jupiter	4/ 8	8/12	12/14	13/15		50/35	22/20	17/16
Saturn	2/ 6	6/10	8/10	8/11	54/30		40/29	25/22
Uranus	2/ 6	4/ 8	5/ 8	5/ 8	23/21	44/28		60/42
Neptune	2/ 5	2/ 7	4/ 8	4/ 8	16/17	25/21	63/33	

TABLE 3. Total number of harmonics for the system of 8 major planets

ε	10^{-6}	10^{-7}	10^{-8}	10^{-9}	10^{-10}	10^{-11}	10^{-12}	10^{-13}	10^{-14}
N_ρ	153	267	388	524	667	826	994	1173	1361
N_τ	255	364	472	576	680	780	880	975	1069
N_ρ/N_τ	0.60	0.73	0.82	0.91	0.98	1.06	1.13	1.20	1.27
Δ_ρ/ε	1.0	1.2	1.4	3.6	2.3	2.1	2.5	3.0	2.9
Δ_ρ/ε	0.87	0.63	0.93	0.81	0.68	0.55	0.58	0.60	0.67

References

Brumberg, V.A. (1994) General Planetary Theory in Elliptic Functions, *Celes. Mech.*, **59**, 1

Brumberg, V.A. (1996) Theory Compression with Elliptic Functions. In: S. Ferraz-Mello, B. Morando, J.E. Arlot (eds.), Dynamics, ephemerides and astrometry of the solar system, Kluwer, Dordrecht

PULSARS AND SOLAR-SYSTEM EPHEMERIDES

J.F. CHANDLER

Harvard/Smithsonian Center for Astrophysics
60 Garden Street, Cambridge, MA 02138

Abstract. The analysis of pulsar time-of-arrival data is intimately bound up with planetary ephemerides. Highly accurate ephemerides are required for Earth and Moon and, to a lesser degree, for the other planets, in order to make full use of the timing data for millisecond-class pulsars. These data, in turn, present an opportunity for improving planetary ephemerides in a variety of ways. Fitting the Earth and Moon orbital parameters to the timing data is the obvious first step, though it is less valuable in the short term for many applications than using the current accumulation of spacecraft-tracking and lunar laser ranging data. By themselves, the pulsar timing data convey no information on the orientation of Earth's orbit, since each pulsar's position on the sky must be determined from those same data. However, independent pulsar position measurements by VLBI, in combination with the timing-derived positions, can serve to fix the orientation of Earth's orbit with respect to the radio reference frame and thereby link the planetary and radio frames. In the long run, the acquisition of timing data over increasing time spans and with improving precision should prove to be an important factor in determining the shape, as well as the orientation, of Earth's orbit. In addition, pulsar timing over a sufficiently long span can directly measure a planet mass through the reaction of the rest of the solar system. The effect must be observed for a major fraction of the orbital period of the planet in question so that the signature can be separated from that of the ordinary spin-down of each pulsar. Finally, pulsar timing can be used to probe gravitational physics, a field with far-reaching consequences and a basic part of the framework for constructing the ephemerides.

1. Introduction

In this paper, I discuss contributions of pulsar timing to the development of planetary ephemerides and *vice versa*. Pulsars, especially millisecond-class pulsars, present a unique opportunity to track the motion of Earth while nominally observing a distant astronomical object. Although the accepted mechanisms for pulsed emission have not been confirmed by direct observation, the extreme regularity of the pulses has been abundantly demonstrated over years of timing observations. Any object that emits signals regularly is, in effect, a clock and is potentially useful as a standard for comparison with other clocks. Even "young" pulsars, which are characterized by large pulse period derivatives and occasional timing "glitches" attributed to the sudden relaxation of rigid structures stressed by changing spin rates, generally show regular behavior between glitches. Unfortunately, the unpredictability of such behavior in young pulsars renders them useless for long-term comparisons.

Millisecond pulsars, however, have not only shorter periods than ordinary pulsars, but also slower spin-down rates and an apparent lack of glitches. These characteristics place them among the best clocks known to us in the Universe and give them the potential for high-precision measurements of Earth motion.

2. Modeling pulse arrival times

In order to account for pulsar timing observations, it is necessary to deal simultaneously with properties of both the pulsar and the observer. Let us first consider the train of pulses emitted by the pulsar. Although individual pulses display considerable variability in amplitude and even in shape, it is possible for the observer to form an average pulse profile by integrating over time spans as short as a few minutes and to determine the arrival time of a reference pulse within each span. Such averaging yields a timing signal characterized by remarkably few parameters. The model of pulse emission must include the pulse period and period derivatives, as well as an epoch for the arrival of one particular pulse (or, equivalently, an initial pulse phase and the first few time derivatives of the phase). These are the only required parameters intrinsic to the pulsar. In addition, of course, the pulsar's position on the sky and proper motion must be included because the observing platform is moving. For a nearby pulsar, even the distance must be included in calculating the annual variations in pulsar-Earth path length; the approximation of projecting the Sun-Earth vector onto the mean line of sight is not sufficiently accurate. The distance is also important in the sense that interstellar dispersion of the pulsar signals increases with distance. Indeed, the dispersion is one of the most striking features of pulsar signals, since it must be determined before the pulsing can be detected in the first place. If the pulsar has a known binary companion, the (time-variable) orbital elements must also be included. However, the fastest known pulsar (and the best for the purposes of solar-system dynamics) is PSR B1937+21, a solitary object.

These few parameters adequately describe the emitted pulse train. For the intrinsic pulsar parameters, there is no need to assign a physical significance, nor (in the absence of a binary companion) any real interest from the point of view of solar-system dynamics. They are simply numbers to be determined. Similarly, the interstellar dispersion, once known well enough to permit detection of the pulses, is nothing more than a time-variable calibration factor that can be readily determined from dual-frequency observations of the pulsar signals.

We turn, then, to the effects of the solar system on pulse arrival times seen by a terrestrial observer. The most obvious is the annual variation due to Earth's orbit around the Sun, but there are many others, and pulsar timing depends critically on a detailed and accurate picture of these effects. If planetary ephemerides did not already exist, they would need to be devised for and from the timing of pulsar signals. In fact, the process would be analogous to that of detecting and characterizing planets in orbit around a pulsar, except that our own planets affect the signals from every pulsar at once and are therefore much more easily characterized. In any event, planetary ephemerides *do* exist, and pulse timing interpretation can take advantage of them in describing the four-dimensional geometry of the observer. Table 1 shows a summary of the effects important for that description.

Most, but not all, of these effects are simply variable displacements, like that of an observatory from Earth's center, of Earth from the Earth-Moon barycenter, and so on. The leap second, of course, is a purely human artifact that affects only the time scale and not the pulse arrival times themselves. Similarly, the item called "local proper time" is simply the variation in rate of Earth-borne clocks in a relativistic sense, due to the changing velocity and gravitational potential. The items marked "orbit" (aside from Earth and Moon) do not directly affect the pulse arrival times, since they are the offsets of the Sun from the solar-system barycenter, but they serve as markers for the corresponding perturbations on Earth's orbit. In a numerically integrated planetary ephemeris, such as those used for the analysis of pulsar data, these perturbations are not presented as separate terms, but are simply included in the Earth orbit. The amplitude of the Shapiro delay in the table is calculated for a "typical" pulsar; it would be larger for a pulsar close to the ecliptic.

TABLE 1. Ephemeris contributions to pulse timing variation

Term	Period	log(amplitude) (sec)
Earth orbit	1 yr	3
Earth eccentricity	1 yr	1
Jupiter orbit	12 yr	0
Saturn orbit	29 yr	0
Leap seconds	1-2 yr	0
Neptune orbit	165 yr	0
Uranus orbit	84 yr	-1
Earth rotation	1 day	-2
Earth precession	long	-2
Moon orbit	27 days	-2
Earth proper time	1 yr	-3
Venus orbit	0.6 yr	-3
Pluto orbit	248 yr	-4
Mars orbit	2 yr	-4
Mercury orbit	0.2 yr	-5
Shapiro delay	1 yr	-5
Earth nutation	19 yr	-6
Local proper time	1 day	-6
Ceres orbit	5 yr	-6

3. Ephemeris frames

One result of the use of pre-computed planetary ephemerides for the analysis of pulse timing data, as opposed to a combined analysis of pulsar and planetary data, is the dependence of the final results upon the choice of ephemeris. Such a dependence would be a serious drawback if the differences induced by switching from one ephemeris to another produced significant changes in interpretation. Fortunately, the ephemerides used for pulse timing are nearly interchangeable, aside from offsets in the orientation and mean motion of Earth's orbit. Since the annual signature of Earth motion dominates the variations of pulse arrival times (see Table 1), each ephemeris defines, in effect, its own reference frame via the specification of coordinates of the Earth-Moon barycenter as a function of time. That reference frame then provides the context for the pulsar positions and proper motions obtained using that ephemeris.

We can confirm the near-equivalence of separate ephemerides by direct comparison of the coordinates. Obviously, these coordinates differ systematically, and the differences can largely be described by a simple, linearly varying orientation offset between the corresponding reference frames. In practice, the residual differences, though still systematic, can be neglected or treated merely as ephemeris "noise" because they are so small.

Among the ephemerides used for pulse timing are a set from MIT and more recently from the Center for Astrophysics: PEP311 (1969), PEP740 (1984), and PEP740R (also 1984). The latter was produced, solely for the convenience of pulsar observers, to agree with the orientation of PEP311 at epoch 1982.9, by rotating all the coordinates by a fixed amount. Another series of ephemerides has been distributed from JPL, including DE96, DE118, and DE200 (see Standish 1982). The latter is the only one of these that is nominally aligned to the equator and equinox of J2000; the others are aligned to B1950. That discrepancy is merely one more offset of orientation and therefore makes no difference to the effective interchangeability of ephemerides.

The ephemerides were compared directly, one to another, using a linearized least-squares fit of tabulations of the cartesian coordinates at four-day intervals over a span of twelve years. The process involved solving for an overall scale factor, an offset of orientation, and a steady rotation. The solutions for these parameters are not intrinsically interesting, except insofar as they indicate the transformation from one ephemeris frame to another. The key point is the postfit RMS residual deviation between the two ephemerides after applying the transformation. In all cases, *i.e.*, for the ephemerides mentioned above, considered pairwise, this was about 10^{-10} AU. A further check on this technique was a series of one-year fits covering the twelve-year span for each pair of ephemerides solving only for the offset of orientation. The resulting set of angular offsets was consistent with the offset and rotation rate determined by the overall fit, to within the indicated "noise" level. For further discussion of this fitting procedure and the resulting transformations, see Bartel *et al.* (1996).

4. Applications of pulse timing

In the near future, pulse timing analysis is not competitive with the more conventional techniques for refining solar-system ephemerides. Even the best available millisecond pulsar, PSR B1937+21, displays timing noise on the order of 0.3 μs = 100 m (Kaspi *et al.* 1994). By comparison, the set of spacecraft-tracking data from interplanetary missions includes range measurements between Earth and Mars with an accuracy of about 10 m (the Viking landers). Further, the lunar laser ranging data have reached an accuracy of a few cm between laser stations on Earth and the retroreflectors on the Moon. These lunar data, being all Earth-based, are much less sensitive to the orientation of the lunar orbit than to its shape, but they have the advantage over pulse timing data of sensing the Moon directly, rather than through the reaction of Earth. Pulse timing can therefore make no significant contributions in the foreseeable future to the Moon ephemeris. Potentially, the long-term accumulation of pulse timing data for more and better pulsars could shift the balance away from the spacecraft-based data now available, but the latter remain for now the primary determinant for the Earth ephemeris.

The one contribution pulsars are uniquely equipped to make is in the overall orientation of the planetary ephemerides. This orientation is only weakly determined by the conventional high-precision techniques, which rely on topocentric distance measurements for objects all in the same system. Though the spinning Earth provides an inertial reference, these techniques tie that reference to the planetary system only by triangulation over terrestrial baselines. Pulse timing measurements, on the other hand, can make use of baselines across Earth's orbit, and the angular accuracy for a 100 m distance resolution is therefore better than 1 mas. However, the position of each pulsar on the sky is customarily determined from the pulse timing, and it is necessary to establish the pulsar position independently before the timing data can be used to set the orientation of the ephemerides. The technique of Very Long Baseline Interferometry (VLBI) is well suited to the task of measuring pulsar positions with respect to an independent reference frame, namely, that of the extragalactic radio sources. PSR B1937+21 has been determined with respect to that frame with an accuracy of about 3 mas (Bartel *et al.* 1990 and 1996). The combination of VLBI and pulse timing, then, can relate the planetary ephemeris frame to the radio reference frame to about that level.

Another possible application of pulse timing is the determination of planet masses. Such a determination is sensitive to correlations between the pulsar model and planet orbital elements, and requires a long time base to break degeneracies, but it shows promise for some of the planets. Table 2 gives the results of a case study assuming 20 years of timing data twice per month with 0.3 μs uncertainty, along with the current state of knowledge for planet orbital elements, and ignoring uncertainties in pulsar spin-down, *etc.* For comparison, the table also shows the current uncertainties in the planet masses, from a combination of spacecraft and planetary data. For the terrestrial planets, even 20 years of pulse timing is insufficient to match the accuracy of existing techniques, and 20 years is also too short a time span for the planets outward of Saturn, but Jupiter and Saturn both fall into the intermediate range. Certainly, the results of the Galileo mission are expected to improve the uncertainty in Jupiter's mass, but Saturn remains a possible target for improvement.

TABLE 2. Planet mass uncertainties in solar mass units

Planet	Current uncertainty (log)	Possible uncertainty from pulse timing over 20 years (log)
Mercury	-11.0	-10.0
Venus	-11.7	-10.0
Mars	-11.5	-10.5
Jupiter	-9.3	-10.7
Saturn	-8.2	-11.1

Finally, pulse timing can be used to study gravitational physics through a variety of effects. As shown in Table 1, the Shapiro delay due to the Sun's gravitational potential plays a role in pulse timing residuals. In order to make a useful contribution, this effect must be magnified by the discovery of a suitable pulsar near the ecliptic, so that observations of the delay near conjunction could be made, and, even there, this technique would still have to compete with the corresponding spacecraft experiments (primarily the Viking mission to Mars). The Shapiro delay due to propagation of pulsar signals in a binary system or the effect of gravitational radiation in such a system (Taylor and Weisberg 1982), or even the possible timing variations due to background gravitational radiation (see, for example, Davis et al. 1985), may also be useful, but these are not directly relevant to solar-system dynamics. An interesting possibility is the use of pulse timing, in combination with VLBI measurements, to study the gravitational redshift of Earth-borne clocks (Shapiro 1986, Chandler 1990). This effect has already been confirmed to a fractional accuracy of 10^{-4} within the potential well of Earth (Vessot et al. 1980), but nothing approaching that accuracy has been achieved in the solar potential. One or two suitably placed millisecond pulsars whose positions can be measured by VLBI to the level of 1 mas would yield a fractional redshift test of 10^{-3}.

Thus, we see that solar-system dynamics and pulse timing serve each other in several ways. Highly accurate planetary ephemerides are essential to the proper modeling of pulse arrival times. Pulse timing, in turn, offers an opportunity to determine the orientation of the planetary ephemeris frame, refine some of the parameters in the solar-system model, and even to test the underlying gravitational physics.

5. Acknowledgments

This work was supported in part by NASA under grant NAGW-3666. Travel funds for presenting this paper at IAU Symposium 172 were supplied through the Organizing Committee.

References

Bartel, N., Cappallo, R., Whitney, A., Chandler, J., Ratner, M., Shapiro, I., and Tang, G. 1990, in *Workshop on Impact of Pulsar Timing on Relativity and Cosmology* (Berkeley).
Bartel, N., Ebbers, A., Pan, R., Cappallo, R.J., Chandler, J.F., Ratner, M.I., and Shapiro, I.I. 1996, in preparation.
Chandler, J.F. 1990, in *Workshop on Impact of Pulsar Timing on Relativity and Cosmology* (Berkeley).
Davis, M.M., Taylor, J.H., Weisberg, J.M., and Backer, D.C. 1985, *Nature*, **315**, 547.
Kaspi, V.M., Taylor, J.H., and Ryba, M. 1994, *Ap. J.*, **428**, 713.
Shapiro, I. 1986, in *Proceedings of the Workshop on the Arecibo Upgrade* (Cornell).
Standish, E.M. 1982, *A & A*, **114**, 297.
Taylor, J.H. and Weisberg, J.M. 1982, *Ap. J.*, **235**, 908.
Vessot, R.F.C., *et al.* 1980, *Phys. Rev. Lett.*, **45**, 2081.

THE NEED FOR MORE ACCURATE 4000-YEAR EPHEMERIDES, BASED ON LUNAR AND SPACECRAFT RANGING, ANCIENT ECLIPSE AND PLANETARY DATA

KEVIN D. PANG
(Fax: USA code+818 952 1371)
AND
KEVIN K. YAU
Jet Propulsion Lab., 230-101, Pasadena, CA 91109, USA

1. Introduction

Long planetary and lunar ephemerides like the JPL DE102 and LE51 (Newhall *et al.*, 1983) and the Bureau des Longitudes VSOP (Bretagnon, 1982) and ELP (Chapront-Touze and Chapront, 1983) have enabled more positive ancient eclipse, planetary and cometary identifications, which have in turn refined ephemerides, *e.g.*, the reconstruction of the orbit of comets Halley and Swift-Tuttle (Yeomans and Kiang, 1981; and Yau *et al.*, 1994). The data used to initialize DE102 are pre-1977. Much more observational data have been collected since. The lunar ephemeris has also been improved. The secular lunar acceleration, $\dot{n}_{moon}$, from laser ranging, is $-25.9 \pm 0.5''/\text{cen}^2$ (Williams *et al.*, 1992). We can now uniquely solve for ΔT, the clock error, from ancient eclipse records. The lack of ΔT values before 700 B.C. has left the early timescale of the ephemerides unconstrained (Morrison, 1992). Our solution of this problem is outlined here.

2. Earth's Rotation Rate Deduced From Ancient Eclipse Records

Since $\Delta T = ct^2$ (t = centuries before 1800), the oldest data have the most weight. Sunrise and sunset eclipses are most valuable, as they can be retrospectively timed. The *Bamboo Annals*, entombed in 299 B.C. and unearthed in A.D. 281, states that "in the first year of King Yi of the W. Zhou dynasty (1100-771 B.C.) the day dawned twice at Zheng (34.5°N,

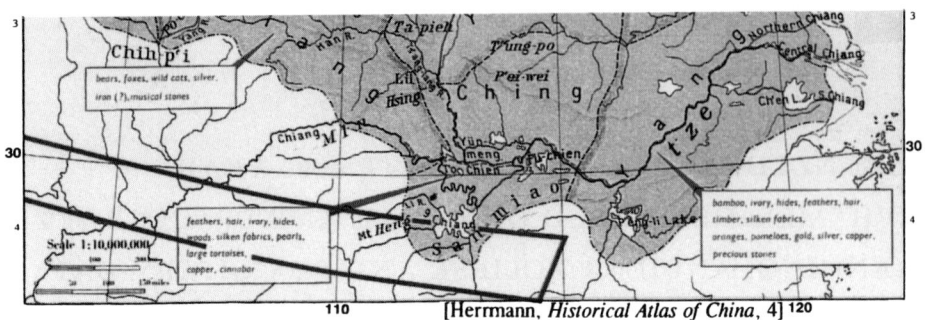

Figure 1. The path of annularity of the September 24, 1912 B.C. solar eclipse. The "double sunset" eclipse was seen in the land of the *san* Miao during the reign of King Yu, founder of the first dynasty Xia. Map from Herrmann (1966).

109.8°E)." *Kaiyuan zhanjing* (Siddhartha, A.D. 724) cites this record and adds that "in the 2nd (*actually 12th*) year of Sheng Ping reign period of King Shang (*actually King Xi*) the day began twice at Zheng." Matching these records with the April 21, 899 B.C. and April 4, A.D. 368 sunrise eclipses gives ΔT values of 5.8 ± 0.15 and 1.7 ± 0.1 hr, respectively (Pang et al., 1988; Pang and Yau, 1992; Pang et al., 1996).

The brightness changes for the magnitudes 0.95–0.97 and 0.991–0.998 annular eclipses were greater than for the January 4, 1992 "double sunset" annular eclipse over S. California (magnitude 0.91–0.92). Levy (1992), *e.g.*, noted that "...as annularity ended. Sunset had come and gone, but the sky began to brighten not darken. For almost 15 minutes it continued to brighten until the onrushing shadow of Earth took over and darkness fell again." A "double sunset" eclipse will occur at Zheng on August 1, 2008.

Analysis of our data gives $\Delta T = (30 \pm 2.5)t^2$. Results from analyzing 14–12th century B.C. Shang dynasty oracle bone eclipse records are included, but not discussed, here (Pang et al., 1989 and 1996). Recent astronomical dating of these unique records has converged to a set of common matching dates with only minor differences (Xu et al., 1995). From these results, we get an $\dot{\omega}/\omega$ of $-(19 \pm 1.6) \times 10^{-11}$/yr. Subtracting a tidal $\dot{\omega}/\omega$ of -27.8×10^{-11}/yr (Lambeck, 1980) gives a nontidal $\dot{\omega}/\omega$ of $(9 \pm 1.6) \times 10^{-11}$/yr ($\equiv \dot{J}_2$ of $-(4.5 \pm 0.8) \times 10^{-11}$/yr). The historical $\dot{J}_2$ and the present $\dot{J}_2$ from satellite laser ranging, -3×10^{-11}/yr (Cheng et al., 1989), are consistent with postglacial rebound from an upper mantle of viscosity 10^{21} Pa s, and lower mantle of $(2-4) \times 10^{21}$ Pa s, deformed by Pleistocene ice sheet loading (Peltier, 1985). The bounceback to its less oblate interglacial shape makes the Earth spin faster, overcoming a third of the tidal braking. The net effect has been lengthening the day by 1.64 ± 0.14 msec/cen. We now test our model with still earlier records.

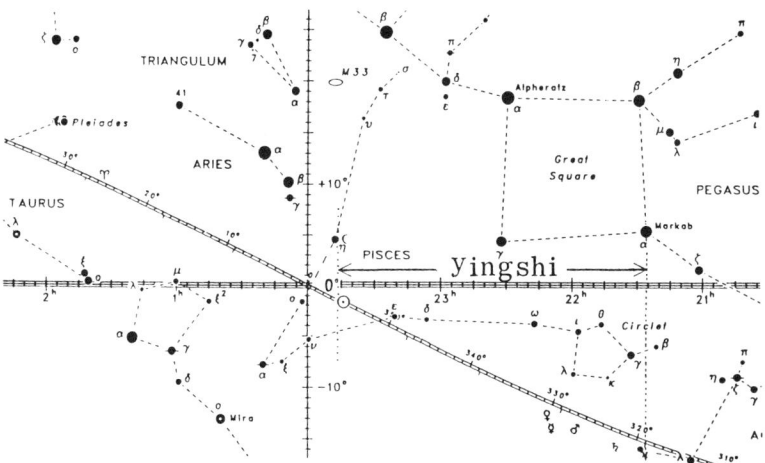

Figure 2. The March 5, 1953 B.C. conjunction of the Sun, Moon and five planets in Yingshi (the equatorial space 35 arc degrees east of α Pegasus, within the two vertical dotted lines). The Moon is just west of η Pisces. The Sun is on and the planets just below, the ecliptic. The reference frame is 1 B.C., close to Liu Xiang's lifetime (77 to 9 B.C.). From *Sky and Telescope* chart SC004

3. Results Obtained From Still Earlier Observations

Philosopher Mozi (*ca.* 468-382 B.C.) wrote: "In ancient times, the *san* (three) Miao tribes were in disarray. Heaven ordered their destruction. The Sun rose in the evening ... King Yu, founder of Xia, first dynasty, vanquished them..." The *Bamboo Annals* adds: "When the *san* Miao perished ... the Sun disappeared by day and reappeared at night ..." Astronomically verified *Bamboo Annals* Xia chronology puts Yu's official reign in 1914-1907 B.C. (Nivison and Pang, 1990). We have found that a "double sunset" eclipse did occur over south-central China on September 24, 1912 B.C. (year 3 of Yu's reign). The *san* Miao domain was south of the Yangzi River, east of Mount Heng and west of Lake Pengli (Herrmann, 1966). A ΔT of 12.2 hr or c of 32 sec/cen^2 would put the sunset eclipse (magnitude 0.97-0.99) right there (28°N, 115°E) (Fig. 1), consistent with results from analyzing third millennium B.C. Mesopotamian lunar eclipse records (Huber, 1987; Pang and Yau, 1995). We conclude that our model holds back to at least 2300 B.C. Analysis of an even earlier planetary record follows.

Sirius' heliacal rise (with Nile floods) in the summer "dog days" regulated the ancient Egyptian Sothic calendar. Yingshi's (Pegasus') heliacal rise initiated spring in the ancient Zhuanxu calendar, *e.g.*, The *Book of Rites* (1000 B.C.) states that "in the first lunar month, the Sun is in Yingshi." The lunisolar calendar was invented in Mesopotamia in mid-third millennium B.C. and has been used by the Chinese since about 2000 B.C.

Liu Xiang's (77–9 B.C.) *Hong Fan Zhuan, e.g.*, states that "the original Zhuanxu calendar began on cyclic day 6, month 51, year 51 (modulo 60) at the start of spring when the Sun, Moon and five planets met in Yingshi, 5°." Liu's interpolated date is wrong, and its errors can be reproduced mathematically. Some even consider it an imaginary epoch when the day, lunation, year and planetary cycles last came together. Using VSOP85 and ELP2000-85, with the new $\dot{n}_{moon}$ and $\Delta T(t)$ values, we have uniquely matched Liu's record with circumstances of the sky computed for early 1953 B.C. (Fig. 2). On February 26, the five planets were visible before dawn like "a pearl necklace," spanning $< 5°$. On March 5, the Sun, new Moon and five planets were all in Yingshi (= lunar mansions 13+14, with an old width of 35° in RA, east of α Peg). From time immemorial, the Chinese have been using the Xia dynasty calendar, which starts the year with the second new moon after winter solstice. In 1953 B.C., winter solstice fell on January 5, so spring began on March 5. All of Liu's conditions are thus satisfied.

4. Conclusions

Whereas the initial conditions of the current planetary ephemerides are better known, the Moon's secular acceleration rate improved by laser ranging, and the history of the Earth's rotation determined for 4300 years, it is timely to produce more accurate 4000-year ephemerides, valuable for analyzing ancient astronomical records. We hope to stimulate such production, and the discussion of old astronomical records. We thank J. Bangert.

References

Bretagnon, P. (1982) *Astron. Astrophys.* **114**, 278
Chapront-Touze, M. and Chapront, J. (1983) *Astron. Astrophys.* **124**, 50
Cheng, M.K. et al. (1989) *Geophys. Res. Lett.* **16**, 393
Herrmann, A. (1966) *Historical Atlas of China*, Edinburgh-Aldine Publishing Co., p.4
Huber, P. (1987) *Acta Historica Scientiarum Naturalium et Medicinalium* **39**, pp.3
Lambeck, K. (1980) *The Earth's Variable Rotation*, Cambridge Univ. Press, pp.337
Levy, D.H. (1992) *Sky and Telescope* **83**, 695
Morrison, L.V. (1992) *Observatory* **112**, 289
Newhall, X.X., Standish, E.M. and Williams, J.G. (1983) *Astron. Astrophys.* **125**, 150
Nivison, D.S. and Pang, K.D. (1990) *Early China* **15**, 87
Pang, K.D. et al. (1988) *Vistas Astron.* **31**, 833
Pang, K.D. et al. (1989) *Bull. Amer. Astron. Soc.* **21**, 753
Pang, K.D. and Yau, K.K.C. (1992) *Eos* **73**, No. 43, 62
Pang, K.D. and Yau, K.K.C. (1995) *Eos* **76**, No. 46, F62
Pang, K.D., Yau, K. and Chou, H.H. (1996) *Pure Appl. Geophys.* **145**, No. 3
Peltier, W.R. (1985) *J. Geophys. Res.* **90**, 9411
Williams, J.G., Newhall. X.X. and Dickey, J.O. (1992) *Eos* **73**, No. 43, 126
Yau, K., Yeomans, D. and Weissman, P. (1994) *Mon. Not. R. astr. Soc.* **266**, 305
Yeomans, D.K. and Kiang, T. (1981) *Mon. Not. R. astr. Soc.* **197**, 633
Xu, Z. T., Stephenson, F.R. and Jiang, Y.T. (1995) *Quart. J. Roy. astr. Soc.* **36**, 397.

REVIEWING THE THEORIES OF MOTION OF THE SATELLITES OF SATURN

L. DURIEZ
Laboratoire d'Astronomie de l'Université de Lille
1, Impasse de l'Observatoire, F 59000 Lille, France

Abstract. The state of knowledge of the motions of all Saturn's satellites is presented (excluding however the rings and their relating shepherding satellites). In particular, it appears that the theory of motion of the major satellites is now more precise than the available Earth-based observations, allowing to expect new progress with the next observations from mutual events and then with those from the Cassini mission.

1. Introduction

The saturnian satellite system is certainly the richest and the most complex in the solar system, specially because of the presence of many resonances: There are 3 resonances between 6 of the major satellites (resonance 1:2 between Mimas and Tethys and between Enceladus and Dione, and resonance 3:4 between Titan and Hyperion), 1 resonance 1:1 between the 2 small coorbitals Janus and Epimetheus, and 3 resonances 1:1 between Tethys and its 2 Lagrangian companions Telesto and Calypso and between Dione and its companion Helena. Then, these Lagrangian satellites are also in resonance 1:2 with Mimas and Enceladus respectively. This resonant feature is unique in the solar system. This present status may have been driven by tidal dissipation (by modifying both mean motions and precession rates of nodes and pericentres) and then locked on the observed resonances. According to Dermott *et al.* (1988), in comparison with the uranian satellites system where no resonance exists now, the resonances in the saturnian system seem to be more regular and more stable because of the larger oblateness of Saturn.

But, before describing the real story of the system, it is important to know accurately the present status of the motions, specially because of spatial observations of the Saturn's satellites from the past missions (Pioneer and Voyager) or from the planed Cassini mission, it will be necessary to have theories of motions as accurate as a few kilometers.

Hence, because of the lack of place in the proceedings, the present review tries to emphasize only the most recent developments of theories and of observations in the restricted field of the major saturnian satellites. However, an extended full TeX version of this review is available on ask from the author by e-mail at : `duriez@gat.univ-lille1.fr` ; it includes a review of both major and small satellites, some historical aspects and an extended bibliography.

2. The recent works about major satellites

All recent works about modelling the motions of the major satellites attempt to reduce the residuals at all costs, attacking the problem of improving both observations and theories (by numerical and analytical methods as well).

2.1. OBSERVATIONS

Strugnel & Taylor (1990) have compiled a catalogue of 51000 ground-based observations made between 1874 and 1989, all being now given in the B1950 reference frame, all times being reduced to UTC only. The quality of each observation is indicated by $(o - c)$ computed from the theories quoted in Taylor & Shen (1988) but using the mean motions and the libration parameters obtained by Dourneau (1987). The catalogue contains most of the published observations made since 1966 (mainly photographic ones) while other data concern a selection of older, mostly visual, observations. To examine the possibility of combining old and recent observations, Harper *et al.* (1989) have analyzed the micrometric observations of Titan, Hyperion and Iapetus made from 1874 to 1923, fitting them to a numerical integration of their motions. The method is the same as that used by Sinclair & Taylor (1985) with the 1967-1985 observations of these satellites. It results in particular that visual observations of angular separation from the Saturn's centre are rather prone to error (systematically large by about $0''.5$). However the order of magnitude of residuals is found to be the same for visual intersatellite observations and for astrometric observations analyzed by Sinclair & Taylor, allowing to plan a next fitting of all the data set. It is hoped that future observations be published in the standardized form given by Strugnel & Taylor.

Other works show that observations are now revitalized by new techniques: improvement of astrographic measurements, emergence of promising CCD observations and opportunity of photometric observations of mutual events during the 1995-1996 opposition.

According to Veillet et al. (1988), by using refined astrographic techniques, the precision of the best ground-based observations may be estimated at about 0″.05, that is about 350 km on positions in the saturnian system. Following the same approach as Sinclair (1974, 1977) to derive corrections to the ephemeris of Saturn as well as corrections of the orbits of the satellites, Pascu & Schmidt (1990) have reduced again their photographic observations of the Saturn's satellites since 1974. They have shown that these observations have an expected precision better than 0″.2 which would suffice to give, as a by-product, better positions of Saturn than by transit circle observations which are not more accurate than 0″.5. The same improvement of the Saturn ephemeris is also deduced by Taylor et al. (1991) from Carlsberg and Bordeaux meridian circle observations of Titan and Iapetus made in 1987 and 1989.

Besides, many observers are now ready to use CCD cameras to obtain images of the Saturn's satellites. In spite of the smallness of the field, astrometric measurements on these images could be now more accurate than large field photographic astrometry: Le Floch (1995) describes the technique used at Bordeaux observatory to obtain astrometric positions of major satellites on CCD fields, using the globular cluster M15 to calibrate the images (a catalogue of hundreds of stars in this cluster have been elaborated by Le Campion et al. (1992) specially for small field astrometry); the precision of the reduction of CCD images of M15 is 0″.025 and that of satellite's images may reach the same value. However, Beurle et al. (1993), Beurle (1994) and Harper (1995) describe an analogous work made at La Palma Observatory, but they predict an accuracy of only 0″.1 on inter-satellite positions. In fact, Pascu (1995) thinks that CCD astrometric observations must be made for faint satellites only, because small fields on CCD chips do not allow to determine the orientation with the same precision as on photographic plates; moreover, if used in astrometric measures, Pascu thinks that the CCD chip must be calibrated in the same manner as the plate-measuring machines (size, position and alignment of each pixel on the chip). Nevertheless, the results obtained by Rapaport & Le Campion (1995) in fitting theories to CCD inter-satellite positions made at Bordeaux and calibrated with M15, show r.m.s. residuals between 0″.06 and 0″.08, that is almost as good results as with the best photographic observations.

At last, observations of mutual events between the Saturn's satellites are planed in 1995-1996. Arlot & Thuillot (1993), Aksnes & Dourneau (1994) and Emel'yanov et al. (1994) have published predictions of possible observations for both eclipses and occultations between the major satellites: Colas (1995) describes the method to apply to observe these events by CCD cameras. Arlot & Thuillot (1995), Thuillot (1995) and Thuillot & Arlot (1995) give details on the organization of the next campaign PHESAT95. Stavinschi & Bocsa (1995) and Zamarashkin et al. (1995) explain how they will organize these observations at Bucarest and Pulkovo observatories respectively. It is hoped that many observers will participate to the campaign PHESAT95, because if the timing of events is obtained at the precision level of some tenths of second, the expected precision on inter-satellite angular distance may be as good as $0''.01$ (that is 70 km on positions!).

Spatial observations exist already which may be used also more extensively: From the tracking data collected during the Pioneer and Voyager missions, Campbell & Anderson (1989) have deduced new values for the constants of the gravity field of Saturn. In particular, their J_2 value ($0.016\,298 \pm 0.000\,010$, with an equatorial radius of 60300 km) is now the most accurate determination.

2.2. THEORIES

Recent progress concern separately the theory of Hyperion and the theory of all other satellites:

For Hyperion, Message (1989) has proposed a new analytical theory of motion, using formal manipulations on computer to expand the disturbing function of the Titan-Hyperion problem in the neighbourhood of the libration. Second-order terms are computed for the long-periodic part of the motion, the influence of the short-period terms of first order being computed by semi-numerical quadratures. Message (1993) shows two fits of this theory to observations collected in the two periods 1875-1922 and 1968-1984, giving two dynamically consistent sets of parameters and coefficients, one for each period; however, these two solutions are slightly different and not exactly compatible each with other, since in particular they correspond to two different values for the Titan's mass; that makes to much regret the total lack of observation of Hyperion during more than forty years, from 1923 to 1967.

Gerasimov & Mushailov (1991) have also constructed an analytical resonant solution of the external case of the circular restricted 3-body problem and accounting for the Saturn's oblateness; however, being independent of the Titan's eccentricity, this solution does not include the important oscillations whose period is 18.8 years.

Other works on Hyperion try to represent numerical integrations by various functions of t, giving semi-analytical solutions:

Duriez (1992) gives a first synthetic theory of motion for Titan-Hyperion, obtained iteratively from the representation of Taylor et al. (1987) by substituting numerically a solution in the Lagrange equations and then by using a multivariable Fourier transformation to reconstruct trigonometrical developments with arguments as combinations of 3 fundamental angles (libration, pericenter and synodic angle). This process allows to construct extensively the long and short period terms in the planar problem, up to the kilometer level. Many new terms exceeding 500 km are found which were not present in Taylor et al. (1987). Some long-period perturbations of Titan by Hyperion are also found which would be non negligible at the 10 km level (assuming the Hyperion's mass equal to half that of Mimas).

Taylor (1992) gives another synthetic theory of motion for Hyperion, obtained from a numerical integration of Titan-Hyperion over ±25 years around 1973 and fitted to recent observations: Then, Taylor represent the resulting time-series by fitting to them, by least-square method, given series of trigonometrical terms with adjusted amplitudes and with arguments like in Taylor et al. (1987) or Duriez (1992). That allows to construct all long-period terms greater than $0\rlap{.}''07$ (as seen from earth) and all short-period terms greater than $0\rlap{.}''02$. Compared to 1967-1983 observations, such a representation gives r.m.s. residuals of $0\rlap{.}''28$ for the Titan-Hyperion intersatellite positions, that is comparable to residuals in Taylor et al. (1987).

In fact, the difficulty to lower the residuals comes from the very bad convergence of the short-period terms: Duriez (1994) gives a new synthetic theory for Hyperion including all long-period terms greater than 1 km and all short period term greater than 5 km; more than 500 terms are thus computed but the maximum difference over one century between the numerical integration and its representation remains as large as about 300 km due essentially to the short period part of the solution. Unlike Duriez (1992), this new representation is constructed from a numerical integration of the 3D Titan-Hyperion problem, over 1500 years, with numerical filtering to obtain only long-term motions; then, a discriminating frequency analysis allows to represent the time-series as sums of long period terms (periods between 45d and 700y; after subtracting this representation from the unfiltered time-series the same method of frequency analysis allows to reconstruct short period terms (periods found between 3d and 45d). Recently, partial derivatives of this representation with respect to masses and initial conditions have been computed, and a preliminary adjustment to recent observations has given r.m.s. residuals of $0\rlap{.}''22$ (not yet published).

At last, Kirsanov (1995) presents new elements of Hyperion as polynomials of t, obtained from a numerical integration of Titan-Hyperion, fitted to 1967–1981 observations; ten years from 1995 are described by series of polynomials, each one being valid over 300 days.

Concerning the other satellites, there are both a new theory of motions and new fits to ground-based observations:

Duriez & Vienne (1991) and Vienne & Duriez (1992) present a new

TABLE 1. – r.m.s. residuals of some major sets of observations (in arcseconds) obtained in the 3 recent adjustments over one century by Dourneau (1993), by Harper & Taylor (1993) [H&T in the table] and by Vienne & Duriez (1995) [TASS in the table].

observations	number	Dourneau	H&T	TASS
Struve (1898)	1780	0.15	0.14	0.12
USNO (1954)	2520	0.23	0.22	0.20
Sinclair (1974,1977)	860	0.18	0.12	0.13
Pascu (1982)	2330		0.13	0.13
Aksnes & al (1984)[*]	14	0.025	0.018	0.015
Dourneau et al (1989)	950	0.27	0.28	0.27
Veillet & Dourneau (1992)	300	0.19	0.18	0.11
Veillet & Dourneau (1992)	1400	0.11	0.13	0.12
all sets	~ 30000	0.20	0.17	0.15

[*] residuals taken from Viateau & Rapaport (1995)

global and analytical theory of the motions of all major satellites (except Hyperion, seen above). The aim of this work is to give a coherent representation for the motions of all satellites, considered all together, in the J2000 reference system, with a precision compatible with spatial observations. The method is the same as in a general planetary theory: The Lagrange's equations are constructed analytically up to degree six in eccentricities and inclinations and to order 2 in the masses and in the oblateness coefficients of the Saturn's figure; all terms greater than 1 km have been retained in the short period perturbations as well as in the critical system which collects all secular, resonant and solar perturbations. Then, a numerical integration of this system is made over 1200 y for the four inner resonant satellites (step 4 d), and over 9000 y for Rhea, Titan and Iapetus (step 100 d); after that, frequency analysis is applied to the resulting time-series, allowing to represent the corresponding mean elements as sums of periodic terms whose arguments are then identified as integer combinations of some fundamental arguments. That leads finally to a semi-numerical representation of all orbital elements. The internal precision of this representation is only a few kilometers for all satellites except Iapetus (for which it is only 100 km over one century). At last, the partial derivatives of this representation with respect to all physical parameters and to initial conditions have also been computed; the resulting theory is called TASS. The main characteristic of this solution comes from the ability to produce a coherent determination of all dynamical parameters for the complete Saturn's system, because each coefficient, each frequency, each phase in the resulting series is expressed analytically with respect to the masses, to the oblateness coefficients and to the initial conditions of motions.

TABLE 2. − r.m.s. residuals by satellites (in arcseconds), obtained by Dourneau (1993) and by Vienne & Duriez (1995).

	1	2	3	4	5	6	8
Dourneau	0.23	0.20	0.20	0.20	0.19	0.20	0.25
TASS1.6	0.18	0.14	0.13	0.13	0.12	0.13	0.19

Vienne & Duriez (1995) present a fit of TASS to observations made between 1874 and 1989, and give formulas to compute ephemerides in the J2000 reference system. A few time before, Harper & Taylor (1993) have produced a new fit of their theory to almost the same observations (from the Strugnell & Taylor catalogue, completed by observations made by Pascu and by Veillet & Dourneau). Hence, with Dourneau (1993), there are now 3 representations of the major satellites which have been adjusted to more than one century of observations.

Table 1 gives the main results from these three fits, showing that the r.m.s. residuals are now between 0″.20 and 0″.15 (for the set of all satellites except Hyperion). The 30000 observations concern TASS but there are 8000 observations more in Harper & Taylor (analyzed in Harper & Taylor 1994) and 8000 observations less in Dourneau. We see also that the old observations by Struve seem as good as those from Veillet & Dourneau. The 14 observations from Aksnes concern mutual events made during the previous disappearance of the Saturn's rings in 1980 : The very small residuals, (computed by Rapaport 1995), reveal the great interest of such observations, and show that the theories of the satellites concerned by these observations (Enceladus to Rhea from Nov. 1979 to Apr. 1980) seem now as good as these residuals.

Table 2 gives the residuals obtained for each satellite : They are still greater for Mimas and Iapetus, requiring probably new developments for their theory. Moreover, Rapaport & Le Campion (1995) have compared the three solutions given by Dourneau, Harper & Taylor et Vienne & Duriez in the 1960–2000 period, showing also that the differences between theories are the greatest for Mimas and Iapetus.

At last, Table 3 gives the values of some dynamical parameters obtained by adjustment of each theory to ground-based observations; there is a good agreement except for the mass of Enceladus from TASS, but this mass gives a mean density closer to 1, like that of Mimas or Tethys. Finally, TASS appears to give now the smallest residuals and probable errors, but space observations would still improve the theories.

TABLE 3. – Adjustments of some physical parameters, where C&A = Campbell & Anderson (1989) (from tracking data of Pioneer and Voyager missions), with in parenthesis the probable error expressed in the units of the last digits.

parameters × 10^6	Dourneau	H&T	TASS	C&A
m_1	0.0648 (21)	0.0646 (11)	0.0634 (8)	
m_2	0.206 (55)	0.213 (46)	0.069 (21)	
m_3	1.088 (31)	1.076 (18)	1.060 (13)	
m_4	1.954 (58)	1.916 (36)	1.963 (21)	
m_5			4.320 (380)	4.059 (53)
m_6				236.368 (8)
m_8	3.7 (1.7)		3.10 (50)	2.79 (8)
J_2	16326 (54)	16298 (38)	16285 (5)	16298 (5)

3. Conclusion

A great deal of work has been spent to improve both theories of motion and ground-based observations of all satellites. For the major satellites, we can hope that the new theories will be improved soon by taking account of new observations of mutual events in 1995-1996. For the other small satellites, some progress has still to be made to obtain the same precision in both theories and observations. However, we must soon take up the challenge of the CASSINI mission, which will require theories able to represent the positions of all satellites with a precision perhaps better than a few kilometers. In effect, the spacecraft will be put in orbit around Saturn, and during the few years of the mission, several tens of close encounters with Titan will be necessary to produce, by gravitational assistance, all the various orbits needed to explore the all satellite's system; to plan these encounters, the motion of Titan will have to be known with a precision of a few kilometers. All that will be very important to know accurately the present state of all motions, that will then allow to extent them with confidence to past and future long-term evolution.

References

Aksnes K., Franklin F., Millis R., Birch P., Blanco C., Catalano S., Piironen J.: 1984, 'Mutual phenomena of galilean and saturnian satellites in 1973 and 1979/1980' *Astron. J.* **89**, 280–286.
Aksnes K., Dourneau G.: 1994, 'Mutual events of Saturn's satellites in 1995-1996' *Icarus* **112**, 545–548.
Arlot J.E., Thuillot W.: 1993, 'Eclipses and mutual events of the first eight saturnian satellites during the 1993-1996 period', *Icarus* **105**, 427–440.
Arlot J.E., Thuillot W.: 1995, 'The former campaigns of observations of mutual events', to appear in the proceedings of the 'Atelier de travail PHESAT95', Bucarest, Sept. 1994.

Beurle K., Harper D., Jones D.H.P., Murray C.D., Taylor D.B., Williams I.P.: 1993, 'Preliminary analysis of CCD observations of Saturn's satellites', *Astron. Astrophys.* **269**, 564–567.

Beurle K.: 1994, 'CCD observations of Saturn's satellites', in Royal Greenwich Observatory Workshop: Galactic and solar system optical astrometry, p312–317.

Campbell, J.K., Anderson, J.D.: 1989, 'Gravity field of the saturnian system from Pioneer and Voyager tracking data', *Astron. J.* **97**, 1485–1600.

Colas F.: 1991, 'Nouvelles observations CCD astrométriques pour l'étude dynamique des satellites des planètes: Application au mouvement de satellite Thebe de Jupiter' *Thesis, Paris Observatory.*

Colas F.: 1995, 'Observation CCD des phénomènes des satellites de Saturne', to appear in the proceedings of the 'Atelier de travail PHESAT95', Bucarest, Sept. 1994.

Dermott S.F., Malhotra R., Murray C.D.: 1988, 'Dynamics of the Uranian and Saturnian Satellite System: a chaotic route to melting Miranda ?', *Icarus* **76**, 295–334.

Dourneau G.: 1987, 'Observation et étude du mouvement des huit premiers satellites de Saturne', Dr.thesis, Université de Bordeaux I, Bordeaux.

Dourneau G.: 1993, 'Orbital elements of the eight major satellites of Saturn determined from a fit of their theories of motion to observations from 1886 to 1985', *Astron. Astrophys.* **267**, 292–299.

Duriez L.: 1992, 'A synthetic theory of motion for Titan-Hyperion', in *IAU Symposium 152: Chaos, resonance and collective dynamical phenomena in the solar system* p209–214.

Duriez L., Vienne A.: 1991, 'A general theory of motion for the eight major satellites of Saturn. I : Equations and method of resolution', *Astron. Astrophys.* **243**, 263–275.

Duriez L., Vienne A.: 1994, 'Modélisation du mouvement d'Hyperion', in 'Notes Scientifiques et Techniques du Bureau des Longitudes', S50, 115–120.

Emel'yanov N.V., Gasanov S.A., Nasonova L.P.: 1994, 'Mutual events in Saturn's satellite system from 1995 to 1996' *Astron. Reports* **38**, 708–717.

Gerasimov V.P., Mushailov B.R.: 1991, 'Evolution of orbits of Hyperion' , *Sov. Astron.* **35** 202–205.

Harper D., Taylor D.B., Sinclair A.T.: 1989, 'Analysis of the orbits of Titan Hyperion and Iapetus by numerical integration', *Astron. Astrophys.* **221**, 359–365.

Harper D., Taylor D.B.: 1993, 'The orbits of the major satellites of Saturn', *Astron. Astrophys.* **268**, 326–349.

Harper D., Taylor D.B.: 1994, 'Analysis of ground-based observations of the satellites of Saturn 1874-1988', *Astron. Astrophys.* **284**, 619–628.

Harper D.: 1995, 'Astrometry of Saturn's satellites from La Palma' to appear in the proceedings of the 'Atelier de travail PHESAT95', Bucarest, Sept. 1994.

Kirsanov N.O.: 1995, 'A theory of the motion of Hyperion' to appear in the proceedings of the 'Atelier de travail PHESAT95', Bucarest, Sept. 1994.

Le Campion J.F., Geffert M., Dulou M.R., Colin J.: 1992 'An astrometric catalogue of stars in the region of M15', *Astron. Astrophys. Sup. Ser.* **95**, 233–247.

Le Floch J.C.: 1995, 'L'observation des satellites de Saturne à l'observatoire de Bordeaux' to appear in the proceedings of the 'Atelier de travail PHESAT95', Bucarest, Sept. 1994.

Message J.P.: 1989, 'The use of computer algorithms in the construction of a theory of the long-period perturbations of Saturn's satellite Hyperion' *Celest. Mech.* **45**, 45–53.

Message J.P.: 1993 'On the second order long-period motion of Hyperion', *Celest. Mech.* **56**, 277–284.

Pascu D., Schmidt R.E.: 1990, 'Photographic positional observations of Saturn' *Astron. J.* **99**, 1974–1984.

Pascu D.: 1995, 'CCD observations of planetary satellites at the U.S. Naval Observatory', to appear in the proceedings of the 'Atelier de travail PHESAT95', Bucarest, Sept. 1994.

Rapaport M., Le Campion J.F.: 1995, 'Observations CCD du système des satellites de Saturne faites à Bordeaux. Comparaisons avec les théories' to appear in the proceedings of the 'Atelier de travail PHESAT95', Bucarest, Sept. 1994.

Rohde J.R., Pascu D.: 1993, 'Astrometric observations of Helene, Telesto and Calypso' *Bull. Amer. Astron. Soc.* **25**, 1235.

Rohde J.R., Pascu D.: 1994, 'CCD Astrometry of Helene, Telesto and Calypso: 1993 observations' *Bull. Amer. Astron. Soc.* **26**, 1024.

Sessin W.: 1992, 'Hori auxiliary system for Mimas-Tethys' in *IAU Symposium 152: Chaos, resonance and collective dynamical phenomena in the solar system* p223–226.

Sinclair A.T.: 1974, 'A theory of the motion of Iapetus', *Month. Not. Roy. Astrom. Soc.* **169**, 591–605.

Sinclair A.T.: 1977, 'The orbits of Tethys, Dione, Rhea, Titan and Iapetus', *Month. Not. Roy. Astrom. Soc.* **180**, 447–460.

Sinclair A.T., Taylor D.B.: 1985, 'Analysis of the orbits of Titan, Hyperion, and Iapetus by numerical integration and by analytical theories', *Astron. Astrophys.* **147**, 241–246.

Soma M.: 1992, 'Eclipses of Iapetus in 1993', *Astron. Astrophys.* **265**, L21–L24.

Stavinschi M., Bocsa G.: 1995, 'The Romanian tradition in the observations of Jupiter's and Saturn's satellites', to appear in the proceedings of the 'Atelier de travail PHESAT95', Bucarest, Sept. 1994.

Strugnell P.R. Taylor D.B.: 1990, 'A catalog of ground-based observations of the eight major satellites of Saturn 1874-1989' *Astron. Astrophys. Sup. Ser.* **83**, 289–300.

Taylor D.B., Shen K.X.: 1988, 'Analysis of astrometric observations from 1967 to 1983 of the major satellites of Saturn', *Astron. Astrophys.* **200**, 269–278.

Taylor D.B., Sinclair A.T., Message P.J.: 1987, 'Corrections to the theory of the orbit of Saturn's satellite Hyperion' *Astron. Astrophys.* **181**, 383–390.

Taylor D.B., Morrison L.V., Rapaport M.: 1991, 'Saturn's position derived from meridian circle observations of Titan and Iapetus' *Astron. Astrophys.* **249**, 569–573.

Taylor D.B.: 1992, 'A synthetic theory for the perturbations of Titan on Hyperion', *Astron. Astrophys.* **265**, 825–832.

Thuillot W.: 1995, 'Les éphémérides pour les observateurs PHESAT95', to appear in the proceedings of the 'Atelier de travail PHESAT95', Bucarest, Sept. 1994.

Thuillot W., Arlot J.E.: 1995, 'La prédiction des phénomènes des satellites de Saturne pour la période de 1995', to appear in the proceedings of the 'Atelier de travail PHESAT95', Bucarest, Sept. 1994.

Veillet C., Bois E., Oberti P.: 1988, '12 ans d'astrométrie photographique des satellites de Saturne, Uranus et Neptune', journées de planétologie, 457–460.

Veillet C., Dourneau G.: 1992, '1980–1985 astrometric observations of the first eight satellites of Saturn at Pic-du-Midi, ESO and CFH', *Astron. Astrophys. Sup. Ser.* **94**, 291–297.

Viateau B., Rapaport M.: 1995, 'Observations astrométriques à l'Observatoire de Bordeaux', in the proceedings of the IAU symposium 172, Paris July 1995.

Vienne A., Duriez L.: 1991, 'A general theory of motion for the eight major satellites of Saturn. II : Short-period perturbations', *Astron. Astrophys.* **246**, 619–633.

Vienne A., Duriez L.: 1992, 'A general theory of motion for the eight major satellites of Saturn. III : Long-period perturbations', *Astron. Astrophys.* **257**, 331–352.

Vienne A., Duriez L.: 1995, 'TASS1.6: Ephemerides of the major saturnian satellites', *Astron. Astrophys.* **297**, 588–605.

Zamarashkin K.N., Kirsanov N.O., Kisseleva T.P., Kisselev A.A., Batrakov J.V.: 1995, 'CCD astrometric observations of the saturnian satellites system on the 26-inch refractor at Pulkovo to be made in 1995-1996 campaign and their comparison with parallel photographic observations', to appear in the proceedings of the 'Atelier de travail PHESAT95', Bucarest, Sept. 1994.

FOUNDATIONS OF A THEORY OF THE MOTION OF THE ORBIT PLANE OF HYPERION

P.J. MESSAGE
University of Liverpool,
L69 3BX, U.K.
sx20@liverpool.ac.uk

Abstract. The principles are set out for the construction of a theory of the motion of the orbit plane of Hyperion, using the mixed set of angle parameters, using different reference planes for different angles, which it has proved convenient to use. It is found that this leads to additional terms, which have not been shown in previous published theories. The theory is developed in general principles exactly, and in detail as far as is needed to enable comparison to be made with the observational data at present available, and, from parameters which have been derived from opposition means from the period 1875 to 1922, the co-efficients of some of the larger long-period terms are computed.

1. Introduction

It has become usual, when developing theories of the motion of Hyperion *in* its orbit plane, including the effects of the very close resonance of orbital period with Titan, to use longitudes, including the longitude of the apse, referred to the Ecliptic and Equinox (*i.e.* First Point of Aries), of course of some specified date. However, when dealing with the motion *of* the orbit plane, the governing equations are very much simplified by using parameters which refer the orientation of the plane to the equator plane, or ring plane, of Saturn, since the orbit planes of Hyperion and Titan (and in fact of all the satellites except Iapetus) are inclined at quite small angles to that plane, and the differential equations for the rectangular-type orbital plane parameters may be treated as linear, for any precision of the theory which has been so far required.

2. The parameters employed

Let us now examine the effects of using this mixed set of parameters, referred to two different reference planes, in work on the motion of Hyperion.

Let us use the following notation:

i for the inclination of the orbit plane to Saturn's equator plane,

h for the longitude of the ascending node of the orbit, on Saturn's equator plane, measured from the ascending node of Saturn's equator on the Ecliptic,

I for the inclination of the orbit plane to the Ecliptic,

Ω for the longitude of the ascending node of the orbit, on the ecliptic, measured from the Equinox,

I_e for the inclination of the equator plane of Saturn to the Ecliptic,

Ω_e for the longitude of the ascending node of Saturn's equator on the Ecliptic, also measured from the Equinox.

A consistent canonical set of orbital parameters may be constructed by using Saturn's equator plane as the reference plane, which has the advantage of being effectively fixed in orientation, since the gravitational couple on Saturn is so small. One way to remove one source of complication from the task of comparing results with those obtained by the use of more usual reference systems, would be to measure all longitudes from the equinox, along the Ecliptic to the to the ascending node of Saturn's equator on the Ecliptic, and then along Saturn's equator to the ascending node of the orbit on Saturn's equator, and then (except for the longitude of the node itself) along the orbit. (Then longitudes so defined would be subject to precession, almost entirely due to the precessional motion of the equinox along the ecliptic.) A canonical set of orbital parameters may be set up in which all longitudes are defined in this way, thus avoiding any complication arising from the use of different reference planes in the construction of a perturbation theory (including the use of a Lie series transformation to separate the long-period effects from those of short period.) Let us denote by $\hat{\psi}$ a longitude defined on this basis (*i.e.* using the Ecliptic, equator plane of Saturn, and orbit plane).

Since, however, most reduction of observational data proceeds on the basis of longitudes measured in the more conventional way, i.e. from the Equinox along the Ecliptic to the ascending node of the orbit on the Ecliptic, it is necessary, in interpreting longitudes predicted by such a canonically consistent theory as is considered above, to relate the two systems of longitudes. Let us denote by ψ a longitude defined in the conventional way (*i.e.* using only the Ecliptic and orbit planes). Usually the theory will involve differences of longitudes of the two satellites, of the type $\hat{\psi}_H - \hat{\psi}_T$ (or corresponding differences of their apse longitudes, etc.). Such a difference

THE MOTION OF THE ORBIT PLANE OF HYPERION 129

will differ from the corresponding difference $\psi_H - \psi_T$ by quantities of the second order in the inclinations i_H and i_T of the satellites' orbits to Saturn's equator plane, and that will, in most aspects of the motion, lead to negligible consequences in the predictions to any precision which has so far been required. But, as observational data of finer precision are acquired, it will become necessary to take these differences into account.

3. The equations for the perturbations

As mentioned above, the differential equations for the motion of the orbit plane will be approximately linear if expressed in terms of the rectangular-type parameters

$$p = \sin i \sin h = \sin I \sin (\Omega - \Omega_e) \qquad (1)$$
$$q = \sin i \cos h = \sin I \cos I_e \cos (\Omega - \Omega_e) - \cos I \sin I_e$$

The Lagrange equations for the mixed set of orbital parameters ($\lambda, \varpi, a, e, q, p$), with the disturbing function, R, expressed in terms of this same set, are found to be, without approximation,

$$\frac{da}{dt} = \frac{2}{na} \frac{\partial R}{\partial \lambda},$$
$$\frac{de}{dt} = \frac{Y}{na^2} \frac{\partial R}{\partial \lambda} - \frac{X}{na^2} \frac{\partial R}{\partial \varpi},$$
$$\frac{d\lambda}{dt} = n - \frac{2}{na} \frac{\partial R}{\partial a} + \frac{Y}{na^2} \frac{\partial R}{\partial e} - \frac{Z}{nab} P,$$
$$\frac{d\varpi}{dt} = \frac{X}{na^2} \frac{\partial R}{\partial e} - \frac{Z}{nab} P,$$
$$\frac{dq}{dt} = -\frac{\cos i}{nab} \frac{\partial R}{\partial p} - \frac{Z}{nab}(q \cos I + \sin I_e)\left\{\frac{\partial R}{\partial \lambda} + \frac{\partial R}{\partial \varpi}\right\},$$
$$\frac{dp}{dt} = +\frac{\cos i}{nab} \frac{\partial R}{\partial q} - \frac{Z}{nab} p \cos I \left\{\frac{\partial R}{\partial \lambda} + \frac{\partial R}{\partial \varpi}\right\},$$

where

$$X = \frac{\sqrt{1 - e^2}}{e},$$
$$Y = X - \frac{1 - e^2}{e},$$
$$Z = \frac{1}{1 + \cos I},$$

$$\mathcal{P} = p\cos I \frac{\partial R}{\partial p} - (q\cos I + \sin I_e)\frac{\partial R}{\partial q},$$

and we note that

$$\cos I = \cos i \cos I_e - q \sin I_e$$

and

$$\cos i = \sqrt{1 - (q^2 + p^2)}.$$

4. The terms from the perturbations by Titan

The most important part of the disturbing function, R, is that corresponding to the effect of Titan:

$$R_T = Gm_T \left\{ \frac{1}{\Delta} - \frac{r_H}{r_T} \cos \mathcal{S} \right\},$$

in which

G is the constant of gravitation,
m_T is the mass of Titan,
Δ is the distance between Hyperion and Titan,
r_H is the distance between Hyperion and Saturn,
r_T is the distance between Titan and Saturn,
$\mathcal{S}$ is the angle subtended at the centre of Saturn by Hyperion and Titan,

so that

$$\Delta^2 = r_H^2 + r_T^2 - 2r_H r_T \cos \mathcal{S}.$$

Now let us put

$$R_T = R_0 + \delta R,$$

in which R_0 is R_T as evaluated with $\mathcal{S}$ replaced by $\psi_H - \psi_T$, the difference between the true longitudes of Hyperion and Titan, and with Δ replaced by Δ_0, which is Δ also evaluated with $\mathcal{S}$ replaced by $\psi_H - \psi_T$. Thus R_0 is that part of R_T giving the main part of the perturbations in the orbit plane, and δR contains all of the terms in R_T involving the parameters of the orbit plane. Then we find, taking proper account of the use of the different reference planes used in the definitions of the various angles, that, to second order in q_H and p_H (the values of q and p, respectively, for Hyperion), we obtain

$$\delta R = G m_T \left\{ r_H r_T \left\{ \frac{1}{\Delta_0^3} - \frac{1}{r_T^3} \right\} \mathcal{F} \right.$$
$$\left. + \frac{3}{4} \cdot \frac{r_H^2 r_T^2}{\Delta_0^5} \tan^2 \left(\frac{I_e}{2} \right) (p_H - p_T)^2 \{ 1 - \cos 2 (\psi_H - \psi_T) \} \right\},$$

where

$$\mathcal{F} = \frac{1}{4} \left\{ -\left\{ (q_H - q_T)^2 + \left\{ 1 + 2 \tan^2 \left(\frac{I_e}{2} \right) \right\} (p_H - p_T)^2 \right\} \cos (\psi_H - \psi_T) \right.$$
$$+ \{ (q_H - q_T)^2 - (p_H - p_T)^2 \} \cos (\psi_H + \psi_T - 2\Omega_e)$$
$$+ 2 (q_H - q_T)(p_H - p_T) \sin (\psi_H + \psi_T - 2\Omega_e)$$
$$+ 2 \left\{ (p_H q_T - q_H p_T) + 2 \tan \left(\frac{I_e}{2} \right) (p_H - p_T) \right.$$
$$\left. \left. + \tan^2 \left(\frac{I_e}{2} \right) (p_H q_H - p_T q_T) \right\} \sin (\psi_H - \psi_T) \right\}.$$

Substituting these terms into the Lagrange equations for the rates of change of q_H and p_H, we find some cancellation of terms, leading to some simplification in the terms of lowest order (as the comments at the end of section 2 lead us to expect), and that $\frac{dq_H}{dt}$ has to first order, in fact the terms

$$n_H m' \mathcal{K} \left\{ (p_H - p_T) \{ \cos (\psi_H - \psi_T) + \cos (\psi_H + \psi_T - 2\Omega_e) \} \right.$$
$$\left. + (q_H - q_T) \{ \sin (\psi_H - \psi_T) - \sin (\psi_H + \psi_T - 2\Omega_e) \} \right\},$$

and that $\frac{dp_H}{dt}$ has, also to first order,

$$n_H m' \mathcal{K} \left\{ (q_H - q_T) \{ -\cos (\psi_H - \psi_T) + \cos (\psi_H + \psi_T - 2\Omega_e) \} \right.$$
$$\left. + (p_H - p_T) \{ \sin (\psi_H - \psi_T) + \sin (\psi_H + \psi_T - 2\Omega_e) \} \right\},$$

where

$$\mathcal{K} = \frac{1}{2} a_H r_H r_T \left\{ \frac{1}{\Delta_0^3} - \frac{1}{r_T^3} \right\},$$

and

$$m' = \frac{m_T}{m_S},$$

where m_S is the mass of Saturn. The extra terms arising from the use of mixed reference planes do not cancel out in the expressions for $\dfrac{d\lambda_H}{dt}$ and $\dfrac{d\varpi_H}{dt}$, even to first order, and, to this order, both have the terms

$$n_H m' \mathcal{K} \tan\left(\frac{I_e}{2}\right)\left\{(q_H - q_T)\left\{-\cos(\psi_H - \psi_T) + \cos(\psi_H + \psi_T - 2\Omega_e)\right\}\right.$$

$$+(p_H - p_T)\sin(\psi_H + \psi_T - 2\Omega_e)$$

$$+\left\{p_T + \left\{\cos I_e \sec^2\left(\frac{I_e}{2}\right) - 2\tan\left(\frac{I_e}{2}\right)\right\}p_H\right\}\sin(\psi_H - \psi_T)\right\}$$

$$+\frac{3}{4}\frac{r_H r_T}{\Delta_0^5}\tan^2\left(\frac{I_e}{2}\right)\cos I_e p_H\{1 - \cos 2(\psi_H - \psi_T)\}.$$

Now from the results of the theory of the motion in the orbit plane (Message, 1989, 1993), we find the expressions, in which we indicate by "$< \mathcal{F} >$" the result of averaging a quantity "$\mathcal{F}$" over $\lambda_H - \lambda_T$, to isolate the long-period and critical terms,

$$< \mathcal{K} \cos(\psi_H - \psi_T) > = \sum_i \sum_j \mathcal{A}_{i,j} \cos(i\tau - j\zeta),$$

$$< \mathcal{K} \sin(\psi_H - \psi_T) > = \sum_i \sum_j \mathcal{B}_{i,j} \sin(i\tau - j\zeta),$$

$$< \mathcal{K} \exp\{\iota(\psi_H + \psi_T - 2\Omega_e)\} > = \sum_i \sum_j \mathcal{C}_{i,j} \exp\{\iota(i\tau - j\zeta)\},$$

where τ is the argument of the free libration (of about 21 month period), and ζ is the linear part of the argument $\varpi_H - \varpi_T$ (of about $18\frac{3}{4}$ year period), and ι is $\sqrt{-1}$.

5. The terms from other perturbations

From the solar perturbations, the main term in R is

$$\frac{1}{2}n_0^2 r_H^2 (3\cos^2 \mathcal{S}_0 - 1),$$

where n_0 is the mean motion in the relative motion of the Sun and Saturn, and $\mathcal{S}_0$ is the angle subtended at the centre of Saturn by the Sun and Hyperion. The largest long-period parts of this term are

$$\frac{3}{8} n_0^2 r_H^2 \Big\{ 1 + \cos^2 I_e - q_H \sin 2I_e - q_H^2 \cos 2I_e + p_H^2 \cos^2 I_e$$

$$+ \Big\{ \sin^2 I_e + q_H \sin 2I_e + q_H^2 \cos 2I_e + p_H^2 (\cos^2 I_e - 2) \Big\} \cos 2(\lambda_0 - \Omega_e)$$

$$+ 2\Big\{ p_H \sin I_e + q_H p_H \cos I_e \Big\} \sin 2(\lambda_0 - \Omega_e) \Big\},$$

from which the largest long-period solar terms in $\dfrac{dq_H}{dt}$ are

$$\frac{3 n_0^2}{4 n_H} \Big\{ 1 + \frac{1}{2} e^2 \Big\} \Big\{ \big\{ \cos^2 I_e + \{\cos^2 I_e - 2\} \cos 2(\lambda_0 - \Omega_e) \big\} p_H$$

$$+ \big\{ \sin I_e + q_H \cos I_e \big\} \sin 2(\lambda_0 - \Omega_e) \Big\},$$

and the largest long-period solar terms in $\dfrac{dp_H}{dt}$ are

$$\frac{3 n_0^2}{4 n_H} \Big\{ 1 + \frac{1}{2} e^2 \Big\} \Big\{ -\frac{1}{2} \sin 2 I_e - q_H \cos 2I_e + p_H \cos I_e \sin 2(\lambda_0 - \Omega_e)$$

$$+ \frac{1}{2} \{ \sin 2I_e + 2 q_H \cos 2I_e \} \cos 2(\lambda_0 - \Omega_e) \Big\}.$$

The largest long-period term from the effect of the figure of Saturn in $\dfrac{dq_H}{dt}$ is

$$\frac{3}{2} n_H^2 R_e^2 J_2 p_H,$$

and the corresponding term in $\dfrac{dp_H}{dt}$ is

$$-\frac{3}{2} n_H^2 R_e^2 J_2 q_H.$$

Here R_e is the radius of Saturn's equator, and J_2 is the co-efficient of the second zonal harmonic in the external gravitational field of Saturn.

6. The solution of the equations.

To proceed to a solution of the equations for q_H and p_H, introduce the complex variable

$$\mathcal{Z} = \kappa(q_H - q_T) + \frac{\iota}{\kappa}(p_H - p_T),$$

where κ is a constant to be chosen. The equations may then be written, to the precision to which we have been working,

$$\frac{d\mathcal{Z}}{dt} = \iota n_H \left\{ -\alpha \mathcal{Z} - \alpha' \bar{\mathcal{Z}} - \delta_s + \sum_j \beta_j \exp(\iota u_j) \bar{\mathcal{Z}} \right.$$

$$- \sum_j \gamma_j \exp(\iota w_j) \mathcal{Z}$$

$$\left. + \sum_j \rho_j \exp(\iota v_j) \right\},$$

where α, α', δ_s, β_j, γ_j, and ρ_j are constants, and u_j, w_j, and v_j are linear functions of the time (corresponding to the various terms in the equations for $\frac{dq_H}{dt}$ and $\frac{dp_H}{dt}$). The constant κ is chosen so that α' takes the value zero. This is found to require that, approximately,

$$\kappa^4 = 1 - \frac{3 n_0^2 \sin^2 I_e}{4 n_H^2 \, m' \, \mathcal{A}_{0,0}},$$

which gives κ the value 0.999565..... The constant α is given by

$$\alpha = m' \mathcal{A}_{0,0} + \frac{3}{2} R_e^2 J_2 + \dots \approx 1.403 m',$$

and

$$\delta_s = \frac{3 n_0^2}{8 n_H^2} \sin 2 I_e \approx 0.00000118.....$$

We note that, if the β_j were all zero (as would be true in the presence only of those terms which Woltjer (1928) took), then the linear equation for $\mathcal{Z}$ would be solvable exactly. However, some of the β_j are in fact significant. The solution may be written in the form

$$\mathcal{Z} = c \exp(-\iota v) - \frac{\delta_s}{\alpha}$$

$$+ \sum_j a_j \exp(\iota v_j)$$

$$+ \sum_j b_j \exp\{\iota(u_j + v)\}$$

$$+ \sum_j b'_j \exp\{\iota(w_j - v)\}$$

$$+ \sum_j \sum_k c_{j,k} \exp\{\iota(u_j - v_k)\}$$

$$+ \sum_j \sum_k c'_{j,k} \exp\{\iota(w_j + v_k)\}$$

$$+ \sum_j \sum_k d_{j,k} \exp\{\iota(u_j - u_k - v)\}$$

$$+ \sum_j \sum_k \sum_\ell e_{j,k,\ell} \exp\{\iota(u_j - u_k + v_\ell)\}$$

$$+ \text{etc} \ldots$$

in which the constants of integration are the amplitude, c, of the free oscillation, and the phase of the linear argument, v, of the free oscillation. Substituting this solution into the differential equation, and equating the co-efficients of each of the (infinite number of) periodic terms, leads to an array of algebraic equations which may be solved by iteration to give the values of the amplitudes a_j, b_j, b'_j, $c_{j,k}$, $c'_{j,k}$, $d_{j,k}$, $e_{j,k,\ell}$, etc., ..., of the forced terms, and the rate of change, χ, say, of the argument v of the free oscillation.

7. Identification of some of the major terms.

Let us now identify some of the main forced terms in the motion of the orbit plane, beginning with those of type $a_j \exp(\iota v_j)$.

Corresponding to $j = 1$ let us set the term arising from the precession of the orbit plane of Titan. This has a period of about 690 years and the relevant argument is $v_1 = 41.4° - 0.5213°(t - 1880.25)$, with t in years. Then the solution gives $a_1 = 0.041°$ and the main contribution to q_H is $0.333° \cos v_1$ and to p_H is $0.333° \sin v_1$.

Corresponding to $j = 2$ let us set a term with argument $v_2 = \Omega_0$, the node of the orbit of the relative motion of the Sun and Saturn. Since the mean motion of this is about 6 seconds of arc per century, it is effectively constant in this context. The contribution to q_H is $-0.745°$ and that to p_H is $-0.037°$.

Corresponding to $j = 3$ let us set a term with argument $v_3 = 2\lambda_0 - \Omega_0 + \pi$. For this we find that $a_3 = -0.018°$; the contribution to q_H is $-0.018° \cos(2\lambda_0 - \Omega_0)$ and that to p_H is $-0.018° \sin(2\lambda_0 - \Omega_0)$.

A significant term of the type $b_j \exp\{\iota(u_j + v)\}$ is associated with that term in the disturbing function which has argument $8\lambda_H - 6\lambda_T - 2\Omega_H$ and appears in the present theory with $u_1 = 2(\zeta + \varpi_T - \Omega_e) + \pi$, which has period 10.3 years. The theory gives $b_1 = -0.013°$ and the contribution to q_H is

$$0.013° \sin\{2(\zeta + \varpi_T - \Omega_e) - v\},$$

and that to p_H is

$$0.013° \cos\{2(\zeta + \varpi_T - \Omega_e) - v\}.$$

It remains, to the precision to which we are at present working, to consider the free oscillation. A fresh analysis of the values of the opposition means of orbital parameters which derive from the observational data from the time interval 1875 to 1922 gives for the amplitude $c = 0.521° \pm 0.012°$ and, for the argument v, $94.91° \pm 1.41° - (2.651° \pm 0.097°) \cdot (t - 1900.0)$. This rate gives an estimate for the mass of Titan, in terms of that of Saturn, of $(2.73 \pm 0.11) \cdot 10^{-4}$, but since the data span only a fraction of the period of this free term, about 136 years, this estimate cannot be considered to have very high weight.

Work is in hand to make an analysis of all the available observational data in one solution in comparison with this theory, which it is hoped will improve the estimates of the parameters. To make comparison with more precise observational data, as may become available for example from the "Cassini" mission, would require the retaining in the theory of more terms than we have considered in section 7, and perhaps also of terms of higher than second order in q_H and p_H in the expression for δR in section 4 above. Also it might possibly be necessary to include, in the theory of the motion *in* the orbit plane, those terms of second order in the mass of Titan resulting from the effect of terms in δR on λ_H and ϖ_H (and perhaps also a and e), though such terms will be very small indeed, however, having as factors the square of Titan's mass and also the very small angle of inclination of the orbit plane to Saturn's equator plane.

References

Message, P.J., 1989, "The use of computer algorithms in the construction of a theory of the long-period perturbations of Saturn's satellite Hyperion". *Celest. Mech. Dyn. Astron.* **45**, 45-53.

Message, P.J., 1993, "On the second-order long-period motion of Hyperion". *Celest. Mech. Dyn. Astron.* **56**, 277-284.

Woltjer. J., 1928, "The motion of Hyperion". *Annalen van der Sterrewacht Leiden* **XVI**, Part 3.

THE MOTION OF HYPERION

ON THE ACCURACY OF THE OBSERVATIONS

NIKOLAI O. KIRSANOV
Institute for Theoretical Astronomy,
Kutuzova 10, St. Petersburg, Russia
kirsanov@ita.spb.su

Abstract.
93 observations of the seventh satellite of Saturn, Hyperion, with 473 observations of Titan—Saturn's sixth and the most massive satellite and the one nearest to Hyperion—are considered. The observations were made in 1967–1981 at several observatories. The values of (O–C) across and along the orbits are obtained. The normality of the distributions of (O–C) is studied.

1. Introduction

Since its discovery, Saturn's seventh satellite Hyperion has attracted the attention of scientists working in the field of celestial mechanics by peculiarities of its motion, which are related to the strong influence of Titan [2, 5]. Being rather dark (magnitude near 14^m) Hyperion is a difficult object for observations. In this sense, it is very interesting to investigate the accuracy of existing photographic astrometric observations comparing them with a numerical theory of motion.

2. Observations

I used photographic astrometric observations (α, δ) made at different observatories between 1967 and 1981 (a total of 473 observations for Titan and 93 observations of Hyperion). The observations were used also in [4] in order to improve the nonsingular orbital elements. The observations are made at the following observatories: Pulkovo, Table Mountain, McDonald, McCormick, Greenwich, US Naval Observatory, Bordeaux, European Southern Observatory.

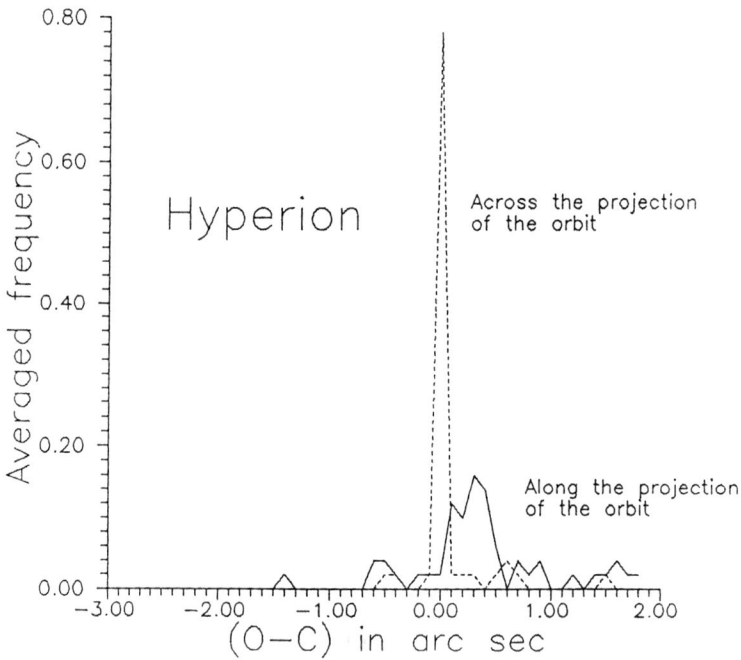

Figure 2. Distribution of (O–C) for Hyperion

Using the known formula for dispersion, I obtained the values of the root mean square errors of distributions of (O–C). They are: for Hyperion, 0.0255 across the orbit and (formally) along, 0.855; for Titan, 0.0236 across the orbit and (formally) 0.571 along the orbit.

The numbers above mean that there is a small systematical error which is to be eliminated by taking into account, in the numerical theory, the omitted factors, e. g. the influence of the errors in the ephemerides of Saturn.

References

1. Bulirsh R., Stoer J. (1966) "Numerical treatment of ordinary differential equations by extrapolation methods", *Numerische Mathematik*, **8** 1.
2. Burns J. A. (ed.). (1977) *Planetary Satellites,* University of Arizona Press, Arizona.
3. Kulikov D. K., Batrakov Yu. V. (1960) "Method for improving orbits of artificial satellites of the Earth, using observations when time is only approximately known", *Bull. ITA*, **7**(7), 554.
4. Kirsanov N. O. (1995) The Orbits of Hyperion and Titan from the 1967–1981 Observations, *Astronomy Letters*, **21**(2), 231.
5. Woltjer J. (1928) The motion of Hyperion, *Annalen van de Sterrewacht te Leiden* **16**(3).

REDETERMINATION OF THE ORBIT OF IAPETUS

SHEN KAIXIAN AND QIAO RONGCHUAN
United Laboratory for Optical Astronomy,
Chinese Academy of Sciences, Shaanxi Observatory,
P.O.Box 18, Lintong, Shaanxi, China

Abstract. This paper gives a test of the reliability of TASS from Vienne & Duriez via the comparison of the theory of Iapetus given by Harper & Taylor. From fitting it to photographic observations, we derived a redetermination of the orbit of Iapetus

A new general theory of the motion of the eight satellites of Saturn – TASS (Théorie Analytique des Satellites de Saturne), has been built by Vienne & Duriez 1991, 1992). Harper & Taylor (1993) have completed also a new determination of the orbits of the major satellites of Saturn. In their paper a set of new analytical formulae of the perturbation for Iapetus was given.

The intention of our work is to construct a feasible analytical representation of the perturbed motion of Iapetus, based on TASS. But this paper only gives a test of the reliability of TASS 1.6 (Vienne & Duriez, 1995) via a comparison with the theory of Iapetus given by Harper & Taylor. We selected only those terms greater than 50 km, which are derived from TASS 1.6 and should be compatible in precision with the terms obtained by Harper & Taylor. In our computations, we have used the same variables used in TASS: p, q, z, ζ and the following values: $\dot{\omega} = \dot{\phi} = 0.°1131$, $\dot{\Omega} = \dot{\Phi} = -0.°1103$, $\dot{\omega}_T = \dot{\phi}_T = 0.°5118$, $\dot{\Phi}_T = -0.°5117$. We easily obtain the expression for the short perturbations of Iapetus. For example from $\Delta z = \Delta h + i\Delta k$, we may deduce $\delta e = \Delta h \cos\omega + \Delta k \sin\omega$, so

$$\begin{aligned}
10^5 \delta e = \ & 59.38 \cos(\ell_T + \omega_T - \omega - 63.°12) \\
& +27.39 \cos(2\ell_T - \ell_T + \omega - \omega_T + 63.°12) \\
& +25.33 \cos\ell + 10.49 \cos(-\ell + 2\ell_S - 2\omega - 188.°25) \\
& +2.06 \cos(3\ell_S - \ell - 2\omega + 160.°78) \\
& +4.89 \cos(3\ell_S - 2\omega - 261.°33) \\
& +37.89 \cos(2\ell_S - 2\omega + 187.°53) \\
& +1.72 \cos(2\ell_T + \omega_T - \omega + 112.°29).
\end{aligned}$$

in which ℓ, ℓ_T, ℓ_S have the same value as used in H&T, because they are just the difference between the longitudes and do not depend on the adopted reference frame. In computing long-period, resonant and secular terms, we use all those terms of TASS 1.6 given by Vienne and Duriez (1995; tables 8a, 8b, 8c).

The epoch and reference frame adopted by us are the same used by Harper & Taylor. The starting epoch is 1930 January 24^d 0^h Ephemeris Time (JED-2426000.5). The mean ecliptic and equinox of B1950.0 was used as the fixed reference system. This is related to two reasons: (1) It will be convenient for comparison of our results with those of Harper & Taylor; (2) We still can use the consistent partial derivatives with them. But, in TASS, the reference plane was assumed to be the equatorial plane of Saturn. The longitudes of node and pericenter are measured in the equatorial plane of Saturn. The origin of the longitudes is the node of this plane on the mean ecliptic J2000 frame. Therefore, it remains to do the rotations corresponding to the inclinations and the longitude of node of the equatorial plane in this frame (i.e., respectively, $28.°0752$ and $169.°5084$). Finally, the conversion from the J2000.0 reference system to B1950 was done with the aid of Lieske's (1994) formulae. In fact, this small adjustment is insignificant. As a test, we have fitted the two theories to 1763 photographic observations from 1967 to 1988. The obtained r.m.s. residuals are $0''.2213$ for TASS and $0''.2215$ for H&T. The preliminary results seem to indicate that it will be possible for the new theory to lead to some small improvements in precision in the determination of the orbit of Iapetus. Table I gives a set of new elements of the orbit of Iapetus corresponding to the date JED 2426000.5. In effect, no distinct differences between the two sets were found. The more refined part of this work is still in progress.

TABLE 1. Orbital elements of Iapetus

Elements Units	a_0 A.U.	λ_0 deg.	e_0	ϖ_0 deg.	i_0 deg.	Ω_0 deg.
H& T	0.0238117	216.99743	0.0288367	357.824	18.02066	141.475
S& Q	0.0238099	216.99519	0.0290824	357.856	18.02032	141.457

References

Duriez, L. and Vienne, A.: 1991, *Astron. Astrophys.* **243**, 263.
Harper, D. and Taylor, D.: 1990, *Astrophys.* **268**, 326. (H& T)
Lieske, J.H.: 1994, *Astron. Astrophys.* **281**, 281.
Vienne, A. and Duriez, L.: 1991, *Astron. Astrophys.* **246**, 619.
Vienne, A. and Duriez, L.: 1992, *Astron. Astrophys.* **257**, 331.
Vienne, A. and Duriez, L.: 1995 *Astron. Astrophys.* **297**, 588.

THE MIMAS-TETHYS AND ENCELADUS-DIONE SYSTEMS

A. VIENNE, L. DURIEZ AND S. CHAMPENOIS
Laboratoire d'Astronomie, Université de Lille 1

We have recently built a coherent theory of the motion of the satellites Mimas, Enceladus, Tethys, Dione, Rhea, Titan and Iapetus. The final form of the "Théorie Analytique des Satellites de Saturne" (TASS1.6) is presented in Vienne & Duriez (1995). The internal precision of TASS is a few kilometers over three years and some tens kilometers over one century. The root-mean-square residuals of the adjustment of TASS over one century of Earth based observations reach $0''.12$ for the best data sets, until $0''.015$ for the few mutual phenomenas of 1981.

In the solution of Mimas and Tethys, nearly all arguments have been recognized as integer combinations of fundamental arguments which are :

ϕ_1, so that $\phi_1 - (\lambda_{o1} - 2\lambda_{o3})$ is close to the proper pericenter of Mimas
ϕ_3, so that $\phi_3 - (\lambda_{o1} - 2\lambda_{o3})$ is close to the proper pericenter of Tethys
Φ_1, so that $\Phi_1 - (\lambda_{o1} - 2\lambda_{o3})$ is close to the proper node of Mimas
Φ_3, so that $\Phi_3 - (\lambda_{o1} - 2\lambda_{o3})$ is close to the proper node of Tethys
ω_1 is the libration argument of the resonance $\Phi_1 + \Phi_3 = 0$.

where λ_{o1} and λ_{o3} are respectively the mean mean longitude of Mimas and Tethys. Among the arguments there are some very long periods :

$\alpha_1 = \phi_3 + 2\Phi_1$ (period: 200 years)
$\alpha_2 = \phi_1 - \phi_3 + 2\Phi_1$ (period: 700 years)

In the model of Dermott *et al.* (1988) the semi major axis of the satellites grows up due to tidal dissipation on the planet. For example, Mimas and Tethys have crossed the resonance $2\Phi_1$ 100 million years ago (without locking inside it), and the system has evolved until to reach the present resonance $\Phi_1 + \Phi_3$. Using the formulas of Dermott *et al.* and their value of the coefficient of the tidal dissipation, we have found that the commensurability $\dot{\alpha}_1 = 0$ has been reached 450 000 years ago, and that $\dot{\alpha}_2 = 0$ will be reached in 190 000 years. Of course, these "datations" are somewhat approximative but give an order of magnitude. In both cases, the semi-major axis are shifted of 1 to 3 km only, with $a_1 = 186\,009$ km and $a_3 = 294\,958$ km.

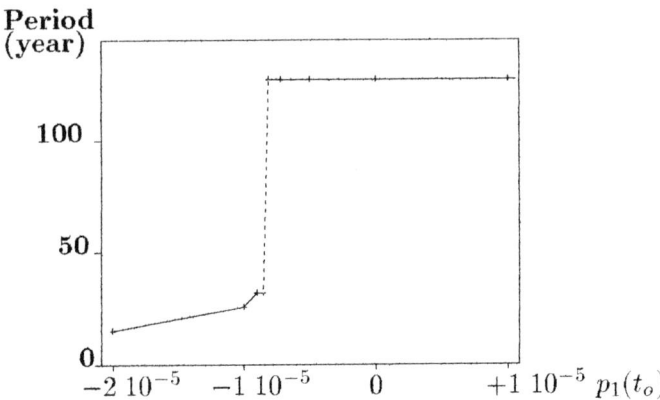

Figure 1. Variation of the frequency of α_1 with respect to the initial contidion $p_1(t_o)$ (a variation of p_1 of $+1\ 10^{-5}$ corresponds to $-1,24$ kilometer in a_1).

We have studied the argument α_1 only. It appears to be very sensitive to initial conditions. We have observed a skip in its frequency for a very small variation in the semi-major axis of Mimas (Fig 1). This skip is suspect. Furthermore, the frequency analysis failed for the integrations inside the skip (dot line).

As it was shown by Laskar (1990, 1992), the measure of the variation of the fundamental frequencies of a dynamical system can be related to the measure of chaos. We have computed the frequency $\dot{\omega}_1$ aver 200 000 years with 130 evaluations of this frequency over 6000 years (intervals are overlaping at 75 %). This fundamental frequency shows significant variations (10^{-5} rad per year), but more computations are necessary to give a definitive conclusion. A similar study has shown that no significant variations exist for the fundamental frequencies of the Enceladus-Dione system.

These results are still preliminary. More computations and studies are necessary to confirm and understand a chaotic feature of the system Mimas-Tethys. Nevertheless, it is clear that these studies can not be neglected to understand the origin of the resonance under the effects of tital dissipation.

Dermott S.F., Malhotra R., Murray C.D.: 1988, *Icarus* **76**, 295–334

Laskar J.: 1990, *Icarus* **88**, 266–291

Laskar J., C. Froeschlé, A. Celletti: 1992, *Physica D.* **56**, 253–269

Vienne A., Duriez L.: 1995, *Astron. Astrophys.* **297**, 588–605

NEW CONSTANTS FOR SAMPSON-LIESKE THEORY OF THE GALILEAN SATELLITES OF JUPITER FROM MUTUAL OCCULTATION DATA

R. VASUNDHARA
Indian Institute of Astrophysics, Bangalore 560034, India
AND
J.E. ARLOT AND P. DESCAMPS
Bureau des Longitudes, URA 0707 CNRS,
77 avenue Denfert-Rochereau, 75014 Paris, France

Abstract. New constants for Sampson-Lieske theory of the Galilean Satellites have been derived using 6360 individual photographic positions (1891-1990) and 438 pseudo astrometric positions from mutual occultations during 1973, 1979, 1985 and 1991. Using these new sets of constants, significant improvement is noticed in the O-C values of the sky plane coordinates of the mutual event data set and residuals in longitude for Io and Europa are found to improve. Problems concerning the inclusion of mutual event data in attempting evaluation of secular variations of the mean motions of the satellites are discussed.

1. Introduction

The constants of motion of the Galilean Satellites are efficiently updated using Lieske's (1974, 1977) technique by computing the corrections (ε, β) to the basic sets of constants. Lieske progressively updated these constants from E1 (Lieske, 1978), using the eclipse data of 1878-1903, to E2 (Lieske, 1980), by supplementing visual eclipses between 1903 and 1972, photographic data between 1967 and 1978 and mutual events of 1973. The E2x3 (Lieske, 1987) ephemeris was derived by further additions of more mutual event pairs from 1973 and 1979, 183 pairs of data from Voyager optical navigation images and 15711 classical eclipses from 1652-1983. Arlot (1982) used 8856 individual photographic observations between 1891 to 1978 to derive the G5 ephemeris. The present investigations relate to determinations of new sets of constants using photographic data between 1891 to 1990 and astrometric positions derived from mutual occultations during 1973, 1979, 1985 and 1991.

2. The observational material

Most of the photographic data that were used by Arlot(1982) were utilized in this study. Additional excellent unpublished photographic material covering the period between 1986 and 1990 (Pascu, private communication) have been included. Table 1 gives the details of the observations and includes the mutual occultation data set containing the astrometric positions derived from the mutual occultation light curves of 1985 and 1991 assuming Hapke's light scattering law for a macroscopically rough surface (Hapke, 1984), using the Hapke's parameters derived by Descamps and Thuillot (1993). The published positions of the 1985 series by the Galilean Satellite Observers (GSO and Franklin, 1991), which were derived using Lambert's law, were reconstructed to adopt Hapke's law and included in our studies (Vasundhara, 1993 and 1994). Published positions of 1973 (Aksnes and Franklin 1976) and 1979 (Aksnes et al. 1984) were also used after accounting for the phase corrections (Aksnes, Franklin and Magnusson 1986). Table 1 gives the statistics for all the data that we used.

3. Corrections to the constants

The method developed by Arlot(1982) has been used in the present study. The starting ephemeris was E1 (Lieske 1987) and 3 more iterations were carried out by progressively adjusting the (ϵ, β) values. The constants that were updated were the mean motions $(\epsilon_6 - \epsilon_8)$, the primary eccentricities $(\epsilon_{16} - \epsilon_{19})$, the primary sine inclinations $(\epsilon_{21} - \epsilon_{24})$, the mean longitudes $(\beta_1, \beta_2, \beta_4)$, the proper perijoves $(\beta_6 - \beta_9)$, the proper nodes $(\beta_{11} - \beta_{14})$, the libration amplitude (ϵ_9) and the libration phase angle (β_5).

Let us remind that the G5 ephemeris was based only on photographic observations from 1891 to 1978. By adding the recent photographic data made from 1986 to 1990 (Pascu, private communication) and the mutual occultations data (see table 1) with a weight of 50, we obtained the ephemeris I32, and I33 using a weight of 10. The ephemeris G6 was constructed using only the photographic data of Table 1 and the updated ε and β values for Jupiter related constants and mass of the satellites introduced by Lieske (1987) in his E2x3 ephemerides. The values of the constants for all these ephemerides are available using FTP anonymous on

TABLE 1. Details on the data

Observers	Codes[1]	Year	No. of positions				r.m.s. of the residuals (arcsec)			
			J1	J2	J3	J4	G5	G6	I32	I33
Renz	H,P	1891-1898	171	175	184	174	0.117	0.111	0.110	0.111
Balanovsky	P	1904-1910								
Chevalier	Zo-Se	1917/1918	132	132	133	121	0.148	0.150	0.151	0.151
De Sitter	G	1918/1919								
De Sitter	C	1924								
Petrescu	B	1934	85	104	106	118	0.217	0.222	0.223	0.222
Petrescu	P	1936								
Van Biesbr.	Y	1961-1963								
Gorel	N	1962-1966								
Soulié	Bo	1966-1974	284	282	318	307	0.390	0.390	0.390	0.390
Gorel	N	1973-1974								
Debehogne	RJ,U,LS	1977/1978								
Ianna et al.	Mc-C	1977/1978	70	100	95	108	0.104	0.104	0.104	0.104
Pascu	Mc-C	1967/1968	88	87	89	95	0.095	0.095	0.095	0.095
"	USNO	1973	72	65	62	67	0.111	0.110	0.111	0.111
"	"	1974	107	115	120	123	0.082	0.085	0.085	0.085
"	"	1975-1977	107	116	109	119	0.085	0.086	0.086	0.086
"	"	1977/1978	59	59	53	59	0.102	0.103	0.103	0.103
Pascu[2]	"	1986	84	74	82	80	0.077	0.074	0.074	0.074
"	"	1987	116	133	129	133	0.074	0.070	0.071	0.071
"	"	1988/1989	52	54	60	60	0.059	0.056	0.056	0.056
"	"	1990	79	72	84	96	0.068	0.066	0.067	0.067
AkFr[3]	-	1973	45	48	3	0	0.017	0.014	0.014	0.014
Ak[4]	-	1979	4	7	3	0	0.010	0.014	0.011	0.011
Ar[5],GSOF[6]	-	1985	59	62	73	26	0.024	0.024	0.024	0.024
Ar[7]	-	1991	45	53	9	1	0.036	0.024	0.028	0.030

1. Observatory codes. H:Helsingfors, P:Pulkovo, G:Greenwich C:The Cape, P:Paris, Y:Yerkes, N:Nicolaiev, Bo:Bordeaux, RJ:Rio de Janeiro, U:Uccle, B:Bucarest, LS:La Silla, Mc-C:Mc Cormick, USNO:U.S. Naval Observatory, Washington D.C.
2. Private communication
3. Aksnes and Franklin, 1976
4. Aksnes et al., 1984
5. Arlot et al., 1992
6. GSO and Franklin, 1991
7. Arlot et al., 1996

TABLE 2. Error in Longitude (minutes of time)

Data set	J1	J2	J3	J4	Ephem.
Ph	0.060 ± 0.011	0.009 ± 0.012	0.053 ± 0.015	0.073 ± 0.020	G5
Ph+ME	0.058 ± 0.010	−0.021 ± 0.011	0.041 ± 0.014	0.076 ± 0.019	G5
Ph	0.019 ± 0.011	0.038 ± 0.012	0.058 ± 0.015	0.207 ± 0.020	G6
Ph+ME	0.019 ± 0.010	0.038 ± 0.012	0.058 ± 0.015	0.207 ± 0.019	G6
Ph	0.019 ± 0.011	0.039 ± 0.012	0.078 ± 0.015	0.201 ± 0.020	I32
Ph+ME	0.018 ± 0.010	0.017 ± 0.011	0.067 ± 0.014	0.203 ± 0.019	I32
Ph	0.021 ± 0.011	0.040 ± 0.012	0.059 ± 0.015	0.204 ± 0.020	I33
Ph+ME	0.021 ± 0.010	0.039 ± 0.012	0.059 ± 0.014	0.205 ± 0.019	I33
Ph+ME	1508	1568	1624	1660	data points
Ph+ME	1508 + 153	1568 + 170	1624 + 88	1660 + 27	data points

TABLE 3. Secular accelerations of the Galilean satellites J1, J2, J3 in $\dot{n}/n \times 10^{11}\,\mathrm{yr}^{-1}$ units

Authors	J1	J2	J3
De Sitter (1928)	25	27	−16
Brouwer & Clemence (1961)	32	27	16
Greenberg (1986)	32 ± 8	−16 ± 4.5	−16 ± 4.5
Lieske (1987)	−0.74 ± 0.87	−0.82 ± 0.97	−0.98 ± 1.53
Goldstein (1996a)	45.4 ± 9.5	-	-
Goldstein (1996b)	70 ± 75	56 ± 57	28 ± 20
This paper/G5	24.6 ± 7.3	−12.7 ± 8.4	−0.22 ± 10.7
This paper/I32	22.7 ± 7.9	−6.1 ± 9.3	+10.6 ± 10.6

ftp.bdl.fr/pub/ephem/satel/theories or may be sent upon request near the authors.

4. Results and discussion

The r.m.s. of the residuals (table 1) shows a small improvement after using the mutual occultations data to fit the theory. More, it is remarkable that G6, which is fitted only on photographic data, shows very small residuals for the mutual occultations data. In any case, these data are of high accuracy. In table 2, we tried to determine a longitude shift from the residuals after different fits of the theory. This shift may be interpreted either as an error on the longitude at the origin of time, or as an error on the mean motion. It is puzzling to see that this error is decreasing when including recent observations to fit the theory (G6, I32, I33) for J1, stable for J3 but increasing slightly for J2 and much more for J4. On the other hand, the comparison between G6 and I32 shows that the mutual occultations data do not improve the ephemeris more than the photographic data for the period 1986-1993. This could be explained by the larger number of photographic data (1388 points) than the one of mutual occultations data (328 points). Table 3 gives the secular accelerations calculated from the residuals of all the data presented in table 1 calculated using G5 and I32. Our results are significant only for J1. Note that the methods of calculations of the other authors are completely different and that we give their data only to appreciate the scale of values of these accelerations.

In conclusion, the residuals of the mutual occultations data look very good but these data are not yet sufficient to improve the ephemerides. We look forward to using data from mutual eclipses and from the forthcoming mutual event opportunity in 1997.

Acknowledgements

We would like to thank Dr.J.H.Lieske from JPL for the computer codes for calculating the revised Sampson theory. We are indebted to Dr.D. Pascu from USNO for allowing us to use his yet unpublished excellent data of 1986-1990.

References

Aksnes, K., Franklin, F. 1976, *Astron. J.*, **81**, 464.
Aksnes, K., Franklin, F., Millis, R., Birch, P., Blanco, C., Catalano, S., Piironen, J. 1984, *Astron. J.*, **89**, 280.
Aksnes, K., Franklin, F., Magnusson, P. 1986, *Astron. J.*, **92**, 1436.
Arlot, J.E. 1982, *Astron. Astrophys.*, **107**, 305.
Arlot, J.E. et al., 1992, *Astron. Astrophys. Supp. Series*, **92**, 151.
Arlot, J.E. et al., 1996, to be published in *Astron. Astrophys. Supp. Series*.
Brouwer, D., Clemence, G.M. 1961, in Planets and Satellites, Kuiper and Middlehurst ed., Chicago University Piers, p. 88.
Descamps, P., Thuillot, W. 1993, *Icarus*, in press.
de Sitter, W., 1928, Leiden Annals, **16**, 92.
GSO and Franklin, F., 1991, *Astrophys.* **102**, 806.
Goldstein, S.J., Jacobs, K.C. 1996, On the Evolution of the galilean satellites, in press.
Goldstein, S.J., Jacobs, K.C. 1996, A recalculation of the secular acceleration of IO, in press.
Greenberg, R., Goldstein, S.J., Jacobs, K.C. 1986, Nature, **323**, 789.
Hapke, B.W. 1984, *Icarus* **59**, 41.
Lieske, J.H., 1974, *Astron. Astrophys.* **31**, 137.
Lieske, J.H., 1977, *Astron. Astrophys.* **56**, 333.

Lieske, J.H., 1978, *Astron. Astrophys.* **65**, 83.
Lieske, J.H., 1980, *Astron. Astrophys.* **82**, 340.
Lieske, J.H., 1987, *Astron. Astrophys.* **176**, 146.
Vasundhara, R. 1993, *Ph. D. Thesis.*
Vasundhara, R. 1994, *Astron. Astrophys.* **281**, 565.

THE EFFECT OF STEP-SIZE ON THE NUMERICAL INTEGRATION OF SATELLITE ORBITS

K.G. HADJIFOTINOU
Department of Mathematics, Faculty of Science, Aristotle University of Thessaloniki, 540 06 Thessaloniki, Greece

AND

D. HARPER
Astronomy Unit, School of Mathematical Sciences, Queen Mary and Westfield College, Mile End Road, London E1 4NS, UK

1. Introduction

This work is a continuation of our study of the efficiency of two well-known methods for the numerical integration of the equations of motion of planetary satellites together with the variational equations of the system. The methods are the 10th-order Gauss-Jackson backward-difference method described in [4, 6] and the Runge-Kutta-Nyström RKN12(10)17M [1].

In our recent work [3] we determined the step-size which achieves the best combination of speed and accuracy for the Gauss-Jackson method. The Runge-Kutta-Nyström method incorporates adaptive step-size control; for this method, we investigated the optimum strategy for local error control. In both cases, we used the two-body Saturn-Mimas system and the corresponding variational equations as our test problem.

Here we present further analysis of the above results and generalise our study to more complicated systems.

2. Study of the Gauss-Jackson method

In [3] we presented some approximate bounds for the behaviour of the partial derivatives of the coordinates with respect to the satellite's initial conditions in the two-body case. These results are confirmed by use of Floquet theory [2] which predicts linear growth in the amplitude of all the partial derivatives. However, we have shown that if the step-size of the Gauss-Jackson method is slightly increased, the system becomes unstable. This is in accordance with the findings of Milani & Nobili [7] for the instability of multistep methods.

It is necessary to emphasise that the instability is observed only in the partial derivatives and not in the satellite's coordinates, since the error in the energy integral at the end of the integration remains small. Furthermore, we have shown that the introduction of the corrector cycle into the integration of the partial derivatives eliminates the instability and allows the use of larger step-sizes.

Investigation of the relationship between step-size of the Gauss-Jackson predictor-corrector method and the accumulation of error in the energy integral shows that it is not possible to use a step-size much greater than $T/76$.

3. Study of the RKN12(10)17M method

For comparison with the Gauss-Jackson predictor-corrector method, we also considered the accumulation of the relative error in the energy integral with the mean step-size used by the RKN12(10)17M method when applied to the same system of equations. It was found that an acceptable relative error can be achieved with step-sizes as large as $T/30$.

4. Generalisation of results

The above experiments have been repeated using a more generalised force model which includes the oblateness of Saturn and the mutual perturbations of Mimas and Tethys. Two-body formulae for the partial derivatives cannot be applied to such a system, but the behaviour of the partial derivatives can still be predicted by Floquet theory when the orbits are at least quasiperiodic [5]. The stability of the two numerical integration methods was found to be the same as in the two-body case.

References

Dormand J.R., El-Mikkawy M.E.A. and Prince P.J., 1987, *IMA J. Num. Analysis* **7**, 423
Hadjidemetriou J.D., 1988, *Periodic Orbits and Stability*, ERASMUS ICP-88-0016-GR, Thessaloniki
Hadjifotinou K.G. and Harper D., 1995, *Astron. Astrophys.* **303**, 940
Herrick S., 1972, *Astrodynamics*, Vol. 2, ed. Van Nostrand Reinhold
Jorba A., Ramirez-Ros R. and Villanueva J. (Preprint from mp_arc@math.utexas.edu # 95-14.)
Merson R.H., 1974, RAE TR 74184
Milani A. and Nobili A.M., 1988, *Celest. Mechan.* **43**, 1

FROM TELESCOPE TO MPC: ORGANIZING THE MINOR PLANETS

B.G. MARSDEN
Harvard-Smithsonian Center for Astrophysics
Cambridge, MA 02138
U.S.A.

1. Introduction

Somewhat more than a century after its introduction for the purpose of discovering minor planets, photography is now rapidly giving way to the CCD as the technology of choice for observing these bodies. A CCD has been used in scanning mode in the University of Arizona's 'Spacewatch' program for the discovery of minor planets since as long ago as 1984 (Gehrels 1984, Gehrels *et al.* 1986), while a CCD in stare mode was first applied as a matter of routine to an established observing program for astrometric follow-up in 1989—that at the Oak Ridge Observatory in Massachusetts (McCrosky 1990). After its initial 1984–1986 success, Spacewatch was modified with the help of a larger CCD and improved computer software and with the adoption of the particular mission of searching for NEOs, or minor planets (and comets) that pass close to the earth (Rabinowitz 1991, Scotti 1994). The Oak Ridge program utilizes a 1.5-m reflector, and the first CCD observations were reduced using the *Astrographic Catalogue*, the mainstay of the Oak Ridge photographic program back to its inauguration in 1972, as well as of other older photographic programs in which the fields observed were significantly less than 1° across. Within months, the availability, on CD-ROMs, of the STScI *Guide Star Catalogue* (Villard 1989) effectively consigned the venerable *AC* to the scrap-heap, and the rapid development of ready-made and relatively inexpensive CCD systems (e.g., di Cicco 1992) has recently increased the volume of CCD astrometry considerably, allowing it to be conveniently and reliably carried out, even by amateur astronomers. At the present time, very nearly 50 percent of the astrometric observations, typically 6000, published each month in the Minor Planet Center's *Minor Planet Circulars* are obtained by means of a CCD.

2. The growth of discoveries

In an earlier review, prepared for the *Asteroids II* conference held early in 1988 (Bowell *et al.* 1989), there was some discussion of the exponentiality of the growth of discoveries of minor planets. Table I in that review showed that, except for one enormously high and four very low anomalies, the doubling time between 1854, when there were 2^5 known minor planets, and 1979, when the number that had been given designations reached 2^{15}, was consistently in the range 6–18 years. From recent patterns it was predicted that the number would not reach 2^{16} until 1993. In fact, this number was achieved in 1991, the latest doubling interval being close to the mean of all the doubling intervals.

At present the number of designations is 93 032. This is the total number of post-1925-style MPC provisional designations, augmented by those in the Palomar-Leiden and three Palomar-Leiden Trojan surveys, as well as by the 1053 objects that had received their permanent numbers prior to 1925—although the number needs to be reduced by 552 for designations that have been removed either before or after publication. This suggests that 2^{17} may be reached by the end of the century, but the change from photography to CCD and the possibility that some extensive 'Spaceguard' program (e.g., Morrison *et al.* 1992) will be initiated to search for NEOs render this very uncertain.

In any case, the count is becoming meaningless because of the new practice of generally not giving designations to objects observed on a single night. One of the proposals made in the *Asteroids II* paper was to downgrade the status of these 'one-night stands'. The initial plan was that they would simply be held over in anticipation that other observers might independently record the objects on two or more nights and thus justifiably claim credit for a more useful discovery. If there were no independent discoveries over four to six months, say, the one-night stands would then receive designations.

In practice, after 1988 this waiting period got longer and longer, and since mid-1992 it has become infinite. Single-night observations of Spacewatch discoveries have generally not been given designations since the start of this project, and they are currently being made at a rate *in excess of 60 000 per year!* All the single-night data do automatically remain in the MPC files, and from time to time they resurface as isolated additional observations of objects newly identified at more than one opposition.

3. Numbered minor planets

In the *Asteroids II* review there was the remark that "it is our premise that the aim of asteroid discovery is to augment the set of numbered asteroids, and that this set should comprise asteroids with orbits that are sufficiently well determined, not only to provide a satisfactorily detailed and reasonably unbiased delineation of the structure of the whole asteroid belt, but also to permit the ready observation of a specific member at any subsequent time." At that time there were 3774 numbered minor planets, 90 percent of which were routinely predictable to 4 arcsec or better; two of the numbered planets, (719) Albert and (878) Mildred, were lost. Now there are 6465 numbered minor planets, well over 95 percent of which are predictable to 2 arcsec or better, and (878) Mildred has been found (Williams 1991).

The mean doubling rate for the numbering of minor planets has been 15.7 years, although variations from the mean have been rather greater than for the discovery growth as a whole. The prediction that there would be 2^{12} numbered minor planets by 1989.8 required correction by -0.4 year, giving a doubling interval of only 11.2 years, the smallest since the 6.4 years for doubling from 2^5 to 2^6. Despite ever more stringent requirements for the numbering of a minor planet, it is now anticipated that the next doubling will take less than nine years, with 2^{13} perhaps being attained by the end of 1997. In the batch of *MPC*s for April 1995 there were a record 77 new numberings, followed by 76 more a month later.

The lion's share of the numbered minor planets has come from the photographic search program carried out since 1963 by N.S. Chernykh and his associates with the 0.4-m astrograph at the Crimean Astrophysical Observatory. With 832 numbered objects, the CrAO program took the lead from the old Heidelberg program, with 809 numbered objects since its inception in 1891, near the end of 1994. It was suggested (Marsden 1994a) that the numbered discoveries at Palomar (which consist mainly of the Palomar-Leiden surveys of Gehrels and the van Houtens with the 1.2-m Schmidt and the NEO patrols by Helin and the Shoemakers, mainly with the 0.46-m Schmidt) would move into first place "rather soon" after CrAO had surpassed Heidelberg. This has not yet happened, the Palomar score currently being 782, but it should be noted that, during the two years since that earlier review was prepared, Heidelberg's total has increased by 0.9 percent, CrAO's by 9.5 percent and Palomar's by as much as 34.4 percent.

After Palomar takes the lead, it is unlikely that there will be new contenders for some considerable time. Fourth-place Lowell Observatory is considerably behind the first three, and essentially all of the principal photographic patrols have now terminated or are in the process of winding down their operations. The Heidelberg and Lowell programs ended in 1959 and 1988, respectively, and there has been little activity from the CrAO program since 1990. The various Palomar-Leiden surveys took place during 1960–1977, the Shoemaker program stopped in late 1994, and Helin is curtailing her program in favor of developing a CCD survey in Hawaii.

The leading nineteenth-century visual search program had been surpassed by the Heidelberg photographic program after the latter had been in operation for little more than a decade, and within another five years more numbered minor planets had been discovered photographically than visually. However, the influence of the CCD revolution is—at least for the present—less extreme. The small fields of the current CCD programs, even Spacewatch, mean that new numberings are going to be dominated by the backlog of identifications of older photographic discoveries for some considerable time to come. There are already 4792 multiple-opposition objects, almost all of them photographic discoveries, that can be numbered when enough additional observations have been made that their orbit solutions are considered to qualify. At present there are 33 CCD discoveries among the numbered minor planets—still a small number, but one that has shown a fourfold increase during the past two years.

4. Qualification for numbering

The introduction of a system of provisional designations in 1892 was intended simply to relieve the pressure caused by the rapid increase in discovery rate brought about by the introduction of photography. Almost all of the objects designated did get numbered, only a short time later, and usually even in the order the provisional designations had been applied. Occasionally the number was given before it was appreciated that the object was already known and had a number, so the new number was then simply reused. From time to time an identification with an earlier object, numbered or unnumbered, was recognized while the new object still just had a provisional designation, but in general it was felt that an object could be numbered if there were an orbit solution from observations made over a month or two. Interestingly, the orbits of the three famous objects of the 1930s that were found to cross the orbit of the earth were not judged good enough to permit numbering. Earnest attempts were made to recover numbered minor planets at their next favorable oppositions, but this became progressively more difficult, and by the time of World War II some 6 percent of the 1500 or so numbered objects had *still* been observed *only* at their discovery oppositions, often decades earlier.

On establishing the MPC in 1947 then-director Paul Herget declared a moratorium on new numberings, while efforts were directed instead to reobserving those already numbered. Numbering was resumed after the IAU

General Assembly in 1948, when it was resolved that objects observed at a single opposition could be numbered only if they passed closer to the sun than the aphelion of Mars, but the orbit solutions for Apollo, Adonis and Hermes were still considered too weak to allow them to qualify. Actually, the concession for Mars crossers was made only three times—for (1565) Lemaître, (1566) Icarus and (1580) Betulia—because Herget wanted to ensure that further numbered minor planets would never be lost, an aim the MPC has so far completely met. Largely unwritten further tightening of the rules later meant that two oppositions became the minimum for a Mars crosser, while for more distant objects the basic requirement was usually three good oppositions or two good and two poor oppositions.

In 1991 the MPC adopted a more quantitative guideline for deciding when a minor planet should be numbered. Among the most important components in the decision are the number N of oppositions at which observations are utilized in the orbit solution and the total span of time covered by them. This time span is represented by the number M of completed decades between the first and last observations, although isolated observations on a single night at the first opposition are ignored, and M is restricted to a maximum of 4. Next it is necessary to consider the distribution of the observations at each opposition. This is done by assigning, for each opposition, an integer between 0 (for single-night observations) and 8 (essentially for observations on two or more nights in each of three lunations); the sum of the scores at the individual oppositions is denoted by L. The distribution of observations *around* the orbit is then considered by looking for the longest consecutive run of calendar months in which there are no observations at any of the oppositions used in the computation of M; the relevant quantity K ranges between 0 and 6, the extremes being the cases when all the observations are in the same month and when there are observations in every month. The sum $J = K + L + M + N$ is then formed. Numbering is considered if *all* of the following are true: (a) the residuals are reasonable, the determination of J of course being modified if critical observations are not good enough for inclusion in the orbit computation; (b) the latest observations in the solution are no more than two years old; (c) the contribution to L at the latest opposition is 4 or more; (d) $N \geq 4$ (although $N = 3$ is allowed if *each* contribution to L is 6 or more, and $N = 2$ is allowed if, in addition, the perihelion distance < 1.3 AU); and (e) $J \geq 21$, critical discussion being necessary if $J < 23$.

The above criterion has the advantage that the initial selection of objects for consideration for numbering can be made automatically in the absence of orbital solutions. Muinonen and Bowell (1993) devised a more sophisticated criterion, based on Bayesian probability, which has a generally similar outcome, except for objects that pass very close to the earth, and they did not consider that the latest observations should have been made quite recently. They did remark that they did not think (5209) 1989 CW_1 should have been numbered. This object was a Trojan that had been well observed at three oppositions and scored $J = 23$. The point is that the observations therefore covered only one-sixth of an orbital revolution, and the motion can thus scarcely be considered very extrapolable. This was known to be a weakness of the MPC system, but if the Muinonen-Bowell criterion is followed, more than a century will elapse before any of the transneptunian objects found during the past few years can be numbered!

Nevertheless, the MPC has recently supplemented its criterion to address the Trojan case more satisfactorily. It has done this by equating the formal mean error in the determination of the orbital semimajor axis to the error in the position in the heliocentric orbit after one decade, this being expressed by another integer U on a logarithmic scale such that an error of less than 1 arcsec gives $U = 0$ and that $U = 9$ means that the error approaches half a revolution (and more). Numbering is considered only if $U \leq 2$. Further, if $N < 4$, numbering requires $U \leq 1$. This effectively restricts three-opposition numberings to NEOs, a very occasional one qualifying at only two oppositions.

It is useful to introduce checks from time to time to ensure standards in the set of numbered minor planets. A "critical list" is maintained of objects that have not been observed during the past ten years or at fewer than four oppositions in all. Two recent categories added in the version issued by the MPC indicate objects observed on only a single night during the past ten years and objects for which new orbit solutions show that they would fail the current stringent numbering standards.

5. Processing the observations

Although photography may be on the way out, some of the photographic surveys for minor planets provide excellent examples of efficient ways to discover and follow-up main-belt minor planets.

Ideally, wide-field exposures are made of an ecliptic field on two neighboring nights shortly before opposition. Images of the same minor planets are recognized on the two nights and their measurements e-mailed to the MPC utilizing a unique observer-defined designation for each object. It is not necessary for the observer actually to identify the object. Obvious identifications are automatically made at the MPC, and orbit improvements are computed as necessary. Unidentified two-night detections are given new MPC designations, and the discovery information is recorded.

Väisälä orbits are next computed and filed as convenient representations of these objects, available for use in processing the next set of observations to be reported. Such orbits, which involve the assumption that the objects are at perihelion (or at aphelion) can be established for a variety of presumed distances, although it is reasonable to adopt one suitably representative example for each object. The concept, devised for use with the old observing program at Turku (Väisälä 1939), is arguably the most important twentieth-century aid to the study of unperturbed orbits. The night-pairs of unidentified objects can be compared with all the existing orbits (even quite poor orbits) computed from observations at single oppositions in the past, searching for cases where a simple shift in the mean anomaly gives a close match. Attempts to verify these suggested identifications can then be made, with genuine new multiple-opposition orbits thereby emerging for probable eventual numbering.

The observer then obtains similar exposures on two nights a month later of the appropriately shifted field, makes measurements of all minor planets found and communicates them as before. Although many of the objects recorded the previous month are expected to be in the field (and at about the same brightness), the observer should again only indicate which observations belong to particular objects observed on the two nights. The MPC makes the identifications using the orbits in its files, including now the Väisälä orbits computed the previous month. Possible matches between the observations of new objects in the two months can be tested for linkage. The resulting single-opposition orbits are then tested against all the remaining night-pairs of unidentified *objects* at previous oppositions, again looking for matches corresponding to a simple shift in mean anomaly. Possible identifications are then examined in detail, resulting new multiple-opposition orbits again going into the file. Remaining unlinked night-pairs are compared with the one-opposition orbit database as before.

The above procedure of what might be termed *contingent follow-up* is elegant in its efficiency and pre-planning. Of course, some potential discoveries are lost, because they happen to be outside the field, either on the confirming nights of the pairs, or from one month to the other. This is particularly true when the observations are made with small-field telescopes. For small-field work, notably that nowadays involving CCD observations, new discoveries are—in the absence of identifications found using the orbit files—specifically followed up on several nights throughout the first month and into the second. A good single-opposition orbit should then be available for possible identification with past night-pairs of observations. This *targeted follow-up* may be more likely to lead to success with some particular object, but it is generally less efficient, given that the aim is to produce multiple-opposition orbits for eventual numbering. Targeted follow-up is further necessary to record well past opposition the objects for which the month-arc orbits do not yield identifications. If the arc over which an object is observed extends over the several months to evening quadrature (bearing in mind that there will be fading during this time), there is the chance that further targeted observations can directly yield the recovery of the object at a future opposition.

6. Near-earth objects

Discoveries of NEOs have traditionally been subject to targeted follow-up. This is understandable, because most of these NEOs were already quite near the earth at discovery, and their apparent motions thus tended to be quite different from those of other minor planets in the field. The proposed Spaceguard survey (Morrison et al. 1992) was distinctive in that it was designed to use large-aperture, large-field telescopes to find NEOs near opposition in the outer parts of their orbits (where they spend most of their time) and to rely mainly on contingent follow-up.

Recently, the Shoemaker Committee, charged by the U.S. Congress to update the Spaceguard proposal, has pointed out that the forthcoming availability of large-format, high-quantum-efficiency, fast-readout CCDs is making it feasible to carry out the Spaceguard task more quickly by means of monthly all-sky surveys for NEOs with a relatively small telescope in each hemisphere (Shoemaker et al. 1995). Certainly, all-sky coverage will make it possible to record NEOs that have their aphelia at or only slightly beyond the earth's orbit. By concentrating on opposition searches, it seems likely that objects with aphelia at 1.2 AU have been undersampled by some 10 percent (Marsden 1995). The disadvantage is that all the follow-up, and probably even confirmation of the initial detection, will have to be targeted. This means that there must be extensive coordination of the activities, and the total communication effort could be quite horrendous.

A useful modification of the two-night contingent confirmation and follow-up procedure discussed in the previous section is to observe long before and long after midnight on one of the nights in order to get some parallax information (Marsden 1992). To attempt parallactic observations may be fine near opposition, but in an all-sky survey much of the sky cannot be observed for very long each night. An added problem with observations far from opposition is that there are likely to be two completely different solutions to the orbit-determination problem. Although an NEO enthusiast might be tempted to select the solution that puts the object moderately far away with, say, its aphelion at or near the orbit of the earth, chances are that the correct solution is the one that makes it a perfectly ordinary minor planet in the main belt and even farther away (Marsden 1991, 1995).

Muinonen and Bowell (1992) have addressed the special problem of orbital uncertainty when one is attempting to recover an NEO observed at only a single opposition. As briefly noted elsewhere (Marsden 1994b), their procedure suggests that the object 1989 ML could be found in 1992 within about 1° of the nominal prediction. Searches for the object proved fruitless. In fact, an apparent new discovery of an NEO, designated 1992 WA, turned out to be 1989 ML, but it was some 28° from the prediction!

TABLE 1. 1989 ML = 1992 WA

1989	UT	R.A.	Decl.	R.A.	Decl.
June	6.39	0.3−	0.4+	0.1−	0.4 +
	6.41	0.3+	0.4−	0.0	0.4 −
	29.38	1.6−	0.8+	0.7−	1.9 +
	29.41	0.3+	3.0−	0.6+	2.0 −
	30.37	0.1−	1.3+	0.6−	0.2 −
	30.40	2.0+	1.7+	0.8+	0.1 +
July	3.34	1.3−	0.6+	(7.1−	9.0−)
	3.37	0.6+	1.3−	(5.9−	11.0−)

Table 1 shows the (O−C) residuals of the 1989 observations, which were pairs made on four nights. The first set of residuals is that of a "standard" fit, treating each observation of equal weight. The 3″ declination residual on June 29.41 is not unusual for photographic measures of trailed images of an NEO on the night of its discovery. Straight application of the Muinonen-Bowell theory would be naïve, however, for while it may seem fortunate to have the June 6 prediscovery observations, they are quite isolated in time from the others. From the point of view of error assessment, it would therefore be quite appropriate completely to discount them. In that case there would be only a four-night arc of observations, and it would be inappropriate to expect the object to be found in 1992.

The second set of residuals in Table 1 is from a fit to the 1989 and 1992 observations combined. This fit shows that the problem was not in fact with the June 6 observations, but with those on July 3, at the end of the four-day post-discovery arc. This outcome may seem a little unexpected, for these observations were not really isolated in time. But this is again a symptom of the isolation of the June 6 measures, even though those measures turned out to be very good. Independent observations on a second night in early June would immediately have shown *something* to be wrong, and with only a little experimentation, attempts to fit those early-June data with the observations on two of the nights during June 29–July 3 would have shown, not only that the problem was on July 3, but also have provided confidence that the solution from the other night-pairs was dependable.

7. Transneptunian objects

The transneptunian objects (TNOs) discovered since 1992 have been no brighter than 22nd magnitude, and the difficulty of obtaining the necessary follow-up in order to determine definite orbital solutions is extreme. The first two years of this effort, involving 13 objects, has been reviewed elsewhere (Marsden 1994c). While bearing in mind that there is always the possibility that some of the transneptunians are objects in unstable planet-crossing orbits detected near aphelion, it seemed that the discoveries were almost equally divided into two groups. One group consisted of low-eccentricity, low-inclination objects with semimajor axes a = 42–46 AU, their essential orbital stability ensured by the impossibility of approaches within 10 AU of Neptune. The second group seemed to contain objects discovered much closer to Neptune's orbit but prevented from encountering Neptune, again generally within 10 AU, because their orbits are librating about the 2:3 resonance with Neptune—i.e., they are like Pluto and have a = 39–40 AU; some of these orbits have rather moderate eccentricities and inclinations, up to 0.32 (for 1993 SB) and 17° (for Pluto), and approaches within only some 8 AU of Uranus are possible.

Table 2 lists the 28 TNOs discovered between mid-1992 and mid-1995. H_R is the absolute red magnitude, and the next six columns give the usual orbital elements (in standard notation) for the common epoch 1995 Mar. 24. The column labeled 'Arc' shows the span in days between the first and last observations, an asterisk indicating that there are observations at additional oppositions. The general pattern noted previously still holds, although the fraction suspected as being 2:3 Neptune librators is now somewhat less than half; only one more multiple-opposition object, 1994 JR$_1$, has been added to the three previously established (more or less) as being in the resonance.

A few more features have developed in the distribution. One is represented by 1995 DA$_2$, an object found opposite Neptune but only just beyond its orbit. It was immediately recognized that at this could not be a stable 2:3 librator, so the possibility of 1:2, 3:4 and even 3:5 librations was considered instead. Although there can be no proof from an orbit with an orbital arc of only 37 days, a 3:4 solution was adopted that seems to keep the object more than 10 AU from Neptune. Rather more conclusive is the case of 1994 JS, initially suspected as being a 2:3 librator, a possibility that has clearly been *disproven* by the observations in 1995. It seems very probable that 1994 JS is instead a 3:5 librator, its minimum distance from Neptune being 10 AU. An attempt at forcing 1995 DA$_2$ to be a 3:5 librator gave a similar minimum distance.

Of the 13 TNOs discovered by mid-1994 two, 1993 RP and 1994 JV, failed to be observed at their next opposition. Both have been considered

TABLE 2. Transneptunian objects

Object	H_R	M	ω	Ω	i	e	a	Arc	MPC
1995 DA_2	7.6	309.38	62.35	127.49	6.58	0.1156	36.3447	37	25184
1993 RO	8.0	357.03	187.46	170.30	3.72	0.1982	39.2782	*	24241
1993 RP	9.0	2.22	180.64	192.09	2.57	0.1135	39.3289	2	23493
1993 SB	7.5	319.73	79.12	354.81	1.93	0.3216	39.3897	*	24408
1994 JV	7.0	1.25	179.99	28.15	16.48	0.1247	39.4570	25	unpub.
1995 GA_7	7.2	63.86	100.20	20.97	3.54	0.1192	39.4551	2	25186
1993 SC	6.3	33.90	319.22	354.64	5.16	0.1795	39.4708	*	24763
1995 KK_1	7.8	21.54	328.35	228.09	9.25	0.1898	39.4748	1	25315
1995 HM_5	7.9	3.26	354.78	186.70	4.60	0.1775	39.5337	33	25315
1994 TB	6.5	31.41	340.17	316.89	11.98	0.2917	39.5656	37	25184
1994 JR_1	7.1	5.73	91.98	144.72	3.80	0.1277	39.8261	*	25341
1994 TH	6.9	0.7	356.62	12.12	16.07	0.0	40.9404	3	24084
1995 GY_7	7.5	0.0	203.47	34.50	0.94	0.0	41.3469	1	25408
1994 TG	6.7	0.6	353.02	15.51	6.76	0.0	42.2544	3	24084
1995 FB_{21}	7.5	0.0	209.63	28.36	0.68	0.0	42.4259	9	25408
1994 TG_2	7.7	0.4	358.86	353.28	2.25	0.0	42.4479	32	24884
1994 JS	6.8	323.68	238.18	56.33	14.03	0.2343	42.8408	*	25341
1995 GJ	6.5	359.96	180.29	338.93	22.93	0.0909	42.9072	1	25185
1994 EV_3	6.8	165.26	13.81	19.51	1.68	0.0381	43.0703	*	25228
1995 KJ_1	6.1	359.8	180.40	47.96	3.80	0.0	43.2345	1	25315
1994 VK_8	5.9	0.3	349.03	72.83	1.43	0.0	43.4504	86	24715
1995 DB_2	6.4	0.10	0.16	128.64	4.27	0.0671	43.4935	34	25031
1994 GV_9	6.8	20.84	335.24	176.85	0.55	0.0420	43.7308	*	25228
1993 FW	6.6	322.72	43.26	187.92	7.74	0.0521	43.8644	*	25227
1992 QB_1	6.7	5.80	357.35	359.40	2.19	0.0698	43.9389	*	24408
1994 JQ_1	6.5	334.80	211.40	25.65	3.74	0.0267	44.1922	*	25228
1995 DC_2	6.0	0.1	358.78	154.29	2.11	0.0	45.2076	42	25184
1994 ES_2	7.6	279.84	99.95	154.73	1.05	0.1333	45.9596	*	25341

as 2:3 librators, although, following recognition of the altered status of 1994 JS, searches—the outcome of which is still not entirely conclusive—for 1994 JV were also made in 1995 on the assumption that its orbit is in 3:5 resonance. A recovery of the 14th TNO, 1994 TB, has recently been made (Gladman 1995), but further observations will be needed before one can state whether this object is a 2:3 librator, a 3:4 librator or has an orbit that is unstable.

Table 2 shows two weak circular orbits around $a = 41$ AU. It is likely that these objects, 1994 TH and 1995 GY_7, are "outer group" objects near perihelion. Duncan *et al.* (1995) have expressed surprise that there seem not to be objects in non-resonant, low-eccentricity orbits with $a = 37\text{--}38$ AU that could be stable. Certainly, no objects have yet been established as in such orbits, although it is not impossible that 1995 GA_7 is a candidate.

References

Bowell, E., Chernykh, N.S. and Marsden, B.G., 1989. Discovery and follow-up of asteroids. In *Asteroids II*, eds. R.P. Binzel, T. Gehrels and M.S. Matthews, University of Arizona Press, Tucson, pp. 21–38.

di Cicco, D., 1992. The ST-6 Imaging Camera, *Sky Tel.*, **84**, 395–401.

Duncan, M.J., Levison, H.F. and Budd, S.M., 1995. The dynamical structure of the Kuiper Belt, in press.

Gehrels, T., 1984. Observations made with the Spacewatch Camera 0.91-m telescope at Kitt Peak, *Minor Planet Circ.*, No. 9198.

Gehrels, T., Marsden, B.G., McMillan, R.S. and Scotti, J.V., 1986. Astrometry with a scanning CCD, *Astron. J.*, **91**, 1242–1243.

Gladman, B., 1995. 1994 TB. *Minor Planet Electronic Circ.*, 1995-M07.

Marsden, B.G., 1991. The computation of orbits in indeterminate and uncertain cases, *Astron. J.*, **102**, 1539–1552.

Marsden, B.G., 1992. Comments on search programs for near-earth objects. In *Observations and physical properties of small solar system bodies (30th Liège Internat. Astrophys. Colloq.)*, eds., A. Brahic, J.-C. Gérard and J. Surdej, Université de Liège, pp. 251–252.

Marsden, B.G., 1994a. Asteroid and comet surveys. In *Astronomy from Wide-Field Imaging (IAU Symposium No. 161)*, eds. H.T. MacGillivray, E.B. Thomson, B.M. Lasker, I.N. Reid. D.F. Malin, R.M. West and H. Lorenz, Kluwer Academic Publishers, Dordrecht, pp. 385–399.

Marsden, B.G., 1994b. Astrometric observations of minor planets and comets: present and future needs. In *Galactic and Solar System Optical Astrometry*, eds. L.V. Morrison and G.F. Gilmore, Cambridge University Press, pp. 263–275.

Marsden, B.G., 1994c. Searches for planets and comets. In *Inventory of the solar system (Astron. Soc. Pacific Conference Series)*, in press.

Marsden, B.G., 1995. Overview of orbits. In *International conference on Near-Earth Objects*, ed. J. Remo, N.Y. Academy of Sciences, New York, in press.

McCrosky, R.E., 1990. 801 Oak Ridge, *Minor Planet Circ.*, No. 15630.

Morrison, D. *et al.*, 1992. *The Spaceguard Survey*, NASA, Washington.

Muinonen, K. and Bowell, E., 1992. Orbital uncertainties for earth-crossing asteroids, *Bull. Am. Astron. Soc.*, 24, 942.

Muinonen, K. and Bowell, E., 1993. Asteroid orbit determination using Bayesian probabilities, *Icarus*, **104**, 255–279.

Rabinowitz, D.L., 1991. Detection of earth-approaching asteroids in near real time, *Astron. J.*, **101**, 1518–1529.

Scotti, J.V., 1994. Computer aided near earth object detection. In *Asteroids, Comets, Meteors 1993 (IAU Symposium No. 160)*, eds. A. Milani, M. di Martino and A. Cellino, Kluwer Academic Publishers, Dordrecht, pp. 17–30.

Shoemaker, E.M. *et al.*, 1995. *Report of the near-earth object survey working group*, NASA, Washington.

Väisälä, Y., 1939. Eine einfache Methode der Bahnbestimmung, *Mitt. Sternw. Univ. Turku* No. 1.

Villard, R., 1989. The world's biggest star catalogue, *Sky Tel.*, **78**, 583–589.

Williams, G.V., 1991. (878) Mildred, *IAU Circ.*, No. 5275.

NUMERICAL MEAN ELEMENTS FOR ASTEROID ORBITS

ANNE LEMAITRE
Département de Mathematique - FUNDP
8 Rempart de la Vierge B5000 Namur Belgique
E-mail: alemaitr@math.fundp.ac.be

1. Introduction

In the context of the search of asteroid families (i.e. identification of minor planets as potential fragments of an old bigger body), the calculation of proper elements plays an important role. They are quasi-invariants of the motion, obtained by a double averaging process of the restricted N-body problem; firstly the osculating elements are averaged over the short periodic terms (namely the longitudes of the asteroid and of the perturbing planets) so to get the *mean elements*, and secondly, the mean elements are averaged over the long periodic terms (longitudes of the pericenters and of the nodes of the asteroid and of the perturbing planets) to obtain the *proper elements*.

The calculation of proper elements for asteroid orbits presently refers to two different computations; a powerful analytical theory, developed by Milani and Knežević (1990, 1992, 1994) and a semi-numerical approach due to Lemaitre and Morbidelli (1994). The analytical theory, based on series expansions in eccentricity and inclination, is particularly suitable for low inclination and low eccentricity orbits, while the semi-numerical method is convenient for the orbits with either large eccentricities or large inclinations. The overlapping region has been tested and is situated about 17° in inclination (see Knežević et al 1994).

Up to now even the semi-numerical approach has used the analytical mean elements as initial conditions for the computation of the proper elements; those mean elements, also obtained by series expansions by Milani and Knežević, could be sources of errors in the calculation of the proper ones for highly inclined or eccentric orbits.

The purpose of this paper is to compare, at least for the numbered asteroids, the *analytical* mean elements with a corresponding set of *numerical* mean ones. For significative differences in the results, proper elements are then recalculated with the numerical initial conditions and compared to their previous values.

2. Numerical integration

We perform a numerical integration by ORBIT8V (implemented by Milani) using the digital filtering techniques (Carpino et al 1987) to average over the short periodic terms. Because of the presence of high eccentricities and high inclinations in our data set, the fixed step integrator has been replaced by a variable step integrator, namely RA15 (Everhart 1985). The integration of 6158 asteroids (1994 catalogue) covers 2000 years and takes into account the perturbations of Jupiter, Saturn, Uranus and Neptune on the motion of the minor planet. Outputs are listed every 100 years and by choice of a filter $n°100$, the filtered data, corresponding to the elimination of the periods between 2 and 200 years, are available between 600 and 1400 years. The reference epoch has been arbitrarily chosen at 1000 years.

To test the validity of the averaging process a triple index (p_a, p_e and p_i) is given for each planet, corresponding to the RMS values of a constant fit of the filtered numerical values of the semi-major axis a and of quadratic fits for the filtered values of the eccentricity e and sine of the inclination $\sin i$. Let us remind that the elimination of the longitudes in the Hamiltonian function theoretically corresponds to a constant Delaunay momentum $L = \sqrt{\mu a}$, which leads to a constant value of a; the mean Hamiltonian so obtained should contain, as main terms in a series expansion, a quadratic term in e and in i, which is checked by two quadratic fits for those elements. The RMS is given on a logarithm scale with respect to 10^{-5}; for example, $p_a = -1$ means that the RMS value of a with respect to a constant is about 10^{-6} and $p_e = +2$ means that the RMS value of e with respect of a quadratic function is about 10^{-3}. This allows the user to extract several subsets from the main data set, by changing the precision of each of the three mean elements. Results about the triple index are given in table 1

3. Comparisons with analytical mean elements

With the result of the numerical integration given at the epoch 1000 as initial condition, an analytical mean element has been calculated by means of SHORT6.3 software. As well as for the numerical mean elements a quality code is given (Milani and Knežević 1994), indicating special situations such as closeness to the mean motion resonances or extremal values of the semi-major axis. 4263 asteroids have a quality code 0; combining the two criteria ("good" triple index and "good" analytical quality code) 4023 planets are retained for the comparison.

NUMERICAL MEAN ELEMENTS

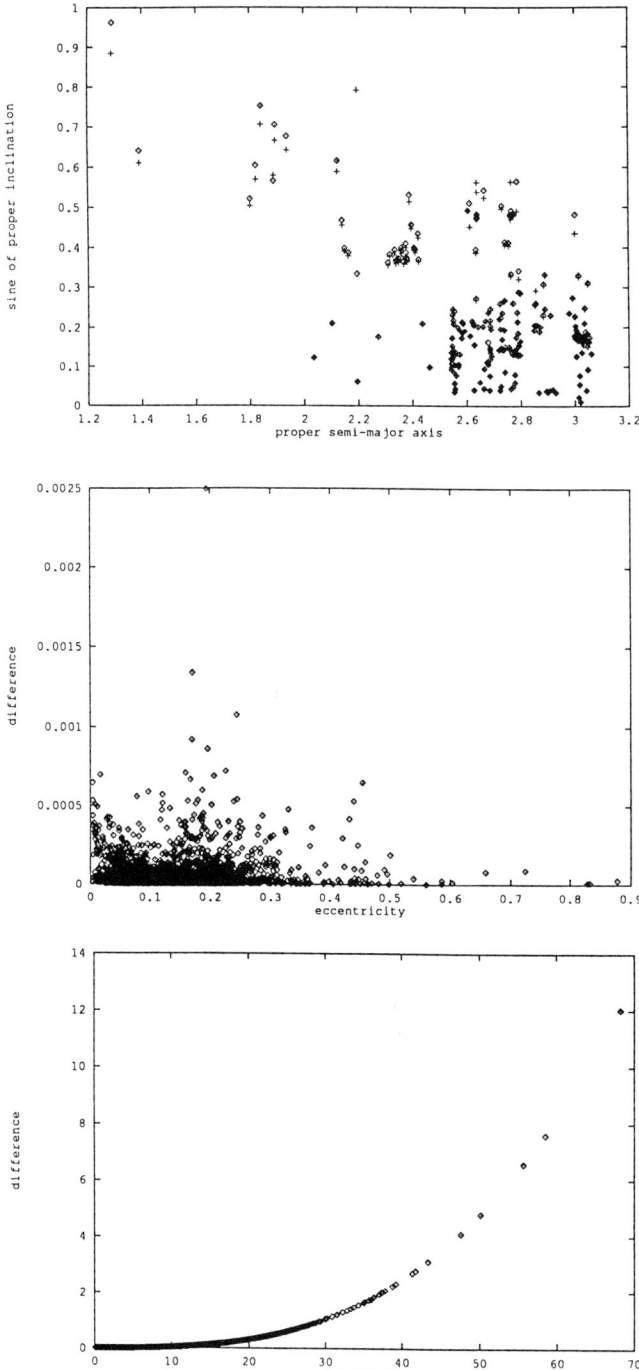

Figure 1. Differences in semi-major axis, eccentricity and inclination (in degrees) for comparable (good quality codes for both elements) analytical and numerical mean elements (4023 planets are plotted)

TABLE 1. Number of asteroids corresponding to a given value of the index, respectively for a, e and i; for the extremal values, the asteroid is identified

Element	-4	-3	-2	-1	0	1	2	3	4
a	1 (2340)	11	127	807	4200	754	163	91	4 (944,5145) (5201,6144)
e	0	0	4 (2100,2102) (4034,4232)	259	5001	563	256	75	0
$\sin i$	0	0	103	1518	4038	458	40	1 (5201)	0

The figures 1 give the differences observed in semi-major axis, eccentricity and inclination (in degrees) between the two sets of mean elements; it is obvious that the differences in a and e are much smaller than in i, where the curve is a quadratic function of the inclination. Even in the interval between 15° and 30° the differences exceed 1° i.e. 10^{-2} for $\sin i$. This result confirms the now classical result obtained by Kozai (1962) who showed the difference of topology between the mean phase space in the circular restricted problem between low and high inclinations; the analytical mean phase space does not take this peculiar topology (with apparition of a critical curve) into account (see Lemaitre and Morbidelli 1994 for a larger discussion).

4. Consequences for proper elements

Out of the 4023 planets, 279 have been isolated, showing either differences in a larger than 10^{-4}, or differences in e larger than $5\,10^{-4}$ or differences in i larger than 1°. Corresponding proper elements have been calculated for both sets of mean elements and the results are plotted in figure 2 for the inclinations and in figure 3 for the frequencies.

A systematic shift in the frequency space as well as in the (a, i) space can be noticed; it means that for family determination the two sets of proper elements are equivalent, because only the mutual distances between the planets are important; for other purposes, it is obvious that differences of a few arcseconds in the frequencies could be very important, in particular for the location or the proximity of the secular resonances.

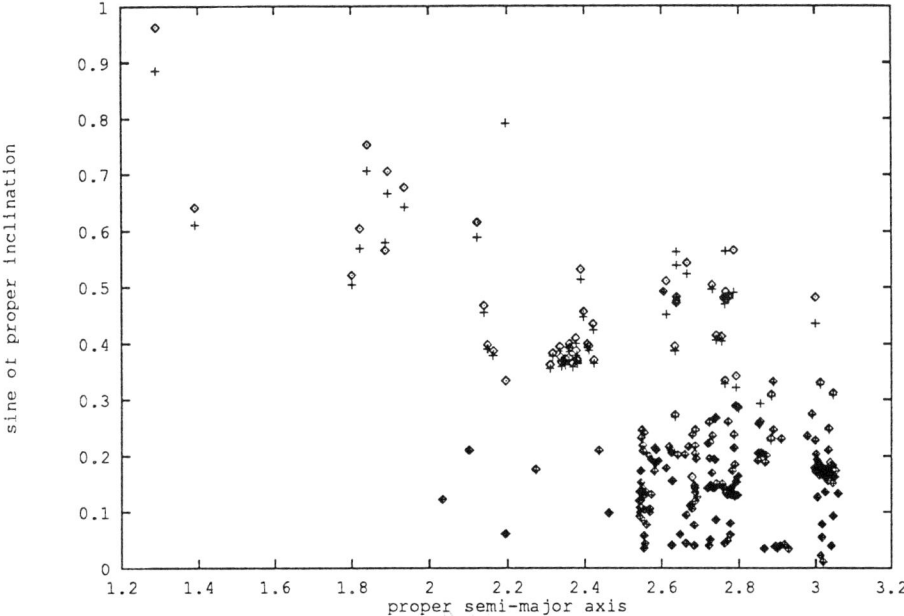

Figure 2. Differences between the proper inclinations calculated from analytical mean elements (+) and the proper inclinations calculated from numerical mean elements (◊) for 279 planets

5. Conclusions

For inclinations lower than 15°, analytical mean elements are equivalent to the numerical ones whatever the semi-major axis or the eccentricity could be; for mean inclinations between 15° and 30°, differences are of the order of 1°; this difference is reported in proper inclination in a linear way, which allows the user to predict the difference in proper elements, knowing the difference in mean elements. For inclinations higher than 30°, numerical mean elements should be used, but in a separate file, the worst thing being to mix up two sets of data. If the reader is interested in some result of the present paper, please contact the author by Email or check the public domain database ftp.dm.unipi.it, where osculating, mean and proper elements are available.

Acknowledgements

I would like to thank Andrea Milani for useful comments and discussions about this contribution, as well as for "public domain" software ORBIT8V which I have intensively used for these results.

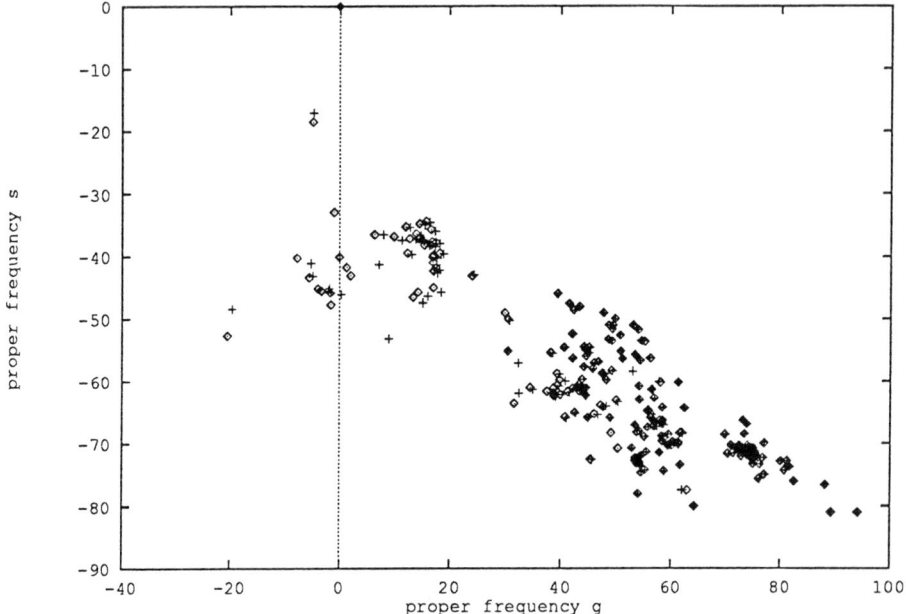

Figure 3. Proper frequencies g and s, given in arcseconds per year, calculated from analytical mean elements (+) and the proper frequencies calculated from numerical mean elements (◇) for 279 planets

References

Carpino, M., Milani, A., Nobili, A.: 1987, "Long term numerical integrations and synthetic theories for the motion of the outer planets", *Astron. Astroph.*, 181, 182–194

Everhart, E.: 1985, "An efficient integrator that uses Gauss-Radau spacings", *Dynamics of Comets: their origin and evolution*, Eds A. Carusi and G.B. Valsecchi, D. Reidel, Dordrecht, 185–202

Knežević Z., Froeschlé Ch., Lemaitre A., Milani A., Morbidelli A.: 1994, "Comparison of two theories of asteroid proper elements", *Astron. Astroph.*, 293, 605–612

Kozai, Y.: 1962, "Secular Perturbations of asteroids with high inclinations and eccentricities", *Astron. J.*, 67, 591–598

Lemaitre A., Morbidelli A.: 1994, "Calculation of Proper Elements for high inclined asteroidal orbits", *Celest. Mech.*, 60, 29–56

Milani, A., Knežević, Z.: 1990, "Secular Perturbation Theory and Calculation of Asteroid Proper Elements", *Celest. Mech. and Dyn. Astr.*, 49, 347–411

Milani, A., Knežević, Z.: 1992, "Asteroid proper elements and secular resonances", *Icarus*, 98, 211–232

Milani, A., Knežević, Z.: 1994, "Asteroid proper elements and the dynamical structure of the asteroid belt", *Icarus*, 107, 219–254

ON THE DYNAMICS OF TROJAN ASTEROIDS

B. ÉRDI
Department of Astronomy
Eötvös University, Budapest, Hungary

Abstract. The author's theory of Trojan asteroids (Érdi, 1988) is developed further. The motion of the Trojans is considered in the framework of the three-dimensional elliptic restricted three-body problem of the Sun-Jupiter-asteroid system including also the secular changes of Jupiter's orbital eccentricity and the apsidal motion of Jupiter's elliptic orbit. An asymptotic solution is derived, by applying the multiple-timescale method, for the cylindrical coordinates of the asteroids and for the perturbations of the orbital elements. This solution is used for the analysis of the long-time dynamical behaviour of the perihelion and the eccentricity of the Trojans.

1. Introduction

The Trojan asteroids have very complex dynamical behaviour, and the need for understanding their motions has stimulated many research efforts. In a series of papers Garfinkel (1985, and references therein) studied the planar motion of the Trojans in the model of the circular restricted three-body problem using the Hori-Lie perturbation method. His theory was extended to the spatial case by Zagretdinov (1986). Significant acheivements were obtained by Bien and Schubart (1987) who derived proper elements for 41 Trojans: proper eccentricity, proper inclination, and a parameter which characterizes the amplitude of libration around the Lagrangian points L_4 or L_5. Milani (1993) extended this work and computed proper elements for 174 Trojans by integrating the orbits for 1 million years and considering the perturbations of the four outer major planets. The proper elements, being almost constant for a very long time, allowed Milani to search for families of Trojan asteroids and also to establish the dynamical structure of the Trojan cloud. Despite all these efforts many problems of the Trojans have remained open. Such unsolved problems are (Milani, 1994): the origin of chaos in the orbits of Trojans, the question of the stability boundaries of the Trojan cloud, the origin of the orbital distribution (high inclinations) of the Trojans.

Unfortunately, existing analytical theories cannot give answers for these questions, they only contribute to the understanding of some basic features of the dynamics of the Trojans. I developed an approximate analytical theory (Érdi, 1981, 1984, 1988) which has been useful in analyzing some long periodic perturbations. This theory is based on the model of the three-dimensional elliptic restricted three-body problem of the Sun-Jupiter-asteroid system with Jupiter moving in a fixed elliptic orbit. Since in the motions of the Trojans there are important perturbations coming from the other major planets too, it would be necessary to develop an analytical theory which accounts for these perturbations. In this paper a step is made in this direction, a part of the effects of the other planets is taken into consideration. Namely, it is assumed that the orbit of Jupiter is secularly changing due to the perturbations of the other planets. Thus the motion of the Trojans is considered in the framework of the three-dimensional elliptic restricted three-body problem of the Sun-Jupiter-asteroid system with Jupiter moving in an elliptic orbit whose eccentricity and perihelion are secularly changing.

2. A Theory of Trojan Asteroids

The following equations of motion can be derived for the Trojans:

$$\frac{d^2r}{dv^2} - r\left(\frac{d\alpha}{dv}\right)^2 - 2r\frac{d\alpha}{dv} - 2\beta r\left(\frac{d\alpha}{dv}+1\right) = \frac{1}{1+e_J\cos v}\left[r - \frac{1-\mu}{R_1^3}r + \mu\left(\frac{\cos\alpha - r}{R_2^3} - \cos\alpha\right)\right],$$

$$\frac{d}{dv}\left(r^2\frac{d\alpha}{dv}+r^2\right) + 2\beta r\frac{dr}{dv} = \frac{\mu r\sin\alpha}{1+e_J\cos v}\left(1-\frac{1}{R_2^3}\right), \qquad (1)$$

$$\frac{d^2z}{dv^2} + z = \frac{z}{1+e_J\cos v}\left(1-\frac{1-\mu}{R_1^3}-\frac{\mu}{R_2^3}\right),$$

$$R_1 = (r^2+z^2)^{1/2}, \quad R_2 = (1+r^2-2r\cos\alpha+z^2)^{1/2}.$$

Here r and α are the polar coordinates of the asteroid in the orbital plane of Jupiter (r is the distance from the Sun, α is the angle between the asteroid and Jupiter), z is the perpendicular distance of the asteroid from the orbital plane of Jupiter, v is the true anomaly of Jupiter which serves as independent variable, e_J is the eccentricity of Jupiter's orbit, $\mu = m_J/(m_S+m_J)$, m_S, m_J being the mass of the Sun and Jupiter. The coordinates r and z

are dimensionless, the instantaneous Sun-Jupiter distance serving as unit distance. The parameter β is connected with the precession of the orbit of Jupiter: $\beta = \dot{\varpi}_J/n_J$, $\dot{\varpi}_J$ is the secular rate of the motion of the perihelion of Jupiter's orbit (the dot means differentiation with respect to the time) and n_J is the mean motion of Jupiter. At the derivation of Equations (1) it has been assumed that the orbit of Jupiter is a precessing ellipse for which the relations

$$R_J = \frac{a_J(1 - e_J^2)}{1 + e_J \cos v}, \quad R_J^2 \dot{v} = \sqrt{k^2(m_S + m_J)a_J(1 - e_J^2)}$$

are valid with slowly changing e_J eccentricity (a_J is the semi-major axis of the orbit of Jupiter, k is the Gaussian gravitational constant). Equations (1) are only approximate, terms depending on $\dot{\varpi}_J^2$, $\ddot{\varpi}_J$, $\dot{e}_J/n_J$ have been neglected. Equations (1) are accurate up to the third order of the small parameter $\varepsilon = \sqrt{\mu} = 0.030885$. The parameter β is of order ε^3, since the mean value of $\dot{\varpi}_J$ is 4.2"/year and $n_J = 299.1$"/day.

The solution of Equations (1) was studied earlier (Érdi, 1981, 1984, 1988) for the case $\beta = 0$, $e_J = \text{const} = 0.048$. Thus the question now is how do the new assumptions ($\beta \neq 0$, e_J changing) modify the previous solution.

Since the basic periods of the Trojans are well separated, Equations (1) are suitable for the application of the method of the multiple-variable expansions. A two-variable solution for the planar motion of the Trojans in the circular restricted three-body problem was first derived by Kevorkian (1970). According to the paper (Érdi, 1984), the solution of Equations (1) can be assumed in the form of a four-variable asymptotic expansion

$$r = 1 + \sum_{n=1}^{N} \varepsilon^n r_n(v, u, \tau, w) + O(\varepsilon^{N+1}),$$

$$\alpha = \alpha_0(u, \tau, w) + \sum_{n=1}^{N} \varepsilon^n \alpha_n(v, u, \tau, w) + O(\varepsilon^{N+1}), \quad (2)$$

$$z = \varepsilon^{1/2} \left[\sum_{n=0}^{N} \varepsilon^n z_n(v, u, \tau, w) + O(\varepsilon^{N+1}) \right],$$

where $u = \varepsilon(v - v_0)$, $\tau = \varepsilon^2(v - v_0)$, $w = \varepsilon^3(v - v_0)$ and v_0 is the epoch. The four variables correspond to the time-scales of the motions of the asteroids, that is to the revolution around the Sun, to the libration around the Lagrangian points, to the motion of the perihelion, and to the motion of the orbital plane. In addition to (2), it will be assumed that in Equations (1)

$$e_J = \varepsilon e_1(w), \quad \beta = \varepsilon^3 \beta_1(w), \quad (3)$$

where e_1 and β_1 are not very large in absolute value compared to unity, and they are changing on the slowest time-scale.

The substitution of (2) into Equations (1) result in a system of partial differential equations for the unknown functions r_n, α_n, z_n with independent variables v, u, τ, and w. In the solution of this system arbitrary functions of the independent variables appear. These are specified according to the principle that the solution for r, α, z be bounded. The solution to $O(\varepsilon^2)$ was derived in Érdi (1981), and the solution in the third order was studied in Érdi (1984).

I have recalculated this earlier solution with the new terms in Equations (1) and with the conditions (3). The terms depending on β modify the solution starting from the third order. The changing of e_J has effects already in the first order. From the solution for the coordinates r, α, z, the perturbations of the orbital elements can be derived by means of the relations of the two-body problem. In the following, perturbations of the eccentricity e and of the longitude of the perihelion ϖ of the Trojan asteroids are discussed.

3. Long Term Perturbations of e and ϖ

Without the terms depending on the two fastest variables v and u, the main perturbations of $O(\varepsilon)$ in e and ϖ are:

$$e\cos(\varpi - \varpi_J) = (b - c\cos\chi)(1 + \tfrac{1}{4}\kappa + \tfrac{5}{8}\kappa\cos\gamma) + (a - c\sin\chi)\tfrac{5}{8}\kappa\sin\gamma, \quad (4)$$

$$e\sin(\varpi - \varpi_J) = (a - c\sin\chi)(1 + \tfrac{1}{4}\kappa - \tfrac{5}{8}\kappa\cos\gamma) + (b - c\sin\chi)\tfrac{5}{8}\kappa\sin\gamma,$$

where

$$a = -e_J\frac{A_2}{A_0}, \quad b = -e_J\frac{A_1}{A_0}, \quad c = \varepsilon\rho_{11}, \quad \kappa = \varepsilon\lambda_0^2,$$

$$\chi = A_0\tau + \psi_{11}, \quad \gamma = A_3\tau + 2A_4w - 2\varpi_J - 2\nu_{02},$$

and ρ_{11}, ψ_{11}, λ_0, ν_{02} are constants. The letters A_i denote known expressions, given in (Érdi, 1981, 1984), of a parameter l which characterizes the amplitude of libration around L_4. The parameter κ is connected with the orbital inclination i, $i = \sqrt{\kappa} + O(\varepsilon)$ ($i = 0$ for $\lambda_0 = 0$).

The long-term variations of e_J and ϖ_J can be described by the equations (Bretagnon, 1974)

$$e_J\cos\varpi_J = \sum_{j=1}^{8}\lambda_{Jj}M_j\cos(g_jt + \beta_j), \quad e_J\sin\varpi_J = \sum_{j=1}^{8}\lambda_{Jj}M_j\sin(g_jt + \beta_j),$$

where λ_{Jj}, M_j, β_j are constants and g_j are the secular fundamental frequencies.

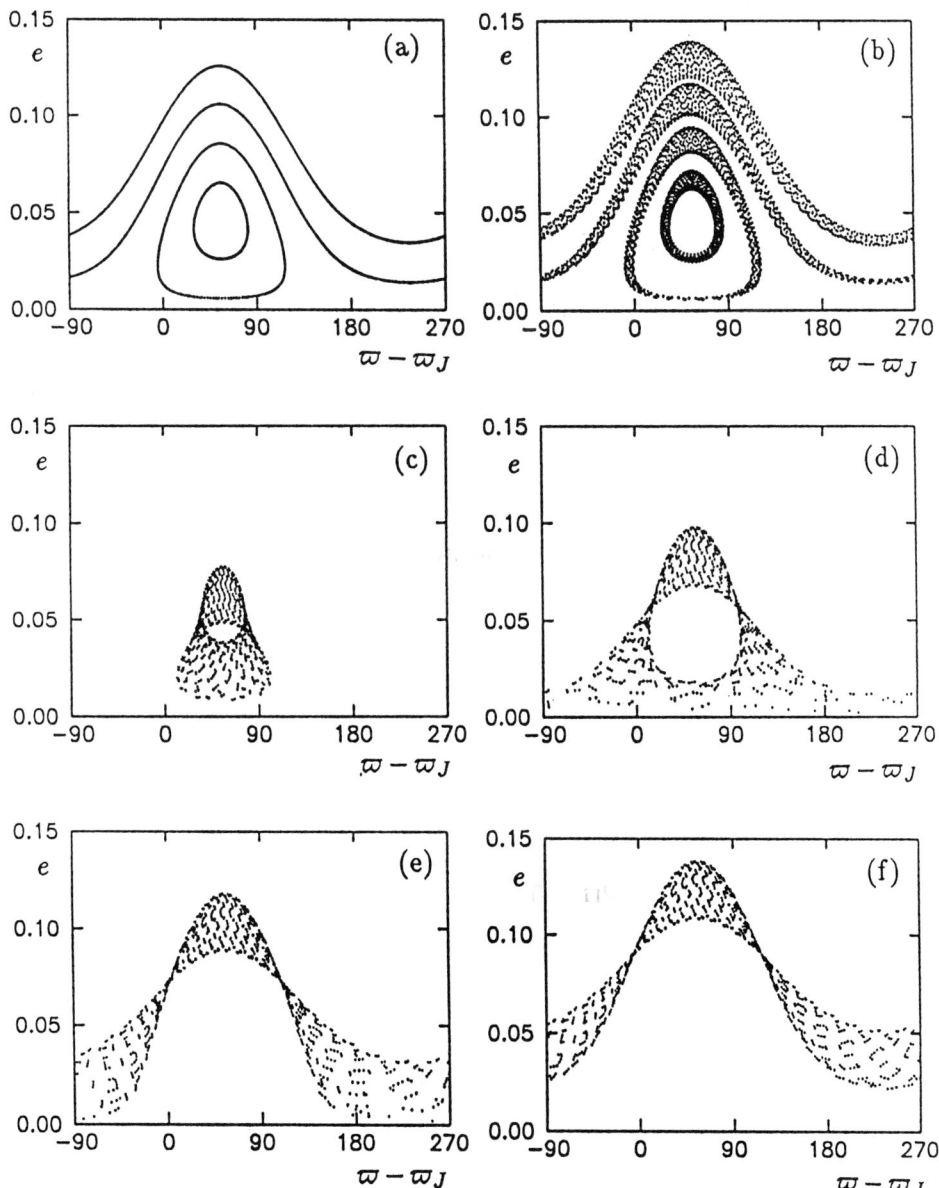

Figure 1. Variation of e versus $\varpi - \varpi_J$. (a): zero inclination orbits, e_J constant; (b): inclined orbits, e_J constant; (c)-(f): zero inclination orbits, e_J secularly changing.

Fig.1 shows e versus $\varpi - \varpi_J$ for several values and cases of the parameters of Equations (4). In all examples $l = 0.3$ (libration amplitude 17.2^0 around L_4), $\psi_{11} = 0$, $\nu_{02} = 0$, and the investigated time interval is 150000 years.

Fig. 1(a) is obtained for $\lambda_0 = 0$ ($i = 0$), $e_J = 0.048$, and for several values of c: $c = 0.02, 0.04, 0.06, 0.08$. The inner closed curve corresponds to $c = 0.02$, and c is increasing from orbit to orbit outwards. In this case there is only a free oscillation of the perihelion with proper frequency $\varepsilon^2 A_0 n_J$, and e and $\varpi - \varpi_J$ change with periods of the order of 3600 years. Note that the centre $e = e_J$, $\varpi - \varpi_J = 60^0$ corresponds to Lagrange's equilibrium solution. Fig. 1(a) is in good agreement with the result of Yoshikawa (1989, Fig. 10).

Fig. 1(b) is obtained for $\lambda_0 = 2$ ($i \approx 19^0$), $e_J = 0.048$, and for the same values of c as in Fig. 1(a). The inclusion of the inclination dependent terms in Equations (4) brings in several long periods and the result is the broadening of the former simple curves of Fig. 1(a).

Figs. 1(c)-1(f) are obtained for $\lambda_0 = 0$ ($i = 0$) and e_J changing: $c = 0.02$ for Fig. 1(c), $c = 0.04$ for Fig. 1(d), $c = 0.06$ for Fig. 1(e), $c = 0.08$ for Fig. 1(f). The secular variation of e_J broadens the limits of the variation of the eccentricity and may cause the alternation of libration and circulation of the perihelion (Fig. 1(d)). When $\lambda_0 \neq 0$, that is for inclined orbits, the regions of Figs. 1(c)-1(f) have further broadening (not shown here), however the dominant effects in the long-term variation of the eccentricity of the Trojans come from the secular change of the eccentricity of Jupiter.

References

Bien, R. and Schubart, J.: 1987, *Astron. Astrophys.* **175**, 292.
Bretagnon, P.: 1974, *Astron. Astrophys.* **30**, 141.
Érdi, B.: 1981, *Celest. Mech.* **24**, 377.
Érdi, B.: 1984, *Celest. Mech.* **34**, 435.
Érdi, B.: 1988, *Celest. Mech.* **43**, 303.
Garfinkel, B.: 1985, *Celest. Mech.* **36**, 19.
Kevorkian, J.: 1970, In *Periodic Orbits, Stability and Resonances* (G. E. O. Giacaglia ed.) D. Reidel Publ. Co. Dordrecht, Holland, 286.
Milani, A.: 1993, *Celest. Mech.* **57**, 59.
Milani, A.: 1994, In *Asteriods, Comets, Meteors 1993* (A. Milani et al. eds.), 1994 IAU Printed in the Netherlands, 159.
Zagretdinov, R. V.: 1986, *Kinematika i Fizika Nebesnih Tel* **2** N 3, 68.
Yoshikawa, M.: 1989, *Astron. Astrophys.* **213**, 436.

ON THE HECUBA GAP

S.FERRAZ-MELLO
Instituto Astronômico e Geofísico, Universidade de São Paulo, Caixa Postal 9638, 01065-São Paulo, SP, Brasil.
sylvio@vax.iagusp.usp.br

Abstract. An asteroid captured in the Hecuba gap (2/1 resonance with Jupiter) may remain there for a long time before escaping. However, the study of the diffusion of orbits in the gap indicates an escape timescale in the range $10^7 - 10^9$ years. The short-period perturbations of Jupiter's orbit play a determinant role in the creation of the stochasticity responsible for the escape.

1. The Hecuba gap

The Hecuba gap is the minimum in the asteroid distribution located at the place where the asteroids have a mean-motion resonance with Jupiter in the ratio 2:1. An asteroid captured in this resonance may remain there for a long time before escaping. However, the study of the diffusion of orbits in the gap, indicates an escape timescale of the order of 10^8 years or, more precisely, in the range $10^7 - 10^9$ years, providing evidences linking the Hecuba gap with the global stochasticity of the 2/1 asteroidal resonance (Ferraz-Mello, 1994a,b; Franklin, 1994; Ferraz-Mello *et al.*, 1995) This scenario contradicts previous ideas in several respects. First of all, it is well known that the inner chaotic region of the 2/1 resonance, in the planar elliptic restricted three-body problem, is shielded from the outside by an extended bunch of regular motions (see the Poincaré maps in Ferraz-Mello 1994a,b). It is also known that, when the long-period perturbations of Jupiter's orbit are taken into account, these regular barriers are not destroyed (Morbidelli and Moons, 1993). The assumptions extending these facts to the exact model led to accept the impossibility of a chaotic diffusion able to drive these orbits off this region in a time of the order of the age of the asteroid belt, making necessary ad-hoc cosmogonic hypotheses to explain the almost absence of asteroids in the gap. But all Lyapunov times computed in the resonant region with an exact Sun-Jupiter-Saturn-asteroid model (Ferraz-Mello, 1994a; Franklin, 1994) point to a global stochasticity whose origin may be searched in the Chirikov regime created by the complex overlap of low- and high-order secondary resonances.

The most serious criticisms to this scenario cames from the scarcity of precise simulations actually showing the diffusion. In fact, only a few results of numerical integrations of fictitious asteroids in the 2/1 resonance are available. They are 5 early numerical integrations reported by Wisdom (1987), one by Scholl (see fig. 7 in Ferraz- Mello, 1994b) and some more recent ones obtained by Henrard et al. (1995). Notwithstanding the limited timespan of these integrations (the longest reaching only 12 Myr), half of them show important diffusing orbital processes, and the later one shows a possible path, through moderate inclinations, leading from the inner chaotic region ($e \sim 0.1$) to high eccentricities. Some more long runs were done by Franklin (1994) in the frame of a planar model. In these runs most of the solutions escaped, but three solutions remained in the resonance zone for more than 120 Myr.

The aim of this communication is to present some results concerning the stochasticity of the 2/1 resonance obtained by means of a planar symplectic mapping and show that the failure of the existing averaged models in showing a stronger chaotic behaviour cames from the fact that the long-period perturbations of Jupiter's orbit are not able, alone, to produce such chaos in a planar model. The situation changes drastically when we add, to the model, the short-period perturbations whose arguments are $\cos(2\lambda_{Jup} - 5\lambda_{Sat})$ and $\cos(\lambda_{Jup} - 2\lambda_{Sat})$.

2. The mapping

The mapping used is a modified form of the symplectic mapping introduced by Hadjidemetriou (1988, 1991). The equations of this mapping are those of the canonical transformation spanned by the Jacobian generator $W = \mathcal{I} + \tau H$ where H is the given *averaged* Hamiltonian, $\mathcal{I}$ the generator of the identical transformation and τ is the map step. The transformation is such that the mapping has the same fixed points – with the same stability characteristics – as the surface of section of the averaged system.

The original Hadjidemetriou formulation of the mapping was modified. The classical Laplacian expansion of the averaged potential of the disturbing forces, problematic in high eccentricities, was substituted by the asymmetric expansion of Ferraz-Mello and Sato (1989) which gives a good representation of it even in high eccentricities provided that the motion is

TABLE 1. Main long- and short-period oscillations in $e_1.e^{i\varpi_1}$

Term		Amplitude	Frequency	
g_5	+	0.0441872	4.257493	arcsec/yr
g_6	−	157002	28.245530	
$-g_5 + 2g_6$	−	5735	52.233567	
$2g_5 - g_6$	+	142	−19.730544	
g_7	+	18139	3.086756	
$-g_5 + g_6 + g_7$	+	1982	27.074793	
$g_5 + g_6 - g_7$	+	1936	29.416267	
$-\lambda_1 + 2\lambda_2 + \varpi_1$		0.000646	$\varpi_1 - 21264.4$	arcsec/yr
$-2\lambda_1 + 5\lambda_2 + \varpi_1$		364	$\varpi_1 + 1467.2$	

a libration of moderate amplitude.

We consider, generically, an asteroid in a resonance $(p+q):p$ with Jupiter and moving in the same plane as the planet. We introduce the long-period angular variables:

$$\psi = \frac{p+q}{q}\lambda_1 + \frac{p}{q}\lambda, \qquad \sigma = \psi - \varpi$$

and their canonical conjugate actions, respectively P and J. The basic function of Hadjidemetriou's mapping is the Jacobian generator

$$W(J_{n+1}, P_{n+1}, \sigma_n, \psi_n, t_n) = \sigma_n J_{n+1} + \psi_n P_{n+1} + \tau \mathcal{H}(J_{n+1}, P_{n+1}, \sigma_n, \psi_n, t_n)$$

where τ is the map step and H is the averaged Hamiltonian

$$\mathcal{H} = -\frac{\mu^2}{2L^2} - L\frac{p+q}{q}n_1 - R;$$

$L = \sqrt{\mu a} = -(J+P)$, μ is the square of the Gaussian constant, n_1 is the mean motion of Jupiter and R is the disturbing potential averaged over the synodic period. All results in this paper were obtained taking the step τ equal to the initial synodic period of the asteroid with respect to Jupiter.

When the orbit of Jupiter is kept fixed, the only possible secular resonance is associated with the angle $\varpi - \varpi_1$ (with $\varpi_1 = $ const.). This secular resonance is responsible for the corotation zone seen in the middle of the resonance, at high eccentricities (see Ferraz-Mello, 1994a). When the actual perihelion of Jupiter is introduced, the corotation zone remains almost the same since its motion is very slow. However, the resulting increase in the number of degrees of freedom may give rise to chaotic regions which were not apparent in the restricted model (Morbidelli and Moons, 1993). Table I gives the amplitudes of the long- and short-period perturbations in the complex quantity $e_1.e^{i\varpi_1}$ used in this investigation, according with, respectively, Nobili et al. (1989) and Simon and Bretagnon (1975). In this table g_5, g_6 and g_7 represent the proper perihelion of Jupiter, Saturn and Neptune, respectively; λ_2 is the mean longitude of Saturn. For the sake of allowing a direct comparison of the long- and short-period perturbations of Jupiter's perihelion, the perturbation equations of the eccentricity and perihelion given by Simon and Bretagnon (1975), which are actually used in the mapping, were slightly modified.

We consider, also, the most important perturbations in the mean longitude of Jupiter, corresponding to the near commensurability 5:2 of its mean-motion with Saturn's According with Simon and Bretagnon (1975), these perturbation are given by

$$\delta\lambda_1 = -5.164 \times 10^{-4} \cos(2\lambda_1 - 5\lambda_2) - 51.502 \times 10^{-4} \sin(2\lambda_1 - 5\lambda_2).$$

3. Scaled models

In order to go over the barriers that a slow chaos puts to investigation, we may introduce artificial parameter variations able to accelerate that mechanisms. We have used scaled models related to the real problem by a mass gauge. In the case of the restricted model, it is simply a factor multiplying the mass of Jupiter. The main frequencies of the planar model – the frequency of the perihelion motion and the libration frequency – are expected to be roughly multiplied, respectively, by this gauge and by its square root. The structure of the resonance web will be altered. Thus, in order to keep this alteration small and avoid new secondary resonances to appear and create artificial domains with different diffusion patterns, we limit ourselves to gauges close to 1.

In what concerns the motion of Jupiter, the frequencies of the long-period perturbations are amplified in the same proportion while the amplitudes remain the same. A mass gauge is likely expected to amplify only the amplitudes of the short-period perturbation of Jupiter's orbit; however, in order to preserve the resonance web, as much as possible, also the frequencies are linearly amplified in the scaled model.

One must kept in mind that no perfect scaling is possible in a non-linear problem. Therefore, results got with a scaled model cannot be considered as true before verified with non-scaled models. This rule is followed in this investigation.

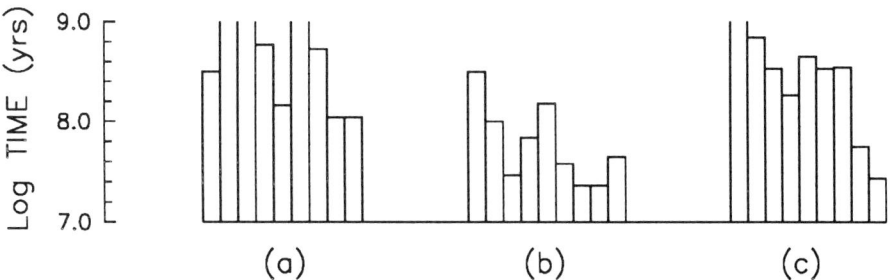

Figure 1. Median times to reach $e = 0.56$ in three cases: (a) Jupiter orbit with short-period perturbations only; scaled $M_g = 1.4$. (b) Jupiter orbit with short- and long-period perturbations; scaled $M_g = 1.4$. (c) Jupiter orbit with short- and long-period perturbations; non scaled ($M_g = 1.0$). In each histogram the initial eccentricity varies from 0.11 to 0.35, from left to right, with a 0.03 step.

TABLE 2. Solutions reaching the limit value $e = 0.56$ ($M_g = 1.4$)

Perturbations in Jupiter's Motion	Limit reached in less than 10^9 yrs	Limit reached in less than 10^8 yrs
Long-period only	15%	10%
Short-period only	68%	32%
both	95%	72%

4. Results and Conclusions

Thanks to the speed of the mapping, we have been able to made a great deal of different experiments. A more systematic study was done selecting a grid of initial conditions delimited by initial eccentricities in the interval 0.11–0.35 and semi-major axes in the interval 3.272–3.284 AU (the middle of the resonance lies close to 3.276 AU). Initial values of σ and $\varpi - \varpi_1$ were fixed at 0°. Scaled models were computed in a net of 40 points in this domain with many different combinations of the perturbations included in Jupiter's orbit. Table II summarizes results obtained using a mass gauge 1.4 (that is, the mass of Jupiter and the frequencies of the perturbations introduced in Jupiter's orbit were multiplied by 1.4) and including all perturbations of table 1.

Table 2 shows that, when the model adopted for Jupiter's orbit include both short- and long-period perturbations, the transition allowing its eccentricity to increase happened in 95% of the studied cases. The limit $e = 0.56$ was set arbitrarily (it corresponds to a perihelion distance 1.44 AU). The results show clearly that, in the long run, in the planar reduction of the

2/1 resonance, the perturbations of Jupiter's orbit whose frequencies are $5n_2 - 2n_1$ and $2n_2 - n_1$ are more important than the long-period perturbations of Jupiter's orbit. The distribution of the times necessary to reach the limit $e = 0.56$ as a function of the initial eccentricity, from 0.11 to 0.35 in steps of 0.03, are shown in Fig. 1 (a) and (b).

The important role played by the short-period perturbations of Jupiter's orbit are confirmed when the experiments are done without any scaling (mass gauge $M_g = 1.0$). When both short- and long-period perturbations of Jupiter's orbit are considered, 38% of the solutions reached $e = 0.56$ in less than 10^8 years and 87% reached it in less than 10^9 years. The distribution of the times necessary to reach the limit $e = 0.56$ as a function of the initial eccentricity, from 0.11 to 0.35 in steps of 0.03, is shown in Fig. 1(c).

These results show that the global stochasticity verified with numerical 4-body models, including Saturn, does not came from the direct action of Saturn, as it was sometimes suspected, but from the main short-period perturbations of Jupiter's orbit.

References

Ferraz-Mello, S.: 1994a, "The convergence domain of the Laplacian expansion of the disturbing function". *Celest. Mech. Dyn. Astron.* **58**, 37-52.

Ferraz-Mello, S.: 1994b, "Dynamics of the 2:1 asteroidal resonance". *Astron. J.* **108**, 2330-2337.

Ferraz-Mello, S., Dvorak. R. & Michtchenko, T.A. (1994). "Chaos and the depletion of asteroids in resonant orbits", In *From Newton to Chaos*, (A.E.Roy and B.Steves, eds.) Plenum Press, New York, pp. 157-169

Ferraz-Mello, S. and Sato, M. 1989, "A very-high-eccentricity asymmetric expansion of the disturbing function near resonances of any order". *Astron. Astrophys.* **225**, 541-547.

Franklin, F. (1994). "An examination of the relation between chaotic orbits and the Kirkwood gap at the 2:1 resonance" *Astron. J.* **107**, 1890.

Hadjidemetriou, J.D.: 1988, "Algebraic mappings near a resonance with an application to asteroid motion". In *Long-Term Dynamical Behaviour of Natural and Artificial N-Body Systems* (A.E.Roy, ed.), Kluwer, Dordrecht, 257-276.

Hadjidemetriou, J.D.: 1991, "Mapping models for Hamiltonian systems with applications to resonant asteroid motion". In *Predictability, Stability and Chaos in N-Body Dynamical Systems* (A.E.Roy, ed.), Plenum Press, New York, 157-175.

Henrard, J., Watanabe, N. and Moons, M.: 1995, "A bridge between secondary resonances and secular resonances inside the Hecuba gap". *Icarus* **115**, 336-346.

Morbidelli, A. and Moons, M.: 1993, "Secular resonances in the mean motion commensurabilities: The 2/1 and 3/2 cases". *Icarus* **102**, 316-332.

Nobili, A.M., Milani, A. and Carpino, M.: 1989, "Fundamental frequencies and small divisors in the orbits of the outer planets". *Astron. Astrophys.* **210**, 313-336.

Simon, J.L. and Bretagnon.P.: 1975, "Results of first-order perturbations of the four large planets" *Astron. Astrophys. Suppl.* **22**, 107-160.

CHAOTIC ASTEROIDAL TRAJECTORIES EXHIBITING MULTIPLE BURSTS OF ECCENTRICITY: A STATISTICAL ANALYSIS

IVAN I. SHEVCHENKO
Institute of Theoretical Astronomy,
Russian Academy of Sciences, St. Petersburg, Russia

AND

HANS SCHOLL
Observatoire de la Côte d'Azur,
Department G. D. Cassini, Nice, France

Abstract. For chaotic asteroidal trajectories exhibiting multiple bursts of eccentricity in the 3/1 Jovian resonance, we derive the distribution function of time intervals between such bursts. At the onset, the resulting distribution decays exponentially. In the tail, an algebraic decay is observed, with the power-law index for the integral distribution in the range -2 to -1.

1. Introduction

Wisdom (1982) found a now-well known behaviour of chaotic asteroidal trajectories in the 3/1 Jovian resonance, namely bursts in the time evolution of orbital eccentricities. During a long period of time (hundreds of thousands of years) the eccentricity of an asteroidal trajectory may remain small (< 0.1), then it jumps suddenly to a high value (> 0.3) and after a comparatively short period jumps back to small values. During this mode, an asteroid is planet-crossing. The bursts in eccentricity repeat and appear to be separated randomly in time. In the following, we call such orbits with chaotic transitions between two modes, intermittent. We recover the shape of the distribution of lengths of time intervals between eccentricity bursts of intermittent orbits. Especially we are interested in the shape of the long time part of this distribution.

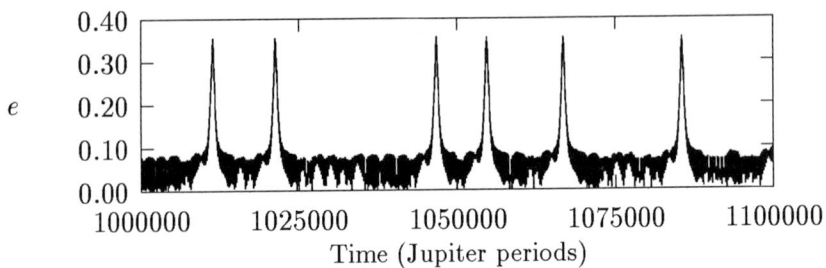

Figure 1. Eccentricity versus time for an intermittent trajectory.

2. Statistical analysis of intermittent trajectories

We used Wisdom's mapping (1983) for the planar-elliptic problem Sun–Jupiter–asteroid in order to obtain intermittent trajectories at the 3/1 mean motion commensurability. There exists a class of intermittent orbits for which the eccentricity bursts are self-similar in shape and the time interval which an asteroid spends in the high eccentricity mode is short as compared to the time spent in the low eccentricity one (see Fig. 13 in Wisdom, 1983).

Intermittent trajectories with various values for Jupiter's eccentricity and initial conditions were analysed. We show the results for a trajectory with the following starting values: the semimajor axis $a = 0.4806$, the eccentricity $e = 0.05$, the longitude of perihelion $\varpi = 0$, and $\varphi \equiv \ell - 3\ell_J = \pi$, where ℓ and ℓ_J are the mean longitudes of an asteroid and Jupiter. For Jupiter, $e_J = 0.05$, $a_J = 1$. Fig. 1 shows the bursts of eccentricity. For all other studied orbits they look very similar, the maximum eccentricity not exceeding $e = 0.3 \div 0.4$. The integral distribution for the length of time intervals between bursts is shown in Fig. 2 in logarithmic scales. The duration D of interburst intervals is measured in Jupiter periods. The quantity F at the vertical axis is the relative fraction of interburst intervals with duration greater than D in the total set of observed intervals for a given orbit. The computation of the orbit spanned 10^9 Jupiter periods. This provided statistics for ≈ 50000 eccentricity bursts. Even with Wisdom's mapping, the computation time needed to obtain these statistics for a single trajectory is several days for a Sun 4 Sparc station. The irregular drop in the distribution function near $\log D = 6$ in Fig. 2 is due to insufficiency of the statistics at high values of D.

For all orbits, the tail fits a power law D^a with a between -2 and -1. For smaller time intervals, not exceeding the limit $D \approx 10^5$, the distribution is of Poisson type, i.e. the decay is exponential. Theoretical interpretation of these dependences will be given elsewhere.

The duration of interburst intervals can be measured e.g. by the number

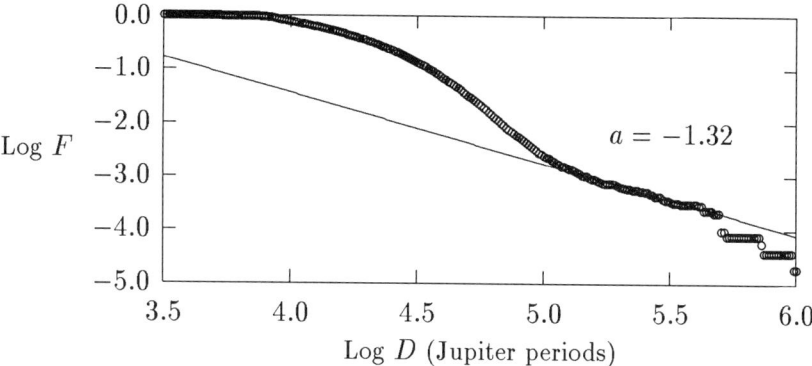

Figure 2. Integral distribution of intervals between eccentricity bursts in decimal logarithmic scales. A best-fit straight line for the tail is plotted, and the corresponding power-law index is indicated. The trajectory is the same as in Fig. 1.

N of periods of rotation of the longitude of the asteroid's perihelion. The resulting distribution of lengths is qualitatively the same as in the case of measuring in real time units: it is a junction of an exponential decay and a slow algebraic one. The transformation from exponential into algebraic behaviour takes place at $N = 50 \div 100$. An influence of the choice of units is that the distribution of intervals measured in Jupiter periods starts to decay beginning from $\log D > 3.5$ (see Fig. 2), while in the case of measuring in perihelion revolutions the distribution starts to decay immediately at $N = 1$.

A somewhat complicating factor, inflicting the very onset of the distributions, occur when an intermittent orbit exhibits not only single random eccentricity bursts, but also bursts forming periodic sequences. For the trajectory with starting values given above the presence of such sequences is negligible, but for some orbits they are abundant. It is easy to exclude them from the analysis, because the quantity N for an interval between bursts in a periodic sequence is less than one. Recovering the shape of distributions of lengths of such sequences constitutes an interesting separate problem.

3. Conclusions

The distribution of the lengths D of time intervals between eccentricity bursts follows a power-law decay for $D > 10^5$ Jupiter periods. For shorter intervals, $D < 10^5$, we find a distribution of Poisson type.

The qualitative character of the over-all distribution is independent of the choice of the time variable (real time or e.g. perihelion revolutions).

The value of the power-law index for the integral distributions for $D > 10^5$ is usually confined between -2 and -1.

Acknowledgements

This research was supported by an exchange program between CNRS, France, and the Russian Academy of Sciences, Grant 3.10 SDU.

References

Wisdom, J.: 1982, *Astron. J.* **87**, 577.
Wisdom, J.: 1983, *Icarus* **56**, 51.

ORBITAL STABILITY OF HIGH INCLINATION ASTEROIDS

N. A. SOLOVAYA
Celestial Mechanics Department
Sternberg State Astronomical Institute
University Prospect 13, 119 889 Moscow, Russia

AND

E. M. PITTICH
Astronomical Institute of the Slovak Academy of Sciences
Dúbravská cesta 9, 842 28 Bratislava, Slovak Republic

Abstract.
The orbital evolutions of fictitious asteroids with high inclinations have been investigated. The selected initial orbits represent asteroids with movement, which corresponds to the conditions of the Tisserand invariant for $C = C(L_1)$ in the restricted three body problem. Initial eccentricities of the orbits cover the interval 0.0–0.4, inclinations the interval 40–80°, and arguments of perihelion the interval 0–360°. The equations of motion of the asteroids were numerically integrated from the epoch March 25, 1991 forward within the interval of 20,000 years, using a dynamical model of the solar system consisting of all planets. The orbits of the model asteroids are stable at least during the investigated period.

1. Introduction

The analytical theory of secular perturbations of asteroids with high inclinations and eccentricities shows that the changes of these elements are limited within certain small domains (Kozai, 1962). The theory was built with one disturbing body, Jupiter, on a circular orbit. The calculations of Gerasimov and Solovaya (1989) have shown that the outer part of the asteroid belt does not contain asteroids with high inclinations. Their orbits are not Hill's stable in this region. Indeed, we do not know of asteroids from the Hilda group having high inclinations.

In this paper we have investigated the evolution of the group of fictitious asteroids which are located near of the boundary of the Hill's stability region. This boundary in the restricted three body problem is limited by the Jacobian constant, which is equal to the constant for the interior stationary point L_1. The semi-major axes of the orbits of the model asteroids were selected such that for arbitrary inclinations and eccentricities the condition $C = C(L_1)$ is always fulfiled. The question is whether the orbits of these asteroids are stable when they move in the dynamical system consisting of more perturbating bodies than in the three body dynamical system.

2. Model orbits

We took a dynamical model of the solar system, consisting of all planets and massless fictitious asteroids. We traced the 20,000 years orbital evolution of these asteroids using the numerical integration program with the RA15 integrator (Everhart, 1985). This period represents approximately 1770 revolutions of Jupiter. The initial asteroid semi-major axes a_0 were calculated from the equation for the integral of the type of Jacobi when $C = C(L_1)$. In the Sun-centered siderical coordinates has the following form:

$$C = (\frac{1}{2a} + \sqrt{p}\cos i) + \frac{1}{2}\mu^2 + \mu(\frac{2K(\kappa_2)}{\pi\nu} - R\cos b \cos(l - l_j)) \quad (1)$$

where $\nu^2 = (\rho^2+3\alpha^2)^2$, $\alpha = (1-\mu)e_j$, and $\kappa_2^2 = (4\alpha^2-(\rho^2+3\alpha^2-\nu^2)/2)/\nu^2$. We used a notation similar as Solovaya et al. (1992), where R is the Sun–asteroid distance, ρ is the Jupiter–asteroid distance, $\mu = 0.0009538$ is the ratio of the Jupiter's mass to the sum of the masses of Sun and Jupiter, e_j is the eccentricity of Jupiter's orbit, l and b are the longitude and latitude of an asteroid, l_j is the longitude of Jupiter, a, i and p are the osculating elements and the parameter of an asteroid orbit, and $K(\kappa_2)$ is the elliptic integral. When $e_j = 0$ then $K(\kappa_2) = \pi/2$.

For the eccentricity of Jupiter $e_j = 0.062$ the integral of the type of Jacobi $C = 1.538$. The equation (1) was solved for i_0 equal $40°$, $60°$, and $80°$, and for e_0 equal 0.0, 0.2, and 0.4. The initial value of argument of perihelion ω_0 of the asteroids was selected from the equation $\cos\Theta_\pi = \cos\omega\cos l_{j_\pi} + \sin\omega\sin l_{j_\pi}\cos i$ at $\Theta_\pi = 180°$ and varied with the step of $60°$ from $0°$ to $360°$. The initial value of the ascending node $\Omega_0 = 90°$. The initial epoch of the numerical integration was March 25, 1991 UT. For this moment l_j was taken from Abalakin (1989). The extreme values of some orbital elements obtained from their evolution within the investigated period of the groups of asteroids with initial $\omega_0 \in \langle 0°, 360°\rangle$ for initial values of $i_0 \in \langle 40°, 80°\rangle$ and $e_0 \in \langle 0.0, 0.4\rangle$ are given in Table 1. The evolution of the osculating eccentricity e, inclination i, perihelion distanace q and aphelion distance Q of the model asteroids with the initial inclinations 40° and 80° is plotted on Figure 1 and 2.

3. Results of orbital integration

The 20,000 years orbital evolution of the model asteroids shows a general behaviour of individual orbital elements in dependence on the initial i_0 and e_0. Within the investigated period all orbital elements of the model asteroids, except of a, change more or less periodically. The changes of a are in all cases negligible. It means that all studied orbits have been stable at least for 20,000 years. The model orbits have more or less smaller i after some time than at the beginning of the investigated evolution.

In the case of $i_0 = 40°$ and $e_0 = 0.2$, some i decline below 30° after seven millenia, or when $e_0 = 0.4$ after four milenia. These changes repeat in periods of ~ 14 milenia in the first case, and ~ 7 milenia in the second one. For $e_0 = 0$, there is also the indication of the periodicity of the inclination change, but it must be more than 20,000 years. The evolution of i of the asteroids with $i_0 > 40°$ has similar behaviour except that the changes are relatively higher than in the previous case. The inclination decreased close to 30° (see Figures and Table 1) for all investigated i_0.

The longitudes of perihelia of all model orbits change slowly. After some milenia the orbits with $e_0 = 0$ become similar for all initial arguments of perihelia, and there is no spread anymore. Within the investigated period, e vary for all i_0 and e_0. The changes develop similarly as those of i (see Figures and Table 1). The eccentricity of some orbits increased. Generally, the increase of e is higher for higher i_0 and e_0.

These model asteroids is not only the change in the shape of their orbits, but also a movement of their orbits relative to the Sun. Of asteroids with $i_0 = 40°$, only some with $e_0 = 0.4$ enter the region between the Sun and the Earth. But practically all asteroids with $i_0 \geq 60°$ enter this region during the 20,000 years. The exceptions are those with $e_0 = 0°$. The perihelion distance of the asteroids reaches 0.68 AU for $i_0 = 40°$ and $e_0 = 0.4$. For $e_0 \geq 0.2$ and $i_0 = 40°$ the asteroids approach the Sun to the distance of 0.42 AU and in the case of $i_0 = 80°$ to 0.04 AU. Generally, the values of q are smaller for higher i_0 and e_0 (see Figures and Table 1).

A group of the investigated model asteroids moves a little further from Jupiter. The aphelion distance of these asteroids with the initial inclination 40° is not higher than 4.26 AU, with 60° $\leq$ 3.82 AU, and with 80° $\leq$ 3.58 AU. Therefore, and with regard to high inclinations, the motion of the investigated asteroids is disturbed with the Earth type planets.

Some of these asteroids are delivered from the main asteroid belt to the near-Mars space and then to the near-Earth space. These orbital changes

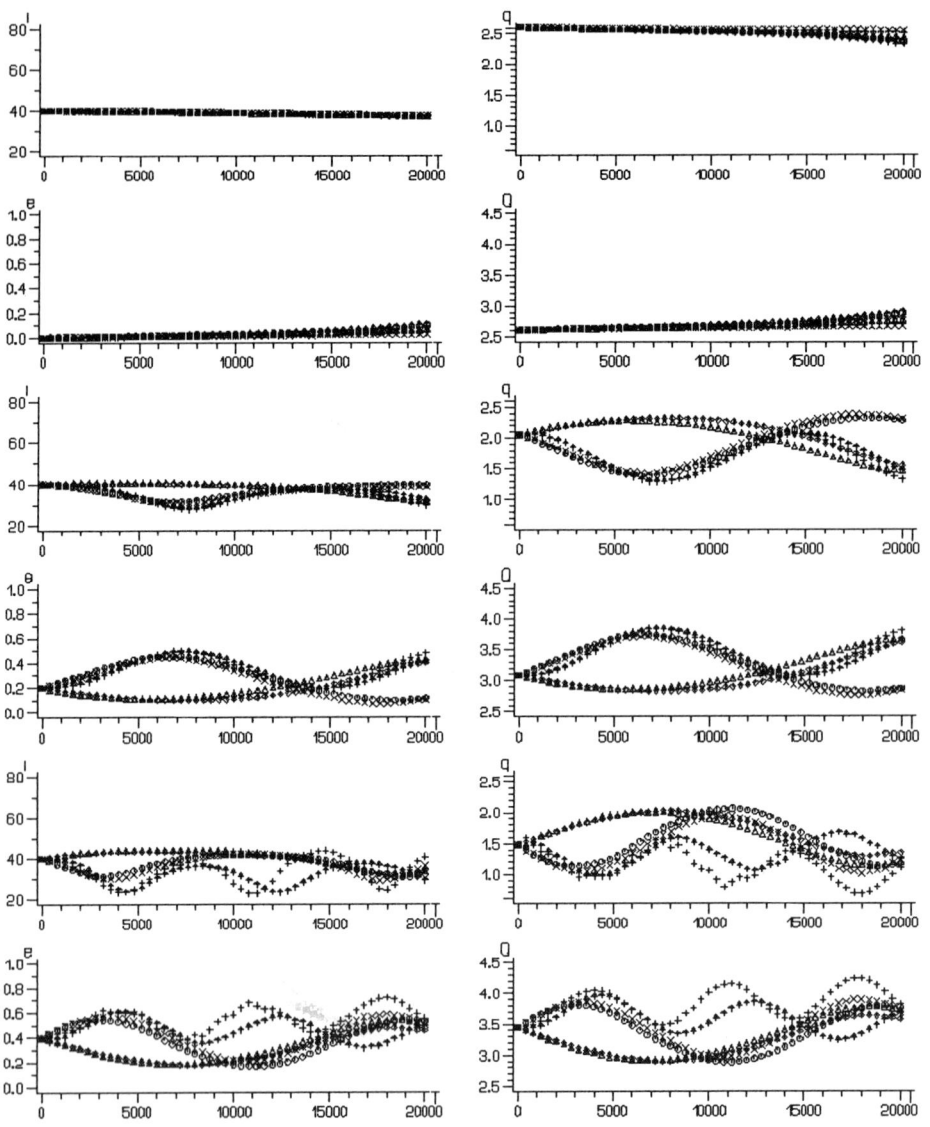

Figure 1. Evolution of orbital elements of the model asteroids within the period of 20,000 years. The initial asteroid inclination $i_0 = 40°$, the eccentricity $e_0 = 0.0$ (top), $e_0 = 0.2$ (middle), and $e_0 = 0.4$ (bottom), and the longitude of perihelion ω_0: ☉ — $\omega_0 = 0°$, △ — $\omega_0 = 60°$, + — $\omega_0 = 120°$, × — $\omega_0 = 180°$, ◇ — $\omega_0 = 240°$, and ✢ — $\omega_0 = 300°$.

are more frequent for higher i_0 and e_0 than for lower ones. They are not much depended on ω_0. Thus, of the asteroids with $i_0 = 40°$, about 20% might be delivered to the near-Earth region, of those with $i_0 = 60°$ about 50%, and of the asteroids with $i_0 = 80°$ about 90% are delivered to this region within twenty millenia.

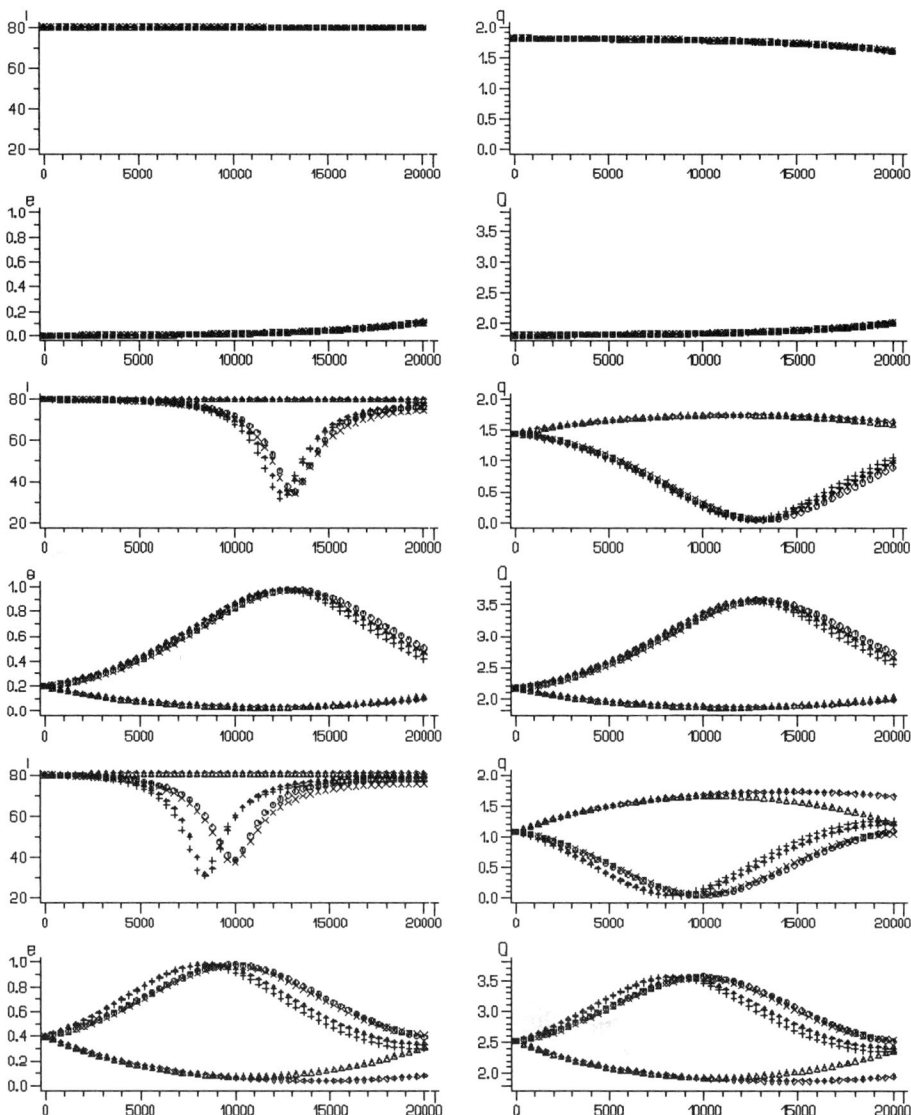

Figure 2. Evolution of orbital elements of the model asteroids within the period of 20,000 years. The initial asteroid inclination $i_0 = 80°$, the eccentricity $e_0 = 0.0$ (top), $e_0 = 0.2$ (middle), and $e_0 = 0.4$ (bottom), and the longitude of perihelion ω_0: ○ — $\omega_0 = 0°$, △ — $\omega_0 = 60°$, + — $\omega_0 = 120°$, × — $\omega_0 = 180°$, ◊ — $\omega_0 = 240°$, and ✦ — $\omega_0 = 300°$.

4. Conclusion

The behaviour of asteroidal orbits with high inclinations were considered in 20 000 years interval and presented in plots. This time span is small for the discovery of secular resonances but sufficient for the investigation of the Hill's stability. The initial semi-major axes of the asteroidal orbits were

TABLE 1. Time evolution of orbital elements of the model asteroids.

i_0 [°]	e_0	i [°]	e	a [AU]	q [AU]	Q [AU]
40	0.0	37–40	0.00–0.11	2.611–2.613	2.32–2.61	2.61–2.91
	0.2	28–40	0.07–0.49	2.568–2.579	1.30–2.39	2.76–3.81
	0.4	24–44	0.18–0.70	2.455–2.538	0.68–2.02	2.90–4.26
60	0.0	57–60	0.00–0.32	2.135–2.137	1.46–2.14	2.14–2.82
	0.2	32–60	0.12–0.81	2.113–2.124	0.43–1.86	2.38–3.82
	0.4	28–62	0.25–0.86	2.052–2.092	0.42–1.56	2.61–3.82
80	0.0	80	0.00–0.12	1.812–1.812	1.60–1.81	1.81–2.02
	0.2	33–80	0.03–0.96	1.806–1.812	0.04–1.76	1.86–3.58
	0.4	31–81	0.40–0.98	1.799–1.805	0.04–1.73	1.87–3.56

taken so that the asteroids with e_0 and i_0 were located at the boundary of the Jacobi's sphera in the restricted three body problem. The chosen maximum value of e_j makes this region smaller. It is interesting to follow the behaviour of the asteroids when the perturbations of all planets are taken into account. The Hill's stability or unstability of an asteroid could be found already at the order of 50 revolutions of Jupiter. The results of integration showed, that e and i change fluently, without large jumping and these changes are limited. The asteroids may be delivered to the near-Mars and to the near-Earth region, but after some period will be returned to the initial belt if the phenomena of resonanses will have not take place.

Acknowledgements

This work was supported by the Slovak Academy of Sciences Grant No. 2/1050/1995 (E. M. Pittich).

References

Abalakin, V.K. (1989) *Astronomicheskij Ezhegodnik na 1991 god.* Nauka, Leningrad.

Everhart, E. (1985) An efficient integrator that used Gauss-Radau spacing, *Dynamics of Comets: Their Origin and Evolution*, (A. Carusi and G.B. Valsecchi, Eds.), pp. 185–202, Reidel, Dordrecht.

Gerasimov, I.A. and Solovaya, N.A. (1989) Evolution of the resonant group of minor planets of the outer part of the asteroidal belt, *Asteroids Comets Meteors III*, (C.-I. Lagerkvist, H. Rickman, B.A. Linblad, and M. Lindgren, Eds.), pp. 91–94, Uppsala University, Uppsala.

Kozai, Y. (1962) Secular perturbations of asteroids with high inclination and eccentricity, *Astronomical Journal*, **Vol. no. 67**, pp. 591–598.

Solovaya, N.A., Gerasimov, I.A., and Pittich, E.M. (1992) 3–D Orbital Evolution Model of Outer Asteroid Belt, *Asteroids, Comets, Meteors 1991*, (A.W. Harris and E. Bowell, Eds.), pp. 565–568, Lunar and Planetary Institute, Houston, Texas.

NONLINEAR OPTIMISATION AND THE ASTEROID IDENTIFICATION PROBLEM

M. EUGENIA SANSATURIO
E.T.S.I.I. University of Valladolid, Spain
AND
ANDREA MILANI AND LUISA CATTANEO
Space Mechanics Group. University of Pisa, Italy

Differential correction procedure allows us to improve orbits for which new observations are available; however, it only works provided the original orbit is within the convergence domain of the pseudo–Newton method. Given the strong nonlinearity of the problem, this only occurs when the residuals of the new observations with respect to the old orbit are quite small.

On the contrary, if a single opposition asteroid, observed only on a short arc, is "lost", i.e. not recovered for several years, it can be difficult to identify it with a newly observed one. There are now $\simeq 20,000$ lost asteroids with poorly determined orbits; if the proposed *Spaceguard survey* will be realized, the problem of identifying millions of new discoveries within catalogues of comparable size will be one of the main challenges. We have begun experimenting with algorithms of orbit determination, to perform both positive and negative identification of asteroids lost for many years.

1. Algorithms

Let us recall the definition of the classical algorithm of *differential corrections* for orbit improvement. If the *residuals* are ξ_i, $(i = 1, m)$ and the *weights* are $W = Diag[\sigma_i^{-2}]$, the *target function* is

$$Q = \frac{1}{m}\xi^T W \xi = \frac{1}{m}\sum_{i=1}^{m}\frac{\xi_i^2}{\sigma_i^2}.$$

If y are the solve for variables, such as the orbital elements at some epoch time, then the least square solution is $\bar{y}$ such that $Q(\bar{y}) = \min Q(y)$.

$$D = \left[\left(\frac{a_1 - a_2}{a_1 + a_2}\right)^2 + (\Delta h)^2 + (\Delta k)^2 + (2\Delta p)^2 + (2\Delta q)^2 + \left(\frac{\lambda_1 - \lambda_2}{10}\right)^2\right]^{1/2},$$

where a, h, k, p, q are equinoctial elements (Broucke and Cefola, 1972) and λ_1, λ_2 are the mean longitudes reduced to the same epoch by two-body propagation. The use of metrics of this kind is discussed by several authors in different contexts (Muinonen and Bowell, 1993; Milani et al., 1994; Zappalà et al., 1990) and it would be worthwhile to experiment also with other ones. We then computed the distance between all the couples of asteroids in a large catalogue with $\simeq 26,321$ records (essentially all the orbits available from MPC and other sources as of Dec. 1994). We sorted the couples by distance and tested for possible identifications the closest ones. This procedure is easy to automatize and is now used also by the MPC; of course it is not good to detect the most difficult cases, those in which the two sets of elements are far apart because of poor determination over too short arcs.

Problem 1: Positive identifications. Which criteria should be used to confirm that two single opposition asteroids are the same?

If a minimum for the target function is found with $Q < 6$ (weights are such that Q is in $arcsec^2$) and each of the two arcs has at least 3 observations over at least 3 days, the identification is confirmed. As a byproduct of this test of the algorithms and the software, we have found 10 new identifications, later accepted by the MPC, and other 3 already known.

However, in some cases, there are observations to be discarded (normally labeled as such in the MPC files). If this is not done, a larger Q can result even from a good identification. If the arcs either have less than 3 observations, or are less than 3 days long, fake low minima can occur, to the point of generating crazy identifications (two asteroids, in different positions on the same plate, identified with a third one). Searching strategies in surveys must take this into account; if not, they could produce useless data.

If LANCELOT gives the same results as DC, as in most of our tests, there is no point in using a more computationally intensive procedure. However, for the reasons explained above, our tests were not too difficult, because the orbital elements solved from each of the two arcs were close.

Problem 2: Reliability of the identification procedure. Can a low minimum, therefore a positive identification, be missed?

The pure DC algorithm is very sensitive to the initial guess chosen. The convergence domain is very small, thus, the iterative procedure can be divergent, even when the low minimum exists. LANCELOT uses a more robust optimisation algorithm, however it is much more expensive.

An effective solution is to use DC, even in the pseudo-Newton formulation, but with an initial guess for the common orbit containing the two arcs which results in moderately large initial residuals. One such procedure is also used at the MPC, but the algorithm is not documented. We have developed our own algorithm to generate an initial guess for the mean longitude and semimajor axis of such a common orbit. It is obtained by computing a two-body orbit such that the mean longitude coincides with those of the two arcs at their epoch times. If the number of revolutions performed by the common orbit in the time span between the two epochs is known, there is only one semimajor axis satisfying this condition. The number of revolutions is not known a priori, but can be guessed by two-body extrapolation from each of the two short arcs; if the two extrapolations differ by one or more revolutions, the only way is to try several different initial guesses for a, obtained for each possible number of intervening revolutions.

With this initial guess we have obtained by DC all the identifications confirmed by LANCELOT; this would not be so with other simple initial guesses, such as the mean of both sets of elements. We have also tested a set of 25 identifications proposed by the MPC, and found convergence to a low minimum by DC in all cases. We intend to perform a large scale test of such an identification procedure, including much more difficult cases.

Problem 3: Can the target function Q have stationary points other than the absolute minimum?

In theory, there is no mathematical proof that Q cannot have many saddles and even many local minima. In practice, we have found one example 3024PL=93 OO7 where DC, with initial guess at the mean of the two sets of elements, finds a stationary point with $Q = 2547.$, while LANCELOT finds a minimum at $Q = 0.67$; the same minimum can also be obtained by DC with the initial guess computed as explained above. This is likely to be a quite rare case, but it is important to know that a saddle can occur. To test whether a stationary point is a minimum or a saddle we would need to have full information on the second derivatives of Q, that is to know the second derivatives of the solution of the N-body problem; this is possible, but computationally expensive. The use of LANCELOT is other alternative, since a nonlinear optimisation method does not converge towards saddles, but its computational cost is high too. Cases with multiple local minima can also occur when the observations are taken far from opposition.

Problem 4: Negative identifications. What happens when the two arcs do not belong to the same asteroid?

In some cases, a stationary point with $Q > 200$ can be found by DC. However, is it a minimum (Problem 3)? In a large fraction of the cases, LANCELOT can find a minimum with $Q > 200$. In this case, the identification should be refused (unless there is a wrong observation). The advantage of a robust nonlinear optimisation algorithm is obvious in these cases; however, the computational cost is too high to propose a brute force searching method for identifications (such as testing all possible couples with "similar" orbital elements). Moreover, in some cases neither DC nor LANCELOT can find a minimum: the iterations diverge ($e > 1$). In these cases the identification can neither be accepted nor be refused. A method to refuse identifications which is totally reliable has not been found yet.

TABLE 1. Statistics of the results

	DIFF. CORR.	LANCELOT
Positive identifications	28.9 %	28.9 %
Negative identifications	33.3 %	53.3 %
Divergent cases	33.3 %	17.8 %

In a set of 45 tests, we have found the results summarized in Table 1. It is apparent that DC should be used first, followed by an additional investigation of both negative identifications and divergent cases.

Problem 5: Which method will be suitable when the number of observations will increase by a factor of about 100?

If the number of asteroids observed only over a very short arc (1-2 days) is very large, the problem of asteroid identifications may become computationally intractable. If the orbits are good enough, because the observing strategy is such that they are all observed for a longer time, we believe it is possible to develop a fully automated algorithm for positive and negative identification. Such a method must take into account all the possible pitfalls, including multiple minima, saddles, divergent cases which result in neither a positive nor a negative identification, and initial extrapolations wrong by more than one revolution. We still cannot present such an algorithm, but we are working to clarify its mathematical foundations.

Acknowledgements

We thank the MPC, in particular B. Marsden, for their prompt answer to our queries and proposed identifications, Ph. Toint for his assistance in defining a formulation of the problem suitable for LANCELOT. This work has been supported by EEC research network CHRX-CT94-0445.

References

Broucke, R. and Cefola, P: 1972, *Celes. Mech.*, **5**, 303–310.
Conn, A.R., Gould, N.I.M. and Toint, Ph.L.: 1992, *LANCELOT: a Fortran package for large-scale nonlinear optimization*. Springer, Berlin.
Everhart, E.: 1985, in *Dynamics of Comets: their origin and evolution*, A. Carusi and G.B. Valsecchi eds., Reidel, 185–202.
Milani, A., Bowell E., Knežević Z., Lemaitre A., Morbidelli A. and Muinonen, K.: 1994, in *Asteroids Comets Meteors 1993*, A. Milani, M. Di Martino and A. Cellino eds., Kluwer, 467–470.
Muinonen,K., and Bowell,E.i: 1993, *Icarus*, **104**, 255-279.
Zappalà,V., Cellino,A., Farinella,P. and Knežević,Z.: 1990, *Astron. J.* **100**, 2030–2046.

A TIME-DEPENDENT EXTENSION TO BROUWER'S METHOD FOR ORBITAL ELEMENTS CORRECTION

F. J. MARCO, J. A. LOPEZ AND M. J. MARTINEZ
Departamento de Matemáticas.
Universitat Jaume I. Castelló. Spain

1. Introduction

One of the most popular methods for orbital elements correction, by means of O-C calculus, is based on Brouwer's method [2], which is very well adapted for integrating short periods of time. We propose a general method to integrate over long intervals of time, when we have good observations, based upon a time-dependent functional relation between the derivatives of all elements with respect to all variations in the initial elements. First, we verify the truth of the unrestricted hypothesis by means of the proposed analytical method and a numerical derivation. In [6], we have incorporated a correction frame model jointly with this general method and, then, we have constructed the residual function which is minimized by the least squares method. But, as we are going to see later, there are correlations between the parameters frame correction model and the initial elements involved in the adjustment. They are obtained here, as well as an expression for the equinox correction from these frame parameters. Finally, by means of a simply weighted method, with the observations from MPC's magnetic tape, in FK4 system, we obtain an estimation for these frame parameters and an equinox correction which is in great accord with the adopted value (see [3]).

2. Analytical model for differential perturbation calculation

Let σ_i^o ($1 \leq i \leq n$) be the initial orbital elements for a body in the Solar System in elliptic motion and let $\triangle \sigma_i^o$ be the desired correction at the same time. The variation induced in right ascension is, at first order,

$$\alpha(\sigma_i^o + \triangle \sigma_i^o; t) = \alpha(\sigma_i^o; t) + \sum_{m=1}^{6} \left.\frac{\partial \alpha}{\partial \sigma_m^o}\right|_t \triangle \sigma_m^o$$

and analogously for declination. We assert that, the generally held hypothesis about these partial derivatives with respect to the initial elements, in the sense that is $\frac{\partial \sigma_k}{\partial \sigma_j^o}(t) = \delta_{kj}, \forall t$, is not true. We denote with $\sigma_{k,j}^o(t) = \frac{\partial \sigma_k}{\partial \sigma_j^o}(t)$ this time-dependence from perturbed elements. These partial derivatives follow by means of a combined solution for the planetary Lagrange equations and their partial derivatives with respect to the initial elements:

$$\frac{d}{dt}\sigma_j = \sum_{k=1}^{6} L_{j,k} \frac{\partial \Re}{\partial \sigma_k}$$

$$\frac{d}{dt}\left\{\frac{\partial \sigma_k}{\partial \sigma_j^o}\right\} = \sum_{i=1}^{6}\sum_{m=1}^{6}\left\{\frac{\partial L_{k,m}}{\partial \sigma_i}\frac{\partial \Re}{\partial \sigma_m} + L_{k,m}\frac{\partial^2 \Re}{\partial \sigma_i \partial \sigma_m}\right\}\frac{\partial \sigma_i}{\partial \sigma_j^o} \quad (1)$$

where L is the Lagrange matrix, $\Re$ is the perturbation function and all indices considered range between 1 and 6. To carry out the integration of this system, we need to consider the first and second derivatives of $\Re$ with respect to all σ^o's, which are computed by means of the chain rule through rectangular ecliptic coordinates, following Simon [8]:

$$\frac{\partial \Re}{\partial \sigma_i \partial \sigma_m} = \vec{V_i} \cdot (\partial^2 \Re) \cdot \vec{V_m} + \vec{V_{i,m}} \cdot \partial \vec{\Re}.$$

$\vec{V}$ is the rectangular ecliptic position vector, $\vec{V_i}$ its partial derivative with respect to the element σ_i and $\partial \Re$ and $\partial \Re^2$ the gradient and second order matrix of the perturbation function in ecliptic cartesian coordinates. The expressions for these $\Re$ function derivatives are computed through the orbital vector plane position: for the first order one, see [5] and, for the second order one, see [7]. Taking into account these analytic expressions, we have selected the minor planet Ceres and we have integrated the equations (1) backwards from 1991 to 1906. Taking the initial elements from the I.T.A. publication [4] and making use of the planetary theory VSOP 87 [1], we obtained these partial derivatives (see table 1). On the other hand, we have numerically computed these values by means of the expressions $\sigma_{k,j}^o \simeq \frac{\sigma_k(...,\sigma_j^o+\Delta\sigma_j,...)-\sigma_k}{\Delta\sigma_j}$ (see table (2)). The agreement between the analytical computation and the numerical estimation confirms our proposal about the time-dependence of $\sigma_{k,j}^o$.

3. Errors in initial elements from catalog errors

To relate the errors involved in the initial elements from catalog errors, we consider $\vec{h}$, orthogonal to the orbital plane, and $\vec{p}$, in the perihelion direction, determining the orbital plane. The superscript "o" is used to denote their dependence on initial elements. Let $\vec{X}^o$ be whatever of these vectors, in rectangular ecliptic coordinates, and let $\vec{X} = \vec{X}^o + \Delta\vec{X}^o$ be

TABLE 1. Analytical computation

$\sigma^o_{k,j}$	a	e	i	Ω	ω	M
a^o	1.29915	-0.06408	-0.01294	-0.06643	-1.01247	63.91748
e^o	-0.01659	0.99336	0.00208	0.03862	-0.05611	0.42464
i^o	-0.00080	-0.00272	1.00081	-0.00233	0.09184	0.07861
Ω^o	0.00280	0.00000	-0.00050	0.99748	-0.05017	0.06743
ω^o	0.00281	0.00043	-0.00022	0.00041	0.95057	0.06454
M^o	0.00304	0.00023	-0.00005	-0.00078	-0.03126	0.94964

TABLE 2. Numerical estimation

$\sigma^o_{k,j}$	a	e	i	Ω	ω	M
a^o	1.29907	-0.06408	-0.01294	-0.06641	-1.01262	63.91452
e^o	-0.01636	0.99337	0.00205	0.03924	-0.05609	0.42131
i^o	-0.00047	-0.00273	1.00079	-0.00220	0.09170	0.07791
Ω^o	0.00281	0.00000	-0.00050	0.99750	-0.05017	0.06734
ω^o	0.00281	0.00043	-0.00022	0.00041	0.95057	0.06455
M^o	0.00304	0.00023	-0.00005	-0.00078	-0.03126	0.94964

its perturbed expression. To relate $\vec{X}^o$ to $\vec{X}$, we can apply the rotations $R_x(-\epsilon)$ for equatorial coordinates, R for the correction model and finally $R_x(\epsilon)$ to retrieve the incremented vector. ϵ is the ecliptic obliquity. The subscript in the R-matrix means rotation around the signaled axis and the rotation R is given by:

$$R = \begin{bmatrix} 1 & -\triangle\xi & -\triangle\eta \\ \triangle\xi & 1 & -\triangle\epsilon \\ \triangle\eta & \triangle\epsilon & 1 \end{bmatrix}$$

Finally, we apply this relation to $\vec{h}$ and $\vec{p}$ and we obtain the system of functional relations:

$$\begin{bmatrix} \triangle\Omega \\ \triangle\omega \\ \triangle i \end{bmatrix} = \begin{bmatrix} \cos\epsilon + \cos\Omega\cot i\sin\epsilon & \sin\epsilon - \cos\Omega\cot i\cos\epsilon & -\sin\Omega\cot i \\ -\csc i\cos\Omega\sin\epsilon & \csc i\cos\Omega\cos\epsilon & \csc i\sin\Omega \\ \sin\Omega\sin\epsilon & -\sin\Omega\cos\epsilon & \cos\Omega \end{bmatrix} \begin{bmatrix} \triangle\xi \\ \triangle\eta \\ \triangle\epsilon \end{bmatrix} \quad (2)$$

4. The equinox correction induced by the rotation model

A local lifting $\triangle\delta_E = \triangle\eta$ in the equinox E implies a correction in their right ascension, in the following sense: $\triangle\alpha_E = \left(\frac{\partial\alpha}{\partial\lambda}\triangle\lambda + \frac{\partial\alpha}{\partial\beta}\triangle\beta\right)_E$. So, we compute the values $\triangle\lambda_E$ and $\triangle\beta_E$ by means of the equations:

$$R_x(-\epsilon)R_z(-\triangle\lambda_E)\vec{e}_1 = \begin{bmatrix} 1 \\ \triangle a \\ \triangle\delta_E \end{bmatrix} \qquad R_x(-\epsilon)R_y(-\triangle\beta_E)\vec{e}_1 = \begin{bmatrix} 1 \\ \triangle b \\ \triangle\delta_E \end{bmatrix}$$

where $\vec{e}_1$ is the unit vector pointing toward the equinox. Then, we obtain the values $\triangle a = \triangle \delta_E \cot \epsilon$; $\triangle b = \triangle \delta_E \tan \epsilon$; $\triangle \lambda_E = \frac{\triangle \delta_E}{\sin \epsilon}$ and $\triangle \beta_E = -\frac{\triangle \delta_E}{\cos \epsilon}$. From $\frac{\partial \alpha}{\partial \lambda}\big|_E = \cos \epsilon$ and $\frac{\partial \alpha}{\partial \beta}\big|_E = -\sin \epsilon$, we obtain the relation $\triangle \alpha_E = \triangle \delta_E(\cot \epsilon + \tan \epsilon)$, implying the true correction for the equinox:

$$\triangle E = \triangle \xi - \triangle \eta (\tan \epsilon + \cot \epsilon) \qquad (3)$$

5. Numerical Results

We have selected the four first minor planets, because a great number of observations in the FK4 System are provided for the MPC's magnetic tape. After solving the system (1), we obtain the σ_i^o values and then we apply the inverse relation (2), obtaining the following contributions for each minor planet and for each one of $\triangle \xi, \triangle \eta, \triangle \epsilon$.

TABLE 3. Amounts in parameter frame correction

$\triangle \xi$	1".765	1".312	1".202	1".980
$\triangle \eta$	0".543	0".328	0".084	0".619
$\triangle \epsilon$	0".217	-0".357	-0".190	0".376

Weighing these results with respect to the number of observations of each minor planet, we obtain the following results:

$$\triangle \xi = 1".541 \quad \triangle \eta = 0".384 \quad \triangle \epsilon = -0".02$$

and applying (3) we obtain the correction $\triangle E = 0^s.0325$ that agrees with the adopted value.

References

1. Bretagnon, P. and Francou, G.(1988) Variation Séculaire des orbites planétaires. Thèorie VSOP87, *Astron. Astrophys.* **202** 309-315.
2. Brouwer D. and Clemence G. (1961) *Methods of Celestial Mechanics* Academic Press.
3. Fricke W. (1982) Determination of the Equinox and Equator of the FK5, *Astron. Astrophys.* **107**, 13-16.
4. I.T.A. (1991) *Ephemeris of Minor Planets*, Academy of Sciences of the URSS.
5. Levallois, J.J.(1969) *Géodésie Générale*, Vol IV. Éditions Eyrolles.
6. López A., López J.A., López R. and Marco F.J. (1991) Precise ephemeris of minor planets. In: *Primer congreso Hipano-Soviético en Astronomía de posición y Mecánica Celeste*.
7. López A., López J.A. and Marco F.J. (1991) Present status of a new algorithm for orbital elements improvement. In: *Primer congreso Hipano-Soviético en Astronomía de posición y Mecánica Celeste*.
8. Simon, J.L. (1987) Calcul des dérivées premières et secondes des équations de Lagrange par analyse harmonique. *Astron. Astrophys.* **175**, 303-308.

ASTEROID MASS DETERMINATION: (1) CERES

M. CARPINO
Osservatorio astronomico di Brera, Milano

AND

Z. KNEŽEVIĆ
Astronomska opservatorija, Beograd

Recent improvements of observational accuracy, stellar catalogues and efficiency of the procedures for searching for asteroid close encounters (Kuzmanoski and Knežević, 1993) gave rise to an increased interest in the problem of determination of asteroid masses (Bowell et al., 1995).

We have made an attempt to determine the mass of the largest asteroid (1) Ceres by analysing variations due to close approaches to Ceres in the orbits of a total of 9 minor planets. The list of our objects includes all the asteroids used in previous attempts, as well as some additional ones for which potentially favourable close encounters were found by Kuzmanoski and Knežević (1994). The method we used consists in the simultaneous determination of the corrections to the orbital elements of the perturbed asteroid and of the mass of Ceres by employing a standard least square fit. The dynamical model included major planets from Mercury to Neptune, (2) Pallas and (4) Vesta. In particular the inclusion of Vesta proved to be important in some cases. For the planets we used initial conditions and masses from JPL DE200 ephemeris, while for Pallas and Vesta we used masses of 1.14×10^{-10} and $1.33 \times 10^{-10}\,M_\odot$ (Goffin, 1991). Observations were obtained from the Computer Service of the Minor Planet Center.

The results are summarized in Table 1, which is sorted in order of increasing formal RMS error of the derived mass. The RMS errors of the observations with respect to the fit (square root of the average value of $\Delta\alpha^2 \cos^2\delta + \Delta\delta^2$) range from 1.0 to 1.8 arcsec, and depend strongly upon the criterion chosen for outlier rejection. Masses found in the first five cases show a satisfactory mutual agreement: their weighted average amounts to $(4.67 \pm 0.09) \times 10^{-10}\,M_\odot$, which is quite close to the values found in the most recent determinations based on close-approach analysis (Goffin, 1991;

TABLE 1. Summary of the results. The columns contain: asteroid number, date of the close approach with Ceres, product of the close encounter distance ρ and velocity Δv (proportional to the reciprocal velocity change produced by the close approach), number and timespan of observations before (N_-, ΔT_-) and after (N_+, ΔT_+) the close approach, inferred mass (M) and its RMS error (δM)

Ast.	Cl. App. MJD	$\rho\Delta v$ 10^{-5} AU^2/d	N_-	ΔT_- y	N_+	ΔT_+ y	M $10^{-10} M_\odot$	δM
(203)	32785.26	3.9	26	68.9	70	45.4	4.77	0.07
(348)	45944.76	2.0	47	91.8	60	7.9	4.77	0.14
(324)	31180.67	10.9	68	52.0	520	49.0	4.54	0.16
(91)	41937.58	6.3	140	80.5	58	19.5	4.25	0.17
(534)	42769.86	3.6	46	71.7	18	16.3	4.19	0.36
(2572)	41036.47	3.3	2	21.1	31	21.5	3.38	0.43
(32)	42741.21	6.8	144	105.2	62	17.2	5.69	0.50
(3643)	41571.14	1.3	6	34.8	14	17.1	6.50	0.90
(2660)	44343.28	5.7	4	55.5	30	13.9	18.17	2.14

Williams, 1992; Bowell et al., 1995) as well as to the value of $4.64 \times 10^{-10}\, M_\odot$ adopted in the latest JPL DE403 Ephemeris (Standish et al., 1995). The other four cases deviate significantly from this value, exhibiting in the same time larger formal errors, due to different reasons:

(2572) Annschnell: the five critical pre-encounter observations collected in 1950 are of poor quality, and at least three of them have certainly to be discarded. A fit employing the remaining two observations produces the very low mass value given in the table; by discarding another observation we obtain $4.15 \times 10^{-10}\, M_\odot$, in better agreement with other more favourable cases. However, fit residuals alone do not give sufficient information to decide which observation should be discarded.

(32) Pomona: available data are not very sensitive to Ceres' mass, since the encounter was not very close and occured comparatively recently.

(3643) 1978 UN2: although the close encounter appeared very promising, there is only one available pre-encounter observation (made in 1937) which is far enough from the close approach date. Thus, the value of the mass depends critically on this observation, the accuracy of which cannot be assessed by statistical methods.

(2660) Wasserman: the orbit is poorly determined due to unfavourable distribution of observations, especially after the close encounter. This is revealed by high correlation coefficients between the orbital elements and, consequently, between the orbital elements and the inferred mass.

In order to infer which are the observations having larger impact on the

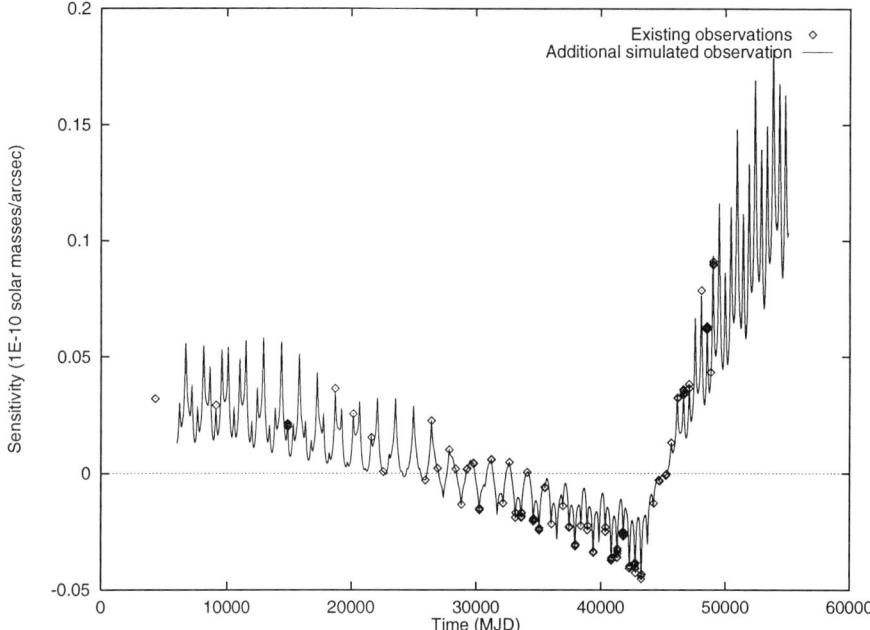

Figure 1. Sensitivity of the solution for the mass of Ceres with respect to the right ascensions of actual observations of (32) Pomona used in the fit (rhombs), and with respect to a fictitious additional observation (solid line)

mass determination, we computed the sensitivity of the fitted mass with respect to each observation, as well as to a possible additional measurement; in the latter case we added a fictitious observation and computed the least square fit using this enlarged data set, repeating the computation for a number of arbitrary instants of time within the time span covered by actual observations. The sensitivity is obtained as the partial derivative of the least square solution for the mass with respect to right ascension and declination. An example of this computation is given in Figure 1 for asteroid (32) Pomona.

The plot shows that post-encounter observations have larger influence on the mass, and hence suggests that additional future observations can significantly improve the result. Since this situation is not uncommon among the cases we have analyzed, we performed a series of simulations in order to establish whether future astrometric measurements (obtained for instance during a dedicated observational campaign) could improve the precision of the mass determination. To this purpose, we repeated the fits by adding fictitious observations distributed around oppositions in the next 10 years (1995–2005). The total number of new observations varied from 104 to 117 (depending on the asteroid). Their assumed accuracy was 0.5 arcsec in right ascension and declination, namely 0.71 arcsec in spherical distance; however, since we assigned equal weights to all the observations, the improvement in the error bar for Ceres mass results from the *number* and *distribution* of new data points rather than from their precision. In the case of (32) Pomona, the RMS error for Ceres mass obtained from the simulation is $0.21 \times 10^{-10} \, M_\odot$ (2.4 times smaller than the one obtained from the analysis of existing observations): this result would place Pomona among the cases which can give some information on the mass of Ceres, though not among the most precise ones. A substantial improvement of the result can be obtained also for (203) Pompeja (the error in the mass decreases by a factor 1.7), (348) May (factor 2.7), (91) Aegina (factor 2.2) and (534) Nassovia (factor 3.9). New observations contribute very little to the mass determination in the case of (324) Bamberga (RMS reduced by a factor 1.1): this is not surprising, since for this object astrometric observations after the date of the close approach with Ceres are already quite numerous. Although a significant increase in the formal precision can be obtained also for (3643) 1978 UN2 (by a factor 3.2) and (2660) Wasserman (factor 6.0), the resulting error bar remains anyway too large for the mass to be usable. As already noticed, a sufficient accuracy in the cases of (2660) Wasserman and (2572) Annschnell cannot be reached only by adding future astrometric data, but requires also good additional pre-encounter observations which could be possibly discovered on archive plates.

The results of the previous simulations can be summarized as follows: if we take into account the increased precision in the determination of the orbits of (203) Pompeja, (348) May, (91) Aegina, (534) Nassovia and (32) Pomona which can be achieved by a dedicated observing campaign lasting 10 years (as assumed above), we estimate an improvement in the *formal* error for the mass of Ceres from the present value of 9.1×10^{-12} to $4.6 \times 10^{-12} \, M_\odot$.

References

Bowell, E., Muinonen, K. and Wasserman, L.H. (1995) Asteroid mass determination from multiple asteroid-asteroid encounters, *Proceedings of Mariehamn*, in press

Goffin, E. (1991) The orbit of 203 Pompeja and the mass of Ceres, *Astron. Astrophys.* **249**, 563–568

Kuzmanoski, M. and Knežević, Z. (1993) Close encounters with large asteroids in the next 50 years, *Icarus* **103**, 93–103

Kuzmanoski, M. and Knežević, Z. (1994) Asteroid close encounters and mutual perturbations, *Planet. Space Sci.* **42**, 297–299

Standish, E. M., Newhall, X X, Williams, J. G. and Folkner, W. M. (1995) JPL Planetary and Lunar Ephemerides, DE403/LE403, *JPL IOM 314.10-127*

Williams, G.V. (1992) The mass of (1) Ceres from perturbations on (348) May, in Harris A.E. and Bowell E. eds., *Asteroids, Comets, Meteors 1991*, Lunar and Planetary Institute, Houston, Texas, 641–643

A METHOD FOR ASTEROID MASS DETERMINATION

M. KUZMANOSKI
Faculty of Mathematics
Studentski trg 16, 11000 Belgrade, Yugoslavia

It is well known that the perturbing effects of any asteroid are negligible at all times, except during a close encounter. The influence of a perturbing asteroid on the motion of a perturbed one can be investigated by means of the differences of the right ascensions and declinations of the perturbed asteroid, as inferred from two numerical integrations, the one with and the other without taking into account the effects of the perturbing asteroid (Kuzmanoski and Knežević, 1994). These effects before reaching the epoch of the close encounter are very small, as expected. Beyond this epoch, the differences increase with time, due to the fact that the orbit is changed on account of the influence of the perturbing body during the close approach, so that perturbations by major planets act on it in a different way.

The essence of this approach is the idea to separate pre–encounter and post–encounter orbits of the perturbed asteroid. In the calculation of these two orbits it is not necessary to know the mass of the perturbing asteroid, because its perturbing effects are practically zero. These two orbits are distinguished by an impulsive change due to the close encounter and can be connected by properly accounting for gravitational effects of the perturbing body. If the pre– and post–encounter orbits are completely precisely determined, the correct mass of perturbing body will give the best fit of the post–encounter observations with the pre–encounter orbit and the pre–encounter observations with the post–encounter orbit. It is essential that no other close approach (or at least not so close to affect the orbit) occurred in the time spans used for the orbits computation.

With such an approach, we have first calculated pre– and post–encounter orbits and examined how large effects one can expect in right ascensions and declinations for some asteroids which had close encounters with (1) Ceres and have already been used for its mass determination, and found that for (203) Pompeja they were the largest. In the time span covered by observations, two very close encounters occurred between (1) Ceres and

(203) Pompeja: the closest one (0.017 AU) in August 1948, and the second one (0.08 AU) in February 1969, while a few more approaches occurred within 0.5 AU. Next, two orbits were calculated: the first one by using the observations from the time span up to the epoch of the close encounter (41 observations at 19 oppositions and RMS residual 2″.01), and the other one by using the ones from the time span after the second close approach (48, 12, 1″.26). Numerical integrations were made by means of a Radau integrator of order 15, developed by Everhart (1985).

With the pre–encounter orbit we have fitted post–encounter observations (in turn used in calculation of the post-encounter orbit), and by means of the post–encounter orbit the pre–encounter observations (used in calculation of the pre–encounter orbit), setting the trial values for the mass of (1) Ceres and computing O–C residuals in the RMS sense. Obtained results are shown below:

TABLE 1. RMS residuals of post–encounter (a) and pre–encounter (b) observations for different trial masses (m)

	$1/m$ (10^9)	2.00	2.10	2.20	2.30	2.40	2.50	2.60	2.70	$m=0$
a)	RMS (″)	6.70	4.89	3.31	2.03	1.45	1.95	2.90	3.91	34.09
b)	RMS (″)	7.75	5.21	3.19	2.33	3.19	4.71	6.31	7.85	50.22

As can be seen, the best representation of post–encounter observations has been obtained for $1/m = 2.4 \cdot 10^9$ (RMS residual 1″.45), and of pre–encounter observations for $1/m = 2.3 \cdot 10^9$ (RMS residual 2″.33). This preliminary result is not in agreement with those derived by Goffin (1991) and Bowell et al. (1994) for the same close encounter. The difference could be to the inaccuracy of initial orbits, neglecting of the other close encounters influence, different integrators used, but also to the different methods for mass determination.

References

Bowell E., Muinonen K., Wasserman L. H. (1994) Asteroid mass determination from multiple asteroid–asteroid encounters, *Proceedings of Mariehamn*, in press.
Everhart, E. (1985) An efficient integrator that uses Gauss–Radau spacings, in Carusi A., Valsecchi G.B. eds, *Dynamics of Comets: Their Origin and Evolution*, 185–211
Goffin E. (1991) The orbit of 203 Pompeja and the mass of Ceres, *Astron. Astrophys.*, **249**, 563–568
Kuzmanoski M., Knežević Z. (1994) Asteroid close encounters and mutual perturbations, *Planet. Space Sci.*, **42**, 297–299

DYNAMICS AND ORBITAL EVOLUTION OF OORT CLOUD COMETS

J.Q. ZHENG, M.J. VALTONEN AND S. MIKKOLA
Tuorla Observatory, University of Turku, Finland

AND

H. RICKMAN
Astronomical Observatory, Uppsala University, Sweden

Investigators generally conjecture a steady flux of new comets from the Oort cloud through the inner Solar system. Due to gravitational perturbations by major planets these objects may escape, become long period comets (LPCs) if their orbital periods P are larger than 200yr or become short period comets (SPCs) when their period is less than 200yr. SPCs are further divided in two types: the Halley type comets (HT, for $P > 20yr$) and the Jupiter family comets (JF, for $P < 20yr$).

The most striking characteristic of the observed SPCs is the great preponderance of low inclination prograde orbits especially for JF comets, in comparison with the roughly random distribution of orbital inclinations of the long period comets. To explain the origin of SPCs some investigators suggest a disk-like source, Kuiper belt or an inner Oort cloud.

The most direct way to attack the problem of orbital evolution of Oort cloud comets is to integrate the orbits of large numbers of comets up to the age of the Solar system. But the required computational effort is overwhelming unless the problem is somehow simplified.

In view of the difficulty with direct integrations, we develop a Monte Carlo method which is outlined by Valtonen et al. (1992). The method is based on cross-sections (energy change distribution) obtained with accurate orbit integrations using a KS-regularization. At the same time, we obtained the distribution functions for changes of orbital elements inclination i and perihelion distance q corresponding to each value of the energy change ΔE and the values of the initial orbital elements i_0 and q_0 (see Zheng, 1994).

We assume that the orbit of a comet in the Solar system follows a constant ellipse for most of the time until occasionally the comet comes close to a planet and changes to a new orbit as a result of the gravitational influence of the planet. Then the comet keeps its new orbit approximately for several tens or hundreds of revolutions before the next close encounter.

In our Monte Carlo method, we first obtain the energy change of the comet using random sampling and the cross-sections which we have previously calculated, and then corresponding to the chosen value of the energy change we use the relative distribution functions of the new orbital elements and choose the new i and the new q of the cometary orbit also by random process.

Our starting point is the isotropic directional distribution of Oort cloud comets and the observed annual flux of new comets. We repeat the encounter process with four major planets and obtain new orbital elements for the comets until they leave the Solar system or become SPCs.

Generally in our simulation an Oort cloud comet makes several hundreds close encounters with major planets before it becomes a JF comet. But we find that the orbital evolution is dominated by a few very close encounters with large energy changes. We calculate the steady state population of the captured objects and record the orbital elements when they are such that the comets might be observed, i.e. if their $q < q_{lim}$, where $q_{lim} = 1.5$ or 2.6 AU in our simulations. Another parameter in the simulations is the number of close approaches to the Sun N_{max} during the active lifetime of a comet. Here 'close approach' means $q < q_{lim}$. We choose $N_{max} = 400, 1000$ or 4000 for different cases in our computations. After N_{max} close approaches to the Sun we assume that a comet becomes a 'dead' comet and it is left out of further calculation.

We have simulated several cases using different values of q_{min} and N_{max}. In each case we calculate $N_{model} = 1000000$ comets with initial perihelion distance q_0 between 0 and 30 AU distributed uniformly or non-uniformly. In our simulations, the uniform q_0 distribution cannot produce enough JF comets. Therefore we suppose that the relative annual number N (per AU) of Oort cloud comets increases with q_0 as $N = 1.0 + 0.014 \cdot q^{1.82}$ when $q < 13$ AU and that it has a constant value of $N = 5$ when $q > 13$ AU. This gives average value of N between 0 – 30 AU of 3.48.

For comparison with observations we have to know the real comet flux. The observed Oort cloud flux is 4.6 comets/AU/yr at the absolute magnitude $H_{10} = 10.8$ and at 1 AU (Bailey and Stagg 1988, Hughes 1988). But in our simulations, we only consider the encounters producing energy changes larger than a value of $\Delta E_{min} = 1/2000 \, AU^{-1}$ which means that only 1/6 of all new comets are captured during the first step. In addition, since the flux is expected to be on the increase at present (Matese et al. 1995), we further divide the observed Oort cloud flux by 1.5 in order to get a representative average flux for the past few million years. Then we get the Oort cloud flux appropriate for comparison with the simulations of $N_{new} = 0.5$ comets/yr/AU at 1 AU and the total flux over the q-interved of 30 AU of $N_{flux} = 30\,AU \cdot 3.48 \cdot N_{new}$.

Finally, the population of JF comets is $N_{flux} \cdot T_{life} \cdot N_{JF}/N_{model}$, where active lifetime of JF comets T_{life} is $P \cdot N_{max} = 8yr \cdot 400 = 3200yr$ (in case of $N_{max} = 400$), and N_{JF}/N_{model} is the probability of an Oort cloud comet to become a JF comet in our simulations.

We obtain $N_{JF} = 3044$ JF comets in our simulation with a non-uniform q-distribution and $N_{max} = 400$. Substituting this to the obove calculation we obtain the population of 508 JF comets with $q < 2.6$ AU. The median value of $\cos i$ is 0.985. Figure 1 shows the comparison between our simulation results and the observed inclination distribution of JF comets. Figure 2 shows the comparison for the q-distributions.

We summarize our results as follows:

(1) We agree with Fernández and Gallardo (1994) that single planet encounters by Jupiter cannot produce JF comets with a low inclination distribution from the Oort cloud comets. But multiple encounters could do that.

(2) The distribution function of energy changes is not symmetrical between negative energy changes and positive energy changes even for small changes. For that reason most comets with near 0° inclinations drift inwards in the diffusion process while only very few comets drift inwards with near 180° inclinations.

(3) Due to the small masses of the outer planets relative to Jupiter, the encounters with the outer major planets produce a strong selection of low inclination distribution of SPCs.

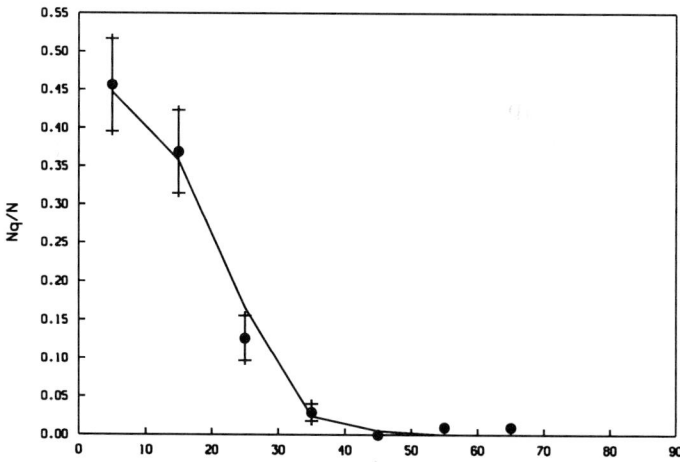

Fig. 1. The distribution of inclination i of Jupiter family comets. Solid line: simulation results; dots with error bars: observations.

(4) No HT comets with inclination less than 18° or larger than 162° have been observed (Marsden and Williams, 1994). It may be impossible for a retrograde comet to become a SP comet near Jupiter's orbital plane. But for low inclination orbit with $i < 18°$, the transfer process to a JF comet is one to two orders of magnitude faster than at the other inclinations. Therefore we may not see these comets when they are in the short-lived HT comets phase. Then we can say that the HT comets form a transfer source of JF comets from the Oort cloud.

In conclusion, we find it entirely possible that the SPCs are a captured subpopulation of the Oort cloud comets, and it is not necessary to assume the existence of additional comet sources to keep up the populations of the Jupiter family and the Halley type comets.

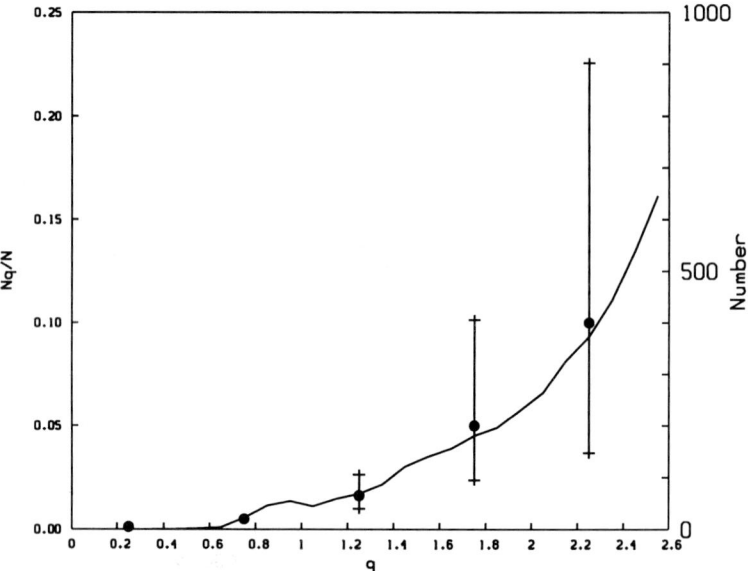

Fig. 2. The distribution of perihelion distance q of Jupiter family comets. Solid line: simulation results; solid circles with error bars: extrapolated numbers from observations by Fernández et al.

References

Bailey, M.E. and Stagg, C.R.: (1988) *MNRAS*, **235**, 1.
Fernández, J.A. and Gallardo, T.: (1994) *A & A*, **281**, 911.
Hughes, D.W.: (1988) *Icarus*, **73**, 149.
Marsden, B.G. and Williams, G.V.: (1994) *Catalogue of Cometary Orbits*, Smithsonian, pp. 80-83.
Matese, J.J., Whitman, P.G., Innanen, K.A. and Valtonen, M.J.: (1995) *Icarus*, **116**, 255.
Valtonen, M.J., Zheng, J.-Q. and Mikkola, S.: (1992) *Celestial Mech.* **54**, 37.
Zheng, J.-Q.: (1994) *A & A Sup.*, **108**, 253.

HYPOTHETICAL EVOLUTIONS OF SHORT-PERIOD COMETS

V. BATLLO
Bureau des longitudes, 0707 CNRS
77 avenue Denfert Rochereau, F.75014 Paris

1. Description of the scenario

The hypothetical model of capture I consider is as follows: a comet with an initial conic orbit, meets close to one of its vertices an outer planet and generates one or several little comets (crossing of the Roche limit) with an elliptic orbit.This initial vertex always remains one of the vertices of the captured orbit by the Solar System. Even if there is a discontinuity of all the orbital elements during a very short time of the capture, I assume that the orbital plane is unchanged during the capture, as indeed was the case for Brooks 2 in 1886 (see Belyaev et al., 1986) or will be for Gehrels 3 in 2300 (see Carusi et al., 1985). After this first decisive step, the "new" comets may be subject to some other captures by jovian planets during their evolution with the same scenario. All these hypotheses allow to find particular numerical results for 142 comets whose current orbital elements were found in the Marsden catalog or given by P. Rocher, but I shall only give several examples in the following section.

2. Numerical results

There are three kinds of comets whose aphelion distance is near a planet since the first encounter, which were, before the capture, nearly-parabolic (Wirtanen: $e=1.001$, $q=5.14$), hyperbolic (Shoemaker Levy 4: $e=1.137$, $q=4.96$) or had an initial aphelion distance beyond the Pluto orbit (Finlay: $e=0.893$, $q=5.53$, $Q=97.96$). Also, there are in this category, two other kinds of comets which were respectively subject to the joint action of the two planets Jupiter-Saturn (Hartley 1, Mueller 3) and Saturn-Uranus (Faye, Shoemaker 4) before the first capture. After the first encounter and the capture by the Solar System, some comets have still changed of group once (Temple 2: Neptune, Jupiter) or even twice (Gunn: Neptune, Uranus, Jupiter).

There are a group of comets which remained in the neighbourhood of a planet during a more or less long time, with a nearly-circular orbit before recovering an elliptic orbit: for Jupiter: Ashbrook Jackson, Russell 1 and Tritton; for Saturn: Kowal Vavrova. There exist some comets which "always" belong to the same planet group: for Jupiter: Borrelly and Wolf; for Saturn: Iras and Tuttle. Finally, let us consider the case of comets which directly came from the neighbourhood of Neptune and Pluto. Hence, they have not the same origin as the other comets since they had neither an hyperbolic or quasi-parabolic initial orbit, nor an initial orbit beyond the Pluto orbit. An explanation can be found in the paper written by S.K. Vsekhsvyatskij and A.S. Guliev (1981). The three comets are Daniel (Pluto, Uranus, Saturn, Jupiter), Grigg Skjellerup (Neptune, Uranus, Jupiter) and Peters (Neptune, Saturn).

After each change of the orbital elements during an encounter with the same planet, three comets become rectilinear. These comets are for Jupiter, Machholtz; for Neptune, Pons Brooks; for Pluto, de Vico.

3. Conclusion

All these numerical results show that there are several possible origins for the short-period comets. They can have an hyperbolic or nearly-parabolic initial orbit (22%). Some others have an initial orbit beyond the Pluto one (29%), that could prove the hypothetical existence of the Kuiper Belt. The study of the previous part also show that some comets were originated in the neighbourhood of a planet or "always" belong to a particular jovian planet group. The joint action of two great planets (Jupiter-Saturn and Saturn-Uranus) generates two groups of comets whose aphelion distance was located between the two respective planets orbits before the first capture (36%). The last case regards the rectilinear comets which also belong to the same planet group. What is the origin of these last three kinds of comets? This question must be studied and developed.

References

Belyaev, N.A., Kresak, L., Pittich, E.M., Pushkarev, A.N. (1986) *Catalog of Short-Period Comets*, A.I.S.A.S. Bratislava.
Carusi, A., Kresak, L., Perrozi, E., Valsecchi, G.B. (1985) *Long-term evolution of short-period comets*, Adam Hilger Ltd. Publsh., Bristol, U.K.
Marsden, B.G. (1995) *Catalog of Cometary Orbits*, C.B.A.T., 4th edit, Cambridge, U.S.A.
Rocher, P. (1995) *Notes Cométaires du Bureau des Longitudes*, server ftp.bdl.fr
Vsekhsvyatskij, S.K., Guliev, A.S. (1981) *Soviet Astronomy*, **25**(3), 358.

TAURID METEOROIDS AND ASTEROID COMPLEX

E. M. PITTICH
Astronomical Institute of the Slovak Academy of Sciences
Dúbravská cesta 9, 842 28 Bratislava, Slovak Republic

AND

J. KLAČKA
Department of Astronomy and Astrophysics
Faculty for Mathematics and Physics
Comenius University, Mlynská dolina, 842 15 Bratislava
Slovak Republic

Abstract.
The orbital evolution of model meteoroids ejected from the comet Encke and from the asteroid Hephaistos was investigated. The particles abandoned the mother body with velocities 40, 150 and 600 ms^{-1} at their perihelia within the interval of 10,000 to 20,000 years, respectively. Their 10,000 to 20,000 years old osculating orbits were numerically integrated forward, using a dynamical model of the solar system consisting of all planets. Forces from solar electromagnetic and corpuscular radiation effecting the particles were considered, too. Orbital dispersions of the model meteoroids, ejected in different epochs, compared with those obtained from observations. It seems probable that the comet Encke and the asteroid Hephaistos are sufficient to produce the observed Taurid meteor complex.

1. Introduction

The long-term orbital evolution of model meteoroids of the comet Encke showed that the observed distribution of the longitude perihelion for the whole Taurid meteoroid complex cannot be explained by this comet alone. It is not possible even in the case when the solar radiation effects are taken into account (Klačka and Pittich, 1994, 1995). It is generally believed that the source of the Taurid meteor complex are both the Taurid asteroid complex and the comet Encke (Asher et al, 1993; see also Klačka, 1995). The meteoroids with the longitude of perihelion higher than 160° had to be ejected from an asteroid or asteroids belonging to the hypothetical Taurid asteroid complex. The aim of the present article is to study the possibility of the Taurid stream creation, to find an asteroid which can explain the observed distribution of the longitudes of perihelia higher than 160°, and to ascertain under which conditions this situation may occur.

2. Computational Model

We traced the 10,000 years orbital evolutions of the model particles ejected from the Encke comet perihelion using the numerical integration with the RA15 integrator (Everhart, 1985). Similar process was applied to the model particles ejected from the Hephaistos asteroid perihelion for a twice longer period. The input data for the integration, ecliptical rectangular coordinates and velocities, were calculated from the Encke comet orbital elements for the epoch 1987 July 24 (Marsden, 1989) and from the Hephaistos asteroid orbital elements for the epoch 1993 August 1 (Batrakov, 1992).

The adopted ejection velocity 40 m s^{-1} corresponds to the value determined by Sykes and Walker (1992) for the comet Encke, 150 m s^{-1} is the value calculated by Gajdošík and Klačka (1995) from the same data as used by the above mentioned authors. The value 600 m s^{-1} seems to be realistic for some comets and asteroids (Harris et al, 1995). The model particles were released from the Encke comet perihelion 10,000 years ago in three directions – tangential to the cometary motion, normal to the comet's orbital plane, and perpendicular to the preceding ones – and for two orientations in each direction. The same procedure was employed in the case of another parent body, the asteroid Hephaistos, except that the model particles were ejected from its perihelion 15,000 and 20,000 years ago. The motions of both the comet Encke and the asteroid Hephaistos were calculated, too.

We have made 10,000 years backward numerical integration of the Encke comet orbit, beginning from the epoch 1987 July 24, and 20,000 years backward numerical integration of the Hephaistos asteroid orbit beginning from the epoch 1993 August 1. Then, the forward numerical integration for the model dust particles – meteoroids – was performed, taking into account the gravitational perturbations of all planets as well as nongravitational effects – solar electromagnetic and corpuscular radiations – acting on the model meteoroids.

Particles with the ratio of the radiation force to the gravity force $\beta = 5.7 \times 10^{-5}/(\varrho s) = 4 \times 10^{-4}$ were taken into account; ϱ is the mass density measured in g cm^{-3} and s is the radius of the particle in cm. For dimensions of particles with such β see Table 1 in Klačka and Pittich (1994).

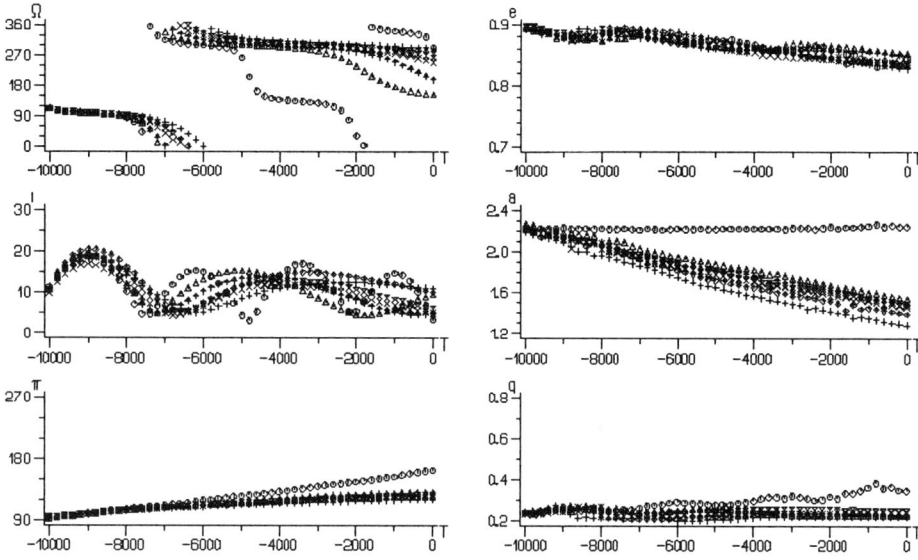

Figure 1. Time evolution of orbital elements of the comet Encke ⊙ — 0 m s^{-1} (ejection velocity), and of the model meteoroids ejected from the Encke comet perihelion 10,000 years ago. The meteoroids were disturbed by gravitational perturbations of planets and solar electromagnetic and corpuscular radiations. For the meteoroids $\beta = 4 \times 10^{-4}$. The particle parameters are: △ — +150 m s^{-1} in the direction of the cometary motion, + — −150 m s^{-1} in the direction of the cometary motion, × — +150 m s^{-1} in the direction normal to the comet's orbit, ◊ — −150 m s^{-1} in the direction normal to the comet's orbit, ⁺ — +150 m s^{-1} in the direction perpendicular to the preceding directions, ⊠ — −150 m s^{-1} in the direction perpendicular to the preceding directions.

3. Results of Orbital Integrations

The results of the orbital integrations for the longitudes of perihelia are given in Table 1. For different cases, the evolution of the osculating elements is plotted on Figures 1–3. The symbols of the orbital elements are standard: ω – argument of perihelion, Ω – ascending node, i – inclination, π – longitude of perihelion, e – eccentricity, a – semi-major axis, q – perihelion distance, Q – aphelion distance. Fig. 1 shows the evolution of these elements for the model particles ejected from the Encke comet perihelion 10,000 years ago with the velocity 150 m s^{-1} in all selected directions. Figs. 2 and 3 show the evolution of the orbital elements of the same model particles ejected from the Hephaistos asteroid perihelion 15,000 and 20,000 years ago, respectively. In both cases the model particles were ejected with the velocity of 150 m s^{-1} in the selected directions.

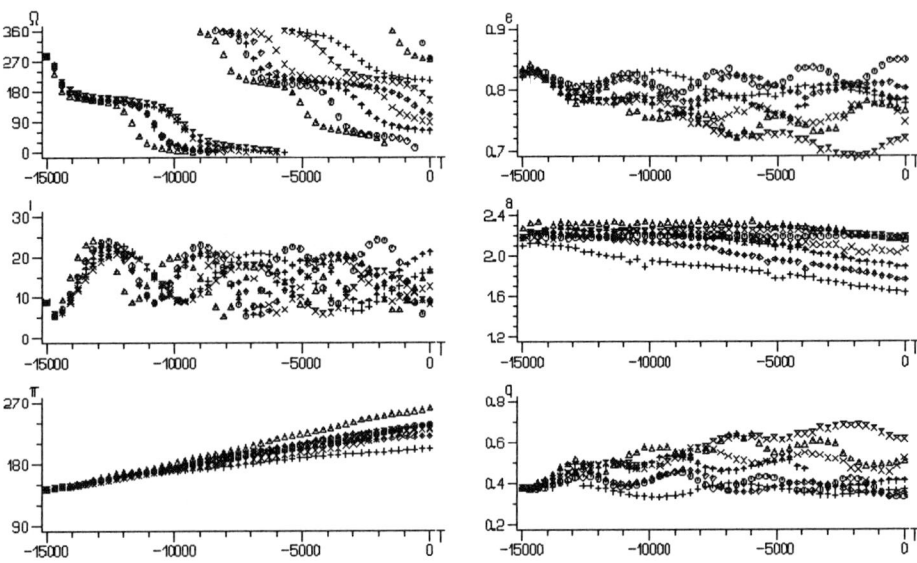

Figure 2. Time evolution of orbital elements of the asteroid Hephaistos ⊙ — 0 m s^{-1} (ejection velocity), and of the model meteoroids ejected from the Hephaistos asteroid perihelion 15,000 years ago. The meteoroids were disturbed by gravitational perturbations of planets and solar electromagnetic and corpuscular radiations. For the meteoroids $\beta = 4 \times 10^{-4}$. The particle parameters are: △ — +150 m s^{-1} in the direction of the asteroid motion, + — −150 m s^{-1} in the direction of the asteroid motion, × — +150 m s^{-1} in the direction normal to the asteroid's orbit, ◇ — −150 m s^{-1} in the direction normal to the asteroid's orbit, ✦ — +150 m s^{-1} in the direction perpendicular to the preceding directions, ✕ — −150 m s^{-1} in the direction perpendicular to the preceding directions.

4. Discussion

The Taurid meteor stream is characterized by the longitude of perihelion $\pi \in (100°, 200°)$ determined from observations already described (Štohl and Porubčan, 1992). From Fig. 1 and Table 1 it can be seen that the observed distribution of the longitudes of perihelia for the Taurid meteoroid complex cannot be explained by the Encke comet alone, if other nongravitational forces, besides those taken into consideration, are neglected. Some other nongravitational forces effecting the distribution of the longitudes of perihelia of model particles heve been dealt with in our earlier paper (Pittich and Klačka, 1994). The evolution of the orbital elements of model particles ejected with smaller velocities from the Encke comet perihelion is described in our preceding papers (Klačka and Pittich, 1994, 1995).

In order to obtain also particles with $\pi > 160°$ we have to take into account some other parent body, or bodies which at present, are probably dormant comets. Our calculations with such a body, the asteroid Hephaistos, show that simultaneous acting in the past both of the Encke comet

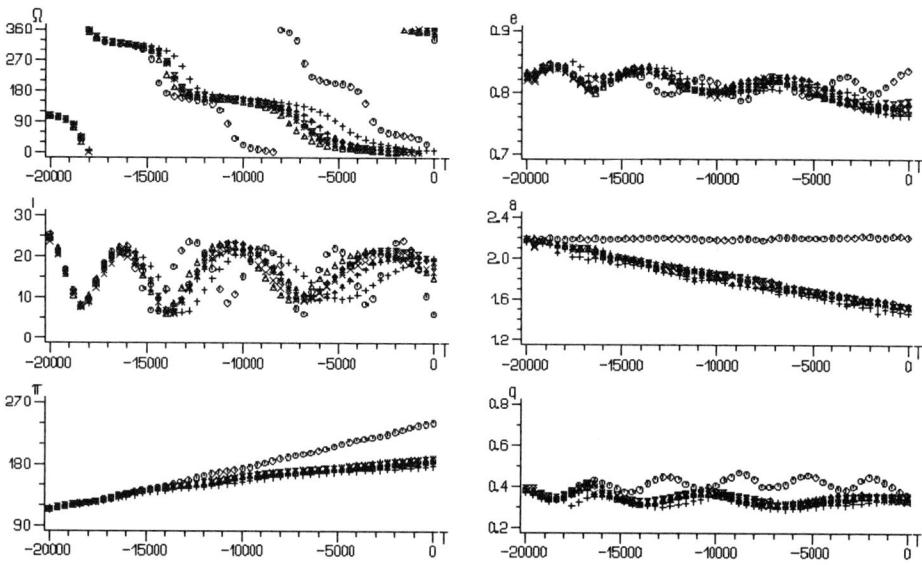

Figure 3. Time evolution of orbital elements of the asteroid Hephaistos ⊙ — 0 m s^{-1} (ejection velocity), and of the model meteoroids ejected from the Hephaistos asteroid perihelion 20,000 years ago. The meteoroids were disturbed by gravitational perturbations of planets and solar electromagnetic and corpuscular radiations. For the meteoroids $\beta = 4 \times 10^{-4}$. The particle parameters are: △ — +150 m s^{-1} in the direction of the asteroid motion, + — −150 m s^{-1} in the direction of the asteroid motion, × — +150 m s^{-1} in the direction normal to the asteroid's orbit, ◊ — −150 m s^{-1} in the direction normal to the asteroid's orbit, ✦ — +150 m s^{-1} in the direction perpendicular to the preceding directions, ✕ — −150 m s^{-1} in the direction perpendicular to the preceding directions.

and this asteroid might explain the distribution of the longitudes of perihelia of the Taurid meteoroids derived from observations. If Hephaistos became a nonactive body 20,000 years ago, the maximum ejection velocity, producing meteoroids with the present 200° longitude of perihelion, would have to be 600 m s^{-1}. The perihelion distance of the Hephaistos asteroid 15,000–20,000 years ago ($q \sim 0.35$ AU) was similar to the longitude of perihelion of the Encke comet at present. Thus, in the first approximation, we can assume similar ejection velocities for particles ejected from the comet Encke now and for those ejected from Hephaistos 15,000–20,000 years ago.

5. Conclusion

The model particles ejected from Hephaistos 20,000 year ago with the ejection velocity 600 m s^{-1} together with the particles of the comet Encke are sufficient, to cover the values of the longitudes of perihelia less than 200° derived from observations (Table 1). The consequence of this result is that

TABLE 1. Longitudes of perihelia of model particles for two parent bodies.

Ejection velocity	Encke 10,000 years ago	now	Hephaistos 15,000 years ago	now	Hephaistos 20,000 years ago	now
m s^{-1}	°	°	°	°	°	°
40	94	125–132	143	213–248	114	184–191
150	94	121–131	143	202–262	114	177–189
600	94	111–152	143	195–349	114	165–198

Hephaistos had to be inactive during the last 20,000 years. All these conclusions hold if nongravitational effects different from solar radiations are negligible.

Acknowledgements

This work was supported by the Slovak Academy of Sciences Grant No. 2/1050/1995 (E. M. Pittich).

References

Asher, D.J., Clube, S.V.M., and Steel, D.I. (1993) The Taurid complex asteroids, *Meteoroids and their parent bodies*, (J. Štohl and I.P. Williams, Eds.), pp. 93–100, Astronomical Institute, Bratislava.
Batrakov, Ju.V. (1992) *Ephemerides of minor planets for 1993*. Institute of Theoretical Astronomy, Sankt Peterburg.
Everhart, E. (1985) An efficient integrator that used Gauss-Radau spacing, *Dynamics of Comets: Their Origin and Evolution*, (A. Carusi and G.B. Valsecchi, Eds.), pp. 185–202, Reidel, Dordrecht.
Gajdošík, M. and Klačka, J. (1995) Cometary Dust Trails and Ejection Velocities from Comets, *Astron. Astrophys.*, in preparation.
Harris, N.W., Yau, K.K.C. and Hughes, D.W. (1995) The True Extent of the Nodal Distribution of the Perseid Meteoroid Stream, *Mon. Not. R. Astron. Soc.*, **Vol. no. 273**, pp 999–1015.
Klačka, J. (1995) The Taurid Complex of Asteroids, *Astron. Astrophys.*,**Vol. no. 295**, pp. 420–422.
Klačka, J. and Pittich, E.M. (1994) Long-term Integration of Dust Particles Released from Comet Encke, *Planetary and Space Science*, **Vol. no. 42**, pp. 109–112.
Klačka, J. and Pittich, E.M. (1995) Orbital Dispersion of Comet Encke's Meteoroids, *Earth, Moon, and Planets*, in press.
Marsden, B.G. (1989)*Catalogue of Cometary Orbits*. Cambridge, Massachusetts.
Pittich, E.M. and Klačka, J. (1994) On the Applicability of the Poynting-Robertson Effect on Meteoroids, *Small Bodies in the Solar System and Their Interactions with the Planets*, Mariehamn, Finland 1994, in press.
Sykes, M.V. and Walker, R.G. (1992) Cometary Dust Trails. I. Survey, *Icarus*, **Vol. no. 95**, pp. 180–210.
Štohl, J. and Porubčan V. (1992) Dynamical aspects of the Taurid meteor complex, *Chaos. Resonances and Collective Dynamical Phenomena in the Solar System*, (S. Ferraz-Mello, Ed.), pp. 315–324, Dordrecht, Holland.

MODEL OF THE COMET OUTBURSTS BASED ON A FRAGMENTATION OF ICE GRAINS IN ITS ATMOSPHERE

D.A. ANDRIENKO AND I.I. MISCHISHINA
Kiev University

Outbursts of the comet's brightness play a significant role in the study of the physical nature and evolution of comets. The causes of cometary outbursts are still not properly understood. It is known, however, that there is correlation between the outbursts and the high-velocity flux of solar wind. The evidences of such causal relationship are the following: the existence of the correlation between the geomagnetic perturbations and the outburst activity of the comets [1]; two-peaks distribution of the comet's outbursts, which depends on the 11-year solar cycle phase with maxima at phases 0.2–0.3 and 0.7–0.8, in coincidence with the distribution of physical characteristics of high-velocity fluxes [2]; the repeated character of the outbursts with main intervals of 7–8, 14–15, 22–24 and 30 days, this corresponds to the four-sector structure of the interplanetary magnetic field [3].

We think that the generation of outbursts requires the existence of some favourable internal conditions in the comet's head besides corpuscular external effects. In our model approximation, the comet nucleus is treated as a conglomerate of different ices and refractory impurities. The surface layer of such nucleus may be destructed periodically being accompanied by ejection of large icy grains into the comet's coma and forming a halo at its periphery. Similar processes of icy grain ejection were observed during comet simulation experiments performed by Kajmakov[4] and the KOSI group [5]. Due to the presence of the structural imperfections such as cracks and refractory impurities, the icy grains are unstable and may be destroyed by the energy particle of solar wind. The process of the halo's icy grain destruction results in increasing of the total reflection of the solar light, giving origin to additional atmospheric dust source and strengthening the extended gas source. The accurate solution of the behavior of the cometary brightness during the process of icy grain halo's destruction is a complex problem. But quite simple numerical simulations of the time dependence of the intensity

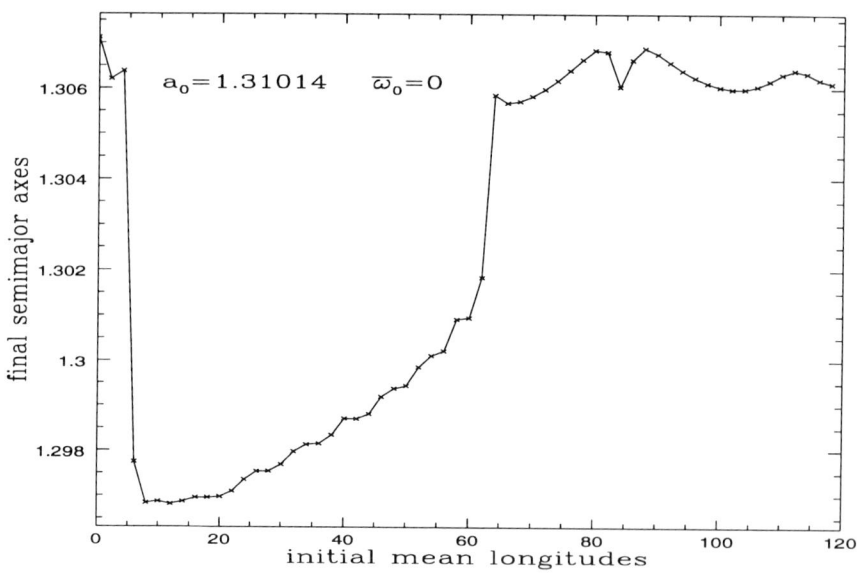

Figure 1.
Initial mean longitudes against semimajor axes after 1200 *years numerical integration of the orbits of* 60 *particles subject to Poynting-Robertson drag* ($\beta = 0.01$), *started just above the 2:3 resonance with the Earth. Other elements are* $a = 1.31014$, $e = 0.05$ *and* $\varpi = 0°$, *for all particles.*

We see that near the resonance there is a continuous range of mean longitudes associated to particles that are trapped (this feature is not however observed in all cases. Some examples show this longitude region split into two). For initial mean longitudes from 120° to 360° this same configuration will repeat itself twice, what is expected from the fact that the resonant angle $\phi = 3\lambda - 2\lambda_T - \varpi$ varies 3 times as fast as λ.

For our next experiment, we slightly change the initial semimajor axis ($a = 1.31023$), other conditions being the same as those of figure 1. Figure 2 shows which particles are trapped for this different initial semimajor axis. The first conclusion we may draw from this figure is that there is no point in taking capture probabilities from the results shown in figure 1, because the number of trapped particles varies considerably with the semimajor axis. This last result almost unavoidably yields our next experiment, which is plotting the number of trapped particles against their initial semimajor axes. This is shown in figure 3. Here we notice peaks of relatively high number of trapped particles between valleys of relatively low number of trapped particles. For an interpretation of this diagram, we first measure

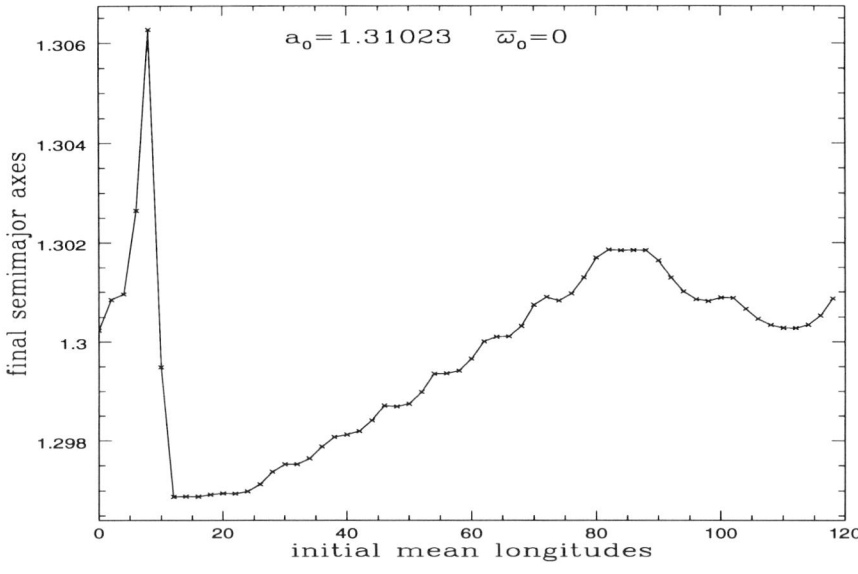

Figure 2.
Like figure 1, for another initial semimajor axis ($a = 1.31023$). The number of trapped particles is quite different from that of figure 1.

the distance between successive peaks. This distance is not constant and it is not difficult to verify that it refers to the variation of the semimajor axis due to the Poynting-Robertson dissipation in a ϕ-period. We postpone a better explanation of this point after we analyze our next experiment.

Figures 4a and 4b show the evolution of the resonant angles ϕ at every 10 years from an initial uniform distribution through 360°. These figures suggest another reason why figure 1 is not suitable to give capture probabilities. In fact, near a resonance, the initial uniform distribution of ϕ on a circle is artificial. After a short time the particles resonant angle accumulate near a point, due to the fact that $\dot{\phi}$ has a relative high variation near $\dot{\phi} = 0$. Figure 4a refers to an initial semimajor axis ($a = 1.31014$) (same as Figure 1), which generated high number of captured particles, whereas figure 4b refers to $a = 1.31023$ (same as Figure 2) and just one particle is captured. These figures suggest that the value of ϕ where $\dot{\phi} = 0$ for the first time is related to whether the particle is going to be captured or not. So these last two conclusions taken from figures 4a and 4b lead us to an explanation of figure 3. The fact that the resonant angles get and run together near a resonance associated to the fact that these angles will have their time derivative vanishing in a region that leads or not to capture gives a good explanation for the peaks and valleys of figure 3.

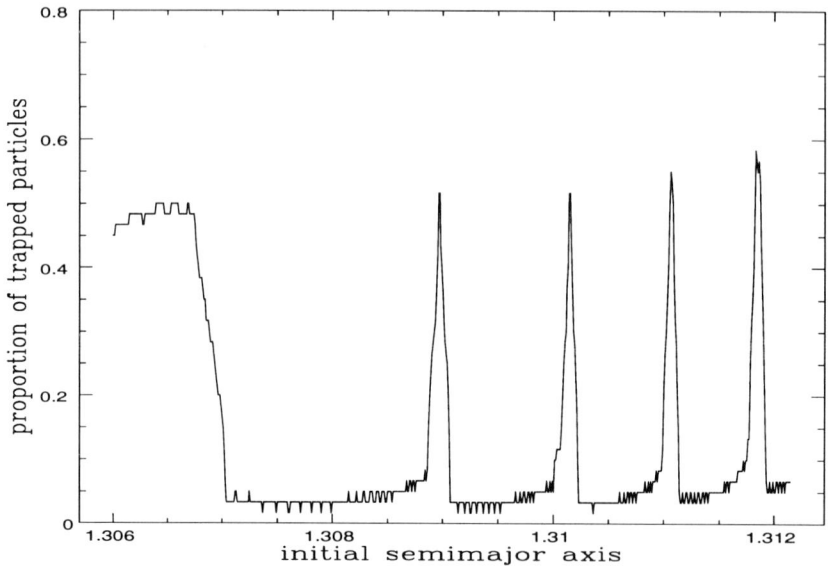

Figure 3.
Proportion of trapped particles against initial semimajor axis, all other initial orbital elements like those of figure 1.

Our last numerical experiment tries to better understand the association of a region of eccentricities and resonant angles where $\dot\phi = 0$ for the first time (return points) with capture or non capture. Figure 5 shows (P, ϕ) points, where $P = na^2(1 - \sqrt{1-e^2})$, associated to the time when $\dot\phi = 0$ for the first time. Larger dots belong to captured particles and the smaller ones belong to non captured particles. We start the integrations with sets of equal eccentricities and different semimajor axes. These elements vary from a minimum value associated to an early separatrix crossing (near 360°) to a maximum value associated to a late separatrix crossing (near 0°), but the separatrix crossing occurs always before a complete turn of the angle ϕ (note that the motion is clockwise). For each eccentricity, the return points lie in approximate arcs of circles as the semimajor axis varies. Here we use several values of β. Smaller β approximate the adiabatic case. We notice here that some general ideas given by the theory for the adiabatic case are also present in the non adiabatic case. Reminding the analytical model (Henrard,1982; Peale,1986), captured particles cross the separatrix early, so that they have time to get the lowest possible energy level before they start to have their resonant angles running in the opposite way. $\dot\phi = 0$ in the lowest levels of energy means $\dot\phi = 0$ for the highest values of ϕ, as found in

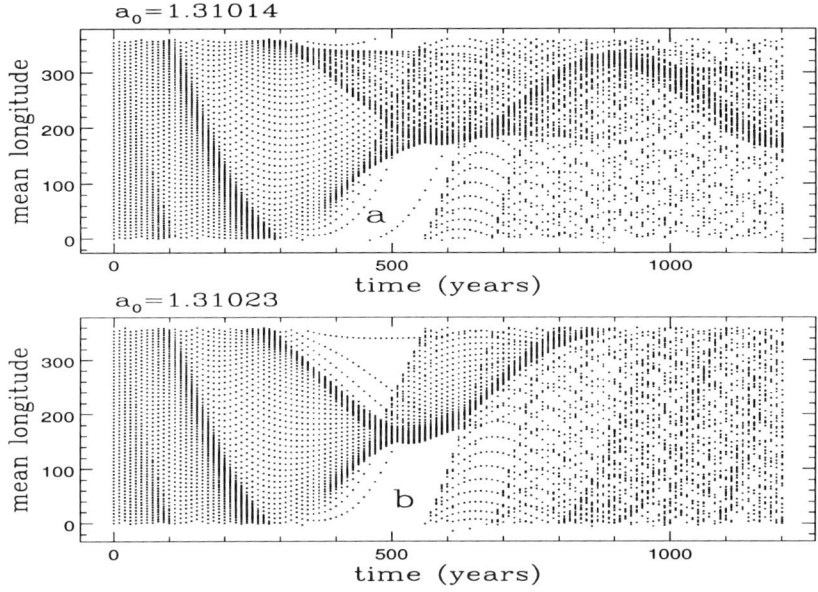

Figure 4.
Evolution of the mean longitudes at every 10 years for 60 particles. (a) the same conditions as in figure 1. (b) the same conditions as in figure 2.

every plotting for low and high β. When we get closer to the adiabatic case (lower β) the return points are closer to 0° as a whole. This is explained by the fact that $\frac{d(H-H^*)}{dt}$, where H is the Hamiltonian associated to the evolving orbit and H^* is the Hamiltonian associated to the separatrix, is proportional to β. Thus, for small β, the return point belongs to a libration curve near the separatrix. These curves have return points near 0°. We also see the variation of the number of captured particles with eccentricity, which for the adiabatic case is higher as smaller is the eccentricity and for faster dissipation there is a peak of highest capture probability away from $e = 0$. This peak is associated to as high e as higher is β, but the height of the peak gets lower itself. These last results confirm (Gomes, 1995).

References

Gomes,R.S.: 1995, 'Resonance trapping and evolution of particles subject to Poynting-Robertson drag: adiabatic and non-adiabatic approaches', *Celestial Mechanics and Dynamical Astronomy* **61**:1, pp. 97-113

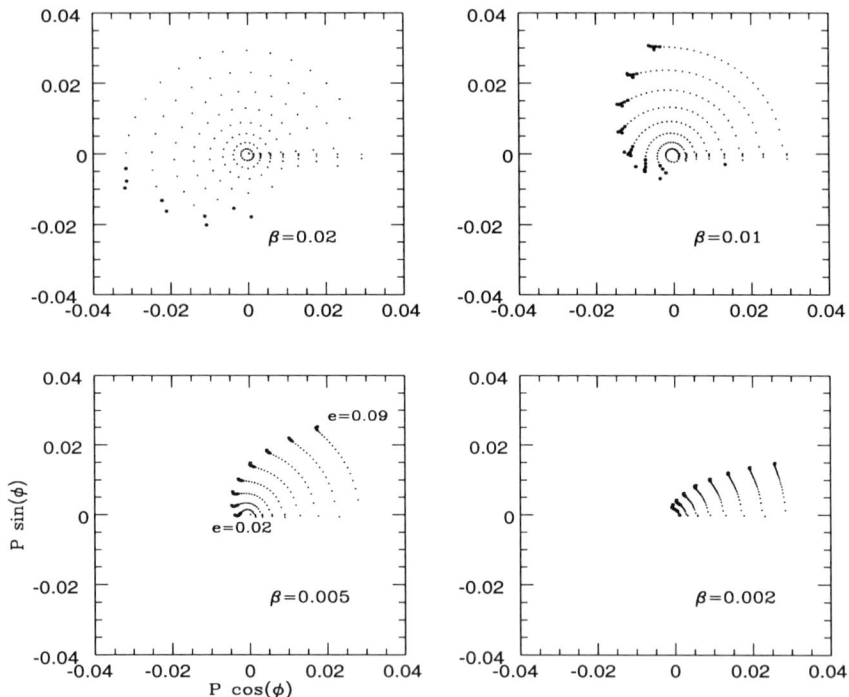

Figure 5.
(P, ϕ) *points where* $\dot{\phi} = 0$ *for the first time, for several initial eccentricities and several* β. *In each figure the eccentricity varies from* 0.02 *to* 0.09. *The return points associated to resonance trapping are printed larger. For each* β *and* e, *the several points located on approximate arcs of circles correspond to different initial semimajor axes. All integrations start with* $\phi = 0$ *(*$\lambda = 0$, $\varpi = 0$ *and* $\lambda_T = 0$*).*

Henrard,K.: 1982, ' Capture into resonance: an extension of the use of adiabatic invariants', *Celestial Mechanics* **27**, pp. 3-22

Lazzaro,D., Sicardy,B., Roques,F., Greenberg,R.: 1994, 'Is there a planet around β Pictoris? Perturbations of a planet on a circumstellar dust disk.', *Icarus* **108**, pp. 59-80

Lemaitre,A.: 1984, 'High-order resonances in the restricted 3-body problem.', *Celestial Mechanics* **32**, pp. 109-126

Marzari,F. and V.Vanzani: 1994, 'Dynamical evolution of interplanetary dust particles.', *Astron. & Astrophys.* **283**, pp. 275-286

Peale,S.J.: 1986, 'Orbital resonances, unusual configurations and exotic rotation states among planetary satellites.' *in Satellites (eds. Burns, J.A. and Mathews, M.S.), University of Arizona Press* pp. 159-223

EJECTA TRANSFER BETWEEN TERRESTRIAL PLANETS

B.J. GLADMAN AND J.A. BURNS
Dept. of Astronomy, Cornell Univ., Ithaca, NY, 14853, USA
H. LEVISON
SWRI, 1050 Walnut St., Boulder, CO, 80302, USA
AND
M.J. DUNCAN
Dept. Astronomy, Queen's Univ., Kingston, K7L 3N6, Canada

As meteorites from the Moon and Mars continue to be discovered, it is increasingly clear that impact fragments can escape from large bodies more easily than previously believed. These escaping fragments are then subject to the gravitational perturbations of the planets, allowing them to be transferred to a body other than their parent. The lunar meteorites and SNC meteorites prove the plausibility of this process. Warren (1994) summarizes cosmic ray exposure ages and other properties of the lunar and martian meteorites. Their existence confirms that lightly shocked material can be launched at greater than the escape speed of the Moon and Mars.

We study the transfer dynamics by means of N-body simulations, considering the evolutions of impact ejecta escaping from the Moon, Mercury, and Mars. Impact ejection from Venus and the Earth, with their massive atmospheres, is less likely to be feasible. Particles were launched off the target's surface at a variety of velocities slightly above the escape speed. The lunar case is somewhat different, since often the material has first to be followed through a number of geocentric orbits until the particles escape to heliocentric space [here most ejecta ($> 80\%$) are boosted by close encounters with the Moon to speeds sufficient to escape from orbits bound to the Earth]. In all cases, we track the material that reaches heliocentric orbit using a regularized mixed-variable-symplectic integration package. The particles are followed for between 2 and 40 Myr, until they impact a terrestrial planet, cross the orbit of Jupiter, or evolve to a Sun-grazing state.

The transfer of lunar meteorites to the Earth has been addressed previously by Arnold (1965) and Wetherill (1969) with a Monte Carlo model

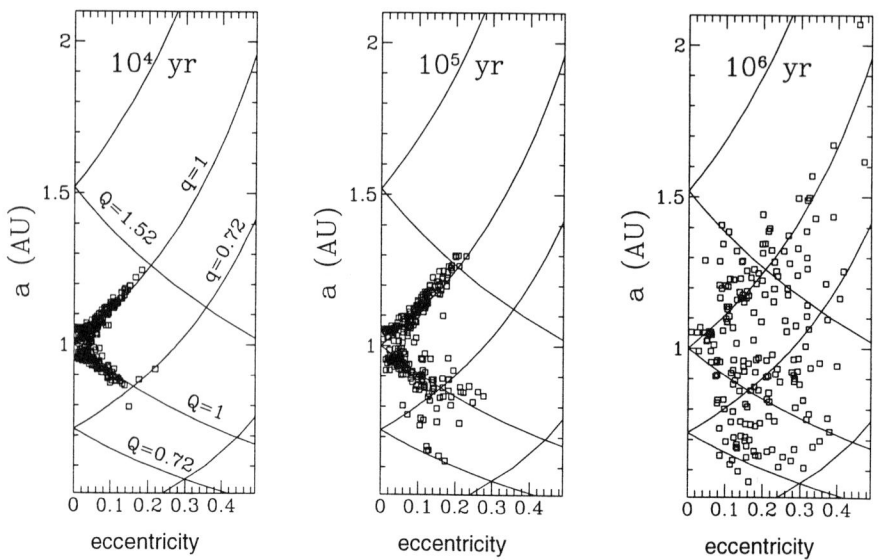

Figure 1. Semimajor axes and eccentricities for particles launched from the Moon at just above the lunar escape velocity. The web of lines marks where the particles' perihelia (q) or aphelia (Q) coincide with the semimajor axes of Mercury, Venus, Earth, and Mars.

using the Öpik collision formulae. Gladman *et al.* (1995) find, as they did, that of the ejecta that just barely escape the Earth/Moon system, about half returns to impact the Earth during 10 Myr. We find that of the Earth impacts that do occur, roughly two-thirds happen within the first 50 000 years, in agreement with the relatively young cosmic ray exposure ages of most recovered lunar meteorites. The early impact rate is very high because the velocities of the escaping particles relative to the Earth are initially low, enhancing the Earth's gravitational cross-section. Within a few Myr, material that does not impact the Earth scatters throughout the inner solar system (see Fig. 1). Over a 10-Myr period the remaining particles continue to collide with the terrestrial planets. Many impacts with Venus, and one with Mars, are observed. After a few Myr, losses commonly occur when orbits become Sun-grazing (with perihelia less than 2 solar radii) as a consequence of entering secular resonances (usually ν_5 or ν_6). Farinella *et al.* (1994) see a similar phenomena for the Near-Earth asteroids.

The transit times of Earth-impacting Martian ejecta may be compared to the age spectrum of the SNC meteorites, which generally have spent less than 15 Myr in space. Yet Monte Carlo simulations (Wetherill 1984) had found a large spread of transfer durations (up to 100 Myr) with only about half being delivered to the Earth in less than 15 Myr. For objects

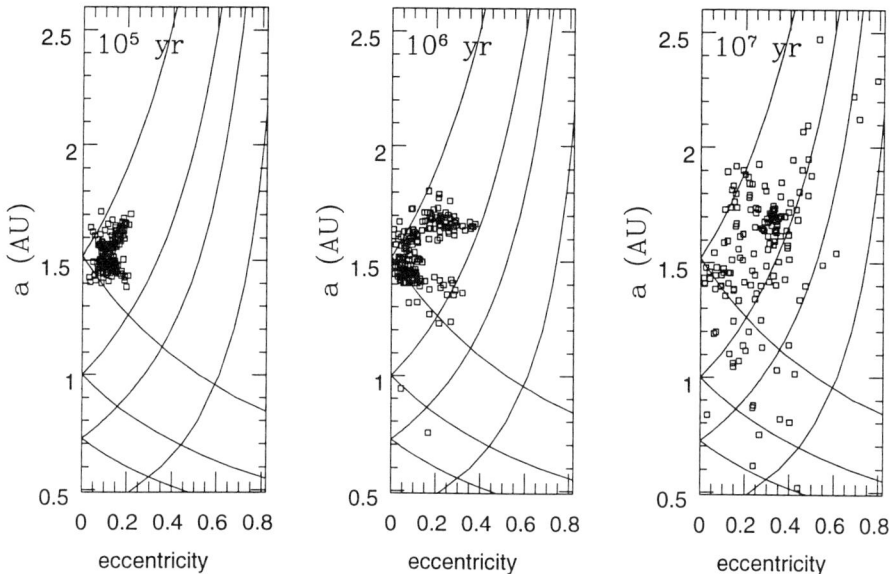

Figure 2. Semimajor axes and eccentricities for test particles launched from Mars at just above the martian escape velocity. The web of lines marks where the particles' perihelia or aphelia coincide with the semimajor axes of Mercury, Venus, Earth, and Mars.

that spend considerable portions of their orbits in the asteroid belt, the Wetherill simulation includes collisional destruction, which might be one way of removing long-lived meteoroids over tens of Myr; however, a decrease in the collisional lifetime by a factor of more than two reduces the yield by only 30%, so apparently collisional destruction is not responsible for the lack of long-lived SNCs. Our numerical integrations (see Fig. 2) show that secular resonances, located throughout most of the inner solar system, are primarily responsible for establishing a relatively short timescale for the delivery of this material to the Earth. They also allow fast transfers even when only very small ejection speeds off of Mars are typical.

We find that, for objects that barely escape Mars (probably the most common case since the amount of ejecta should drop rapidly with increasing velocity), the evolution is divided into three main phases. During the first 2 Myr, roughly 10% of the escaped particles are re-accreted by Mars; after this period few Mars collisions are observed because no longer are the meteoroid orbits sufficiently similar to Mars that the relative velocities (and thus collision time scales) are small. In the second phase, which lasts until about 10 Myr after launch, a variety of removal mechanisms operate: Venus and Earth collisions each absorb 2% of the meteoroids; roughly 1% encounter Jupiter (which we assume removes particles from the system on a short

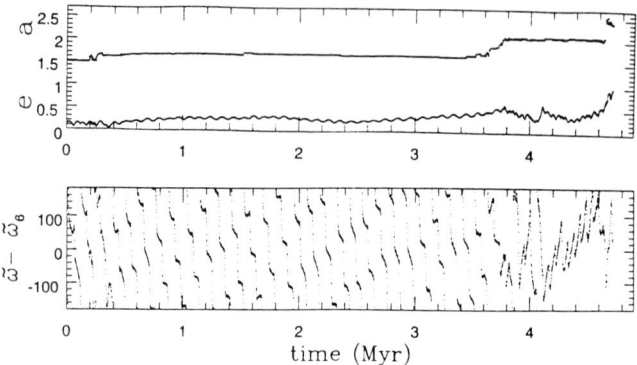

Figure 3. Semimajor axis and eccentricity for a martian ejecta particle that ends its life as a Sun-grazer. Beginning just after 4 Myr from launch the bottom panel shows that the particle is heavily influenced by the ν_6 resonance.

time scale); and the largest fraction (4%) become Sun-grazers (see Fig. 3).

After 10 Myr, the Sun-grazing end-state becomes extremely common, and systematically depopulates the meteoroids, with Jupiter-crossing occurring occasionally, and Earth or Venus collisions happening rarely. By 50 Myr 7%, 7%, and 9% of the initial number of particles particles have struck Venus, Earth, and Mars (respectively), 10% have crossed Jupiter's orbit, and about one third (32%) have become Sun-grazing. Thus perhaps the lack of old SNCs is due to the fact that secular resonances efficiently deplete the remaining meteoroid population by driving them into the Sun.

Three quarters of the material ejected from Mercury is re-accreted by that planet. A small fraction (10%) is driven on Myr timescales to Venus-crossing orbits, where collisions with Venus occur. Transfer of any material to the Earth would require at least 10 Myr, and only a very small fraction, which we estimate should be of order 0.1–1%, of the launched material would arrive (in obvious contrast with the lunar and martian cases). The lack of any discovered mercurian meteorites is thus consistent with what we know about the transfer dynamics.

References

Arnold, J.R. (1965) The origin of meteorites as small bodies; II. The model, *Astrophys. J.*, **141**, 1536-47.

Farinella P., Froeschlé, Ch., Froeschlé, Cl., Gonczi, R., Hahn, G., Morbidelli, A., and Valsecchi, G.B. (1994) Asteroids falling into the Sun. *Nature*, **371**, 314-17.

Gladman, B., Burns, J.A., Levison, H., and Duncan M.(1995) The dynamical evolution of lunar impact ejecta. *Icarus*, **118**, in press.

Warren, P. (1994) Lunar and martian meteorite delivery services. *Icarus*, **111**, 338-363.

Wetherill, G. (1969) Relationships between orbits and sources of chondritic meteorites. in *Meteorite Research*, ed. Millmann, P., pp.573-589.

Wetherill, G. (1984) Orbital evolution of impact ejecta from Mars. *Meteoritics*, **19**, 1-12.

CANONICAL TREATMENT OF DISSIPATIVE FORCES BETWEEN EARTH MANTLE AND CORE

JUAN GETINO
Grupo de Mecánica Celeste
Facultad de Ciencias, 47005 Valladolid. Spain
e-mail: getino@hp9000.uva.es

AND

JOSE M. FERRANDIZ
Depto. Matemática Aplicada
Universidad de Alicante, 03080 Alicante. Spain

Abstract. Dissipative effects arising from the core–mantle interaction are treated in a Hamiltonian framework, using a simple model. Analytical solutions are obtained for free and forced motions. The first show the persistence or damping of the different components. The latter, frequency dependent changes of amplitude and phase. Preliminary numerical values are in acceptable agreement with observational data.

1. Introduction

The advantages of the Hamiltonian formalism are well known in general Celestial Mechanics, where it has been frequently used. The study of the Earth rotation is not an exception, and the theory of Kinoshita (1977) is the most complete and accurate among those for a rigid Earth model. Recently, the Hamiltonian approach has been followed by the authors in order to extend Kinoshita's theory to general models of deformable Earth through a series of papers, first considering an elastic mantle (Getino and Ferrándiz, 1995), then a liquid core (Getino, 1995a, 1995b). We only point out here that the results obtained are completely analytical, and depend on parameters that are provided by the different Earth models available.

In this paper we consider the possibilities of including dissipative effects in the previous formulations, like those arising from friction in the boundary layer between mantle and core. As a first step, we have chosen a fairly well known simplified model, in which the dissipation appears through a torque proportional to the difference of angular velocities (Kubo,1979). Using canonical Andoyer–like variables as in Getino (1995b), it is only necessary to introduce the generalized forces in a straightforward way. By integrating the unperturbed Hamiltonian equations, solutions for the free wobbles are obtained, which in our model represent the CW and the NDFW, using the standard terminology for conservative free oscillations. It is remarkable that only the component terms corresponding to the NDFW exhibit a damping, but not the component associated to CW. That is in good agreement with the difficulties found in the experimental determination of the free nutation itself (Mathews and Shapiro, 1992), and on the other hand is a common feature of general dissipative systems (Birkhoff, 1927), that are to tend towards conservative systems of lower dimension.

Besides the solutions for free wobbles and polar motion, the forced nutations are also obtained. Together with the frequency dependent amplitude variations due to non–rigidity, phase changes caused by the dissipation appear. This last effect is not obtained in other theories like Wahr (1981). As in our previous developments, the analytical expressions depend on parameters provided by quite general Earth models.

2. Canonical expression of the kinetic energy

We consider an Earth model composed of a rigid mantle and a liquid core. Let $OXYZ$ be a non rotating inertial frame, $Oxyz$ the frame of the principal axes of the total Earth rotating with an angular velocity $\vec{\omega}$ with respect to the inertial frame, and $Ox_c y_c z_c$ a core fixed frame rotating with angular velocity $\delta\vec{\omega}$ with respect to the mantle. To formulate canonically the kinetic energy let us use the set of canonical variables $\lambda, \mu, \nu, \Lambda, M, N$ for the total Earth, and $\lambda_c, \mu_c, \nu_c, \Lambda_c, M_c, N_c$ for the core, described in Getino (1995a,b).

We can write the components of $\mathbf{M}$ and $\mathbf{M}_c$ in the $Oxyz$ frame in terms of the canonical variables as follows

$$\mathbf{M} = \begin{pmatrix} A\omega_1 + A_c\delta\omega_1 = K\sin\nu \\ A\omega_2 + A_c\delta\omega_2 = K\cos\nu \\ C\omega_3 + C_c\delta\omega_3 = N \end{pmatrix}, \quad \mathbf{M}_c = \begin{pmatrix} A_c\omega_1 + A_c\delta\omega_1 = K_c\sin\nu_c \\ A_c\omega_2 + A_c\delta\omega_2 = -K_c\cos\nu_c \\ C_c\omega_3 + C_c\delta\omega_3 = N_c \end{pmatrix} \quad (1)$$

where we have put

$$K = (M^2 - N^2)^{1/2} = M\sin\sigma, \quad K_c = (M_c^2 - N_c^2)^{1/2} = M_c\sin\sigma_c. \quad (2)$$

Then, the canonical expression of the kinetic energy is (Getino, 1995b):

$$T = \frac{1}{2(A - A_c)}\left(K^2 + \frac{A}{A_c}K_c^2\right) + \frac{1}{2(C - C_c)}\left(N^2 - 2NN_c + \frac{C}{C_c}N_c^2\right) + \frac{KK_c}{A - A_c}\cos(\nu + \nu_c),$$
(3)

3. Generalized forces and equations of motion

The frictional torque, including electromagnetic coupling and the effects of the viscosity, is proportional to the difference of the angular velocities of the mantle and the core. According to Kubo (1979) this torque is given by

$$\begin{array}{l}\vec{t_m} = \Gamma(\vec{\omega_c} - \vec{\omega_m}) = \Gamma\,\vec{\delta\omega} \\ \vec{t_c} = \Gamma(\vec{\omega_m} - \vec{\omega_c}) = -\Gamma\,\vec{\delta\omega}\end{array}, \text{ with } \Gamma = \begin{pmatrix} \lambda_\perp & 0 & 0 \\ 0 & \lambda_\perp & 0 \\ 0 & 0 & \lambda_\| \end{pmatrix},$$
(4)

$\lambda_\perp$ and $\lambda_\|$ being the coefficients for perpendicular and parallel directions to the rotational axis. The generalized forces due to the dissipative torque are obtained through the dot product:

$$\vec{t}\cdot\vec{d\phi} = \vec{t_m}\cdot\vec{d\phi_m} + \vec{t_c}\cdot\vec{d\phi_c} = \vec{t_c}\cdot\left(\vec{d\phi_c} - \vec{d\phi_m}\right),$$
(5)

where $\vec{d\phi_m}$ and $\vec{d\phi_c}$ are the infinitesimal rotations of the reference frames of mantle (total Earth) and core with respect of the inertial frame. Taking into account the meaning of the canonical variables (see Getino, 1995a), the generalized equations of motion will be

$$\dot{q}_i = \partial T/\partial p_i - W_{p_i}, \quad \dot{p}_i = -\partial T/\partial q_i + W_{q_i},$$
(6)

where W_{p_i} and W_{q_i} are the components of the generalized forces.

4. Solutions for the free motion

Taking into account the relationships (6), the necessary generalized equations for the free motion problem are:

$$\begin{array}{ll}\dot{\nu} = \partial T/\partial N, & \dot{\nu}_c = \partial T/\partial N_c - W_{N_c}, \\ \dot{N} = -\partial T/\partial \nu, & \dot{N}_c = -\partial T/\partial \nu_c + W_{\nu_c}, \\ \dot{M} = -\partial T/\partial \mu, & \dot{M}_c = -\partial T/\partial \mu_c + W_{\mu_c},\end{array}$$
(7)

where

$$W_{N_c} = -\frac{\lambda_\perp}{A_m} \frac{K}{K_c} \sin(\nu + \nu_c)$$
$$W_{\nu_c} = \frac{\lambda_\parallel}{C_m} \left(N - \frac{C}{C_c} N_c \right)$$
$$W_{\mu_c} = \frac{\lambda_\parallel}{C_m} \cos\sigma_c \left(N - \frac{C}{C_c} N_c \right)$$

As a first result, by means of (1) and (7) we have that $\dot{N} - \dot{N}_c = \lambda_\parallel \delta\omega_3$, so that, neglecting second order terms, we obtain $\dot{\omega}_3 = \frac{\lambda_\parallel}{C_m} \delta\omega_3$, while $\delta\omega_3 = e^{-\frac{\lambda_\parallel}{C_m}\frac{C}{C_c}t}$. That is to say, the relative angular velocity of the core with respect to the mantle is damped by the frictional forces. Thus, we can take $\delta\omega_3 \to 0$, and then $\omega_3 =$ constant $= \Omega$. With these simplifications the system of equations (7) can be easily solved performing the change of variables $p = K\sin\nu$, $q = K\cos\nu$, $p_c = K_c\sin\nu_c$ and $q_c = K_c\cos\nu_c$. Notice that these new variables are linear combinations of ω_1, ω_2, $\delta\omega_1$ and $\delta\omega_2$ (see Getino, 1995b). The solution can be written in the form:

$$\begin{aligned} u &= p + iq = \gamma_a\, \alpha\, e^{im_1 t} + \beta\, e^{-dt}\, e^{im_2 t}, \\ v &= p_c - iq_c = \gamma_b\, \alpha\, e^{im_1 t} + \gamma_c\, \beta\, e^{-dt}\, e^{im_2 t}, \end{aligned} \qquad (8)$$

α and β being the arbitrary constants of integration, the coefficients γ_i are

$$\begin{aligned} \gamma_a &= \frac{A}{A_m}\left(\frac{C-A}{A} + \frac{C_c}{A_c}\right) - i\frac{\lambda_\perp}{\Omega A_c}\frac{A}{A_m}, \\ \gamma_b &= \frac{C_c}{A_m} - i\frac{\lambda_\perp}{\Omega A_m}, \\ \gamma_c &= \frac{C_c}{A_c} - i\frac{\lambda_\perp}{\Omega A_c}, \end{aligned} \qquad (9)$$

and the frequencies

$$\begin{aligned} m_1 &= \Omega\,\frac{C-A}{A_m} & &\to \text{CW}, \\ m_2 &= -\Omega\left(1 + \frac{C_c - A_c}{A_c}\frac{A}{A_m}\right) & &\to \text{FCN}, \\ d &= \frac{\lambda_\perp}{A_c}\frac{A}{A_m} & &\to \text{Damping}. \end{aligned} \qquad (10)$$

According to (10), the well known frequencies for an Earth model with liquid core, that is to say, the Chandler Wobble (CW) and the Free Core Nutation (FCN), have been obtained directly from the equations of motion of the Hamiltonian of the system, unlike the majority of approaches which obtain these frequencies from the Liouville equations. On the other hand, from (8) we see that the CW is not affected by the dissipation, while the FCN is damped by the coefficient d. This fact would justify the difficulties found in the experimental determination of this free nutation, which "remains elusive" (Mathews and Shapiro, 1992).

5. Forced nutations

Once the solution for the unperturbed Hamiltonian has been obtained, we can undertake the problem of the perturbed case, taking into account the disturbing potential of Moon and Sun, following the same analytical procedure described by Kinoshita (1977) for the rigid Earth, by Getino and Ferrándiz (1995) for a deformable Earth with elastic mantle, and by Getino (1995b) for a rigid mantle–liquid core conservative Earth model. The most remarkable fact is that, when obtaining the generating function of the canonical transformation corresponding to the analytical perturbation method, a delay appears due to the imaginary coefficients γ_i (9) presents in the solution (8) of the unperturbed case. Thus, when obtaining the nutation series we have in–phase and out–of–phase terms. For the Oppolzer terms of obliquity and longitude of the figure plane, we have respectively the final expressions:

$$\Delta(I_f - I) = k \sum_i \sum_{\tau=\pm 1} \frac{C_i}{n_\mu - \tau n_i} F_a(\tau n_i) \cos \Theta_i -$$
$$- \frac{k}{\sin I} \sum_i \sum_{\tau=\pm 1} \frac{\tau C_i}{n_\mu - \tau n_i} F_b(\tau n_i) \sin \Theta_i ,$$
$$\Delta(\lambda_f - \lambda) = \frac{k}{\sin I} \sum_i \sum_{\tau=\pm 1} \frac{\tau C_i}{n_\mu - \tau n_i} F_a(\tau n_i) \sin \Theta_i +$$
$$+ \frac{k}{\sin I} \sum_i \sum_{\tau=\pm 1} \frac{C_i}{n_\mu - \tau n_i} F_b(\tau n_i) \cos \Theta_i .$$
(11)

In these expressions, the effect of dissipative forces as well as the liquid core is included in the corrector factors F_a and F_b, which are written as follows:

$$F_a(\tau n_i) = \frac{\frac{C}{A}\Omega - \tau n_i}{f_1(\tau n_i)\, f_2(\tau n_i)} \left(f_2(\tau n_i) - \Omega \frac{A_c}{A_m} \right) ,$$
$$F_b(\tau n_i) = \frac{\frac{C}{A}\Omega - \tau n_i}{f_1(\tau n_i)\, f_2(\tau n_i)} \frac{\lambda_\perp}{A_c} \frac{A^2}{A_m^2} \frac{f_2(\tau n_i) - \Omega \frac{A_c}{A}}{f_2(\tau n_i)} ,$$
(12)

with

$$f_1(\tau n_i) = \Omega + m_1 - \tau n_i, \quad f_2(\tau n_i) = \Omega + m_2 - \tau n_i .$$

Note that the corrector factor F_b is responsible of the out–of–phase terms, while the in–phase terms depend on F_a. If we neglect the dissipation ($\lambda_\perp = 0$), the out–of–phase terms disappears, and we get similar expressions as those obtained in Getino (1995b) for the conservative case. Even more, if we particularize to the rigid case ($A_c = 0$), then $F_a = 1$, and we obtain the same expressions as in Kinoshita's theory.

Thus, we can conclude that it is possible to develop a canonical theory for the rotation of an Earth model with liquid core, including the effect of dissipative forces, in a way very similar to that of Kinoshita for the rigid case. Neglecting the presence of liquid core and dissipation, our approach coincides exactly with that of Kinoshita, so that we can say that this model is a more general one, including the theory of Kinoshita as a particular case.

6. Final remarks

The Hamiltonian treatment of dissipative effects, like those arising in the mantle–core boundary, provides a fairly simple, completely analytical solution for the Earth rotation, including free motion as well as forced nutations. In spite of the simplicity of the dissipative model, the solution is qualitatively well–fitted to the experimental evidence, providing amplitude and phase–frequency dependent forced nutations, and a clear theoretical derivation of the free motion showing the damping of NDFW and the persistence of CW. Numerical values of in–phase and out–of–phase nutations are also compatible in magnitude with experimental data. Nevertheless, we insist that the main purpose of this report is not to obtain accurate series for the real Earth, but to show how the canonical approach can be applied to study dissipative phenomena in Earth rotation.

In future steps we expect to progress in the development of a complete unified Hamiltonian theory for the Earth rotation adapted to more accurate Earth models.

Acknowledgements

This work has been partially supported by CICYT, Project No. ESP93-741.

References

Birkhoff, G.D.: 1927, *Dynamical Systems*, A.M.S.
Getino, J. and Ferrándiz , J.M.: 1995, *Celes. Mech.*, **61**, 117–180.
Getino, J.: 1995a, *Geophys. J. Int.*, **120**, 693–705.
Getino, J.: 1995b, *Geophys. J. Int.*, **122**, 803–814.
Kinoshita, H..: 1977, *Celes. Mech.* **15**, 277–326.
Kubo, Y.: 1979, *Celes. Mech.* **19**, 215–241.
Mathews, P.M. and Shapiro, I.I.: 1992, *Annu. Rev. Earth Planet. Sci.*, **20**, 469–500.
Wahr, J.M.: 1981, *Geophys. J. R. Astr. Soc*, **64**, 705–727

NEW SERIES OF RIGID AND NON RIGID EARTH NUTATION. COMPARISON WITH OBSERVATIONS

J. SOUCHAY
Observatoire de Paris/DANOF, URA 1125 du CNRS
61 Avenue de l'Observatoire, 75014 Paris
E-mail: souchay@obspm.fr

Abstract. After analysing the recent developments of the theory of the nutation for a simplified rigid Earth model (Kinoshita and Souchay,1990) with new corrections and new contributions (Williams,1994; Souchay and Kinoshita,1996), we will study the effect of these developments on the calculation of the coefficients of nutation for a non rigid Earth model, based on the transfer function given by Wahr (1979). The relative improvements characterized by the residuals with the observations are explained in the following.

1. Introduction

Because of the high accuracy obtained by the VLBI observations in the recent years, the conventional IAU1980 nutation series (Wahr,1979; Seidelmann,1982) can no more be available, mainly because only the coefficients up to 0.1 milliarcsecond (mas) were retained, whereas the precision of the observational individual determination of nutation coefficients by VLBI (Herring et al., 1991) falls far below this limit.

In view of this remark it was necessary to reconstruct the tables of the nutation for a rigid Earth model by taking into account all the coefficients up to 0.005 mas. This was done both by Zhu and Groten (1989) and Kinoshita and Souchay (1990). For some reasons explained by Souchay (1993), this last work is more complete than the former one, especially because planetary effects (direct and indirect) have been considered. Notice also that some important corrections to large coefficients (among them the corrections at the level of 1 mas of coefficients of argument Ω and 2Ω) were done by Kinoshita and Souchay (1990), which characterize an interaction between the orbital motion of the Moon and the motion of rotation of the Earth.

Williams (1994) as well as Souchay and Kinoshita (1996) pointed out corrections and new contributions which were not included in the series of Kinoshita and Souchay (1990). In the following we present all the improvements to these last series related to the remarks above.

2. Corrections and new contributions in rigid earth nutation

Since the tables of the nutation for a rigid Earth model were established by Kinoshita and Souchay (1990), many authors have shown that a correction to the value of the general precession in longitude p_A, that is to say: $p_A = 5029.0966"/\text{cy}$ (Lieske et al., 1977), was necessary. The values found for the correction δp_A are in agreement one to each other (Herring,1991; Herring et al., 1991; Mc Carthy and Luzum,1991; Miyamoto and Soma,1993; Williams et al., 1993) and close to : $\delta p_A = -0.3"/\text{cy}$. Notice that they are coming from the reductions of various kinds of observations (LLR, VLBI, optical catalogue).

Souchay and Kinoshita(1995), by choosing the value $\delta p_A = -0.3266 "/\text{cy}$, as suggested by Williams (1994), show that this correction leads to a correction to the dynamical ellipticity of the Earth H_d as calculated by Kinoshita and Souchay (1990). Their new value: $H_d = 0.0032737548$, is different from the value in this last paper: $H_d = 0.0032739567$, by a relative amount of 6×10^{-5}. This change in the value of H_d affects in a significant manner the largest trems of nutation as calculated by Kinoshita and Souchay (1990), at the level of 1 mas for the 18.6y component. Notice that the value of H_d recently calculated by Williams (1994) by taking the same correction $\delta p_A = -0.3266"/\text{cy}$, that is to say: $H_d = 0.0032737634$, is close to the new value calculated by Souchay and Kinoshita (1996).

Williams (1994) pointed out a planetary tilt effect on nutation not taken into account in the tables of Kinoshita and Souchay (1990). They are characterized by out of phase components for the terms of argument Ω and 2Ω. Souchay and Kinoshita (1996) calculated the total amplitudes of these terms. They found:

$$\Delta\Psi = -17283.977 \sin\Omega + 0.135 \cos\Omega + 209.077 \sin 2\Omega + 0.005 \cos 2\Omega$$

$$\Delta\varepsilon = 9228.910 \cos\Omega - 0.030 \cos\Omega - 90.360 \cos 2\Omega + 0.003 \sin 2\Omega$$

to be compared with the respective amounts in Kinoshita and Souchay (1990):

$$(\Delta\Psi)_{out-of-phase} = -17285.201 \sin\Omega + 209.095 \sin 2\Omega$$

$$(\Delta\varepsilon)_{out-of-phase} = 9229.578\cos\Omega - 90.368\cos 2\Omega$$

Their value for the new out-of-phase contributions are very close to those of Williams (1994) who found:

$$\Delta\Psi = 0.137\cos\Omega + 0.006\cos 2\Omega$$

$$\Delta\varepsilon = -0.028\cos\Omega + 0.003\sin 2\Omega$$

Moreover, Souchay and Kinoshita (1996) found new terms not included in the tables of Kinoshita and Souchay (1990), due to the oscillations of the true ecliptic with respect to the mean ecliptic of the date. They are characterized by 9 coefficients larger than the level of truncation of the series (0.005 mas).

They also summarized in their tables all the corrections and new contributions explained above, with respect to the results of Kinoshita and Souchay (1990). Notice that they concern the luni-solar part of the nutation for a rigid Earth model, the indirect planetary effects being included. Moreover, their tables include also all the sign errors, omissions or misprints in the tables of Kinoshita and Souchay (1990).

3. VKSNRE95.1 Series for non rigid earth nutation

The conventional IAU80 nutation series (Seidelmann,1982) are based on the rigid Earth nutation series established by Kinoshita (1977) to which the transfer function calculated by Wahr (1979) has been applied, to determine the coefficients for a non rigid Earth model. In order to evaluate the improvement in nutation theory due to the reconstruction of the tables of nutation for a rigid Earth model by Kinoshita and Souchay (1990), the same transfer function as above has been applied to these reconstructed tables. Souchay et al. (1995 a) have compared the nutation given by resulting series, called KSNRE (Kinoshita and Souchay Non Rigid Earth) with VLBI observations (Ma et al., 1994). By comparing the KSNRE nutation with 10 years of VLBI data, they have shown that the weighted rms post fit residuals are 1.16 mas in longitude and 1.18 mas in obliquity. If the comparison with IAU80 series is done, the rms post fit residuals becomes respectively 1.27 mas and 1.44 mas. This improvement is undoubtedly due to the new level of truncation of the KSNRE series (0.005 mas instead of 0.1 mas).

In a similar way, Souchay et al. (1995 b) have shown that after fitting only the linear trend, the FCN (Free Core Nutation) term and 9 among the 10 largest terms of the KSNRE series, which are affected by the geophysical modeling, the residuals after comparison with 15 years VLBI observational data falls below 0.1 mas for the total amplitude of the nutation. This leads them to construct new series, called VKSNRE95.1, consisting in adopting the empirical fitted value for the 9 terms above (both in longitude and obliquity), all the other terms being those of KSNRE analytical series. Among the fitted terms are the 18.6 y, the annual, semi-annual monthly and fortnightly components. Even if there is clearly a need to explain the relatively big differences between the theory and the observations for the 9 fitted terms above, the VKSNRE95.1 series can be of very useful help for users needing to use analytical series characterized by a very good agreement with the observations.

References

Herring,T.A.,1991, Proc. IAU Coll.127 "Reference Systems",Hughes, Smith and Kaplan (eds.),US Naval Observatory, p.157

Herring, T.A., Buffet,B.A., Mathews,P.M., Shapiro,I.I.,1991 *J. Geophys. Res.* **96**,p.8259

Kinoshita,H,1977,*Celest. Mech.* **26**,p.296

Kinoshita,H.,and Souchay,J.,1990,*Celest. Mech.***48**, p.187

Lieske,J.H.,Lederle,T.,Fricke,W.,Morando,B.,1977, *Astron. and Astrophys.* **58**,1

Ma,C.,Gipson,J.M.,Gordon,D.,Caprette,D.S.,and Ryan,J.W.,1994, IERS Technical Note 17,Charlot (ed.), p.R-7.

Mc Carthy,D.D., and Luzum,B.J.,1991, *AJ* **102**,p.1889

Miyamoto, M, Soma, M.,1993, *AJ* **105**,p.691

Seidelmann,P.K.,1982,*Celest.Mech.* **27**, p.79

Souchay,J.,1993,*Astron. Astrophys.* **276**, p.266

Souchay,J.,Feissel,M.,Bizouard,C.,Capitaine,N.,Bougeard,M, 1995a, *Astron. Astrophys.* **299**, p.277

Souchay,J.,Feissel,M., Ma,C., 1995b, *Astron. Astrophys.* (accepted)

Souchay and Kinoshita,1996,"Corrections and new contributions in rigid Earth Nutation Theory", *Astron. Astrophys.*, in Press

Wahr,J.M.,1979,Ph.D. Thesis, University of Boulder, Colorado,USA;

Williams,J.G., Newhall, X.X., Dickey, J.O., 1993, in Contributions of Space Geodesy to Geodynamics: Earth Dynamics, Geophysical Monograph of the American geophysical Union, Vol.24, edited by D.E. Smith and D.L.Turcotte, Washington D.C., p. 83

Williams,J.G.,1994,*Astron.J.* **108**, 711.

Williams,J.G.,1995,"Planetary-induced nutation of the Earth- direct terms" (submitted)

Zhu,S.Y., Groten,E., 1989, *Astron.J.* **98**, 1104.

ABOUT THE APPLICATION OF ANGLE–ACTION VARIABLES TO THE ROTATION OF DEFORMABLE CELESTIAL BODIES

YURI V. BARKIN AND JOSE M. FERRANDIZ
Depto. Matemática Aplicada
Universidad de Alicante, 03080 Alicante. Spain

AND

JUAN GETINO
Grupo de Mecánica Celeste
Facultad de Ciencias, 47005 Valladolid. Spain
e-mail: getino@hp9000.uva.es

Abstract. We begin with the development of a new analytical theory for the rotational motion of deformable celestial bodies. In this work we present a new unperturbed Hamiltonian which includes the main term affecting the Chandler period.

1. Motivation of this paper

The effect of the elasticity of the mantle on the Earth's rotation has been developed by Getino and Ferrándiz (1991,1995) in a Hamiltonian framework, using a set of Andoyer–like canonical variables, $\lambda, \mu, \nu, \Lambda, M, N$ (and the auxiliary parameters σ, I). As described in Getino and Ferrándiz (1995), the main contribution to the Chandler period arises from the term T_r, that is to say, the increment of the kinetic energy due to the centrifugal deformation (caused by the own rotation of the Earth). So that, the principal ideas of this work are, on the one hand, to consider a new expression of the Hamiltonian in such a way that the new unperturbed Hamiltonian includes the main effects affecting the Chandler period, in order to make its study easier, and on the other hand, to develop an analytical theory suitable to be applied to more general celestial bodies, when the principal moments of inertia are in the form $A \neq B \neq C$.

Let us break down the term T_r into two parts, $T_r = T_r^{0)} + T_r^{1)}$. From Getino and Ferrándiz (1991), the main part $T_r^{0)}$ has the expression:

$$T_r^{0)} = \frac{3}{2} D_r \frac{M^2 - N^2}{C} \left[\frac{\sin^2 v}{A} + \frac{\cos^2 v}{B} \right],$$

where D_r is a coefficient depending on the internal structure of the body (for the case of the Earth, $D_r = -2.845379 \times 10^{41}$ c.g.s.), and A, B and C are the principal moments of inertia including the effect of the centrifugal deformation in the form $A \simeq A_0 + D_r$, $B \simeq B_0 + D_r$, $C \simeq C_0 - 2D_r$, where A_0, B_0, C_0 correspond to the body in the absence of deformation (Getino and Ferrándiz, 1991).

Now, if we define new principal moment of inertia $\tilde{A}$, $\tilde{B}$, $\tilde{C}$ in the form

$$\frac{1}{\tilde{A}} = \frac{1}{A} \left(1 + \frac{3D_r}{C} \right), \quad \frac{1}{\tilde{B}} = \frac{1}{B} \left(1 + \frac{3D_r}{C} \right), \quad \frac{1}{\tilde{C}} = \frac{1}{C},$$

the new unperturbed Hamiltonian has the following expression

$$H_0 = T_0 + T_r^{0)} = \frac{M^2 - N^2}{2} \left(\frac{\sin^2 \nu}{\tilde{A}} + \frac{\cos^2 \nu}{\tilde{B}} \right) + \frac{N^2}{2\tilde{C}}, \qquad (1)$$

and we can conclude that:

"The free rotational motion of a celestial body deformed by its own rotation is described in first approximation by the solution of the Euler–Poinsot problem corresponding to a rigid body with special principal moments $\tilde{A}$, $\tilde{B}$, $\tilde{C}$".

Thus, this solution will explain the Chandler period. The analytical integration of (1) by means of action–angle variables, as well as the development of the of the perturbations due to the force function of the Earth–Moon system is in progress, and we hope to present the obtained results soon.

Acknowledgments

This work has been partially supported by CICYT, Project No. ESP93-741.

References

Getino, J. and Ferrándiz, J.M.: 1991, *Celes. Mech.*, 51, 35–65.
Getino, J. and Ferrándiz, J.M.: 1995, *Celes. Mech.*, 61, 117–180.

A NUMERICAL SCHEME TO INTEGRATE THE ROTATIONAL MOTION OF A RIGID BODY

TOSHIO FUKUSHIMA

National Astronomical Observatory
2-21-1, Ohsawa, Mitaka, Tokyo 181, Japan
(Internet) toshio@spacetime.mtk.nao.ac.jp

1. Introduction

Once, we numerically integrated the precession and nutation of a spheroidal rigid Earth (Kubo and Fukushima 1987). As a natural extension, we tried to integrate the rotation of a triaxial rigid Earth numerically and faced a problem: a loss of precision in long-term integration. This is due to the smallness of the characteristic period of the problem: 1 day. Of course, one can integrate the rotational motion in higher precision arithmetics with a smaller stepsize. However, the quadruple precision integration is roughly 30 times more time-consuming than the double precision integration. See Table 1. Therefore, it is desirable if there is a formulation 1) reducing the overall integration error, 2) being independent on the choice of the integrator and 3) requiring no extra computations. The key points to achieve this goal will be to find a set of variables which 1) are efficiently convertible to the physical quantities required finally, say, the orientation matrix in the case of the rotational dynamics, *and* 2) vary with time as smoothly as possible. In this note, we report a discovery of such an example.

2. Scheme

As the basic variables describing the rotational motion of a rigid body, we adopt $(L, L_X, L_Y, L_A, L_B, f)$; the magnitude and X-, Y-, A-, B- components of the rotational angular momentum vector $\vec{L}$, and the angle f measured from the plane containing $\vec{L}$ and the X-axis to the plane containing $\vec{L}$ and the A-axis. The orientation matrix $\mathcal{E}$ is expressed in terms

TABLE 1. Comparison of CPU time

Scheme	Precision	CPU time	
		Evaluation of $\mathcal{E}$	Eq. of Motion
New	Double	117	195
Euler		117	206
Serret-Andoyer		147	250
Euler	Quadruple	3378	7195

Note: The unit of CPU time is μs in HP/9000 715/50MHz.

of these variables as

$$\mathcal{E} \equiv (\vec{e}_A, \vec{e}_B, \vec{e}_C) = \mathcal{Q}_{12}\left(\frac{L_X}{L}, \frac{L_Y}{L}\right) \mathcal{R}_3(-f) \mathcal{Q}_{12}^T\left(\frac{L_A}{L}, \frac{L_B}{L}\right). \quad (1)$$

Here, $\vec{e}_j$ is the unit column vector defining the j-th axis, and

$$\mathcal{Q}_{12}(x,y) \equiv \mathcal{R}_1\left(+\tan^{-1}\frac{y}{z}\right) \mathcal{R}_2\left(-\sin^{-1}x\right) = \begin{pmatrix} w & 0 & x \\ -xy/w & z/w & y \\ -xz/w & -y/w & z \end{pmatrix},$$

where $w = \sqrt{1-x^2}$ and $z = \sqrt{1-(x^2+y^2)}$. Thus, $\mathcal{E}$ is generated by the five successive rotations in the sense of 1-2-3-2-1. In the actual integration, we integrate not these basic variables but their departures from nominal constants, their initial values for the first five and a linear function of time for f. In the language of the Earth rotation, these departures correspond to the variation of LOD, the nutation in obliquity and in declination, the polar motion and the variation of UT1. The equation of motion is simple. Those for the first five are just the translation of the conservation law of $\vec{L}$ in the inertial and body-fixed coordinate systems. That for f is

$$\frac{d\Delta f}{dt} = \frac{\Delta L}{C} + \left(\frac{1}{B} - \frac{1}{C}\right)\frac{LL_B^2}{L^2 - L_A^2} \\ + \frac{L_A(L_C N_B - L_B N_C)}{L(L^2 - L_A^2)} - \frac{L_X(L_Z N_Y - L_Y N_Z)}{L(L^2 - L_X^2)} \quad (2)$$

where $L_Z = \sqrt{L^2 - (L_X^2 + L_Y^2)}$, $L_C = \sqrt{L^2 - (L_A^2 + L_B^2)}$, and $N_j = \vec{N} \cdot \vec{e}_j$ is the j-th component of the torque $\vec{N}$.

3. Comparisons with Other Schemes

Figure 1 illustrates the growth of integration errors in ≈ 90 years of the rotational motion of a rigid Earth perturbed by Moon and Sun in model

NUMERICAL SCHEME TO INTEGRATE RIGID ROTATION 247

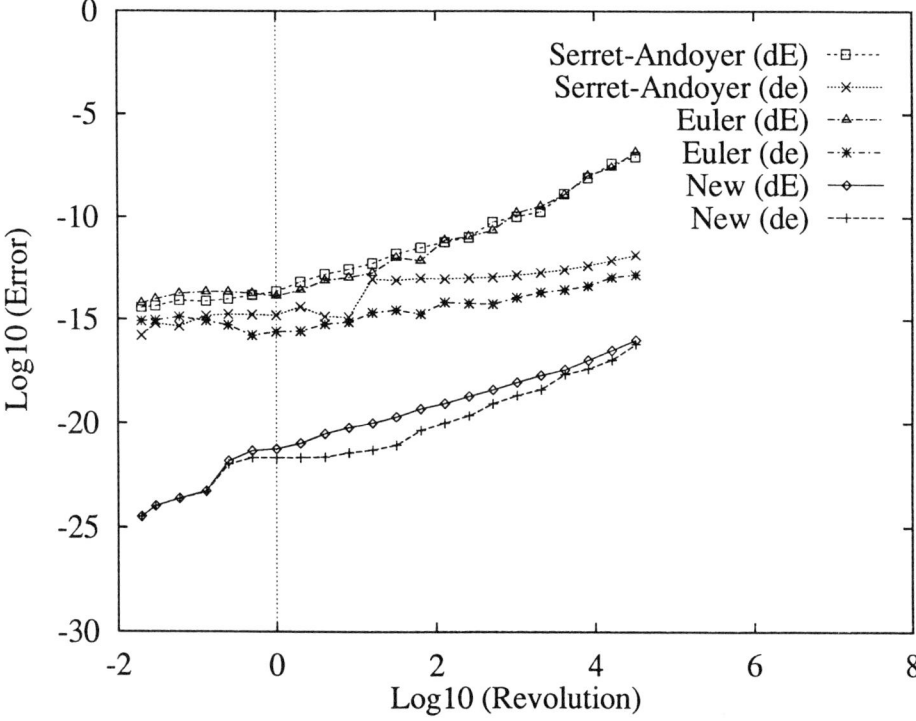

Figure 1. Growth of Integration Error of Earth Rotation

orbits for the new and two other schemes: the well-known Eulerian and the Serret-Andoyer canonical ones. The lines (dE) and (de) denote the errors in $\mathcal{E}$, the total orientation, and in $\vec{e}_C$, the figure axis, respectively. We obtained similar results for the Moon's rotation also. The integrations were done by the 12th order Adams method in 53-bits mantissa arithmetics with the stepsize of 1/128 nominal rotational period. It is clear that the integration error of the new scheme is drastically smaller, say, by 7-9 digits for the overall orientation, and 3-10 digits for the figure axis. The observed large differences in the errors are due to the large differences in the magnitude of the integrated variables, especially those related to the rotation angle. All the integrated variables remain very small in the new scheme, namely less than 10", which is the magnitude of nutation. Also, the growth of error seems smaller, with an almost linear growth for the first 10^4 revolutions. The reason is not clarified yet. On the other hand, Table 1 shows the averaged CPU times for the three formulations. As is seen, there is no actual difference between the Euler and new schemes.

4. Conclusion

A new numerical scheme to integrate the orientation of a rigid body was presented. The adopted basic variables are the magnitude, the X-, Y-, A- and B-components of the angular momentum vector, and the longitude of the A-axis measured from the X-axis along the great circle perpendicular to the angular momentum. Not these basic variables, but the correction to their nominal constants and/or linear motion are integrated. Numerical simulations showed that the new scheme integrates the orientation matrices of the Earth and the Moon 5-9 digits more precisely than the ordinary Eulerian approach or the alternative Serret-Andoyer one does while the required computational time does not change significantly.

We will make two comments on the new variables themselves. First, they are translated to the set of canonical variables $(L, L_X, L_A; f, -\sigma, \xi)$ where the auxiliary angles σ and ξ are defined in Fukushima (1994). This indicates the possibility to develop another symplectic integrator for the rotational motion as was done for the Serret-Andoyer set by Touma and Wisdom (1994). Next, the new variables are closely connected to the concept of the Non Rotating Origin (Guinot 1981). For example, the angle f looks similar to the sidereal angle based on NROs, although the rigorous relation between them remains an open problem. These two facts may imply the possibility to develop an analytical theory of the Earth rotation based on NROs.

Acknowledgements

The author greatly thanks Dr. E.M. Standish, Jr. for his hospitality during the stay at the JPL, and for valuable suggestions. The author also thanks Drs N. Capitaine, J. Lieske, Skip Newhall and J.G. Williams for fruitful discussions.

References

Fukushima, T. (1994) New Canonical Variables for Orbital and Rotational Motions, *Celestial Mechanics and Dynamical Astronomy*, **60**, 57-68.

Guinot, B. (1981) Comments on the Terrestrial Pole of Reference, the Origin of the Longitudes, and on the definition of UT1, in *Reference Coordinate System for Earth Dynamics*, ed. by E.M. Gaposchkin and B. Kolaczek,

Kubo, Y., and Fukushima, T. (1986) Numerical Integration of Precession and Nutation of the Rigid Earth, in *The Earth's Rotation and Reference Frames for Geodesy and Geodynamics*, ed. by G.A. Wilkins and A.K. Babcock, Proc. of IAU Symp. No.128, Reidel, Dordrecht, 331-340.

Touma, J., and Wisdom, J. (1994) Lie-Poisson Integrators for Rigid Body Dynamics in the Solar System, *Astron. J.*, **107**, 1189-1202.

ANALYTICAL INTEGRATION OF A GENERALIZED EULER-POINSOT PROBLEM: APPLICATIONS

R. MOLINA AND A. VIGUERAS
Dpto de Matematica Aplicada y Estadistica
Esc. Politecnica Superior de Cartagena, U. Murcia, Spain

Abstract. We consider a generalized Euler-Poinsot problem for a stationary gyrostat whose first two components of the gyrostatic momentum are null. The problem is formulated in the Serret-Andoyer canonical variables and analytically integrated by means of the Hamilton-Jacobi equation in terms of elliptic functions and integrals. The obtained solutions are just the same as those for rigid bodies if a specific constant is annulled. Finally, two applications are proposed: 1) to obtain the action-angle variables of this problem, and 2) to the problem of the rotation of the Earth, using a triaxial gyrostat as a model.

The problem of the Earth's rotation, using as a model a symmetric gyrostat is studied in Vigueras (1983) and Cid and Vigueras (1990), by assuming that the first two components of the gyrostatic momentum are null and the third component chosen as a constant, in such a way that the free polar motion has a period of 430 days (Chandler's period). To extend this study to the triaxial case (the effects of triaxiality have been considered in some papers of Kinoshita (1977, 1992)), we generalize, in part, previous papers about the problem of the free rotation of a rigid body, due to, amongst other authors, Deprit (1967), Sadov (1970), Kinoshita (1972,1992), and Deprit and Elipe (1993).

In the Serret-Andoyer variables $(\lambda, \mu, \nu, P_\lambda, P_\mu, P_\nu)$, the Hamiltonian for the free motion of a stationary gyrostat is

$$H = \frac{1}{2}\left[(P_\mu^2 - P_\nu^2)\left(\frac{\sin^2(\nu)}{A} + \frac{\cos^2(\nu)}{B}\right) + \frac{P_\nu^2}{C}\right] - \sqrt{P_\mu^2 - P_\nu^2}\left(\frac{a_1}{A}\sin(\nu) + \frac{a_2}{B}\cos(\nu)\right) - \frac{a_3}{C}P_\nu \quad (1)$$

and the corresponding system is integrable because it has three integrals (P_μ, P_λ, H) which are functionally independent and in involution. When $a_1 = a_2 = 0$, we have a generalized Euler-Poinsot problem that reduces to that of free rotation of a rigid body if the constant a_3 is zero. This problem is separable and a complete solution of the Hamilton-Jacobi equation is given by the formula

$$W(\lambda, \mu, \nu, \alpha_3, \alpha_2, \alpha_1) = \alpha_3\lambda + \alpha_2\mu + \int \frac{a_3 + \sqrt{a_3^2 + C[1 - Cf(\nu)][2\alpha_1 - \alpha_2^2 f(\nu)]}}{1 - Cf(\nu)} d\nu \quad (2)$$

being $f(\nu) = \dfrac{\sin^2 \nu}{A} + \dfrac{\cos^2 \nu}{B}$. We have used this generating function to: 1) calculate the Serret-Andoyer canonical variables in function of the time t, and 2) obtain the action-angles variables. In both cases, the solutions can be expressed in terms of elliptic functions and integrals. All these calculations and other necessary in order to be able to study the perturbed motion will be given in a next paper.

Finally, the problem of the Earth's rotation when it is attracted by the Sun and the Moon is formulated using as a model a triaxial stationary gyrostat in the canonical variables $(\pi, \zeta, \nu, P_\pi, P_\zeta, P_\nu)$ introduced by Cid and Correas (1973), and referred to the mean ecliptic of an adopted epoch, in a similar way to Kinoshita (1977). The Hamiltonian can be break down into the sum of one unperturbed part $\mathcal{H}_0$ and one perturbed $\mathcal{H}_1$, in this form $\mathcal{H} = \mathcal{H}_0 + \mathcal{H}_1$, where $\mathcal{H}_0$ is the Hamiltonian of the previous generalized Euler-Poinsot problem, whose solutions we have obtained; and $\mathcal{H}_1$ contains the remaining terms. The Hamiltonian depends on the gyrostatic momentum whose constant components we want to determinate in such a way that the polar motion fits perfectly the observed values.

References

Cid, R. and Correas, J.M., 1973: *Actas I Jornadas Matemáticas Hispano-Lusas*, 439-452.
Cid, R. and Vigueras, A., 1990: Rev. Acad. Ciencias. Zaragoza (Spain), **45**, 83-93.
Deprit, A., 1967: Free rotation of a rigid body studied in the phase plane. *Amer. J. Phys.* **35**, 424-428.
Deprit, A. and Elipe, A., 1993: Complete reduction of the Euler-Poinsot problem. *J. of the Astronautical Sciences*, **41**, 4, 603-628.
Kinoshita, H., 1972: First-order perturbations of the two finite body problem. *Publ. Astron. Soc. Japan*, **24**, 423-457.
Kinoshita, H., 1977: Theory of the rotation of the rigid earth. *Celestial Mechanics*, **15**, 277-326.
Kinoshita, H., 1992: Analytical expansions of torque-free motions for short and long axis modes. *Celestial Mechanics and Dynamical Astronomy*, **53**, 365-375.
Sadov, A., 1970: The action-angle variables in the Euler-Poinsot problem. *PMM*, **34**, 962-964.
Vigueras, A., 1983: Movimiento rotatorio de giróstatos y aplicaciones. Ph. Dissertation. Univ. de Zaragoza, Spain.

ON THE TIDAL FUNCTION BETWEEN TWO REAL BODIES

V. SHKODROV, V. IVANOVA
Institute of Astronomy,
Bulgarian Academy of Sciences
1784 Sofia, Blvd. Tsarigradsko shosse, 72

Abstract. The general expression of the tidal function between two real bodies of arbitrary mass distribution has been derived. The case has been regarded when the distances between the bodies are comparable to their sizes. In this case, the effect of the body figure is substantial.

The relations obtained are utilized for the tidal influence of Jupiter on the surface of Io. It has been shown that Jupiter's shape (with $J_2 = 0.015$) causes a deformation in the sub-Jupiter point of about 15m. This is approximately 10% of the total tidal deformation of Io caused by Jupiter's gravitation.

1. General theory

When the tidal effect of a celestial body A on another celestial body B is determined, the first body is usually treated as a mass point A. This is so because the distances between the bodies are large and because the difference between the real figure of the tide formation body and the spherical symmetry in its mass distribution is negligibly small. In this case, the tidal function of the body A on the body B is given by the usual expression

$$W_0 = \frac{4}{3} G(r) \sum_{n=2} \left(\frac{\rho}{r}\right)^{n-2} P_n(\cos H), \qquad (1)$$

where $G(r)$ is the known Dutson constant [1]

$$G(r) = \frac{4}{3} \frac{fM_A}{r} \left(\frac{\rho}{r}\right)^2,$$

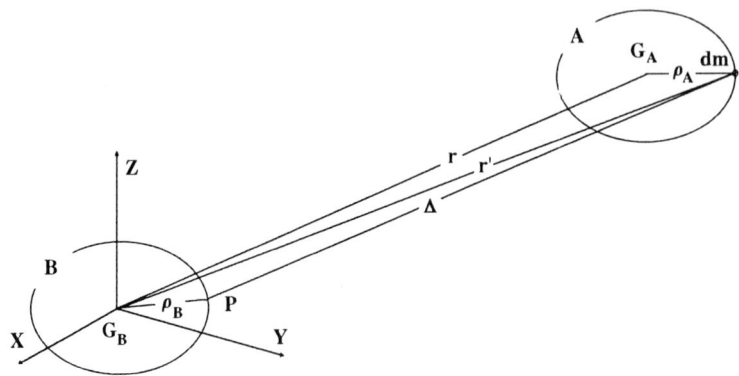

r is the position vector of the point where the mass M_A is located, and $P_n(cos H)$ is the Legendre polynomial with an argument $cos(H) = \mathbf{r}_0 \boldsymbol{\rho}_0$, $(\mathbf{r}_0 = \mathbf{r}/r; \boldsymbol{\rho}_0 = \boldsymbol{\rho}/\rho)$.

In this paper, we analyze the case when the distance R between the bodies A and B is not great and our purpose is to find the tidal influence of the real body A on B (Fig. 1). Here, the tidal function W at point P is determined by

$$W = V_p - V_{G_B} = fM_A \int \left(\frac{1}{\Delta} - \frac{1}{r'}\right) \frac{dm}{M_A}, \qquad (2)$$

where M_A is the mass of the body A, dm is an elementary mass; r' and Δ are the distances from dm to the mass center G_B and to the point P, respectively (Fig. 1). If Δ^{-1} is expressed by volume spherical functions [2], then

$$\Delta^{-1} = \sum_{n=0,m} \frac{1}{2n+1} T_n^{m*}(\boldsymbol{\rho}_p) N_n^m(\mathbf{r}'), \qquad (3)$$

where

$$T_n^{m*}(\boldsymbol{\rho}_p) = \rho_p^n Y_n^{m*}(\tau), \qquad (4)$$

$$N_n^m(\mathbf{r}') = \frac{1}{r'^{n+1}} Y_n^m(\tau').$$

In (4), $Y_n^{m*}(\tau)$ and $Y_n^m(\tau')$ are surface spherical functions whose arguments are the components of the unit vectors τ and τ' in a spherical coordinate system with an origin G. By (*) we denote the intricate conjugated quantity to $Y_n^m(\tau)$.

Hence, for V_p, we obtain

$$V_p = fM_A \sum_{n=0,m} \frac{1}{2n+1} T_n^{m*}(\boldsymbol{\rho}_p) \int_{T_A} N_n^m(\mathbf{r}') \frac{dm}{M_A}. \qquad (5)$$

and, from (1), (4) and (5),

$$W = fM_A \sum_{n=2,m} \frac{1}{2n+1} T_n^{m*}(\rho_p) \int_{T_A} N_n^m(\mathbf{r}') \frac{dm}{M_A}. \quad (6)$$

The last expression is obtained by taking into account that $N_0^0(\mathbf{r}') = -1/r'$ and that the harmonic with an index $n = 1$ does not cause a tidal effect.

To connect the tidal function W with the harmonic coefficients characterizing the distribution of mass in T_A, we shall refer $N_n^m(\mathbf{r}')$ to the origin G_A of T_A. For this purpose, we use the relation (Fig. 1)

$$\mathbf{r}' = \mathbf{r} + \boldsymbol{\rho}_A, \quad (7)$$

and $N_n^m(\mathbf{r} + \boldsymbol{\rho}_A)$ is presented according to the rule described in [2] in the following form:

$$\Omega N_n^m(\mathbf{r} + \boldsymbol{\rho}_A) = \sum_{(L,M)} Q_{LMnm} T_n^{m*}(\boldsymbol{\rho}_A) N_{n+L}^{m+M}(\mathbf{r}'). \quad (8)$$

In (8), Ω denotes the Taylor operator whose action is identical to the translation of the origin of the coordinate system by G_B at G_A, while

$$Q_{LMnm} = \frac{(-1)^L [(2n+1)(n-m+L-M)!(n+m+L+M)!]^{1/2}}{[(2L+1)(2n+2L+1)(n+m)!(n-m)!(L+M)!(L-M)!]^{1/2}}. \quad (9)$$

The action of the operator W on $N_n^m(\mathbf{r}') \in G_B$ leads to the appearance of

$$T_{M*}^L(\boldsymbol{\rho}_A) \in G_A; N_{n+L}^{m+M}(\mathbf{r}') \in G_B.$$

The final expression for the tidal function W of the real body A on B is obtained from (5) and (8) in the form

$$W = fM_A \sum_{n=2,m,L,M} Q_{LMnm} \frac{2L+1}{2n+1} A_L^M T_n^{m*}(\rho_p) N_{n+L}^{m+M}(\mathbf{r}) \quad (10)$$

where

$$A_L^M = \frac{1}{2L+1} \int_{T_A} T_L^{M*}(\boldsymbol{\rho}_A) \frac{dm}{M_A}$$

are the harmonic coefficients characterizing the mass distribution in the body T_A.

Expression (10) generalizes the usual presentation of the tidal potential where the body causing the tides is treated either as a point mass M or as a spherical body with spherical symmetry of its mass distribution. Really, if in (10) the term with $L = M = 0$ is separated, we obtain the expression

$$W = fM_A \sum_{n=2,m} \frac{1}{2n+1} T_n^{m*}(\rho_p) N_n^m(\mathbf{r}) + W_q, \qquad (11)$$

the first part of which is, in fact, the usual presentation (1) of the tidal potential, while the second part may be presented in the following form

$$W_q = \frac{4}{3} G(r) \sum_{nmLM} Q_{LMnm} \frac{2L+1}{2n+1} \left(\frac{\rho_p}{R}\right)^{n-2} \left(\frac{R}{r}\right)^{n+L-2}$$
$$\cdot A_L^{M*} Y_n^{m*}(\theta_p, \lambda_p) Y_{n+L}^{m+M}(\theta, \lambda), \qquad (12)$$

where R is constant, q, l and q_p, l_p are the spherical coordinates of the directions $\mathbf{r}$ and $\mathbf{r}_p$, respectively, and

$$A_L^{M*} = \frac{1}{2n+1} \int_{T_B} \left(\frac{\rho_A}{R}\right)^L Y_L^{M*}(\theta_A, \lambda_A) \frac{dm}{M_A}.$$

The summing is performed according to the scheme

$$\sum_{(n,m,L,M)} = \sum_{(n=2, m=-n)}^{n} \sum_{(L=2, M=-L)}^{L}$$

The relation (12) gives the tidal effect related to the mass distribution of the body.

2. The tidal function of Jupiter

The relations obtained are utilized for the tidal influence of Jupiter on the surface of Io. It has been shown that Jupiter's shape (with $J_2 = 0.015$) causes a deformation in the sub-Jupiter point of about 15m. This is approximately 10% of the total tidal deformation of Io caused by Jupiter's gravitation.

The investigations made for this paper were thanks to the financial support by National Foundation " Scientific Investigations" under contract Ph. 477/94, with Ministry of Education, Science and Technology.

References

1. Melchior P. (1971), *Physique et dynamique, Planetaires*, **Vol.1**
2. Shkodrov V. (1989), *Planet Potential*, Publishing House of Bulgarian Academy of Sciences, Sofia.

SYMPLECTIC MAPPINGS

JOHN D. HADJIDEMETRIOU
University of Thessaloniki, Department of Physics
GR-540 06 Thessaloniki, Greece
e-mail: hadjidem@olymp.ccf.auth.gr

Abstract. This paper reviews the main methods for constructing mapping models for Hamiltonian systems, for the study of motion in the Solar System. The emphasis is given to the relation between the various mapping techniques, the methods to check how close is a mapping model to the original system and the effects of an incomplete model on the evolution of the system.

1. Introduction

The evolution of a subsystem in the Solar System, as for example the motion of an asteroid under the gravitational attraction of the Sun and Jupiter, can be described by a Hamiltonian system. It is known that the flow in the $2n$-dimensional phase space, from an initial position (q_0, p_0) at time $t = t_0$ to the position (q_1, p_1) at the time $t = t_1$ is a canonical transformation, where n is the number of degrees of freedom and q, p are n-vectors. This means that the mapping in phase space from the time t_0 to the time t_1 is a symplectic mapping. [1] An important property of a symplectic map is that it conserves the volume in phase space.

A very useful tool in the study of a Hamiltonian system is the method of Poincaré map on a surface of section. The continuous flow in the $2n$-dimensional phase space is reduced to a map in a $(2n-2)$-dimensional phase space. The Poincaré map is also symplectic. This method is especially useful in systems with two degrees of freedom, where instead of studying the flow

---

[1] A map $T : M \to M$ is symplectic if its Jacobian matrix $L = DT$ satisfies the symplectic property $L^T J L = J$, where J is the symplectic matrix with rows (in block form) $(0, I)$ and $(-I, 0)$, (I and 0 are the $n \times n$ unit and zero matrices, respectively) and L^T is the transpose of L.

in a 4-dimensional phase space we study the map in a 2-dimensional space. The fixed points of the Poincaré map coincide with the periodic orbits of the system and it is clear that their position and stability character determine the topology of the reduced phase space.

By making use of the Poincaré surface of section we do not lose any information: the map describes completely the system. However, in order to find the Poincaré map we must solve the system of differential equations which, in general, cannot be solved in a closed form, because the system is nonintegrable. Thus the only way to obtain the Poincaré map is to use numerical integrations.

2. Construction of mapping models

As mentioned above, the Poincaré map involves the solution of a nonintegrable dynamical system. In order to overcome this difficulty, we can construct mapping models to study the evolution of the Hamiltonian system. There are several methods to do this, that will be explained in the following. All of them are based on a perturbation method and consequently the series may not converge beyond a certain value of a small parameter.

In order for a mapping model to be realistic, its phase space must have the same topology as that of the Poincaré map of the original Hamiltonian system. The *necessary conditions* for a realistic mapping model are:
1. The mapping must be symplectic.
2. The mapping model must have the same fixed points as the Poincaré map.
3. The fixed points must have the correct stability character.

Note that it is not difficult to check the correct position and stability of the fixed points of the mapping model, because they must coincide with the periodic orbits of the original Hamiltonian system and it is relatively easy to compute the main families of periodic orbits of a Hamiltonian system. This provides us with an efficient check of the validity of the mapping model and also of the convergence of the perturbation series involved in the construction of the mapping model.

3. The averaging method

Some important methods for the construction of a mapping model for a nearly integrable dynamical system are based on the method of averaging. For this reason we shall comment on this method and we shall discuss its relation to the original system on one hand and to the corresponding mapping model on the other.

We consider a Hamiltonian system with n degrees of freedom,

$$H = H_0 + \epsilon H_1 \tag{1}$$

where H_0 is the integrable part and ϵ a small parameter. We transform first to action-angle variables, θ, I, of the integrable part and H_0 becomes a function of the actions only, $H_0(I)$. We assume that we are close to a resonance (of H_0) and transform further to new canonical resonant action-angle variables. In this way we have *fast* and *slow* angles. Next, by a new canonical transformation through a suitable generating function, we eliminate the fast angles and we are left with a new, *averaged* Hamiltonian $\bar{H}$ with $(n-1)$ degrees of freedom. It is in this last step that the perturbation method enters and several things could go wrong, because the perturbation series may not converge beyond a certain value of the small parameter (in fact this is always the case).

From the theory of averaging it is known that the fixed points of the averaged Hamiltonian $\bar{H}$ coincide with the periodic orbits of H (or the fixed points of the Poincaré map of H). If H were integrable, the perturbation method would converge and the fixed points of $\bar{H}$ would be identical with the fixed points of the original system (Poincaré map). In a nonintegrable Hamiltonian however the fixed points of $\bar{H}$ may be in the wrong position, or may have the wrong stability, or more fixed points may appear, or some fixed points may be missing. In all these cases the topology of the mapping model is different from that of the real system and consequently the model is not realistic. We shall discuss all these problems in the study of actual dynamical systems in the Solar System that follows.

4. Methods to construct mapping models

We can separate the methods to obtain a mapping model for a Hamiltonian system of the form (1) into two main categories:

A- Solve *approximately* the differential equations of motion and then use the approximate solution to obtain a mapping model.

B- Construct *analytically* the Poincaré map of the integrable part H_0 and express it by a generating function. Then perturb the generating function to obtain the mapping model.

Before we present applications of these methods to the study of actual systems in the solar system, we shall describe the method A, as applied by Wisdom and the method B, as applied by Hadjidemetriou by a simple example, and we shall discuss their similarities and also their relation to the actual system they are designed to represent. Both methods are based on an averaged Hamiltonian $\bar{H}$ of the original Hamiltonian H of the form (1). Other mapping techniques will be also presented. A review on mapping techniques has been presented by Froeschlé (1991).

5. A comparison of the methods A and B by a simple example

5.1. METHOD A

We consider the time dependent Hamiltonian system

$$H = H_0(I) + (K_0/2\pi) \cos(\theta) + H_{hf}(\theta, I, t), \qquad (2)$$

where K_0 is a constant, I is the action, θ the angle and

$$H_{hf} = \sum_{n \neq 0} K_n(I) \cos(\theta - nt) \qquad (3)$$

is a 2π-periodic function of the time that represents the high frequency terms. By making use of the averaging method (see for example Hadjidemetriou 1991), we obtain the averaged Hamiltonian

$$\bar{H} = H_0(\bar{I}) + (K_0/2\pi) \cos(\bar{\theta}), \qquad (4)$$

where $\bar{I}$, $\bar{\theta}$ are the averaged variables. On going from H to $\bar{H}$ we lose information (the time dependence is eliminated with the high frequency terms, which means that in fact we lose one degree of freedom).

We *reintroduce* now the high frequency terms that were eliminated in the averaging process (but not the same terms!). Instead of H_{hf} we add to $\bar{H}$ the high frequency terms (also with period 2π)

$$H_{hf}^* = \sum_{n \neq 0} (K_0/2\pi) \cos(\bar{\theta} - nt) \qquad (5)$$

and we have now the Hamiltonian (dropping overbars)

$$H = H_0(I) + K_0 \cos(\theta) \delta_{2\pi}(t), \qquad (6)$$

that will be used in place of the original Hamiltonian (2), where

$$\delta_{2\pi}(t) = \frac{1}{2\pi} \left(1 + 2 \sum_{n=1}^{\infty} \cos(nt) \right)$$

is the 2π-periodic delta function.

Note that this is equivalent to substituting $\cos(\bar{\theta})$ in (4) by the sum

$$\sum_{n=-\infty}^{\infty} \cos(\bar{\theta} - nt) = \cos(\bar{\theta}) \left(1 + 2 \sum_{n=1}^{\infty} \cos(nt) \right) = 2\pi \cos(\bar{\theta}) \delta_{2\pi}(t), \qquad (7)$$

i.e. we replace $\cos(\theta)$ by 2π-periodic impulses, multiplied by 2π.

The Hamiltonian (6) can be easily solved because it is equal to $H_0(I)$ in all open intervals $(0, 2\pi)$, $(2\pi, 4\pi)$, ... The solution from 0^- to $2\pi^-$ gives the mapping for one period 2π of the forcing high frequency term (3). This mapping is:

$$I_1 = I_0 + K_0 \sin(\theta_0), \quad \theta_1 = \theta_0 + 2\pi\,\omega_0(I_1), \tag{8}$$

where $\omega_0(I) \equiv \partial H_0/\partial I$. It can be readily verified that this mapping is symplectic and can be obtained by the generating function

$$F(\theta_0, I_1) = I_1\theta_0 + 2\pi\left[H_0(I_1) + (K_0/2\pi)\cos(\theta_0)\right]. \tag{9}$$

We ask now the question what is the relation between the map (8) and the original system (2). In order for the map (8) to be a realistic model, it must coincide with the Poincaré map of H, i.e. its fixed points should coincide with the fixed points of the Poincaré map, both in position and in stability properties. We remind now that the fixed points of the Poincaré map coincide with the fixed points of the averaged Hamiltonian $\bar{H}$ (provided that $\bar{H}$ is a good model!). So, finally, the comparison between the fixed points of the map (8) and of $\bar{H}$ will provide us the test whether (8) is a realistic model for the original Hamiltonian H.

The fixed points (θ_0, I_0) of the averaged Hamiltonian $\bar{H}$, given by (4), are obtained from the equations

$$\sin(\theta_0) = 0, \quad \omega_0(I_0) = 0. \tag{10}$$

The stability index of the corresponding periodic orbit of the original Hamiltonian (2), which has the period 2π of the forcing term H_{hf}, defined as the sum of the eigenvalues of the monodromy matrix, is

$$k = \exp 2\pi\sqrt{\beta} + \exp -2\pi\sqrt{\beta}, \tag{11}$$

where

$$\beta = (K_0/2\pi)\omega_{0I}(I_0)\cos(\theta_0), \tag{12}$$

and $\omega_{0I} \equiv \partial\omega_0/\partial I$, (Hadjidemetriou, 1991).

The fixed points of the mapping (8) are also given by the equations (10), but the stability index (defined as the sum of the eigenvalues of the linearized mapping at the fixed point) is different and is given by

$$k = 2 + 2\pi K_0 \omega_{0I}(I_0)\cos(\theta_0). \tag{13}$$

From the above analysis it is clear that the position of the fixed points is the same in both the averaged Hamiltonian and the corresponding mapping, but the stability index is different, in general. However, the two stability indices are close if $K_0 \ll 1$, as can be seen by expanding the exponentials in (11) in powers of $\sqrt{\beta}$:

$$k = 2 + 2\pi K_0 \omega_{0I}(I_0) \cos(\theta_0) + O(K_0^2). \tag{14}$$

Thus, for small K_0 the mapping model is a realistic model for the original system (2).

5.2. THE METHOD B

In this method we solve first the integrable part of $H_0(I)$ of H assuming that K_0 is small, so that the second term in the right hand side of (4) to be considered as a perturbation. The solution is $I = I_0$, $\theta = \omega_0(I_0)t$ and this gives the Poincaré map, at integral multiples of the period 2π:

$$I_1 = I_0, \quad \theta_1 = \theta_0 + 2\pi\omega_0(I_0). \tag{15}$$

This is obtained from the generating function

$$F_0(I_1, \theta_0) = I_1\theta_0 + 2\pi H_0(I_1) \tag{16}$$

through the equations $I_0 = \partial F_0/\partial \theta_0$, $\theta_1 = \partial F_0/\partial I_1$. We perturb now the generating function (16) in such a way that the new map has the same topology as the Poincaré map of the original system. Note that this map is *always* symplectic by its construction. It can be proved (Hadjidemetriou, 1991,1993) that this can be achieved by adding a perturbation term F_1 to F_0, which is the perturbation term of $\bar{H}$, given by (4), multiplied by the period 2π,

$$F_1 = K_0 \cos(\theta). \tag{17}$$

Thus, the generating function is $F = F_0 + F_1$ and we can verify that it coincides with the generating function (9) of the mapping (8).

5.3. COMPARISON OF METHODS A AND B

From the previous two sections we come to the conclusion that both methods A and B are equivalent, for the simple example we used to demonstrate these methods, provided $K_0 \ll 1$. This may not be always the case. Method B gives always a symplectic map, and it can be proved that it gives mapping models that are realistic (i.e they satisfy the necessary conditions of section 2), provided of course that the averaged Hamiltonian on which they are based are realistic. The map however that is obtained with method A may not be symplectic (Henrard, 1995), as can be seen by considering K_0 in (2) as a function of I, $K_0 = K_0(I)$. The mapping obtained by the method A is

$$I_1 = I_0 + K_0(I_0)\sin(\theta_0), \quad \theta_1 = \theta_0 + 2\pi\omega_0(I_1) + K'(I_0)\cos(\theta_0),$$

which is not symplectic, while the mapping obtained by method B is

$$I_1 = I_0 + K_0(I_1)\sin(\theta_0), \quad \theta_1 = \theta_0 + 2\pi\omega_0(I_1) + K'(I_1)\cos(\theta_0),$$

which is symplectic.

Finally, we remark that the above two methods that were demonstrated by a simple example, can be extended to systems with more degrees of freedom, as we shall see in the applications that follow.

6. Application of mapping methods in the Solar System

6.1. RESONANT ASTEROID MOTION: THE WISDOM MAPPING

The method described in section 5.1 to obtain a mapping model for a Hamiltonian system has been applied by Wisdom (1982,1983,1985) for the study of an asteroid at the 3:1 resonance with Jupiter. The underlying dynamical system is the planar elliptic restricted three body problem, with the Sun and Jupiter as primaries. The averaged Hamiltonian used to obtain the mapping model is

$$\begin{aligned} H = & -\frac{\mu_1^2}{2\Phi^2} - 3\Phi + \mu F(x^2 + y^2) \\ & + e_j \mu G x - \mu \left[C(x_2 - y_2) + e_j D + e_j^2 E \right] \cos\phi \\ & - \mu \left[2Cxy + e_j Dy \right] \sin\phi, \end{aligned} \quad (18)$$

where $\mu_1 = 1 - \mu$, $\Phi = \sqrt{\mu_1 a}$, $\phi = l - 3l_j$ and

$$x = \sqrt{2\rho}\cos\omega, \quad y = \sqrt{2\rho}\sin\omega, \quad (19)$$

with $\rho = \sqrt{\mu_1 a}\left(1 - \sqrt{1-e^2}\right)$. The semimajor axis of the asteroid is a, its eccentricity is e, its mean longitude is l, the argument of perihelion is ω, the subscript j refers to Jupiter and μ is the mass of Jupiter. The angle ϕ is the resonant angle (*slow* angle). The averaged Hamiltonian (18) is obtained from the Hamiltonian of the elliptic restricted three body problem by eliminating the high frequency terms, which in this case is the orbital motion of Jupiter in its orbit around the Sun, with period 2π. The high frequency terms are reintroduced by 2π-periodic impulses as

$$\cos\phi \to \cos\phi \times 2\pi\delta_{2\pi}(t), \quad \sin\phi \to \sin\phi \times 2\pi\delta_{2\pi}(t - \pi/2). \quad (20)$$

The new Hamiltonian can now be solved easily from $t = 0$ to $t = 2\pi$ and in this way a mapping is obtained, which is a model for the Poincaré map of the elliptic restricted three body problem at integral multiples of the period 2π of Jupiter's orbit (stroboscobic map). Note that all periodic

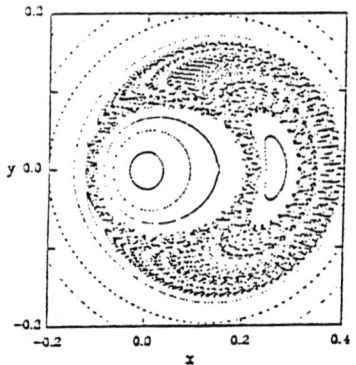 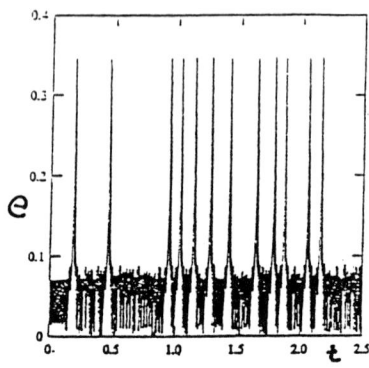

Figure 1. (a):The evolution in the xy plane. The distance from $(0,0)$ is proportional to the eccentricity. (b):The time evolution of the eccentricity (from Wisdom).

orbits in the elliptic restricted three body problem have a period equal to 2π, or a multiple of it. It turns out that this mapping is symplectic. This is a four dimensional map in the space ϕ, Φ, x, y and it is not easy to present geometrically. In this case however, we can further reduce the map to a two dimensional one by taking advantage of the fact that after the averaging we still have a "fast" and a "slow" angle, and we eliminate the fast angle by a new averaging. The fast angle is the resonant angle ϕ, compared to the argument of perihelion ω which is much slower. This is equivalent to taking a "mapping" of the mapping, for example by the "surface of section" H =constant, $\phi = \pi$. In this way we finally obtain a two dimensional map in the space $x\ y$.

The evolution of an asteroid at the 3:1 resonance, obtained by this mapping, is shown in Figure 1. We note that there is a large chaotic region and the asteroid may be locked in a motion with small radius (eccentricity) but suddenly may jump to a motion where the radius takes large values. This is clearly seen in Figure 1b, where the eccentricity jumps to values higher than 0.3 at unpredictable times, showing an intermittent behaviour. In this way Wisdom explained the observed gap in the distribution of the asteroids at the 3:1 resonance, because for $e > 0.3$ the asteroid becomes a Mars crosser and the resulting perturbation will remove the asteroid from the 3:1 resonance region.

The Wisdom method has been applied by Šidlichovský (1988, 1990, 1992, 1993) and by Murray and Fox (1984).

6.2. RESONANT ASTEROID MOTION: HADJIDEMETRIOU'S METHOD

The method discussed in section 5.2 has been applied by Hadjidemetriou (1991,1993) for the study of asteroid motion at the 3:1 resonance. The starting point is also the averaged Hamiltonian of the elliptic restricted three body problem, which is of the form

$$H = H_0(S, N) + \mu H_1(\sigma, S, N) + \mu e_j H_2(\sigma, S, \nu, N), \qquad (21)$$

where

$$H_0 = -\frac{2(1-\mu)^2}{(N-S)^2} - \frac{3}{2}(N-S), \quad H_1 = 2FS - b\frac{S}{N}\cos 2\sigma,$$
$$H_2 = \sqrt{2S}[G\cos(\sigma+\nu) + D\cos(\sigma-\nu)] + 2\mu e_j K \cos 2\nu.$$

The resonance action-angle variables are defined by

$$S = \sqrt{\mu_1 a}\left(1 - \sqrt{1-e^2}\right) \quad \sigma = \frac{1}{2}(3\lambda_j - \lambda) - \omega_j$$

$$N = \sqrt{\mu_1 a}\left(3 - \sqrt{1-e^2}\right) \quad \nu = -\frac{1}{2}(3\lambda_j - \lambda) + \omega_j$$

where $\mu_1 = 1-\mu$, $e_j = 0.048$, λ, ω, a are the mean longitude, the longitude of perihelion and the semimajor axis, respectively, of the asteroid and the corresponding quantities with subscript j refer to Jupiter. This is in fact the same as the Hamiltonian (18) used by Wisdom. The mapping is obtained from the generating function

$$F = \sigma_n S_{n+1} + \nu_n N_{n+1} + 2\pi H(S_{n+1}, N_{n+1}, \sigma_n \nu_n), \qquad (22)$$

though the equations

$$\sigma_{n+1} = \partial F/\partial S_{n+1}, \; S_n = \partial F/\partial \sigma_n, \; \nu_{n+1} = \partial F/\partial N_{n+1}, \; N_n = \partial F/\partial \nu_n.$$

and is evidently symplectic.

We can verify that the first two equations are decoupled from the other two if $e_j = 0$. The mapping in this case (*circular* restricted three body problem) is two dimensional, in the space σ, S, with N as a parameter. For different values of N we have different surfaces of section, but no chaotic regions appear. As soon however as $e_j \neq 0$, the parameter N varies slowly and as a consequence a slow drift appears from one N = constant plane to the next. This is the mechanism by which large scale chaos appears (Hadjidemetriou 1993,1995). This mechanism is also given by Wisdom (1985) and Henrard (1992), using the averaged model. The evolution of an asteroid inside the 3:1 resonance is given in Figure 2 for different initial values of σ and ν. The behaviour is in fact the same as that given in Figure 1. (Note the similarity between Figures 1a and 2a). Not all motions however inside the resonance zone are chaotic. We also have ordered motion, as shown in

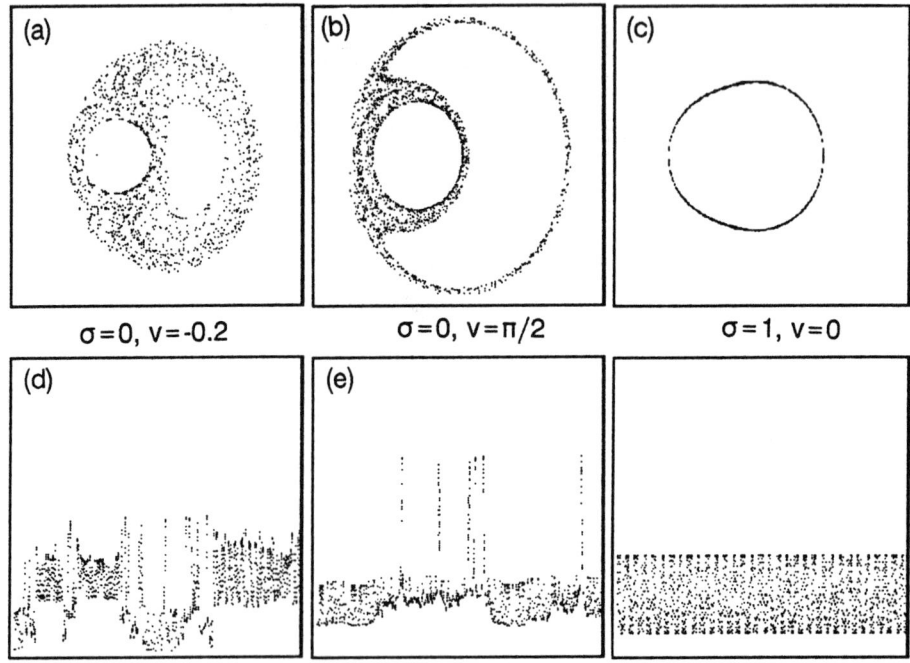

Figure 2. The evolution of an asteroid by the mapping obtained from the generating function (22) for $S = 0.0116$, $N = 1.4$ and σ, ν as indicated: a,b and c give the evolution in the xy plane, and d,e,f the corresponding time evolution of the eccentricity

Figures 2c,f, where we have an oscillation with an amplitude which depends on the initial angles and may be quite small (as is the case with $\sigma = 0$, $\nu = \pi$). This means that we may have a trapping of the asteroid inside the 3:1 resonance.

This method of constructing mapping models has been also used by Ji-Liu et al.(1994) for near conservative systems and by Ferraz-Mello (1995) for the 2:1 resonance.

6.3. REMARKS ON THE PREVIOUS TWO MAPPINGS

The general behaviour of both mappings is the same, as far as the long term evolution is concerned. However both models have their limitations, as we shall explain:
- The averaged model on which these two mapping models are based is valid only for values of the eccentricity smaller than 0.3. This means that the high eccentricity resonances that exist in the real problem (the elliptic restricted three body problem) at eccentricities $e = 0.8$ are missing (Hadjidemetriou 1992). These high eccentricity resonances can be introduced to the system

through a *correction term* (Hadjidemetriou 1993). The corrected mapping behaves in a much different way and the eccentricity can jump to very high values of the eccentricity, 0.9 or even larger. The asteroid can become now an Earth crosser.

-Both models are not realistic for the long time study of asteroid motion, because the orbit of Jupiter is fixed. The elements of Jupiter's orbit vary considerably, as shown by Laskar (1990) and Nobili et al.(1989). If this variation is introduced in the mapping (22), by varying e_j and ω_j, we find (Hadjidemetriou 1993, 1995) that the chaotic behaviour appears for all values of the angles σ and ν, including those cases where the motion was ordered for a fixed orbit of Jupiter, while no appreciable effect appears outside the resonance zone.

Thus, we see that the gravitational effect of Saturn must also be included in the model in order to explain the observed gap in the distribution of the asteroids at the 3:1 resonance.

6.4. OTHER MAPPING TECHNIQUES

6.4.1. *Nearly parabolic orbits*

Mapping models for nearly parabolic orbits have been developed by several people, with the aim to study cometary motion. They are all in the framework of the restricted circular three body problem, with the Sun and Jupiter as primaries. For a nearly parabolic orbit we can assume that the motion is close to the separatrix. Due to the very large time scales (the period of the parabolic orbit is infinite), a Fourier analysis is made and the spectrum is assumed to be continuous. In this way Petrosky and Broucke (1988) transformed the nonintegrable system to an integrable one by embedding the small denominators in an analytic function through a suitable analytic continuation. The corresponding mapping is called *Keplerian map* and is a two dimensional symplectic map in the space P, g, where P is the heliocentric energy of a comet and g the phase angle of Jupiter when the comet is at perihelion. Similar work on the Keplerian map has been made by Chambers (1993), Chirikov and Vecheslavov (1988), Sagdeev and Zaslansky (1987) and Liu and Yi-Sui Sun (1994).

6.4.2. *Nearly circular orbits*

This technique has been developed by Duncan et al.(1989) for the study of a test particle in nearly circular orbit, in the framework of the *circular* restricted three body problem, with the Sun and a planet as primaries. The perturbation on the test particle is assumed to be significant only at conjunction. Under this assumption, the first order perturbation on the elements of the orbit are obtained by Hill's equation. In this way an approximate solution is found, from which a mapping is obtained.

References

Chambers, J. E.: (1993) "A simple mapping for comets in resonance". *Celest, Mech,* **57**, 131.
Chirikov, B. V. and Vecheslavov, V. V.:(1986) "Chaotic dynamics of comet Haley". preprint 86-184, Institute of Nuclear Physics, Novosibirsk.
Duncan, M., Quinn, T. and Tremaine, S.: (1989) "The long term evolution of orbits in the Solar System: A mapping approach". *Icarus* **82**, 402.
Ferraz-Mello, S.: (1995) preprint.
Froeschlé, C.: (1991) "Modelling: An aim and a tool for the study of the chaotic behaviour of asteroidal and cometary orbits". In *Predictability, Stability and Chaos in N-Body Dynamical Systems*, A. E. Roy (ed), Plenum Press, p. 125-155.
Hadjidemetriou, J. D.: (1991) "Mapping Models for Hamiltonian Systems with application to Resonant Asteroid Motion". In *"Predictability, Stability and Chaos in N-Body Dynamical Systems*, A. E. Roy (ed), Kluwer Publ., p. 157-175.
Hadjidemetriou, J. D.: (1992) "The Elliptic Restricted Problem at the 3:1 Resonance". *Celest. Mech.* **53**, 151-183.
Hadjidemetriou, J. D.: (1993) "Asteroid Motion near the 3:1 Resonance". *Celest. Mech.* **56**, 563-599.
Hadjidemetriou, J. D. and Voyatzis, G.: (1993) "Long Term Evolution of Asteroids near a Resonance". *Celest. Mech.* **56**, 95-96.
Hadjidemetriou, J. D.: (1995), "Mechanisms of Generation of Chaos in the Solar System". In *"From Newton to Chaos*, A. E. Roy and B. A. Steves, (eds), Plenum Press, 79.
Henrard, J.: (1995) private communication.
Henrard, J.: (1988) "Resonances in the Planar Elliptic Restricted Problem". In *Long Term Dynamical Behaviour of Natural and Artificial N-Body Systems*, A. E. Roy (ed), 405- 425, Kluwer Publ.
Ji-Lin Zhou, Yan-Ning Fu and Yi-Sui Sun: (1994) " Mapping models for near-conservative systems with applications". *Celest. Mech.* **60**, 471.
Laskar, J.: (1990) "The chaotic motion of the Solar System". *Icarus* **88**, 266-291.
Liu Jie and Yi-Sui Sun: (1994) "Chaotic motion of comets in near-parabolic orbit: mapping approaches". *Celest. Mech.* **60**, 3.
Nobili, A., Milani, A., Carpino, M.: (1989) "Fundamental Frequencies and Small Divisors in the Orbits of The Outer Planets". *Astron. Astrophys.* **210**, 313-336.
Murray, C. D. and Fox, K.: (1984) "Structure of the 3:1 Jovian resonance: A comparison of numerical methods". *Icarus* **59**, 482.
Petrosky, T. Y. and Broucke, R.: (1988) "Area-preserving mappings and deterministic chaos for nearly parabolic orbits". *Celest. Mech.* **42**, 53.
Sagdeev, R. Z. and Zaslavsky, G. M.: (1987), *Nuovo Cimento* **97**, BN2.
Šidlichovský, M.: (1988), "On the Origin of 5/2 Kirkwood Gap" In *Proceedings of the IAU Colloq. 96, The Few Body Problem* (Turku), Valtonen, M. (ed.), p. 117.
Šidlichovský, M.: (1990), "The Existence of a Chaotic Region Due to the Overlap of Secular Resonances ν_5 and ν_6". *Celes. Mech. Dyn. Astr.* **49**, 177.
Šidlichovský, M.: (1992) "Mapping for the asteroidal resonances". *Astron. Astrophys.* **259**, 341-348.
Šidlichovský, M.: (1993) "Chaotic Behaviour of Trajectories for the Fourth and Third Order Asteroidal Resonances". *Celest. Mech.* **56**, 143.
Wisdom, J.: (1982) "The origin of the Kirkwood Gaps". *Astron. J.* **87**, 577-593.
Wisdom, J.: (1983) "Chaotic Behaviour and the Origin of the 3/1 Kirkwood Gap". *Icarus* **56**, 51-74.
Wisdom, J.: (1985) "A Perturbative Treatment of Motion Near the 3/1 Commensurability". *Icarus* **63**, 272-289.

L'ALGÈBRE SYMBOLIQUE EN MÉCANIQUE CÉLESTE

ANDRÉ DEPRIT
*National Institute of Standards and Technology,
Gaithersburg, MD 20899-0001, U.S.A.*

Que dire de l'algèbre symbolique en mécanique céleste si ce n'est que les astronomes furent les premiers sur le chantier. Rien là qui doive surprendre.

LA GRANDE PÉRIODE

D'un calcul qui dépasse les bornes de la patience humaine, on dit qu'il est astronomique. L'expression est heureuse, vous l'admettrez volontiers s'il vous arrive de retirer de la bibliothèque les mémoires de Lagrange, de Laplace, du sieur de Pontécoulant et de Leverrier, à quoi vous pourriez ajouter, s'il vous reste de la place au creux des coudes, une certaine théorie de la lune qui fit histoire par ses formules longues comme des écheveaux sans fin. Nul ne se faisait un nom en mécanique céleste s'il n'était un athlète en calcul algébrique. Hansen, Hill, Airy, Tisserand, Radau, Brown, Von Zeipel consacrèrent des années, voire des décennies, à pousser le rocher de leur théorie au sommet de la précision. A de rares exceptions, pour chacun d'eux la légende se répéta: les observations haussaient la précision de plusieurs ordres, le souffle coupé, Sisyphe lâchait prise et sa théorie tombait à pic dans l'oubli.

De ces travaux d'Hercule, la discipline est sortie apauvrie, rétrécie, écœurée, écœurante même, et probablement à jamais, au goût des astronomes. Il fallait se rendre à l'évidence. Patience et longueur de temps n'en font pas plus que force ni que rage quand il s'agit de réduire la complexité des calculs algébriques que la mécanique céleste se doit d'exécuter.

LE TEMPS DES MOULINETTES

Il en est toutefois du calcul comme de la diplomatie: la fortune ne vous clôt jamais une porte qu'elle ne vous en déclôt une autre aussitôt. Tandis que la mécanique céleste faisait sa dépression, apparaissaient les moulinettes commerciales, des Mercedes et des Brunsviga. L'analyse numérique prit le pas sur l'algèbre symbolique. Comme de bien entendu, la mécanique céleste suivit le mouvement, mais de loin. On se mit à créer les figures qui manquaient aux *Méthodes nouvelles* de Poincaré. George Darwin se choisit les orbites périodiques, Elie Strømgren les familles d'orbites périodiques. Pour leur compte, Georges Lemaître, Sandoval Vallarta et René de Vogelaere s'en prenaient à des sujets délicats comme les exposants caractéristiques et les orbites asymptotiques. Il importe, à notre point de vue, de se rendre compte que ces explorations numériques couvaient une conversion radicale des esprits: à dater de ce temps, le sort de la mécanique céleste est lié à l'avenir des moyens de calcul. Fort de cette conviction, Leslie Comrie introduisit les tabulatrices comptables imprimantes au Nautical Almanach Office de Greenwich. Aussi riche de promesses fut la nomination de Wallace Eckert à la direction du Nautical Almanach Office à Washington.

Eckert était un homme de calcul au double sens du mot. Son initiation à l'algèbre symbolique, il la tenait de Brown; sa formation en analyse numérique, il l'acquit au cours des années que Brown l'employa à réduire les observations de la lune. Au Nautical Almanach Office, Eckert établit les tâches de calcul sur la base des cartes perforées. C'était une innovation.

RÉVEIL

L'effort de guerre d'abord, la course à l'espace ensuite, sans parler de l'impitoyable concurrence commerciale et financière, imprimaient une allure effrénée au développement des calculateurs. Eckert surveillait la tempête. Le moment lui sembla venu en 1958 d'offrir à ses collègues l'occasion de prendre ensemble conscience des conditions dans lesquelles se ferait à l'avenir la recherche en mécanique céleste. Ils le firent à souhait lors de la conférence convoquée à New York, si du moins l'on en croit le procès verbal publié par l'*Astronomical Journal*.

Qu'on se garde toutefois de voir un plan de travail dans ce rapport: sur le thème central des équipements et du logiciel comme aussi des approches et des axes de recherche, le document ne contient rien de précis. Il s'en dégage néanmoins plusieurs éléments de grande conséquence pour l'avenir, à commencer par le fait que le lancement de Sputnik a tiré la mécanique céleste de son état dépressif. Au début du siècle, la discipline s'était réfugiée auprès des gens du Métier des Almanachs; à lire le rapport, on devine qu'elle ne s'est pas encore soustraite à leur emprise. Les circonstances ne le permettaient pas.

Il fallait en effet fournir de toute urgence aux ingénieurs d'aérospatiale de quoi prédire la trajectoire d'un satellite dans le champ de gravité de la terre. La mécanique céleste sortit donc des étagères tout ce qu'elle avait accumulé à propos de modèles dynamiques intégrables et de leurs perturbations. A la conférence d'Eckert, par exemple, tandis qu'Herget plaçait son mot sur la méthode de Hansen, Brouwer confessait qu'il essayait d'appliquer au problème des opérations que Delaunay, un siècle auparavant, avait mises au point dans la théorie de la lune.

Avec l'inclinaison critique Brouwer se prit les pieds dans une difficulté de nature mathématique. La mécanique céleste du XIXe siècle ne l'avait pas rencontrée. Orlov s'y était achoppé quelques années auparavant lorsqu'il misait sur la méthode de Lindstedt pour traiter la trajectoire d'un satellite artificiel à la manière de Poincaré comme la déformation d'un arc pris sur une orbite périodique de première espèce. Les complexes militaires-industriels qui, de part et d'autre du Rideau de Fer, jouaient les mécènes d'une renaissance de la mécanique céleste, n'avaient ni patience ni curiosité pour ce genre d'énigmes. Qu'à cela ne tienne! D'entrée de jeu, il aurait fallu faire face au problème, à tout le moins le marquer auprès des générations montantes comme un sujet de recherche susceptible d'ouvrir la discipline à de nouveaux horizons. Mais les invités d'Eckert ne l'ont pas fait. La mécanique céleste faisait sa rentrée à l'avant-scène avec adresse et assurance, mais à reculons, les yeux fixés sur les décors de son passé.

LE PREMIER PAS

On fera exception toutefois pour ce qui est de l'algèbre symbolique. Herget et Musen avaient ouvert la brèche sur une IBM 650 à Cincinnati. Ils demandaient à une machine de réaliser les trois opérations fondamentales

$$(f, g) \to f + g : \quad \mathcal{F} \times \mathcal{F} \to \mathcal{F},$$
$$(f, g) \to fg : \quad \mathcal{F} \times \mathcal{F} \to \mathcal{F},$$
$$(\alpha, f) \to \alpha f : \quad \mathcal{K} \times \mathcal{F} \to \mathcal{F}$$

dans une algèbre de Fourier $\mathcal{F}$ à plusieurs variables sur le corps des nombres réels. Leurs programmes procédaient d'une représentation triviale des séries de Fourier comme des éléments dans un espace vectoriel. Aux lignes trigonométriques

$$\cos(m_1 x_1 + m_2 x_2 + \ldots + m_p x_p) \quad \text{et} \quad \sin(n_1 x_1 + n_2 x_2 + \ldots + n_p x_p)$$

correspondaient respectivement des listes de la forme

$$(0, m_1, m_2, \ldots, m_p) \quad \text{et} \quad (1, n_1, n_2, \ldots, n_p).$$

Il ne fallait pas être grand clerc pour inventer ce schéma. Mais Herget et Musen entendaient le faire exécuter par une machine rudimentaire dans une langue de programmation primitive.

On a peine, de nos jours, à se représenter les obstacles auxquels Herget et Musen se heurtaient. En voici un exemple parmi beaucoup d'autres. A la sortie d'un calcul qui ne pouvait se faire que sur cartes, chaque vecteur $\{m_j\}$ ou $\{n_j\}$ ainsi que son coefficient était perforé dans des colonnes fixées à l'avance quels que fussent les résultats de l'opération. Ou bien la sousroutine de sortie interceptait les dépassements de capacité—auquel cas elle n'avait d'autre recours que d'arrêter la machine et perdre tout ce qu'elle avait accompli—ou bien elle laissait passer l'indice excédentaire, ce qui déclencherait une perforation multiple, laquelle on ne pouvait déceler qu'en passant le paquet des résultats tout entier à une vérificatrice de cartes perforées. Qu'on pense aussi aux complications de manier des nombres réels avant que n'entrent en exploitation les circuits électroniques opérant sur des nombres en virgule flottante. Bref, si l'entreprise ne se distinguait pas par la hauteur de ses conceptions, elle étonnait par son acharnement à poursuivre une tâche prometteuse sur un équipement de toute évidence insuffisant.

On se serait attendu, pour la démonstration de ces techniques, à des exercices scolaires comme de résoudre l'équation de Képler. Brouwer, au contraire, choisit de les faire appliquer à un problème aussi redoutable à l'époque que la théorie du huitième satellite de Jupiter. La thèse de doctorat de Jean Kovalevsky sur ce sujet fit grand bruit chez les gens de mécanique céleste. Que les machines puissent être programmées pour traiter des formules mathématiques, on en caressait l'idée aussitôt qu'on entra dans l'ère cybernétique. Il n'en reste pas moins que, lorsqu'on attend une surprise et qu'elle survient, elle ne laisse pas pour autant d'émerveiller.

UN COUP DE GRÂCE

Une fois la brèche ouverte, on s'employa sans délai à la rallonge. Tout semblait possible en algèbre symbolique. On ferait tomber toutes ces théories des XVIIIe et XIXe siècles qu'on tenait jusqu'ici pour des citadelles inviolables. Ce n'était qu'affaire de creuser une tranchée de programmation sous leurs fondations. On voyait juste; on se trompait sur le temps qu'il faudrait mettre à investir ces ouvrages. J'en prends Eckert à témoin. Il persévéra dans un rôle que la tradition avait consacré, celui d'Atlas traversant les âges l'échine ployée sous le fardeau d'une théorie. Il commença par prendre en charge la théorie de Hill et de Brown.

Seuls les compagnons du Métier des Almanachs peuvent apprécier le travail d'épuration auquel se livra la première équipe d'Eckert. Selon l'ordre établi, une Théorie se composait de trois formules: la longitude, la latitude et, sinon la distance radiale, du moins une fonction de celle-ci comme, par exemple, un angle de parallaxe. Pour en faire un instrument de calcul, il fallait débiter la théorie en schémas emboîtés de formules trigonométriques intermédiaires, de préférence calculables par logarithmes. En termes mathématiques, cela revenait à composer des fonctions: on remplaçait les arguments d'une fonction par des fonctions de nouveaux arguments indépendants des premiers. En termes pratiques, comme on s'adressait à des employés sans formation mathématique avancée, on présentait les fonctions sous forme de tableaux; les valeurs dans les cases d'un tableau servaient à calculer des valeurs qu'on entrait dans les cases d'un tableau ultérieur. Ainsi s'explique le nom de Tables qu'on donna à ce procédé. C'était, au fond, de l'EXCEL ou du LOTUS avant le temps. Une théorie dépourvue de tables n'avait pas d'intérêt. Mais qu'on ne se leurre pas, la confection des tables était une besogne aussi considérable que l'édification d'une théorie. Brown consacra dix années à démonter sa théorie de la lune pour la mettre à portée des petites mains dans les bureaux nationaux d'éphémérides. Avant lui, Radau à l'Observatoire de Paris en avait fait de même pour la théorie de Delaunay.

Eckert s'était assigné de mettre à jour les valeurs que Brown avait attribuées aux constantes à la base de ses tables. A cette fin, il mit les tables elles-mêmes sur cartes perforées, ce qui lui permit de les manipuler par machine comme des séries de Fourier multiples à coefficients réels. Il obtenait par exemple les corrections en dérivant formellement les coefficients donnés dans la théorie sous formes de séries potentielles tronquées. De fil en aiguille, à retracer le chemin qu'avait descendu Brown de la théorie aux tables, Eckert réalisa qu'on pouvait se passer des tables. Les machines étaient devenues puissantes assez pour évaluer d'emblée les formules d'une théorie. De cette phase de ses recherches, Eckert avec son équipe publia les conclusions dans l'*Improved Lunar Ephemeris*.

Atlas allait-il déposer son fardeau? Il en était libre. Il ne le fit pas. Car les nouvelles éphémérides présentaient des écarts systématiques avec les observations. Manifestement la théorie elle-même était en erreur.

LES SÉRIES DITES DE POISSON

Le travail d'Eckert allait donc se compliquer. Mais les fabricants sortaient de nouveaux modèles de calculateurs électroniques; ils suffiraient à la besogne; on pouvait maintenant compter sur eux pour travailler des heures durant sans craindre une panne de machine. COBOL et FORTRAN ouvraient, en logiciel, l'âge des compilateurs. Grâce à eux, on en vint, en algèbre symbolique, à manier des structures plus riches que les algèbres de polynomes ou les algèbres de séries de Fourier à coefficients réels. On passa aux séries de Poisson.

Ce sont des séries de puissances positives ou négatives à plusieurs variables avec coefficients dans l'algèbre des séries réelles de Fourier. Poisson n'eut jamais rien à faire avec elles; je leur avais donné ce nom pour taquiner Danby qui s'était offert de m'aider à les programmer.

Je proposais de coder en assembleur les opérations algébriques de base comme l'addition et la multiplication des séries ou leurs dérivations partielles. Par ailleurs, je m'en remettais à FORTRAN pour combiner ces opérations à un niveau supérieur. Cette répartition des tâches tenait à deux raisons. La première était qu'un assembleur vous laisse libre d'optimiser un code, ce à quoi un compilateur ne saurait prétendre; par ailleurs, je ne pouvais pas m'astreindre à ne coder qu'en assembleur alors que j'entendais explorer les possibilités de l'algèbre symbolique en dynamique plutôt que de m'appliquer à un problème de type fixe. La seconde raison, d'ordre pratique celle-là, était que je ne pouvais pas prévoir la longueur des séries qui surgiraient dans un problème.

Arnold Rom qui était mon assistant trouva une solution: au coup d'envoi, nos programmes couraient s'emparer de toute la mémoire que leur concédait le système opérationnel. Après quoi, notre système—MAO pour l'appeler par son nom— gérait le territoire annexé. Il le divisait en deux zones à chaque bout du terrain, la "pile" et le "tas". Pour exécuter une opération sur des séries de Poisson, on tirait les opérands du disque pour les entasser sur la pile, après quoi on faisait pousser le résultat sur le tas. Dès qu'il se rendait compte que le résultat en croissance allait empiéter sur la pile, MAO sifflait la fin de la partie, ce qui arrivait très souvent. Les grandes mémoires, celles qui coûtaient les yeux de la tête, ne dépassaient pas 256 K.

On ne pouvait donc pas se permettre, comme on le fait aujourd'hui, d'enrôler les termes au fur et à mesure qu'ils arrivaient. Quand on créait un vecteur partiel, on cherchait tout de suite à le combiner avec son semblable, s'il en avait, dans le résultat en formation. Au niveau du Fortran, on veillait à effacer les résultats intermédiaires aussitôt qu'inutiles. Toujours à l'affût d'économies de mémoire, MAO réduisait les trous dans la pile entre deux séries encore vives. De nos jours, on attend d'un interprétateur qu'il vous rende ces services sans qu'il faille les demander.

Tout malingre qu'il était, MAO fit des imitateurs: au Jet Propulsion Laboratory avec Broucke, au Naval Research Laboratory avec Dasenbroeck. Henrard l'emporta aux Facultés universitaires de Namur. J'en fis part à l'ITA, l'Institut d'astronomie théorique à Saint Pétersbourg. William Jefferys perfectionna le système en donnant aux séries de Poisson la forme de listes à pointeurs, et je m'empressai d'en faire autant.

UNE RÉVISION QUI FINIT BIEN

Encore qu'il n'ait rien dit à ce sujet, on peut croire qu'Eckert appliquait des techniques analogues à refaire les développements de Brown dans la théorie de la lune.

Pour ce qu'il appelait le problème principal, Brown avait mis en place un dispositif récursif de systèmes d'équations différentielles linéaires qui généralisaient l'équation linéaire à coefficients périodiques dite de Hill, et qu'on devait intégrer en forme littérale par approximations successives. Eckert suivit la méthode pas à pas. Il découvrit des petits malheurs, comme des itérations prématurément arrêtées ou des valeurs légèrement différentes attribuées aux constantes, rien là pensait-il qu'il ne fût à même de corriger sans altérer le Grand Dessin de Brown. Il reviendrait à d'autres, comme Jean Chapront, Michele Chapront-Touzé et Thomas Van Flandern, de découvrir insuffisances et omissions beaucoup plus graves, celles qui entâchaient les perturbations planétaires, les effets dûs à l'aplatissement de la terre et ceux dûs à la relativité.

Vous connaissez l'adage: *A tant mettre la main au bénitier le diable finit par se mouiller.* Ainsi s'explique que Wallace Eckert abandonna les fonctions de réviseur algébriste pour créer sa propre théorie de la lune. Il en finit le second ordre; il en prépara la publication. Il mourut. Il laissait au Naval Observatory un manuscrit dont on ne savait que faire.

Las d'attendre une décision qu'on semblait à jamais retarder, Martin Gutzwiller, un physicien de renom, ami personnel d'Eckert et chargé par IBM de liquider son héritage scientifique, me téléphona au National Bureau of Standards. Je mis Gutwiller en rapport avec Dieter Schmidt à l'université de Cincinnati. Ce dernier travaillait, en effet, depuis quelques années à moderniser la solution semi-analytique de Hill et Brown.

Dieter Schmidt et moi, nous avions repris les idées de MAO, mais en PL/1 cette fois. Je me réservais les séries de Poisson, Dieter Schmidt me suivait dans les séries de puissances positives et négatives de plusieurs variables à coefficients complexes.

LE DÉFI DE DELAUNAY

Pourquoi changer de langage? Après tout, MAO avait passé en FORTRAN l'examen le plus difficile que la mécanique céleste offrait en ce temps: la théorie de Delaunay. Jacques Henrard et moi l'avions rhabillée de pied en cap. Barton avant nous s'était perdu à reproduire litéralement les fameuses opérations de Delaunay; il jouait les réviseurs. Pour notre part, nous entendions faire du neuf: nous avancions par transformations de Lie. Comme Delaunay, nous voulions une théorie complètement analytique; en outre, bien au-delà de la précision atteinte par Delaunay, nous cherchions la distance moyenne de la terre à la lune au décamètre près, ce qui nous conduisit par endroits au vingt et unième ordre. L'élimination des termes de courte période dura vingt-huit heures sans interruption sur une IBM 360-44, l'équivalent de nos jours d'une petite station de travail. C'était une gageure.

Une vérification s'imposait. Nous pensions la trouver dans une correspondance terme pour terme dans les parties communes aux deux théories, l'ancienne et la nouvelle. Nous nous trompions. Nous avions perdu de vue que les variables moyennisées de Delaunay étaient différentes des nôtres puisqu'elles ne résultaient pas des mêmes transformations canoniques. En

tensions nouvelles ou de déchiffrer la *Fundamental Theory* d'Eddington.

L'instrument, s'il est souple, invite à prendre des risques. On peut donc se permettre une fois encore de relever le défi de Delaunay. J'ai toujours eu dans l'idée que des opérations préparatoires comme ce que j'ai appelé l'"élimination de la parallaxe" suffiraient à évacuer les anomalies moyennes du soleil et de la lune, et cela sans passer par des développements selon les puissances des excentricités. A Zaragossa, Carlos Osácar et Jesús Palacián sont en train d'en faire la preuve.

Faut-il mentionner les opérations qui sont devenues routines d'exploitation? On n'hésite plus à normaliser des combinaisons d'oscillateurs harmoniques avec couplages dépendant de plusieurs paramètres. Teodoro López vient d'automatiser les calculs que requiert le théorème d'Arnold à propos des perturbations sur des formes quadratiques qui ne sont pas définies positives. Il applique son outil aux satellites stationnaires dans le voisinage d'une planète dont le champ de gravité n'est pas sphérique. Mais son outil est conçu dans un tel esprit d'abstraction et de généralisation qu'il s'applique également à des situations qui ne relèvent pas de l'astronomie. Dans ce sens, on voudra bien admettre que la mécanique céleste sert de prétexte pour développer de l'algèbre symbolique à la machine.

Sous prétexte encore de faire de la mécanique céleste, Alberto Abad et moi avons entrepris de mettre sur machine les *Tables of Elliptic Integrals* de Byrd et Friedman. Loin de nous l'idée de passer cette compilation au scanner pour en faire une base de données à graver sur un disque compact comme on fait couramment aujourd'hui d'un dictionnaire ou d'une encyclopédie. Nous voulons mettre dans les mains d'un ingénieur ou d'un physicien les algorithmes dont Byrd et Friedman se sont servis pour dresser leurs Tables, ce qui permettrait à l'usager non seulement de reconstruire à volonté les formules mentionnées dans les Tables mais encore d'en ajouter selon ses besoins. C'est une œuvre de longue haleine. Dans une phase préliminaire, Abad s'est occupé d'intégrer dans le mode symbolique les produits

$$\operatorname{sn}^\alpha(u,k)\operatorname{cn}^\beta(u,k)\operatorname{dn}^\gamma(u,k)$$

dont les exposants sont des entiers positifs ou négatifs. Il n'en fallait pas moins pour aborder dans une seconde phase les intégrales de type général

$$\int^x R\left(\overline{x}, \sqrt{y(\overline{x})}\right) d\overline{x},$$

R étant une fonction rationnelle de ses arguments tandis que $y(x)$ est un polynome

$$a_0 + a_1 x + a_2 x^2 + a_3 x^3 + a_4 x^4$$

à coefficients numériques, voire même littéraux. Dans la phase présente, Miquel Vallejo s'est empressé d'ajouter aux *Fundamenta Nova Theoriæ*

Functionum Ellipticarum de Jacobi une technique pour développer des intégrales elliptiques comme des séries de Fourier dont les coefficients seraient des fonctions rationnelles de la nome plutôt que des séries dans les puissances du module.

Je pourrrais continuer la liste des services que l'algèbre symbolique par machine rend à la mécanique céleste. Mais je l'arrête ici pour conclure ma leçon.

L'algèbre symbolique tient aujourd'hui la mécanique céleste pour un Jardin des Plantes. On est loin du temps où les plus grands esprits scientifiques se rencontraient en ce jardin. Il convient cependant de rappeler que c'est là que Newton et Leibniz créèrent le calcul différentiel; c'est là encore il n'y a guère que Poincaré sema ses pressentiments en dynamique non linéaire. Il se trouve encore aujourd'hui des astronomes et des mathématiciens qui poussent les grilles du parc et, cheminant par les sentiers usés, trouvent un coin où planter des espèces nouvelles. L'algèbre symbolique aime de faire métier de jardinière dans cet enclos.

Appendix
BIBLIOGRAPHICAL NOTES

The text above is a response to an invitation to deliver an address on the relations between symbolic algebra and celestial mechanics. The author felt bound to honor the page limitation set by the editors of the Proceedings. Within that constraint, he felt he could not report on progress made at the interface between automated algebra and modern mechanics. He opted therefore for a lecture—based on his personal experience—on the evolution of the main ideas that have driven research on this topic within the horizons of celestial mechanics. A lecture of the sort is usually meant to elicit the interest of readers beyond the pale of specialists. General lectures do not include a bibliography because general readers have no use for it. In this case, however, the author underestimated the curiosity of his public, and the editors have asked him to annotate his text, yet without disfiguring it.

For historical references in Section 1 ("The Great Period"), Deprit advises the readers to consult Poggendorff (1863).

The works of Sir George Darwin and Elie Strømgren mentioned in Section 2 ("The era of the cranking machines") are listed in the later editions of the Poggendorff. The contributions of Lemaître and his colleagues, although more significant and far-sighted, are less known; they were published in French before WWII in obscure Belgian journals. The main items are posted in de Vogelaere (1958). A first hand account on the use of punched cards for astronomical calculations by W. Eckert (1957), an eye witness narrative by Duncombe (1988), a detailed and affectionate biography of Leslie Comrie by W. M. H. Graves (1953), these are but a few items picked in a rich documentation covering the first half of this century.

In Section 3 ("Awakening"), Deprit analyzes the proceedings of the Conference on Celestial Mechanics (Brouwer, 1958) held at Columbia University (17–21 March 1958). The problem of the "critical inclination" was raised in 1953 by Orlov; elucidation of it came some thirty years later. The conclusive paper (Coffey *et al*,1986) covers just about all contributions to the subject.

About the problem of Jupiter VIII mentioned in Section 4 ("The first step"), there is no better survey than the one given by the person who solved it (Kovalevsky, 1958). The thesis (Kovalevsky, 1959) defended at the Faculty of Sciences of the University of Paris has been reprinted in the *Bulletin astronomique*.

Concerning the research prior to 1972 which is the background of Sections 6 ("The so-called Poisson series") and 8 ("Delaunay's challenge"), Deprit wants to quote only one piece, but it is in his opinion a masterpiece in the survey genre, namely the progress report commissioned by the Institute of Physics to Barton and his colleague Fitch at the Computer Laboratory of the University of Cambridge (Barton and Fitch, 1972). Their analysis is thorough—about 350 items in their list— yet discriminating, critical yet even-handed. Their compilation, however, still views symbolic algebra by machine in the perspective of von Neumann. The update by Henrard (1988) covers the period 1972–1987, but without the breath and depth of the Barton-Fitch report. Its author should not be blamed for that. Proceedings are meant to reflect what happened at a conference, not what happened in the fifteen years preceding the conference. The book by Brumberg (1995) is a better witness of that period. But, as a critical summary, let it be said that it is biased. Brumberg advertises his preferences for iterative over inductive schemes, for sequential over parallel processing. He is a FORTRAN *aficionado* and, as such, has fallen well behind the times in regard to both Symbolic Algebra and Celestial Mechanics. Graduate students fishing in Brumberg's monograph for a research topic in Celestial Mechanics *cum* Symbolic Algebra should know that.

Eckert's theory of the Moon—the subject of Section 7—combined with the revision and extension of Brown-Hill's solution by Schmidt appeared in (Gutzwiller and Schmidt, 1986).

The Quarrel of the Simplification evoked in Section 9 ("Tailor made processors") is re-enacted in vivid terms in a famous paper by Joel Moses (1971). The subtitle is an irreverent invitation to take with respect to algebraic simplification the attitude that, in his celebrated "Guide of the Perplexed". the great Moses ben Maimon adopted in ontology with respect to the concept of divinity. For the outstanding contributions made by the Bureau des Longitudes, Deprit sends readers to the well-penned reports by the late Bruno Morando in this volume; they will find there directions to the relevant literature, e.g. (Bretagnon and Francou, 1988) and (Chapront-Touzé and Chapront, 1988); there is so much of it!

About Section 10 ("Ascensions in abstraction"), Deprit assumes that on a shelf under their workstation readers have stacked the major manuals: Steele (1984) for LISP, MACSYMA (1995), Char *et al*(1991, 1991a, 1992) for MAPLE, Wolfram (1991) for MATHEMATICA, Jenks and Sutor (1992) for AXIOM. All these systems have raised *pro* and *contra* arguments. Those who have not yet chosen a processor of mathematical texts will find a reliable *consumer guide* in Chapter I of Geddes *et al*(1992). Around MATHEMATICA, the controversies have been somewhat acrimonious, and Deprit hopes readers will balance the hostile analysis by a MACSYMA partisan (Fateman, 1991) with the discerning diagnostics of applied mathematicians—see, e.g., Foster and Bau (1989) or Simon (1990). The ideas exposed in Section 10 took their time to emerge; the tumult finally organized itself into a programming methodology most felicitously articulated in Abelson and Sussman (1986). Symbolic Algebra has definitely broken away from its roots in arithmetic as represented by a most consulted encyclopedia (Knuth, 1969). A word of caution to the apprentice in Symbolic Algebra: do not take a seat to watch the debate about parallel processing until you have read the first chapters of the book of Hillis (1985), the creator of the Connection Machine.

Finally, a few quick notes about Section 11 ("Return to the flower garden"). The Lissajous transformations are defined, refined, exercised and exorcized in a quartet of articles published in *Celestial Mechanics* **51** (1991), pp. 201–302; *ditto* for the KS transformation and its avatars [**58** (1994), pp. 151–201]. The "elimination of the parallax" is explained in *Celestial Mechanics* **24** (1981), pp. 111–153, and generalized into a simplification technique by Lie transformations in the *Journal of Astronautical Sciences*, **37** (1989), pp. 451–463; *ditto* for computer implementations of Arnold's theorem (Deprit and López Moratalla, 1996). Rudiments of a processor for elliptic functions and integrals are outlined by Brumberg (*op. cit.*, pp. 52–60); much more is coming (Abad, 1995). Instead of transcribing the well known Fourier series obtained by Jacobi in the *Nova Fundamenta Theoriae Functionum Ellipticarum* or copying manuals like Byrd and Friedman (1954), Gradshteyn and Ryzhik (1980), Aba, Elipe and Vallejo (1994; also Vallejo, 1995) devised a technique for expanding series of that sort for a large class of elliptic functions and integrals. On the project of scanning optically a handbook of mathematical functions and tables for the purpose of converting it into a data base of mathematical objects to be queried and processed by Symbolic Algebra, re ad the work proposal presented at the latest International Symposium on Symbolic and Algebraic Computation (Fateman and Berman, 1994).

References

Abad Medina, A., Elipe Sánchez, A. and Vallejo Carrión, M.: 1994, "Automated Fourier series expansions for elliptic functions", *Mechanics Research Communic.* **21**, 361–366.
Abad Medina, A.:1995, "Integrales y funciones elípticas", *Grupo de Mecánica espacial*, Universidad de Zaragoza (Spain), in preparation.
Abelson, H. and Sussman, G. J.:1986, *Structure and interpretation of computer programs*, The MIT Press, Boston, MA.
Barton, D. and Fitch, J. P.:1972, "Applications of algebraic manipulative programs in physics", *Reports on Progress in Physics*, **35**, 235–314.
Bretagnon, P. and Francou, G.: 1988, "Planetary Theories in rectangular and spherical variables. VSOP87 solutions", *Astronomy and Astrophysics* **202**, 309–315.
Brouwer, D.:1958, "Celestial mechanics conference", *Astronomical Journal*, **63**, 401–463.
Brumberg, V.:1995, *Analytical Techniques of Celestial Mechanics*, Springer-Verlag, Berlin.
Byrd, P. F. and Friedman, M. D.:1954, *Handbook of Elliptic Integrals for Engineers and Physicists*, Springer-Verlag, Berlin/Göttingen/Heidelberg.
Chapront-Touzé, M. and Chapront, J.: 1988, "ELP 2000-85: a semi-analytical lunar ephemeris adequate for historical times", *Astronomy and Astrophysics* **190**, 342–352.
Char, B. W., Geddes, K. O., Gonnet, G. H., Leong, B. L., Monagan, M. B. and Watt, S. M.: 1991, *Maple V Language Reference Manual*, Springer-Verlag, Berlin.
Char, B. W., Geddes, K. O., Gonnet, G. H., Leong, B. L., Monagan, M. B. and Watt, S. M.: 1991, *Maple V Library Reference Manual*, Springer-Verlag, Berlin.
Char, B. W., Geddes, K. O., Gonnet, G. H., Leong, B. L., Monagan, M. B. and Watt, S. M.: 1992, *First Leaves: A Tutorial Introduction to Maple V*, Springer-Verlag.
Coffey, S. L., Deprit, A. and Miller, B.: 1986, "The critical inclination in artificial satellite theory", *Celestial Mechanics* **39**, pp. 365–406.
Deprit, A. and López Moratalla, T., "Estabilidad de satélites geoestacionarios", *Revista matemática de la Universidad Complutense de Madrid*, accepted for publication.

de Vogelaere, R.: 1958, "On the structure of symmetric periodic solutions of conservative systems, with applications ", in *Contributions to the theory of nonlinear oscillations* vol. IV, ed. S. Lefschetz, *Annals of Mathematical Studies* **41**, 53–84.
Duncombe, R. L.: 1988, "Early applications of computer technology to dynamical astronomy", *Celestial Mechanics* **45**, pp. 1–10.
Eckert, W.: 1957, "Computing in Astronomy", in *The Computing Laboratory*, ed. P.R. Hammer, pp.43–50, The University of Wisconsin Press, Madison, WI.
Fateman, R. H.: 1991, "A review of Mathematica", *J. of Symbolic Comput.* **13**. 353–394.
Fateman, R. H. and Berman, B.: 1994, "Optical character recognition for typeset mathematics", in *Proceedings, ISSAC*, Oxford, UK, July 1994, pp. 348–353.
Foster, K. R. and Bau, H. H.: 1989, "Symbolic Manipulation programs for Personal Computers", *Science* **243**, 679–684.
Geddes, K. O., Czapor, S. R. and Labahn, G.: 1992, *Algorithms for Computer Algebra*, Kluwer Academic Publishers, Boston/Dordrecht/London.
Gradshteyn, I. S. and Ryzhik, I. M.: 1980, *Table of Integrals, Series and Products*, Academic Press, New York/London/Toronto/Sydney/San Francisco.
Graves, W. M. H.: 1953, "Leslie John Comrie", *Monthly Notices of the Royal Astronomical Society* **113**, 294–304.
Gutzwiller, M. and Schmidt, D.: 1986, "The Motion of the Moon as computed by the Method of Hill, Brown, and Eckert", *Astronomical Papers* **23**, Part I.
Henrard, J.: 1988, "A survey of Poisson series", *Celestial Mechanics* **45**, 245–254.
Hillis, W. D.:1985, *The Connection Machine*, MIT Press, Cambridge MA.
Jenks, R. D. and Sutor, R. S.: 1992, *AXIOM: the scientific computation system*, Springer-Verlag, Berlin.
Knuth, D.: 1969, *The Art of Computer Programming*, chapter 4, Addison-Wesley Publishing Company, Menlo Park, CA.
Kovalevsky, J.: 1958, "The problems of the eight satellite of Jupiter", *Astronomical Journal* **63**, 452–456.
Kovalevsky, J.: 1959, "Méthode numérique de calcul des perturbations générales. Application au VIIIe satellite de Jupiter", *Thèses présentées à la Faculté des Sciences de l'Université de Paris* Série A n° 3369, Gauthier-Villars, Paris.
Macsyma: 1995, *Mathematics and system reference manual. Version 15*, Macsyma Inc., Arlington, MA.
Moses, J.: 1971, "Algebraic Simplifications: A Guide to the Perplexed", *Communications of the ACM* **14**, 527–537.
Poggendorff, J. C.:1863, *Biographisch-Literarisches Handwörterbuch zur Geschichte der exacten Wissenschaften*, Verlag von Johann Ambrosius Barth, Leipzig.
Simon, B.: 1990, "Four computer mathematical environments", *Notices of the American Mathematical Society* **37**, 861–868.
Steele, G. L. Jr., Fahlman, S. E., Gabriel, R. P., Moon, D. A. and Weinreb, D. L.: 1992, *Common Lisp: the language*, Digital Press, Burlington, MA.
Vallejo Carrión, M.: 1995, "Series de Fourier de funciones elípticas. Aplicación a la precesión terrestre", Doctoral dissertation, Universidad de Zaragoza, reprinted in *Boletín* **2/95**, Real Instituto y Observatorio de la Armada, San Fernando (Cádiz).
Wolfram, S.: 1991, *Mathematica. A system for doing mathematics by computer*, Addison-Wesley Publishing Co., Redwood City, CA.

PSP: A NEW POISSON SERIES PROCESSOR

T.V. IVANOVA
Institute of Theoretical Astronomy of the Russian Academy of Sciences, St.Petersburg, E-mail: 1197@ita.spb.su

Abstract. A specialized Poisson Series Processor (PSP) is proposed. It is designed for manipulating long Poisson series. The Keplerian Processor and analytical generator of special celestial mechanics functions based on the PSP are proposed as well.

The PSP (Poisson Series Processor) is a typical software for the implementation of analytical algorithms of Celestial Mechanics. It is a new realization and development of Universal Poisson Processor (Babaev et al., 1980). All the procedures of the PSP are general and may be used in other fields of science. It manipulates the Poisson series of the form:

$$S = \sum C_j^i x^i \, {\sin \atop \cos} (jy)$$

Here $x = (x_1, \ldots, x_n)$, $y = (y_1, \ldots, y_m)$ are vectors of power and angular variables respectively; $i = (i_1, \ldots, i_n)$, $j = (j_1, \ldots, j_m)$ are vectors of integer components. Coefficients C_j^i are rational or floating-point numbers. Summation is performed over all integer values of indices i and j.

The most important characteristics of the PSP are as follows:
- The PSP is written in standard FORTRAN-77 language and runs under MS-DOS on IBM PC and under UNIX on a Sun workstation.
- There are two versions of the PSP depending on the representation of the coefficients of Poisson series as rational or floating-point numbers. The range of the representation of the rational coefficients was increased by approximately 7 decimal orders in comparison with the range of the standard representation of computer integer number due to using the double precision floating-point numbers for correct operations on integer numbers.
- The PSP has no restrictions on the number of power and angular variables and on the ranges of the associated indices.

- Each term of the series is characterized by the analytical order of smallness which is defined as the sum of the power indices multiplied by the weight functions of the corresponding variables.
- All the computer operations on the series are formal. The criterion for rejecting a term is its smallness which is determined by the analytical order of the term, the numerical estimation of its coefficient and the values of indices if they fall outside the preassigned limits.
- The PSP allows the user to write his own procedures.
- The hierarchical architecture of the PSP allows rather easily to adapt the system on other computers and to modify it for the objects slightly different from standard Poisson series (for instance, for manipulating the exponents instead of the trigonometric functions or for changing the type of coefficients), or in case of using other storage technique for series.
- The list of basic operations of the PSP includes the standard arithmetic operations on series, the partial differentiation and integration with respect to polynomial and trigonometric variables, the total differentiation and integration with respect to time under the assumption that all the trigonometric variables are linear functions of the time with numerical values for the frequencies, raising to any power, binomial and Taylor expanding up to prescribed order, substituting other series in place of any set of the variables, fast evaluating of the series in fixing values of some variables, converting of the series, different sortings and selectings, etc. The PSP allows to input and output the series in any format or unformatted mode and to type them in the natural mathematical form. An effective algorithm for the most crucial operations of binary searching and inserting of terms into the series was worked out. It takes into account the advantages of table and linked representation of series. The searching of any term demands at most $\ln(N) + 2$ operations of terms comparison and N fast numeric assignment operations (N – the number of terms in the series).

The Keplerian Processor and analytical generator of special celestial mechanics functions based on the PSP are proposed as well. These systems are designed for implementing the expansions of the most important mathematical functions for celestial mechanics, for constructing the expansions of the elliptic motion functions of the unperturbed two-body problem and the expansions of the typical celestial mechanics functions.

The PSP is available on request by electronic address: 1197@ita.spb.su.

References

Babaev, I. O., Brumberg, V. A., Ivanova, T. V., Skripnichenko, V. I., Tarasevich, S. V. and Vasiliev N. N.: 1980, "Universal Poissonian Processor (UPP)", *Internat. Conf. on Systems and Techniques of Analytical Computing and their Applications in Theoretical Physics*, Dubna, 80.

ON GENERATORS OF NEW METHODS OF THE PERTURBATION THEORY

E.A. GREBENIKOV
Institut of high-performance computing systems
RAS, Moscow

Abstract. In this article are discussed classical and modern interpretations of the perturbation theory methods.

1. On the classical perturbation theory

Let us consider an n-dimensional differential equation with small parameter μ

$$\frac{dz}{dt} = Z(z, t, \mu), \qquad z(0) = z_0 \qquad (1)$$

where the vector-function $Z(z, t, \mu)$ is determined and has properties guaranteeing the existence and uniqueness of the solutions of the Cauchy problem (1) in a $(n + 1)$-dimensional domain $G_{(n+1)} = \{z \in G \times R \ni t\}$ of the Euclidean space. Our purpose is to construct this solution. Together with equation (1), we consider an equivalent one

$$\frac{dz}{dt} = \bar{Z}(z, t, \mu) + Z(z, t, \mu) - \bar{Z}(z, t, \mu), \qquad z(0) = z_0, \qquad (2)$$

in which $\bar{Z}(z, t, \mu)$ is an arbitrary function. We write the linear equality

$$z(t, \mu) = \bar{z}(t, \mu) + u(t, \mu) \qquad (3)$$

where $\bar{z}, u$ are some new unknown functions. The solution of a Cauchy problem can be found by solving the following two Cauchy problems:

$$\frac{d\bar{z}}{dt} = \bar{Z}(\bar{z}, t, \mu), \qquad \bar{z}(0) = \bar{z}_0 \in G_n \qquad (4)$$

$$\frac{du}{dt} = Z(\bar{z} + u, t, \mu) - \bar{Z}(\bar{z}, t, \mu), \qquad u(0) = z_0 - \bar{z}_0 \qquad (5)$$

where $\bar{z}_0$ is some new initial point. The equation (4) defines the choice of the initial approximation $\bar{z}(t, \mu)$ and equation (5) defines the total perturbation $u(t, \mu)$. From problem (5), one can see that perturbation $u(t, \mu)$ depends on the choice of function $\bar{Z}(\bar{z}, t, \mu)$, initial point $\bar{z}_0$ and, moreover, its finding is possible only after the solution of a equation (4). Thus, for a Cauchy problem (1), it is possible to construct a set of variants of the perturbation theory with parameters $\bar{Z}$ and $\bar{z}_0$. It is necessary that the solutions of equation (5) be "small" under the norm. We call $\bar{Z}(\bar{z}, t, \mu)$ and $\bar{z}_0$ the generators of the perturbation theory for problem (1) and equation (4) the generating equation for equation (1).

2. New variants of the perturbation theory

Now, we assume that the perturbation u depends on $\bar{z}, t$ and μ, that is, instead of (3) we have an equality

$$z(t, \mu) = \bar{z}(t, \mu) + u(\bar{z}, t, \mu). \qquad (6)$$

Therefore, instead of equations (4) and (5) we shall have equations (4) and (7):

$$\frac{\partial u}{\partial t} + \left(\frac{\partial u}{\partial \bar{z}}, \bar{Z}(z, t, \mu)\right) = Z(\bar{z} + u, t, \mu) - \bar{Z}(\bar{z}, t, \mu), u(0) = z_0 - \bar{z}_0. \qquad (7)$$

The perturbation theory based on equations (4) and (7) differs from the classical perturbation theory in an essential point: the determination of perturbation $u(\bar{z}, t, \mu)$ from the equation (7) does not require the preliminary solving of a generating equation (4).

So, let a problem of classical dynamics be described by a multifrequency system of $(m+n)$-order

$$\begin{cases} \frac{dx}{dt} = \mu X(x, y) \\ \frac{dy}{dt} = \omega(x) + \mu Y(x, y). \end{cases} \qquad (8)$$

where

$$X(x, y) = \sum_{\|k\| \in I} X_k(x) e^{i(k, y)}, \qquad Y(x, y) = \sum_{\|k\| \in I} Y_k(x) e^{i(k, y)}, \qquad (9)$$

$$i = \sqrt{-1}, \quad (k, y) = \sum_{s=1}^{n} k_s y_s, \quad \|k\| = \sum_{s=1}^{n} |k_s|, \quad I = \{0, 1, 2, \cdots\},$$

$$k_s = 0, \pm 1, \cdots$$

We choose, corresponding to (8), a generating system of the form

$$\begin{cases} \frac{d\bar{x}}{dt} = \mu \bar{X}(\bar{x},\bar{y}) + \sum_{k\geq 2} \mu^k A_k(\bar{x},\bar{y}), \\ \frac{d\bar{y}}{dt} = \omega(x) + \mu \bar{Y}(\bar{x},\bar{y}) + \sum_{k\geq 2} \mu^k B_k(\bar{x},\bar{y}), \end{cases} \quad (10)$$

where $\bar{X}, \bar{Y}, A_k, B_k$ are arbitrary functions of their arguments. Let's look for the replacement of variable (6) as formal series

$$x = \bar{x} + \sum_{k\geq 1} \mu^k u_k(\bar{x},\bar{y}), \qquad y = \bar{y} + \sum_{k\geq 1} \mu^k v_k(\bar{x},\bar{y}), \quad (11)$$

with unknown functions $u_k(\bar{x},\bar{y}), v_k(\bar{x},\bar{y})$. We have infinite system of linear partial differential equations of first order

$$\begin{cases} \left(\frac{\partial u_1}{\partial \bar{y}}, \omega(\bar{x})\right) = X(\bar{x},\bar{y}) - \bar{X}(\bar{x},\bar{y}), \\ \left(\frac{\partial v_1}{\partial \bar{y}}, \omega(\bar{x})\right) = \left(\frac{\partial \omega}{\partial \bar{x}}, u_1\right) + Y(\bar{x},\bar{y}) - \bar{Y}(\bar{x},\bar{y}), \\ \left(\frac{\partial u_k}{\partial \bar{y}}, \omega(\bar{x})\right) = F_k(\bar{x},\bar{y}, u_1, v_1, \cdots, v_{k-1}, u_{k-1}, A_2, B_2, \cdots, A_k), \\ \left(\frac{\partial v_k}{\partial \bar{y}}, \omega(\bar{x})\right) = \Psi_k(\bar{x},\bar{y}, u_1, v_1, \cdots, v_{k-1}, u_k, A_2, B_2, \cdots, A_k, B_k), \end{cases} \quad (12)$$

$$k = 2, 3, \cdots$$

The system (12) has a remarkable property: it is possible to integrate it in analytical way for any vector-index k if, for functions $\bar{X}$ and $\bar{Y}$, are chosen some averages of the functions X and Y.

Really, let the generators $\bar{X}(\bar{x},\bar{y}), \bar{Y}(\bar{x},\bar{y})$ be the partial sums of series (9):

$$\bar{X}(\bar{x},\bar{y}) = \sum_{\|k\|\in I'} X_k(\bar{x})e^{i(k,\bar{y})}, \qquad \bar{Y}(\bar{x},\bar{y}) = \sum_{\|k\|\in I''} Y_k(\bar{x})e^{i(k,\bar{y})}, \quad (13)$$

where $I' \in I$ and $I'' \in I$ are "resonance sets". Then

$$\begin{cases} X(\bar{x},\bar{y}) - \bar{X}(\bar{x},\bar{y}) = \sum_{\|k\|\in I\setminus I'} X_k(\bar{x})e^{i(k,\bar{y})}, \\ Y(\bar{x},\bar{y}) - \bar{Y}(\bar{x},\bar{y}) = \sum_{\|k\|\in I\setminus I''} Y_k(\bar{x})e^{i(k,\bar{y})}, \end{cases} \quad (14)$$

The sets $I \setminus I'$ and $I \setminus I"$ are not "resonance sets". It is possible to find the exact solution of the first equations (12):

$$u_1(\bar{x}, \bar{y}) = \sum_{\|k\| \in I \setminus I'} \frac{X_k(\bar{x})}{i(k, \omega(\bar{x}))} e^{i(k,\bar{y})} + \varphi_1(\bar{x}), \qquad (15)$$

$$v_1(\bar{x}, \bar{y}) = \sum_{\|k\| \in I \setminus I"} \frac{Y_k(\bar{x})}{i(k, \omega(\bar{x}))} e^{i(k,\bar{y})} + \left(\frac{\partial \omega(\bar{x})}{\partial \bar{x}}, \sum_{\|k\| \in I \setminus I'} \frac{X_k(\bar{x}) e^{i(k,\bar{y})}}{i^2(k, \omega(\bar{x}))^2} \right)$$

$$+ \left(\left(\frac{\partial u_1}{\partial \bar{x}}, \varphi_1(\bar{x}) \right), \bar{y} \right) + \psi_1(\bar{x}).$$

Here, ψ_1, φ_1 are arbitrary differentiable functions of their arguments $\bar{x}_1, \cdots, \bar{x}_m$.

Integration of equations (12) at $k = 2, 3, \cdots$ is without complicated difficulties. Rather important is the fact that, while determining functions u_2 and v_2, we can use functions $A_2, B_2, \psi_1, \varphi_1$.

The stated analytical algorithm means that we consequently construct the replacement of variables

$$(x, y) \to (\bar{x}_1, \bar{y}_1) \to (\bar{x}_2, \bar{y}_2) \to \cdots \to (\bar{x}_s, \bar{y}_s) \to \cdots.$$

Naturally, for the final construction of the solution of initial equations (8), one should solve the generating equation (10) with new initial conditions $\bar{x}(0), \bar{y}(0)$. In conclusion we want to note once again that, in formulas (11), the functions u_k, v_k are found by analytical methods. The solution of the generating equation (10) can be found with the combination of numerical and analytical methods.

References

Grebenikov E., Ryabov J. 1983, *Constructive methods in the analysis of nonlinear systems* Moscow. Ed.MIR, 324 p.

ANALYTIC-NUMERICAL SOLUTIONS OF RESTRICTED NON-RESONANCE PLANAR THREE-BODY PROBLEM

Y. A. RYABOV
Moscow Automobile & Highway Engineering University
Leningradsky pr.64, Moscow, 125829, Russia

We consider a restricted planar circular three-body problem (Sun–Jupiter–asteroid) in a non-resonance case. There are two new algorithms developed for construction of a quasi-periodic solution in a trigonometric form by means of computer algebra. The first corresponds to classical method of simple iterations leading to series in powers of small mass m_J, the second, to iterations with rapid (quadratic) convergence, but having ordinary type and not involving a successive coordinate transformations. All these iterations require a realization of algebraic operations on trigonometric polynomials with the help of computers of high capacity. It would be interesting to compare the solutions obtained with the two algorithms and to estimate the domain of their practical convergence.

1. Let us consider the following equations of the given problem

$$dp/d\theta = \mu\Phi_1(p,e,G,L), \quad dG/d\theta = 1 - \mu F_1(p,e,G,L),$$
$$de/d\theta = \mu\Phi_2(p,e,G,L), \quad dL/d\theta = 1 - F_2(p,e,G,L), \quad (1)$$

where θ is the longitude of asteroid, p is the orbital parameter, e is the eccentricity, G is the true anomaly, L is the difference between θ and Jupiter's longitude. Φ_1, Φ_2 are uneven and F_1, F_2 even functions of angular variables G, L. Units of mass and time are defined such that $k^2 = 1$ (gravitational constant), $m_S + m_J = 1$, $m_J = \mu$ and Jupiter's semimajor axis and mean motion are $a_J = 1$ and $n_J = 1$. These equations and the expressions of their right-hand sides are well known. In particular, $\Phi_1 = 2p^3[(1 + e\cos G)^{-3} - A^{-3/2}]\sin L$, where $A = p^2 + (1 + e\cos G)^2 - 2p(1 + e\cos G)\cos L$. Introducing vectors $x = (p,e)$, $y = (G,L)$, $\Phi = (\mu\Phi_1, \mu\Phi_2)$, $F(\mu F_1, F_2)$ we will seek the solution of system (1) in the form:

$$x = x_0 + \sum_{||k||=1}^{N} U_k \cos(k,\psi), \quad y = \psi + \sum_{||k||=1}^{N} V_k \sin(k,\psi), \quad (2)$$

where $\psi = (\psi_1, \psi_2)$, $\psi_j = \omega_j \theta + \psi_{j0}, j = 1, 2$, $k = (k_1, k_2)$ is a vector with integer components k_1, k_2. $x_0 = (p_0, e_0)$ and $\omega = (\omega_1, \omega_2)$ are vectors of the mean values p, e and frequencies ω_1, ω_2, respectively; $\|k\| = |k_1| + |k_2|$, $(k, \psi) = k_1 \psi_1 + k_2 \psi_2$, $(k, \omega) = k_1 \omega_1 + k_2 \omega_2$ and the number N is a given maximal (sufficiently high) order of harmonics. Coefficients U_k, V_k (two-dimensional vectors) and ω_1, ω_2 are unknown; p_0, e_0 are also unknown, but we fix their numerical values; vector $\psi_0 = (\psi_{10}, \psi_{20})$ is left arbitrary. We will obtain the solution (2) with numerical coefficients (i.e. with numerical components of vectors U_k, V_k, ω) and we also fix the value of mass μ. Substituting (2) into (1) we obtain the corresponding relations and our main equations in U_k, V_k, ω follows:

$$\omega = 1 - \tilde{F}_0(U, V), \quad -(k, \omega)U_k = \tilde{\Phi}_k(U, V), \quad (k, \omega)V_k = -\tilde{F}_k(U, V), \quad (3)$$

where $1 \leq \|k\| \leq N$ and $\tilde{F}_0 \tilde{F}_k, \tilde{\Phi}_k$ are coefficients of Fourier expansions:

$$\Phi(x, y) = \sum \tilde{\Phi}_k(U, V) \sin(k, \psi)$$

$$F(x, y) = \tilde{F}_0(U, V) + \sum \tilde{F}_k(U, V) \cos(k, \psi). \quad (4)$$

U, V denote vectors whose components are all U_k, V_k respectively. Coefficients $\tilde{\Phi}_k, \tilde{F}_k$ are theoretically certain expressions in different coefficients U_k, V_k.

2. It is essential that we can solve equations (3) by iterations in absence of analytical expressions for $\tilde{\Phi}_k(U, V), \tilde{F}_k(U, V)$.

The zero-approximation (corresponds to known formulae of unperturbed motion): $x^0 = (p_0, e_0)$, $y^0 = (G^0, L^0)$, where

$$G^0 = \omega_1^0 \theta + \psi_{10}, \quad L^0 = \omega_2^0 \theta + \psi_{20} + \sum_{j=1}^{N} S_j^0 \cos(j G^0), \quad (5)$$

$\omega_1^0 = 1$ and ω_2^0, S_j^0 are known expressions in p_0, e_0; hence, $U^0 = 0, V^0 = (0, S_j^0)$. If numbers p_0, e_0 are given, then we obtain corresponding numbers ω_2^0, S_j^0. We assume that there is no acute resonance between frequencies ω_1^0, ω_2^0.

The first approximation

$$\omega^{(1)} = 1 - \tilde{F}_0(U^0, V^0), \quad \psi^{(1)} = \omega^{(1)} \theta + \psi_0,$$

$$U_k^{(1)} = -\frac{1}{(k, \omega^{(1)})} \tilde{\Phi}_k(U^0, V^0), \quad V_k^{(1)} = -\frac{1}{(k, \omega^{(1)})} \tilde{F}_k(U^0, V^0). \quad (6)$$

It is possible to compute components of vectors $\tilde{\Phi}_k(U^0, V^0), \tilde{F}_k(U^0, V^0)$ in the following way. For example, quantities $\tilde{\Phi}_{1k}(U^0 V^0)$ are Fourier coefficients of the function

$$\Phi_1(x^0, y^0) = 2p_0^3 \left[(1 + e_0 \cos G^0)^{-3} - (A_0)^{-3/2}\right],$$

where $A_0 = p_0^2 + (1 + e_0 \cos G^0)^2 - 2p_0(1 + e_0 \cos G^0)\cos L^0$, $G^0 = \psi_1^0$ and L^0 is represented by (5). The algebraic manipulations of Fourier polynomials done with the help of computer algebra lead us to the expansion of form (4) with coefficients $\tilde{\Phi}_{1k}^0$. Similarly, we obtain other components of mentioned vectors. Afterwards, we calculate $\omega^{(1)}, U_k^{(1)}, V_k^{(1)}$ and obtain $x^{(1)}, y^{(1)}$ in form (2) with numerical coefficients.

We can use for construction of subsequent approximations a) simple iterations and b) iterations with quadratic convergence.

<u>Simple iterations</u> $\omega^{(2)} = 1 - \tilde{F}_0(U^{(1)}, V^{(1)})$,

$$U_k^{(2)} = -\frac{1}{(k, \omega^{(2)})} \tilde{\Phi}_k(U^{(1)}, V^{(1)}), \quad V_k^{(2)} = -\frac{1}{(k, \omega^{(2)})} \tilde{F}_k(U^{(1)}, V^{(1)}). \quad (7)$$

The calculations of $\tilde{F}_k(U^{(1)}, V^{(1)}), \tilde{\Phi}_k(U^{(1)}, V^{(1)})$ are reduced to obtaining Fourier expansions of functions $F(x^{(1)}, y^{(1)}), \Phi(x^{(1)}, y^{(1)})$, where $x^{(1)}, y^{(1)}$ are known expansions of form (2) with numerical coefficients. Subsequent approximations are defined similarly.

<u>Iterations with quadratic convergence</u>
In accordance with Newton's method we put in (3)

$$\omega = \omega^{(1)} + \nu^{(1)}, \quad U_k = U_k^{(1)} + u_k^{(1)}, \quad V_k = V_k^{(1)} + v_k^{(1)}$$

and form linearized algebraic equations in $\nu^{(1)}, u_k^{(1)}, v_k^{(1)}$ (corrections to the first approximation) introducing vectors $u = \{u_k\}, v = \{v_k\}$, whose components are sets of all u_k, v_k correspondingly with $1 \leq \|k\| \leq N$. These equations are the following

$$(k, \nu^{(1)})U_k^{(1)} + (k, \omega^{(1)})u_k^{(1)} + \left(\partial \tilde{\Phi}_k / \partial U\right)_1 u^{(1)} + \left(\partial \tilde{\Phi}_k / \partial V\right)_1 v^{(1)} = \Delta \tilde{\Phi}_{(k)}^{(1)}$$

$$(k, \nu^{(1)})V_k^{(1)} + (k, \omega^{(1)})v_k^{(1)} + \left(\partial \tilde{F}_k / \partial U\right)_1 u^{(1)} + \left(\partial \tilde{F}_k / \partial V\right)_1 v^{(1)} = \Delta \tilde{F}_k^{(1)}$$

$$\nu^{(1)} + \left(\partial \tilde{F}_0 / \partial U\right)_1 u^{(1)} + \left(\partial \tilde{F}_0 / \partial V\right)_1 v^{(1)} = \Delta \tilde{F}_0^{(1)}$$

where $1 \leq \|k\| \leq N, \Delta \tilde{\Phi}_k^{(1)} = \tilde{\Phi}_k^0 - \tilde{\Phi}_k^{(1)}$ etc. and $\tilde{\Phi}_k^{(j)} = \tilde{\Phi}_k(U^{(j)}, V^{(j)})$ etc. The derivative $\left(\partial \tilde{\Phi}_k / \partial U\right)_1 = \partial \tilde{\Phi}_k / \partial U \big|_{U=U^{(1)}, V=V^{(1)}}$ is the block matrix $[C_{l1,l2}]$ with $C_{l1,l2}$ being $2 \times 2-$ matrices $(\partial \tilde{\Phi}_k / \partial U_{l1,l2})_1$.

According to the expansion of $\Phi(x^{(1)}, y^{(1)})$ we have

$$\Phi(x^{(1)}, y^{(1)}) = \sum_{\|k\|=1}^{N} \tilde{\Phi}_k(U^{(1)}, V^{(1)}) \sin(k, \psi),$$

and for fixed vector $l = (l1, l2)$

$$\frac{\partial \Phi}{\partial U_{l1,l2}} = \frac{\partial \Phi}{\partial x} cos(l, \psi) = \sum_{||k||=1}^{N} \frac{\partial \tilde{\Phi}_k}{\partial U_{l1,l2}} \sin(k, \psi).$$

Hence, $\left(\partial \tilde{\Phi}_k / \partial U_{l1,l2}\right)_1$ for different vectors k are Fourier matrix-coefficients for the function $\Phi^* = (\partial \Phi / \partial x)_1 \cos(l, \psi)$.

We obtain Fourier expansion for this function by means of manipulations considered above. Others derivatives are calculated similarly. Certainly, these calculations are very cumbersome, but they are feasible by using computers of sufficient capacity.

Having calculated $\nu^{(1)}, u_k^{(1)}, v_k^{(1)}$, we obtain the second approximation for U_k, V_k, ω and the second approximation for x, y in form (2).

Algebraic equations for the corrections $u_k^{(2)}, v_k^{(2)}, \nu^{(2)}$ are formed in the similar way. The left-hand sides of these equations differ from the left-hand sides of equations in $u^{(1)}, v^{(1)}, \nu^{(1)}$ only in their super- or subscripts: (2) instead (1); on the right-hand sides we obtain the following functions:

$$\tilde{\Phi}_k(U^{(2)}, V^{(2)}) - \tilde{\Phi}_k(U^{(1)}, V^{(1)}) - \left(\partial \tilde{\Phi}_k / \partial U\right)_1 u^{(1)} - \left(\partial \tilde{\Phi}_k / \partial V\right)_1 v^{(1)},$$

$$\tilde{F}_k(U^{(2)}, V(2)) - \tilde{F}_k(U^{(1)}, V^{(1)}) - \left(\partial \tilde{F}_k / \partial U\right)_1 u^{(1)} - \left(\partial \tilde{F}_k / \partial V\right)_1 v^{(1)}.$$

We obtain, after calculation of $u_k^{(2)}, v_k^{(2)}, \nu_k^{(2)}$, the third approximation, etc. The computations are ended when differences between two adjacent approximations for ω and for all U_k, V_k are less, in norm, than a given quantity δ. These iterations possess quadratic type of convergence in relation to small mass μ if we leave out of account possible small divisors $(k, \omega^{(j)})$. Taking into consideration results of KAM-theory we could hope that quadratic convergence will compensate above mentioned small divisors in absence of an acute resonance between ω_1^0, ω_2^0.

Certainly, it arises the question about practical effectiveness of proposed algorithms in the course of immediate computations. If results are positive, the algorithm may be complicated for the purpose of considering 3-dimensional three-body problem and also to leave arbitrary some quantities as, for example, the mass μ.

References

Grebenikov E., Ryabov Y. Constructive methods in the analysis of nonlinear systems.- Moscow, ed.MIR, 1983, 324 p.

ON THE MEASURE OF THE STRUCTURE AROUND AN INVARIANT KAM TORUS

ANALYTICAL AND NUMERICAL INVESTIGATION.

C.FROESCHLÉ[1], A.GIORGILLI[2], E.LEGA[1,3], A.MORBIDELLI[1]
[1] Observatoire de Nice B.P.229, 06304 Nice cedex 4
[2] Dipartimento di Fisica dell'Università, Via Celoria 16, Milan
[3] LATAPSES, 250 Rue A.Einstein, 06560 Valbonne

1. Introduction

In a recent paper, Morbidelli and Giorgilli (1995) proved the superexponential stability of invariant tori. As usual in the theory of dynamical systems, the results are rigorously proved assuming that the perturbation is small enough. The numerical experiments show, however, that invariant tori persist up to much larger perturbation magnitudes. Therefore, it is interesting to check numerically if the superexponential stability and the other properties outlined in Morbidelli and Giorgilli's theorem persist up to the value of the perturbation for which the torus actually breaks up. Moreover, one would like to have a numerical indication about the size of the superexponentially stable region existing around a torus. Is the superexponential stability just an asymptotic result, or does it concern a macroscopic region of physical interest?

The relation of this work with the dynamics of the solar system may not appear clear at a first glance. Nevertheless, it is well known that the fundamental problem of the stability of the solar system is connected with the stability of non linear Hamiltonian systems. At the beginning of this century Poincaré had already shown that it was impossible to demonstrate the integrability of the Hamiltonian system representing the motion of the solar system. In 1954 Kolmogorov obtained a partial integrability, for some small values of the perturbing parameter ϵ, for Hamiltonian systems described by:

$$H(p,q) = H_0(p) + \epsilon H_1(p,q) \qquad (1)$$

The next step is due to the work of Nekhoroshev (1977) concerning the diffusion of the actions of the invariant KAM tori. Although the mathematical demonstrations require small perturbations (direct application implies that the Jupiter mass is much smaller than the size of an orange), the numerical experiments (Hénon 1969) have shown the existence of the regions of stability up to larger perturbation magnitudes.

This work has been realized, respect to the Morbidelli and Giorgilli's theorem, in the same spirit of that of Hénon respect to the KAM theorem. The aim is therefore to be in continuity with the researches about the stability of Hamiltonian systems, and hence of the solar system.

Our numerical computations show in a striking way that the description of the dynamics given, for small perturbations, by Morbidelli and Giorgilli's result is true in reality as long as the invariant torus persists. Moreover, the size of the structure described by Morbidelli and Giorgilli around the invariant torus shrinks to 0 like $\exp(-\epsilon_c/(\epsilon_c - \epsilon))$ when the size of the perturbation ϵ tends to the threshold value ϵ_c corresponding to the torus break-up. This implies that, when the perturbation magnitude is a little bit smaller than the break-up threshold, the size of such structure is macroscopic.

In section 2 we recall the result by Morbidelli and Giorgilli and the main ideas of their approach. In section 3 we discuss our numerical experiments and their significance.

2. Superexponential stability of invariant tori

In their investigation of the dynamics in the vicinity of an invariant KAM torus, Morbidelli and Giorgilli started from the so called Kolmogorov normal form (Kolmogorov,1954).

According to Kolmogorov's construction, one can introduce suitable action angle variables P, Q, such that, in the neighbourhood of the invariant torus $P = 0$, the Hamiltonian writes:

$$H(P,Q) = \omega \cdot P + O(P^2)f(Q) .$$

The Kolmogorov normal form shows in an equivocally way that in the vicinity of the invariant torus the significant perturbation parameter is the distance $|P|$ from the torus itself.

Therefore, in the ball $|P| < \rho$ one can introduce new action angle variables J_ρ, ψ_ρ such to reduce the local perturbation to its optimal size, which, assuming analytic Hamiltonians, is exponentially small with $1/\rho$, i.e.

$$H(J_\rho, \psi_\rho) = \omega \cdot J_\rho + H_0(J_\rho) + \epsilon_\rho H_1(J_\rho, \psi_\rho)$$

with H_0 quadratic in J_ρ and $\epsilon_\rho \sim \exp(-1/\rho)$.

At this point, it is enough to remark that, provided ρ is small enough, ϵ_ρ is smaller than the threshold for the applicability of Arnold's version of

KAM theorem (Arnold, 1963) in the ball $|J_\rho| < \rho$. This allows to prove that in the vicinity of the central torus at $P = 0$ there exist an infinity of invariant tori, the volume of the complement decreasing to zero exponentially with $1/\rho$.

On the other hand, provided $H_0(J_\rho)$ is convex in $J_\rho = 0$, if ρ is small enough, the local perturbation parameter ϵ_ρ is also smaller than the threshold for the applicability of Nekhoroshev's theorem. This allows to prove that the diffusion of the actions J_ρ must be bounded by ϵ_ρ^b for all times up to $\exp(1/\epsilon_\rho)$, which, by substitution gives the superexponential estimate $\exp[\exp(1/\rho)]$. The hypothesis of local convexity is a very natural one. It means indeed that on a given energy surface, the torus with given frequency ratios is locally unique.

The picture provided by Morbidelli and Giorgilli's result is therefore the following. The tori given by Kolmogorov's theory are *master* tori, surrounded by a structure of *slave* tori, which accumulate in an exponential way to the central master torus. These slave tori are all n–dimensional Diophantine ones, but they are characterized by a very small Diophantine constant γ (we recall that a frequency ω is said to be Diophantine if it satisfies the relation $|k \cdot \omega| > \gamma/|k|^{n+1}$ for all integer vectors k and some positive γ); for this reason, they could not be found directly by Kolmogorov's construction. Moreover, diffusion among this structure of slave tori is superexponentially slow, so that chaotic orbits can enter in, or escape from, only in a time proportional to $\exp[\exp(1/\rho)]$.

The interest of this result is double. On the one hand, this makes open the set of invariant tori from all practical point of view; this is important for what concerns the compatibility of KAM theorem with the errors in initial conditions of numerical experiments. On the other hand, a direct consequence of the local superexponential stability is that invariant tori can form, even in three or more degrees of freedom, a kind of impenetrable structure which orbits cannot penetrate for an exceedingly long time, very large even with respect to the usual Nekhoroshev's estimates.

3. Numerical measures

In order to test the structure of invariant tori around a chief torus we have taken as a model problem the standard map (Lega & Froeschlé 1995). We recall the set of equations for this mapping:

$$\begin{cases} y_{i+1} = y_i + \epsilon \sin(x_i) & y \in \Re \\ x_{i+1} = y_{i+1} + x_i & mod(2\pi) \end{cases} \quad (2)$$

We have computed for a set of initial conditions on the line $x = 0$ the rotation number associated to each initial condition. As already pointed

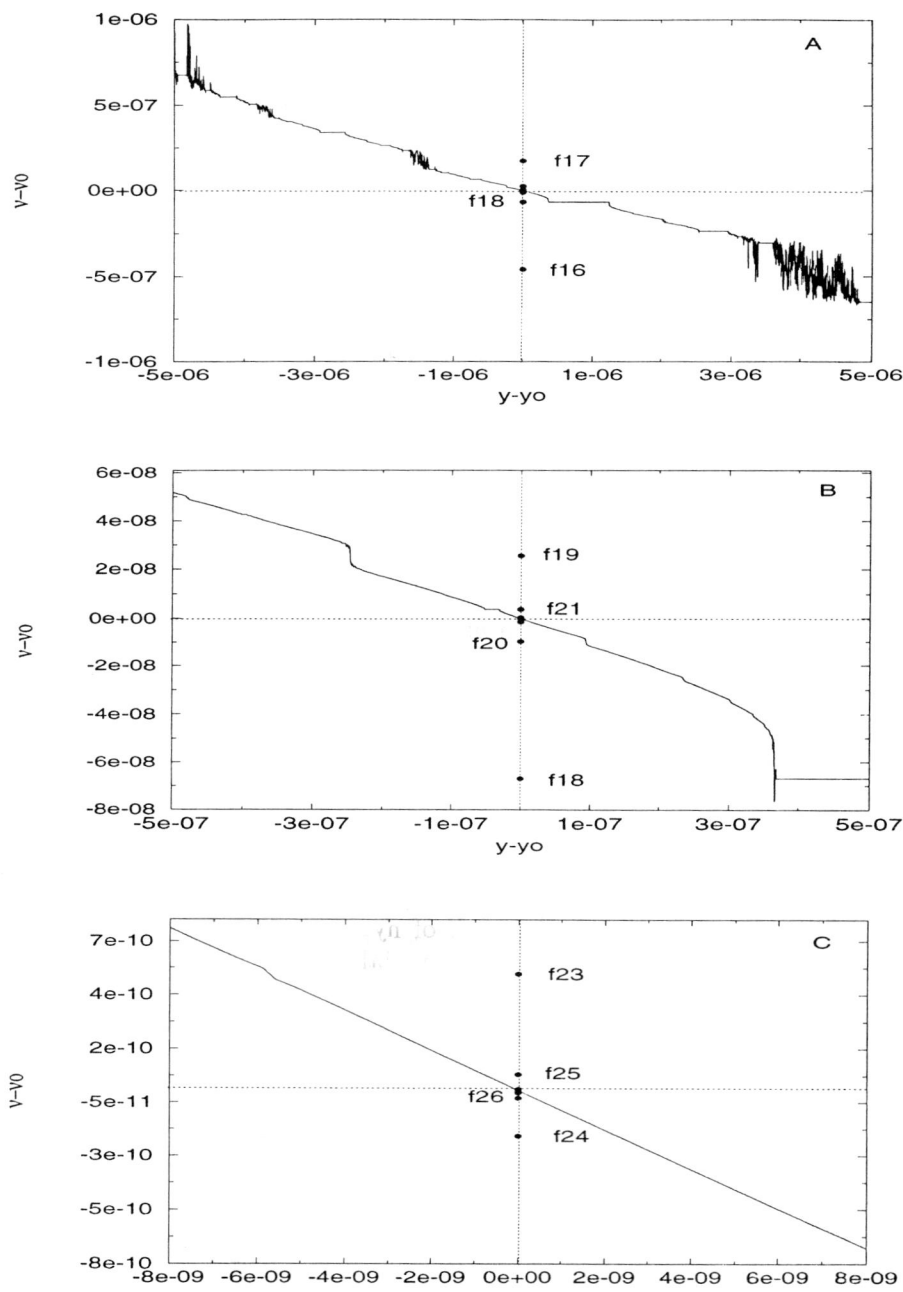

Figure 1. Variation of the fundamental frequency ν for the standard mapping, with $\epsilon = 0.9715$, in the vicinity of the golden rotation number ν_o which corresponds to the origin of the axes. The Fibonacci terms are indicated on each figure by the set of points f_i.

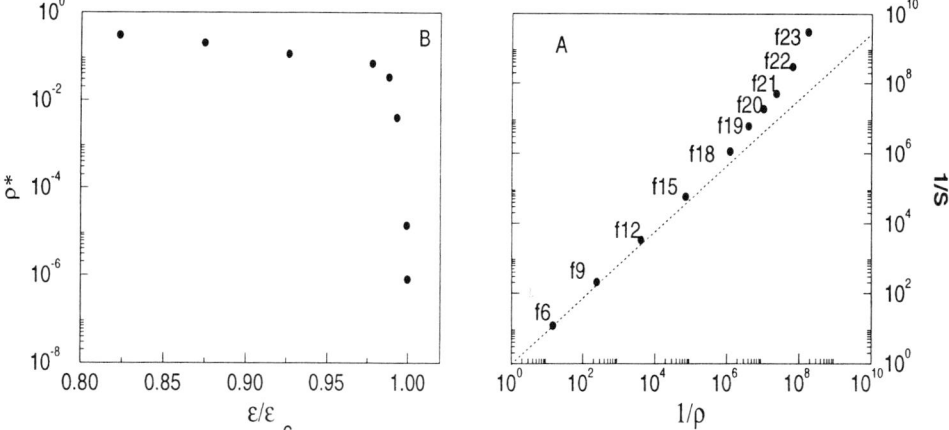

Figure 2. A) Variation of the inverse of the size (S) of the Fibonacci islands as a function of the inverse of the distance ρ to the golden torus. B) Variation of the threshold distance ρ^* as a function of the perturbing parameter ϵ/ϵ_c.

out by Laskar et al. (1992), the existence of KAM tori corresponds to monotonic variations of the rotation number as a function of y. Conversely, islands correspond to constant frequencies and chaotic regions correspond to either noisy or simply non monotonic variations of frequencies. All this features appear clearly in Figure 1. On each plot the origin correspond to the golden torus, i.e. the torus whose rotation number is equal to the golden number $\nu_o = \frac{1}{2}(3 - \sqrt{5})$. We observe that the majority of orbits of Fig.1a correspond to chaotic regions and islands. The situation changes drastically in Fig.1b: the noisy variation of ν corresponding to strong chaotic regions has disappeared, islands and crossing of hyperbolic points are still there, but their relative measure in the action variables is now definitively smaller than the relative measure of tori. This phenomenon is strongly enhanced in the last magnification: up to a step size of $\Delta y = 1.6\,10^{-11}$ (Fig. 1c), we only see one hyperbolic point and a large continuous region of tori. It is clear that the magnifications represented in Figs. 1b,c show a completely different regime and in Fig. 1c only the chief torus and its "slaves" appear with a density which seems to be in full agreement with the prevision of the Morbidelli Giorgilli theorem. We have indicated on each plot the values of frequencies corresponding to the Fibonacci sequence, i.e. the set of the successive terms obtained when developing the golden number through the continued fraction process. In order test the exponential decrease of the volume occupied by the complement of the set of tori V_c as a function of the distance ρ to the chief torus we have measured the size of the Fibonacci islands. Such islands are the largest ones and therefore they fill the major part of V_c.

Figures 2a show the variation of the size of the Fibonacci islands as a function of the distance ρ in a log-log diagram. The distance ρ is the absolute value: $|y_c - y_o|$ where y_c is the action of the center of the island and y_o is the action of the golden torus. In the Morbidelli-Giorgilli regime we have also taken into account the hyperbolic points corresponding to the Fibonacci chain of islands. We have made an estimate of the dimension of the corresponding islands through the jump in frequency which occurs when crossing the hyperbolic point.

On Fig.2a the change of regime is drastic: at the 19th term of the Fibonacci sequence we enter in the Morbidelli-Giorgilli regime where the size of the perturbation decreases exponentially with the distance from the golden torus. Let us emphasize that the measure has been done for a value of the perturbation parameter: $\epsilon = 0.9715$ very close to the critical one: $\epsilon_c = 0.971635$.

Using the same technique of the previous we have estimated the distance $\bar{\rho}^*$ for a set of different values of the perturbing parameter. Fig.2b shows our final result on the variation of $\bar{\rho}^*$ as a function of ϵ/ϵ_c. After a linear decrease of $\bar{\rho}^*$, up to $\epsilon = 0.95$, we observe a sharp drop of $\bar{\rho}^*$ up to $\bar{\rho}^* = 8\,10^{-7}$ for $\epsilon = 0.9715$. It seems therefore that all the slave tori disappear at once when approaching the critical value ϵ_c.

4. Conclusion

The breakthrough provided by the Morbidelli-Giorgilli theorem seems, at first glance, to be out of the range of the numerical experiments. This is still our opinion concerning the super-exponential character of the diffusion. Using the frequency map analysis we have first confirmed the existence of a chief torus surrounded by slaves. Then we have shown that such a structure exist even for values of the perturbing parameter close to the one for which no KAM tori survive. Figure 2b shows the variation of the threshold distance $\bar{\rho}^*$ as a function of the perturbing parameter. The behavior observed for $\bar{\rho}^*(\epsilon)$ might deserve further studies.

References

Arnold, V. I. (1963), *Russ. math. Surv.*, **18**, 9.
Hénon, M. (1969), *Quarterly of Applied Mathematics*, **27**, 291-306.
Kolmogorov, A. N (1954), *Dokl. Akad. Nauk. SSSR*, **98**, 527
Laskar, J., Froeschlé, C. and Celletti, A. (1992), *Physica D*, **56**, 253.
Lega, E. and Froeschlé, C. (1995), *submitted to Physica D*.
Morbidelli, A. and Giorgilli, A. (1995), *J. Stat. Phys.*, **78**, 1607.
Nekhoroshev, N. N. (1977), *Russ. math. Surveys*, **32**, 1-65.

ON TWO–PARAMETER LINEARIZING TRANSFORMATIONS FOR UNIFORM TREATMENT OF TWO–BODY MOTION

LUIS FLORÍA

*GMC, Departamento de Matemática Aplicada a la Ingeniería,
E.T.S. de Ingenieros Industriales, Universidad de Valladolid,
E-47 011 Valladolid, Spain.*

Abstract. Differential changes of time involving two parameters are considered. Universal expressions for dynamical variables of interest in Keplerian motion allow us to reduce the integration of the time transformations to that of integrands depending on an eccentric–like universal anomaly. Elliptic integrals and functions are required to complete the integration.

Key words: reparametrizing transformations, generalized anomalies, uniform treatment of Keplerian systems, universal functions, elliptic functions.

1. Introduction

For the integration of perturbed Kepler problems within the framework of the elliptic–type two–body motion, Ferrándiz & Ferrer (1986) and Ferrándiz et al. (1987) replaced the physical time t by a fictitious time τ as the independent variable. Thus, they generalized previous research by Belen'kii (1981) and Cid et al. (1983) concerning *regularization and linearization* of equations governing perturbed Keplerian systems. They used the method of linearization by time transformations on introducing a *family of generalized anomalies* defined via a differential relation in which the linearizing function depends on *two homogeneous parameters* that can be taken as functions of some orbital elements. For pure Kepler problems, they analytically integrated the time transformations, in closed form, by means of elliptic integrals and functions. Their developments were originally intended to facilitate (on the basis of some radial intermediaries) the analytical treatment of artificial Earth satellite orbits under a zonal model of geopotential, and as a preconditioning of the problem prior to numerical integration.

For analytical step regulation in numerical integration of highly eccentric orbits, E. V. Brumberg (1992) proposed the use of *orbital length of arc* as independent variable, and made to fit his derivation into a pattern resembling that of Ferrándiz and his colleagues. He also pointed out that his transformation is applicable to *any kind of Keplerian motion*.

We aim at a systematic study of changes of time allowing the extension of results established by those authors to a uniform treatment of motion under a *universal formulation* of the two–body problem, in the sense, e. g., of Stumpff (1959), Chapter V, Stiefel & Scheifele (1971), §11, or Battin (1987), §4.5 and §4.6. To be precise, we consider time transformations

$$t \longrightarrow \tau \quad \text{given by} \quad dt = \Phi(r; \alpha_0, \alpha_1) d\tau = \frac{r^\alpha}{\sqrt{\alpha_0 + \alpha_1 r}} d\tau, \quad (1)$$

α_0 and α_1 being parameters of the transformation and α is a real number, combined with the Sundman transformation $t \to s: dt = r\, ds$, introducing a *universal eccentric–like anomaly* (see, e. g., Stumpff 1959, §41, Formulae [V; 35] and [V; 38]; Stiefel & Scheifele 1971, §11, Formula [60].)

To this end, we resort to certain families of special functions, the so–called *universal functions* (Battin 1987, §4.5 and §4.6) and the *Stumpff c–functions* (Stumpff 1959, §37, §41; Stiefel & Scheifele 1971, §11), by means of which the integration of the reparametrizing transformation is reduced to that of some algebraic functions. As expected, when integrating the changes of time parameter at issue, elliptic integrals and functions will enter.

2. Universal Functions and Useful Relations

The *Stumpff c–functions* (Stumpff 1959, §37, §41; Stiefel & Scheifele 1971, §11) are a family of transcendental functions whose first members solve, under a unified treatment, the second–order linear differential equation

$$d^2y/ds^2 + \varrho y = 0 \quad (\text{where } \varrho \text{ is a real parameter.}) \quad (2)$$

After appropriate changes of dependent and independent variables, this is the form to which Kepler problems are reduced, the parameter ϱ being then related to the value of the energy of the two–body system. For $z = \varrho s^2$, the general solution to Eq. (2) is a linear combination of the Stumpff functions $c_0(z)$ and $c_1(z)$. In general, these functions obey the *defining relation*

$$c_n(z) = \sum_{k=0}^{\infty} (-1)^k \frac{z^k}{(2k+n)!}, \quad n = 0, 1, 2 \ldots .$$

The power series are convergent for all complex values of z, and define real–valued functions for real values of z. Some calculations with these functions

are simplified if the *universal functions* U_n introduced by *Battin* are used:
$$U_n(s, \varrho) \equiv s^n c_n(\varrho s^2), \quad n = 0, 1, 2.... \tag{3}$$

Let $L = \mu(1-e)/(2q)$ be the *negative of the energy* of the Keplerian orbit (Stiefel & Scheifele 1971, p. 50.) Take $\varrho = 2L$, and consider (Stiefel & Scheifele 1971, pp. 50–51; Battin 1987, §4.5, §4.6.): $\mu \equiv$ gravitational parameter, $q \equiv$ distance of the pericentre, $e \equiv$ eccentricity, $r \equiv$ modulus of the radius vector, $(x, y) \equiv$ Cartesian coordinates in the orbital plane,

$$r = q + \mu e s^2 c_2(2Ls^2) = q + \mu e U_2(s, 2L); \tag{4}$$
$$x = q - \mu s^2 c_2(2Ls^2) = q - \mu U_2(s, 2L), \tag{5}$$
$$y = \sqrt{\mu q(1+e)}\, s c_1(2Ls^2) = \sqrt{\mu q(1+e)}\, U_1(s, 2L); \tag{6}$$
$$dt = r\, ds \text{ (Sundman transformation)}; \quad t = qs + \mu e U_3(s, 2L); \tag{7}$$
$$\frac{dU_n}{ds} = U_{n-1}, \quad n = 1, 2, 3, \ldots; \tag{8}$$
$$1 = U_0^2 + \varrho U_1^2, \quad U_1^2 = U_2(1 + U_0) = 2U_2 - \varrho U_2^2. \tag{9}$$

3. Universal Two–Parameter Time Transformations

We deal with two–parameter time transformations $t \to \tau$ (e. g., those of Ferrándiz & Ferrer 1986, or Ferrándiz *et al.* 1987) as the change of time in Eq. (1). To extend the results of these authors to universal form, introduce

$$C^2 = \sqrt{\alpha_0^2 + q^2 \alpha_1^2}, \quad a_0 = \alpha_0/C^2, \quad a_1 = \alpha_1 q/C^2, \quad a_0^2 + a_1^2 = 1.$$

We use the *distance of the pericentre* q instead of the semi–major axis or the semi–latus rectum, employed by some authors when dealing with elliptic orbits. Now, the differential relation governing the time transformation is

$$C\, dt = r^\alpha [a_0 + a_1(r/q)]^{-1/2}\, d\tau.$$

We combine it with the Sundman transformation (7), along with Formula (4), to obtain

$$\sqrt{a_0 q + a_1 r}\, r^{1-\alpha} = \sqrt{(a_0 + a_1) q + a_1 \mu e U_2}\, (q + \mu e U_2)^{1-\alpha}.$$

After the *change of integration variable* $s \to v$ given by $U_2(s, 2L) = v$, using Formulae (8) and (9), the integration of the transformation relating the generalized anomaly τ to s is reduced to that of an integrand in v:

$$d\tau = \frac{C}{\sqrt{q}} \frac{\sqrt{(a_0 + a_1) q + a_1 \mu e v}}{(q + \mu e v)^{\alpha-1} \sqrt{v[2 - (2L)v]}}\, dv. \tag{10}$$

The calculations and results hinge on the analysis of the polynomials occurring in (10). Further details and applications will be communicated in a future paper. For pure elliptic motion, Ferrándiz & Ferrer (1986) took $\alpha = 2$, while Ferrándiz et al. (1987) considered $\alpha = 3/2$, and integrated their general time transformations in terms of elliptic integrals and functions.

4. A Universal Approach to Arc Length as Time Argument

Let $d\sigma$ denote the arc element along a Keplerian conic–section orbit. From (5) and (6), universal expressions for the Cartesian coordinates in the orbital plane, by taking into account (8) and (9) we obtain a differential relation between the arc length σ and the universal anomaly s:

$$d\sigma = \sqrt{\mu q(1+e) + \mu^2 e^2 U_1^2(s, 2L)}\, ds. \qquad (11)$$

The change of integration variable $s \to w$ given by $U_1(s, 2L) = w$ yields

$$d\sigma = \sqrt{q}\, \sqrt{\frac{\mu q(1+e) + \mu^2 e^2 w^2}{q - \mu(1-e)w^2}}\, dw. \qquad (12)$$

These developments are intended to extend Brumberg's (1992) study. Concerning the integration of these differential relations, comments like those at the end of Section 3 are in order.

Acknowledgements

This research is partially supported by CICYT of Spain within the National Programme of Space Research (Project ESP. 93–741.)

References

Battin, R. H. (1987) *An Introduction to the Mathematics and Methods of Astrodynamics.* American Institute of Aeronautics and Astronautics Education Series, New York.
Belen'kii, I. M. (1981) A Method of Regularizing the Equations of Motion in the Central Force–Field, *Celestial Mechanics* **23**, 9–32.
Brumberg, E. V. (1992) Length of Arc as Independent Argument for Highly Eccentric Orbits, *Celestial Mechanics and Dynamical Astronomy* **53**, 323–328.
Cid, R., Ferrer, S. and Elipe, A. (1983) Regularization and Linearization of the Equations of Motion in Central Force–Fields, *Celestial Mechanics* **31**, 73–80.
Ferrándiz, J. M. and Ferrer, S. (1986) A New Integrated, General Time Transformation in the Kepler Problem, *Bulletin of the Astronomical Institutes of Czechoslovakia* **37**, 226–229.
Ferrándiz, J. M., Ferrer, S. and Sein–Echaluce, M. L. (1987) Generalized Elliptic Anomalies, *Celestial Mechanics* **40**, 315–328.
Stiefel, E. L. and Scheifele, G. (1971) *Linear and Regular Celestial Mechanics.* Springer-Verlag, Berlin, Heidelberg, New York.
Stumpff, K. (1959) *Himmelsmechanik, I.* VEB Deutscher Verlag der Wissenschaften, Berlin.

RELATIVISTIC EQUATIONS OF MOTION OF CELESTIAL BODIES

MICHAEL H. SOFFEL
Lohrmann Observatory, TU Dresden
Mommsenstr. 13, 01062 Dresden
e-mail: soffel@rcs.urz.tu-dresden.de

Abstract. The problem of relativistic equations of motion for extended celestial bodies in the first post-Newtonian approximation is reviewed. It is argued that the problems dealing with kinematical aspects have been solved in a satisfactory way, but more work has to be done on the dynamical side. Concepts like angular velocity, moments of inertia, Tisserand axes etc. still have to be introduced in a rigorous manner at the 1PN level.

Usually, relativistic equations of motion (EOM) for massive celestial bodies are treated within the so-called post-Newtonian framework, based on a formal expansion of the form

$$\text{EOM} = (\text{EOM})_{\text{Newton}} + \epsilon^2 (\text{EOM})_{\text{1PN}} + \epsilon^4 (\text{EOM})_{\text{2PN}} + \ldots, \quad (1)$$

where

$$\epsilon^2 \sim \left(\frac{v}{c}\right)^2 \sim \frac{GM}{c^2 R}. \quad (2)$$

Here, v denotes a typical orbital velocity of celestial bodies, M a typical mass and R some characteristic distance between the bodies. In the solar system, $\epsilon^2 < 10^{-5}$ and for most purposes the first post-Newtonian approximation will be sufficient. Now, the problem of celestial mechanical equations of motion can be divided into two parts: a kinematical part and a dynamical one. The kinematical problems are related with the theory of relativistic astronomical reference frames. These problems have been solved in a satisfactory manner by Brumberg, Kopejkin and Klioner (see, *e.g.*, Brumberg 1991 and references cited therein) and by Damour, Soffel and Xu (1991, 1992, 1993, 1994; DSX I-IV). The dynamical part deals with the physical interaction of the gravitating bodies. Some basic problems have been solved by Damour, Soffel and Xu in the first post-Newtonian approximation. However, some problems still remain unsolved. These will be mentioned below.

In relativity, the *local* equations of motion that can be written as

$$0 = \nabla_\nu T^{\mu\nu} \tag{3}$$

form the basis for any derivations of *global* equations of motion, *e.g.*, for the center-of-mass of a body. Here, ∇_ν denotes the covariant derivative and $T^{\mu\nu}$ is the energy-momentum tensor of matter. To Newtonian order, in inertial Cartesian coordinates, (ct, x^i) equations (3) take the form

$$\frac{\partial \rho}{\partial t} + \frac{\partial(\rho v^i)}{\partial x^i} = 0 \tag{4}$$

and

$$\frac{\partial(\rho v^i)}{\partial t} + \frac{\partial}{\partial x^j}\left[\rho v^i v^j + t^{ij}\right] = \rho \frac{\partial U}{\partial x^i}. \tag{5}$$

Here, ρ is the matter density, **v** the coordinate velocity of some material element, U is the Newtonian potential and t^{ij} the stress-tensor of matter. Going to higher and higher accuracies, the definition of quantities like center-of-mass, mass-moments, spin-moments and the derivation of useful equations of motion will become increasingly more complicated. For that reason, there is some chance that in the (far) future the local equations of motion (3) might be the only useful ones in which case they would have to be solved by means of techniques from numerical relativity after having specified the visco-elastic properties of matter (*i.e.*, the energy-momentum tensor) in a relativistic framework.

Damour, Soffel and Xu have demonstrated that global Laws of Motion (LOM) can be derived in the first post-Newtonian approximation without specification of the energy-momentum tensor. Let us consider a system of N weakly self-gravitating, rotating bodies of arbitrary shape and composition gravitationally interacting with each other. To describe the time evolution of such a system, we introduce $N+1$ different coordinate systems,

- one global coordinate system $x^\mu = (ct, x^i)$ and
- one local coordinate system $X_A^\alpha = (cT, X^a)_A$, for each of the bodies, comoving with body A.

It is assumed that the global chart extends to spatial infinity. For celestial mechanical problems related with the solar system it will be a barycentric coordinate system. Among the local ones, one will be a geocentric frame, moving with the Earth. Now, in the DSX-formalism, the gravitational interaction is described by two potentials (w, w_i) in the global frame and (W, W_a) in one of the local frames. In Newton's theory of gravity, there is only one scalar potential, the Newtonian potential U. To lowest order, the scalar potential w or W agrees with U but it also has terms proportional to c^{-2}. w is also called the gravitoelectric potential describing the gravitational action of a static matter distribution. On the other hand, the vector potential w^i (or W_a) describes the gravitational effect of moving matter (e.g., rotating bodies), i.e., gravitomagnetic effects. It has no Newtonian analogue.

Let us consider the environment of a body (A) such as the Earth in its own local frame. In a special representation, one finds that the equations satisfied by the potentials $W_\alpha = (W, W_a)$ are *linear* and there is a unique split of the potentials into two pieces:

$$W_\alpha = W_\alpha^+ + \overline{W}_\alpha. \tag{6}$$

Here, the self-part of the potentials, W_α^+, describes the gravitational action of body (A) itself. It can be characterized by two sets of multipole moments, called M_L and S_L. Here, L denotes a Cartesian multi-index, $L = i_1 i_2 \ldots i_l$ and each index i_j takes the values $(1, 2, 3) = (x, y, z)$, so we face components like M_{xx}, or M_{xxyyz} etc. The M_L's are called mass-multipole moments, S_L are the spin-moments of body (A). Both, M_L and S_L are so-called STF (symmetric and trace-free) tensors which are the Cartesian analogues of coefficients in an expansion in terms of spherical harmonics. M_L generalizes the usual Newtonian potential coefficients C_{lm} and S_{lm} to the first post-Newtonian level. S_i represents the total angular momentum (spin) of body (A). M_L and S_L, as functions of local coordinate time T, are fully determined by integrals over the densities

$$\Sigma = \frac{T^{00} + T^{ss}}{c^2} = \rho + O(c^{-2}); \quad \Sigma^a = \frac{T^{0a}}{c} = \rho v^a + O(c^{-2}), \tag{7}$$

i.e., they can formally be obtained without specification of the energy-momentum tensor of matter.

The external part of the potential, $\overline{W}_\alpha$ describes the inertial forces in the local frame and the gravitational effects from the other bodies, i.e., the tidal forces. In the DSX-formalism this external part of the potentials is available either in closed form as function of the mass- and current-moments of the other bodies and of the position and velocity of the origin of the local system in the global frame or as expansion in terms of tidal moments, G_L and H_L. In Newton's theory, the magnetic-type moments H_L play no role and

$$G_i^{\text{Newton}} = \partial U_A^{\text{ext}}(\mathbf{z}_A) - \frac{d^2 z_A^i}{dt^2} \quad (l = 1) \tag{8a}$$

$$G_{i_1 \ldots i_l}^{\text{Newton}} = \frac{\partial^l U_A^{\text{ext}}(\mathbf{z}_A)}{\partial X^{i_1} \cdots \partial X^{i_l}} \quad (l \geq 2). \tag{8b}$$

To derive Laws of Motion we first have to relate the origins of the local frames with the matter distribution of the bodies. Usually, one chooses the origin of the local A-frame to coincide with the center-of mass of body A. This can be achieved by means of

$$M_a = 0. \tag{9}$$

With that condition the potentials (W, W_a) are completely fixed and the trajectory, e.g., of a satellite can be considered as a geodesic in the metric that is determined by the potentials. It is equivalently determined by a Lagrangian of the form

$$\mathcal{L} = \frac{1}{2}\mathbf{v}^2 + W + \frac{1}{c^2}\left[\frac{1}{8}\mathbf{v}^4 - \frac{1}{2}W^2 + \frac{3}{2}W\mathbf{v}^2 - 4W^a v^a\right], \tag{10}$$

where $\mathbf{v}$ is the satellite's coordinate velocity. This leads to an equation of the form

$$\ddot{X}^a = f^a_{\text{self}} + f^a_{\text{tidal}} + f^a_{\text{mixed}}. \tag{11}$$

Here, f^a_{self} is an acceleration resulting from the action of body (A) itself and f^a_{tidal} results from the action of the other bodies. In the DSX-formalism,

$$\begin{aligned} f^a_{\text{self}} &= W^+_{,a} + (c^{-2} - \text{terms}) \\ &= G\sum_{l\geq 0} \frac{(-)^l}{l!} M_L(\partial_{L a} R^{-1}) + (c^{-2} - \text{terms}) \end{aligned} \tag{12}$$

has been obtained fully to post-Newtonian order in terms of M_L and S_L of body (A),

$$f^a_{\text{tidal}} = \sum_{l\geq 0} \frac{X_L}{l!} G_{La} + (c^{-2} - \text{terms}) \tag{13}$$

completely to PN-order in terms of G_L and H_L or in closed, i.e., non-expanded form. Similarly, the mixed acceleration

$$f^a_{\text{mixed}} = -\frac{4}{c^2}\left(\overline{W}W^+_{,a} + W^+\overline{W}_{,a}\right) \tag{14}$$

is completely available to post-Newtonian accuracy. In DSX IV, we have also derived post-Newtonian satellite Laws of Motion in the global barycentric frame which might be useful for high flying satellites or for comparisons with previous results from other authors.

General translational post-Newtonian Laws of Motion for a system of massive bodies have been derived from relation (9). More specifically, using

$$\frac{d^2 M_a}{dT^2} = 0 \tag{15}$$

and the local equations of motion one finds

$$0 = \frac{d^2 M_a}{dT^2} = \sum_{l \geq 0} \frac{1}{l!} M_L G_{La} + (c^{-2} - \text{terms}). \tag{16}$$

Since

$$G_a = -\frac{d^2 z_A^a}{dt^2} + \overline{W}_{,a}|_{X^a=0} + (c^{-2} - \text{terms})$$

equation (16) can be solved for the desired acceleration of the center-of-mass of body (A). In this way, Laws of Motion in the form

$$\frac{d^2 z^a}{dt^2} = \overline{W}_{,a}|_{X^a=0} + (c^{-2} - \text{terms}) \tag{17}$$

have been derived to PN accuracy. For more details, the reader is referred to DSX II. In the case of pure mass-monopoles (*i.e.*, only the masses of bodies different from zero), one recovers the usual Einstein-Infeld-Hoffmann equations of motion.

The problem of rotational motion is even more complex than the translational case. The reason for that being that the spin of a body does not enter the Newtonian potential so only the Newtonian expression for the spin enters the post-Newtonian potentials. A post-Newtonian theory of the spin of a body therefore has to deal with gravitational potentials to 2PN-order. In DSX III, we gave a definition of a post-Newtonian spin of body (A), $S_i^{(A)}$, in the N-body case such that the resulting torque can be expressed entirely in terms of the moments $(M_L, S_L; G_L, H_L)$ in the A-frame. The resulting Law of rotational motion is of the form

$$\frac{dS_i^{(A)}}{dT} = \epsilon_{iab} \sum_{l \geq 0} \frac{1}{l!} \left(M_{aL} G_{bL} + \frac{1}{c^2} \frac{l+1}{l+2} S_{aL} H_{bL} \right) \tag{18}$$

$$+\text{further}(c^{-2} - \text{terms}).$$

For a body with only mass-monopole and spin-dipole this equation reduces to

$$\frac{dS^{(A)}}{dT} = \frac{1}{2c^2} S_a H_b. \tag{19}$$

Now, the local frame can be oriented in space such that $H_b = 0$ and the torque in (19) vanishes. In that case, the local frame is called dynamically non-rotating and the axes to the spatial coordinate line are Fermi-transported along the central worldline (DSX III).

So far, we were just talking about Laws of Motion rather than about Equations of Motion, the reason being that the various quantities M_L and

S_L have not been specified as functions of time in the general case. Note that these multipole moments physically are determined by the visco-elastic properties of matter. To close the whole system of equations, one has to introduce specific models for the time behaviour of the various multipole moments. One might, *e.g.*, start with "rigid models" with

$$\frac{dM_{ab}}{dT} = \epsilon_{adc}\Omega^c M_{db} + \epsilon_{bcd}\Omega^c M_{ad} \qquad \text{etc.} \qquad (20)$$

and

$$S_c = I_{cd}\Omega^d, \qquad (21)$$

where Ω^c is some time dependent vectorial function, but there is indeed the problem if such an ansatz is consistent and compatible with Einstein's field equations. It is known (Thorne and Gürsel 1983) that for one isolated body everything is fine if one restricts to first-order terms in Ω. In that case, problems related with Lorentz-contraction effects will not show up. To higher order in Ω, equations (20) and (21) are only compatible with the field equations for stationary bodies which do not precess and show no nutation.

One can say that concepts like angular velocity, figure axes, Tisserand axes, etc., still present basic problems in the gravitational N-body problem. A good way to proceed has been outlined by Klioner (1996): he introduces some rotating local frame by the action of some vectorial function $\Omega^i(T)$ and then formally specifies the rotating frame by requiring that the resulting rotational equations of motion take a particularly simple form. More work is needed, however, to put these ideas into a more rigorous form.

References

Brumberg, V.A.: 1991, *Essential Relativistic Celestial Mechanics*, Hilger.
Damour,T., Soffel, M. and Xu, C.: 1991-94 *Phys. Rev.* D **43**, 3273 (1991) **45**, 1017 (1992); **47**, 3124 (1993); **49**, 618 (1994). called DSX I - IV in the text.
Klioner, S.: 1996, these proceedings.
Thorne,K. and Gürsel, Y.: 1983 *Mon. Not. Roy. Ast. Soc.* **205**, 809

ANGULAR VELOCITY OF ROTATION OF EXTENDED BODIES IN GENERAL RELATIVITY

S.A. KLIONER
Institute of Applied Astronomy
197042 St.Petersburg, Russia

Abstract. We consider rotational motion of an arbitrarily composed and shaped, deformable weakly self-gravitating body being a member of a system of N arbitrarily composed and shaped, deformable weakly self-gravitating bodies in the post-Newtonian approximation of general relativity. Considering importance of the notion of angular velocity of the body (Earth, pulsar) for adequate modelling of modern astronomical observations, we are aimed at introducing a post-Newtonian-accurate definition of angular velocity. Not attempting to introduce a relativistic notion of rigid body (which is well known to be ill-defined even at the first post-Newtonian approximation) we consider bodies to be deformable and introduce the post-Newtonian generalizations of the Tisserand axes and the principal axes of inertia.

1. Introduction

In the framework of general relativity the spin (angular momentum) of an arbitrarily composed and shaped, weakly self-gravitating deformable body being a member of a system of N arbitrarily composed and shaped, weakly self-gravitating deformable bodies is a well-defined notion (see, e.g., Damour, Soffel, Xu, 1993). However, from the astronomical (observational) point of view not only the spin, but also the angular velocity of rotation of celestial bodies plays an important role. Indeed, it is the angular velocity of the Earth's rotation which is directly related to the observable quantities in the modern geodynamical observational techniques (VLBI, SLR, LLR, GPS/GLONASS). On the other hand, the angular velocity of rotation of pulsars plays the primary role in modelling pulsar timing data. Permanently increasing accuracy of geodynamical and pulsar timing observations makes it very important to have a relativistic definition of angular velocity of real bodies as well as relativistic equations describing time-dependence of the angular velocity.

For the first time, rotational equations of motion of extended bodies in general relativity have been discussed by Fock (1959). Fock succeeded to derive at the first post-Newtonian level the rotational equations of motion of an extended body being a member of a system of N gravitationally interacting extended bodies. Later the equations derived by Fock have been thoroughly investigated in Brumberg (1972) assuming the rigid-body distribution of the velocity of the matter of each body relative to a global reference system. Recently the result has been significantly improved by Damour, Soffel and Xu (1993). In that paper the post-Newtonian rotational equations of motion of an arbitrarily shaped, weakly self-gravitating bodies being members of a system of N bodies have been derived. Considering each body in its own local reference system the authors showed how to define a post-Newtonian spin vector of a body whose local-time evolution can be entirely expressed through bilinear couplings between the Blanchet-Damour (BD) multipole moments of that body and the tidal moments of the external gravitational field it experiences.

In order to introduce the post-Newtonian angular velocity we have to understand how the post-Newtonian spin vector can be split into a product of a tensor of inertia and an angular velocity. This problem, being quite trivial in Newtonian physics, is nontrivial and, so far, unsolved problem in the framework of general relativity (see, Thorne, Gürsel, 1983; Soffel, 1994). Until now virtually all papers dealing with the problem in question were devoted to relativistic generalizations of the rigid-body rotation. It is quite understandable since the coupling between the relativistic effects in the rotational motion of the Earth or a pulsar and the effects of their non-rigidity seem to be much smaller than the relativistic effects themselves which were the primary goal of the investigations. However, in the relativistic framework it is impossible to define rigorously a precessing, rigidly rotating body even at the first post-Newtonian approximation (see, e.g., Thorne, Gürsel, 1983; Soffel, 1994 and references cited therein). That is why, the authors were forced to drop all the terms nonlinear with respect to angular velocity which might be a critical restriction. Moreover, it seems to be unsatisfactory to introduce, into relativistic considerations, a notion which cannot be rigorously defined even in the first post-Newtonian approximation.

In Newtonian case the notion of rigid body plays a fundamental role in discussing the spatial rotation of extended bodies. Rigid-body rotation is not only a first-order (though usually quite accurate) approximation for real celestial bodies, but it also intimately relates to the properties of the absolute Newtonian space which is an Euclidean 3-dimensional manifold from the mathematical point of view. However, it is well known what to do if the body under consideration cannot be considered as rigid: we have to consider velocity distribution of the matter inside the body and split the velocity field into a rigid rotation and a deformation (see, e.g., Moritz, Mueller, 1987). In this way we can introduce the so-called Tisserand axes (a rigidly rotating reference system with respect to which the Newtonian angular momentum of the body vanishes identically) or the principal axes of inertia (a rigidly rotating reference system in which the Newtonian tensor of inertia is diagonal at any moment of time). Both these rigidly rotating reference systems allow one to define the angular velocity of a non-rigid body which has some definite physical meaning. This angular velocity is not in fact directly related to the body. It is the angular velocity of rotation of some rigidly rotating reference system (in which the body appears to be nonrotating in one sense or another) with respect to the corresponding Newtonian inertial reference system.

In analogy to Newtonian physics it is quite natural idea to consider deformable bodies in general relativity. However, the first and probably the only paper dealing with rotation of non-rigid bodies in general relativity is Voinov (1988). The principal idea is to consider relativistic effects in internal motions within the body as additional deformations which can be treated in analogy to Newtonian deformations. Unfortunately, because of several unjustified (or, sometimes, unnecessary) assumptions the paper can be considered only as a preliminary one. Our aim is to follow the principal idea formulated above as rigorously as possible, and generalize the Newtonian concepts of the Tisserand axes and the principal axes of inertia onto the post-Newtonian approximation of general relativity without any attempt to introduce the notion of a post-Newtonian rigid body.

2. Relativistic reference systems

Although it is well known that in the framework of general relativity any reference systems covering the space-time region under consideration are mathematically equivalent, one can prefer one reference system or another to perform actual calculations. The basic reason for the preference is physical adequacy (or convenience) of a reference system for the problem under consideration. In fact, the choice of an adequate reference system is an important part of solving any problem. Recently it has been shown that to describe adequately physics within an isolated system of N gravitating bodies one should use several relativistic reference systems: one global reference system covering all the bodies simultaneously and one local reference system for each of the bodies (see, e.g., Brumberg, Kopejkin (1989), Kopejkin

(1988) and Damour, Soffel, Xu (1991), (1992), (1993)).

The global reference system is suitable for describing relative motion of the bodies. We will designate it as BRS (Barycentric Reference System). The local reference systems are physically adequate for describing local physical processes related to the body under consideration. The local reference systems are proved to be suitable to describe rotational motion of the individual bodies (Damour, Soffel, Xu, 1993). Bearing in mind that the Earth is the primary body whose motion and rotation is to be investigated we will designate the local reference system of the body under consideration as GRS (Geocentric Reference System), although "GRS" can be related not only to the Earth, but to any other body.

We assume general continuous distribution of the matter having the energy-momentum tensor $T^{\alpha\beta}$. According to the usual post-Newtonian assumptions for the matter

$$T^{00} = \mathcal{O}(c^2), \quad T^{0i} = \mathcal{O}(c), \quad T^{ij} = \mathcal{O}(1). \tag{1}$$

We also suppose that the matter is localized in N separate blobs (bodies) and we discuss rotation of one of them.

Not going into complicated details of constructing the global and local reference systems (see, e.g., Brumberg, Kopejkin (1989); Kopejkin (1988); Damour, Soffel, Xu (1991); Klioner, Voinov (1993) for details), let us write down the generic form of the metric tensor which is valid for both the BRS and the GRS

$$g_{00} = 1 - \frac{2}{c^2}W + \frac{2}{c^4}\left(W^2 - \chi_{,00} - a_{,0}\right) + \mathcal{O}(c^{-6}),$$
$$g_{0i} = \frac{1}{c^3}\left(4U^i - a_{,i}\right) + \mathcal{O}(c^{-5}),$$
$$g_{ij} = -\delta_{ij}\left(1 + \frac{2}{c^2}W\right) + \mathcal{O}(c^{-4}). \tag{2}$$

Here, as usual, comma denotes a partial derivative (so that, $a_{,0} = \frac{\partial}{\partial t}a$ and $a_{,i} = \frac{\partial}{\partial x^i}a$), δ^{ij} is the Kronecker symbol and $a = a(t, \mathbf{x})$ is an arbitrary function which parameterizes a class of coordinate gauges allowed by our formalism and can be interpreted as a transformation of time $\tilde{t} = t + c^{-4} a$. This function can be left unspecified since the post-Newtonian equations of motion (both translational and rotational) are well known to be independent of a. Note that $a = 0$ corresponds to the harmonic gauge, while $a = -\chi_{,0}$ leads to the standard PPN (isotropic) gauge. The post-Newtonian potential W, the vector potential U^i, and the superpotential χ can be derived from the Einstein equations reduced with the corresponding gauge conditions

$$W_{,ii} = -4\pi G \sigma, \quad U^i_{,jj} = -4\pi G \sigma^i, \quad \chi_{,ii} = W, \tag{3}$$

where following Damour, Soffel and Xu (1991) we designate

$$\sigma = \frac{1}{c^2} T^{\alpha\alpha}, \quad \sigma^i = \frac{1}{c} T^{0i}. \tag{4}$$

Equations (3) should be solved with the account for specific boundary conditions which are to be chosen differently in the BRS and the GRS (see, e.g., Kopejkin, 1988; Damour, Soffel, Xu, 1991). In both the BRS and the GRS the potentials W, U^i and χ can be split into parts generated by the body under consideration and the external potentials induced by the rest of the matter (and, in case of the GRS, by inertial forces)

$$W = W_{\text{int}} + W_{\text{ext}}, \quad U^i = U^i_{\text{int}} + U^i_{\text{ext}}, \quad \chi = \chi^{\text{int}} + \chi^{\text{ext}}. \tag{5}$$

3. Post-Newtonian rotational equations of motion

Let us consider a reference system $(t, \mathbf{x})$ which can be either the BRS or the GRS. We will call this reference system RS. There are several ways to derive the post-Newtonian rotational equations of motion of an extended body relative to the RS. One can mention the Fock approach and the Landau-Lifshitz one. The former consists in evaluating the following integral over the support V of the body under consideration

$$\varepsilon_{ijk} \int_V (-g)\, x^j\, T^{k\beta}{}_{;\beta} dx^3 = 0, \tag{6}$$

which vanishes due to the local equations of motion of the matter

$$T^{\alpha\beta}{}_{;\beta} = 0. \tag{7}$$

Here semicolon denotes the covariant derivative,

$$-g = 1 + \frac{1}{c^2} 4W + \mathcal{O}(c^{-4}) \tag{8}$$

is the determinant of the metric tensor $g_{\alpha\beta}$ defined by (2), and ε_{ijk} is the fully antisymmetric Levi-Civita symbol ($\varepsilon_{123} = +1$).

The Landau-Lifshitz approach is based on the use of the so-called Landau-Lifshitz pseudotensor $t^{\alpha\beta}$ of energy-momentum of gravitational field defined in such a way that (see, e.g., Landau, Lifshitz (1971))

$$\left((-g)\left(T^{\alpha\beta} + t^{\alpha\beta}\right)\right)_{,\beta} = 0. \tag{9}$$

Then one can derive the rotational equations of motion from the integral

$$\varepsilon_{ijk} \int_V x^j \left((-g)\left(T^{k\beta} + t^{k\beta}\right)\right)_{,\beta} dx^3 = 0. \tag{10}$$

Both approaches mentioned above result in the same post-Newtonian rotational equations of motion which can be written as (details will be published elsewhere)

$$\frac{d}{dt}S^i = F^i + \mathcal{O}(c^{-4}). \tag{11}$$

Here S^i is the post-Newtonian spin which can be written as a well-defined compact support integral

$$S^i = \varepsilon_{ijk} \int_V x^j Q^k dx^3 + \mathcal{O}(c^{-4}), \tag{12}$$

$$Q^k = \sigma^k(1 + \frac{4}{c^2}W) - \frac{1}{2c^2}G\sigma \int_V \sigma^s(t,\mathbf{x}') \frac{7\delta^{ks} + n^k n^s}{|\mathbf{x}-\mathbf{x}'|} dx'^3 + \mathcal{O}(c^{-4}). \tag{13}$$

$$n^i = \frac{x^i - x'^i}{|\mathbf{x}-\mathbf{x}'|}. \tag{14}$$

Now let us assume the following relation between the components of the energy-momentum tensor, which, in fact, is valid under very general assumptions on the matter (see, e.g., Fock, 1959)

$$(-g)\sigma^i = \left(\rho^*\delta^{ij} + \frac{1}{c^2}p^{ij}\right)\dot{x}^j + \mathcal{O}(c^{-4}), \qquad \rho^* = (-g)\frac{1}{c^2}T^{00}, \tag{15}$$

where p^{ij} is the stress tensor of the matter. Then we get

$$Q^k = \rho^*\dot{x}^k + \frac{1}{c^2}\left(p^{ks}\dot{x}^s - \frac{1}{2}G\rho^* \int_V \rho^*(t,\mathbf{x}')\dot{x}'^s \frac{7\delta^{ks} + n^k n^s}{|\mathbf{x}-\mathbf{x}'|} dx'^3\right) + \mathcal{O}(c^{-4}). \tag{16}$$

On the right-hand side of (11) we have a post-Newtonian torque F^i defined as

$$F^i = \varepsilon_{ijk} \int_V \sigma(t,\mathbf{x}) x^j f^k dx^3 + \mathcal{O}(c^{-4}),$$

$$f^k = W_{,k}^{\text{ext}} + \frac{1}{c^2}\left(4\dot{U}_{\text{ext}}^k + \chi_{,0k}^{\text{ext}} - \dot{x}^s(4U_{\text{ext},k}^s + \chi_{,0ks}^{\text{ext}})\right) \tag{17}$$

Note that f^k is proportional to the external potentials, and therefore, in analogy to Newtonian case the post-Newtonian torque F^i vanishes for isolated bodies ($W_{\text{ext}} = 0$, $U_{\text{ext}}^i = 0$, $\chi^{\text{ext}} = 0$).

It is important to note also that casting the integrals (6) or (10) into the form (11) is not unique. For example, the equations of rotational motion which we quoted above do not coincide with those derived in Brumberg (1972) where a part of our post-Newtonian torque F^i have been moved to the left-hand side of (11) and interpreted as a part of the post-Newtonian spin S^i. However, Eqs. (11)–(17) allow us to re-write the post-Newtonian spin density Q^k in the form (13)–(16) which makes it (and, therefore, S^i itself) explicitly proportional to σ^i and therefore to the velocity $\dot{x}^i$ of the matter). The latter circumstance will be important for further considerations. Note also that our form of the rotational equations of motion (11)–(17) being written for a Damour-Soffel-Xu local reference system coincides with Eq. (2.4)–(2.8) of Damour, Soffel, Xu (1993).

4. Rotating reference system

According to the basic idea exposed in Introduction we have to define a relativistic rotating reference system (let us designate it as RS$^+$ and its time and space coordinates as $(\bar{t}, \bar{x}^i)$). The RS$^+$ has to be a post-Newtonian generalization of the Newtonian Tisserand axes (or the Newtonian principal axes of inertia). This implies that $\bar{t} = t + \mathcal{O}(c^{-2})$, and $\bar{x}^i = P^{ij}(t)x^j + \mathcal{O}(c^{-2})$, where $P^{ij}(t)$ is a time-dependent orthogonal matrix. Generally speaking, we could add some post-Newtonian pieces $\mathcal{O}(c^{-2})$ into the transformations. However, considering that it is unclear how to introduce relativistically meaningful macroscopic spatial rotation in curved space-time of general relativity, it is quite reasonable to define the coordinate transformations between the RS$^+$ and the RS as a rigid Newtonian spatial rotation

$$\begin{aligned}\bar{t} &= t, \\ \bar{x}^i &= P^{ij}(t)x^j.\end{aligned} \qquad (18)$$

Here the rotation has the meaning of an Euclidean 3-dimensional time-dependent rotation in the 3-space formed by the spatial coordinates of the RS. Using (18) one can easily derive the metric tensor $\bar{g}^{\alpha\beta}$ of the RS$^+$, the relations between the components of the stress-energy tensor $\bar{T}^{\alpha\beta}$ in the RS$^+$ and those in the RS, etc.

5. Rotational equations of motion relative to the RS$^+$

There are several ways to derive the rotational equations of motion of an extended body relative to the RS$^+$. Let us mention

- the Fock approach (Eq. (6) written in the RS$^+$);
- the Landau-Lifshitz approach (Eq. (10) written in the RS$^+$);
- transforming the rotational equations of motion relative to the RS (11)–(17) into those relative to the RS$^+$ with the aid of the coordinate transformations (18).

All the approaches result in the same rotational equations of motion relative to the rigidly rotating reference system RS+

$$\frac{d}{dt}(\overline{S}^i + \overline{C}^{ij}\overline{\omega}^j) + \varepsilon_{ijk}\overline{\omega}^j(\overline{S}^k + \overline{C}^{ks}\overline{\omega}^s) = P^{ij}F^j + \mathcal{O}(c^{-4}), \quad (19)$$

where

$$\overline{\omega}^i = \frac{1}{2}\varepsilon_{ijk}\dot{P}^{jm}P^{km} \quad (20)$$

is the angular velocity of rotation of the RS+ relative to the RS projected onto the axes of the RS+,

$$\overline{S}^i = \varepsilon_{ijk}\int_V \overline{x}^j \overline{Q}^k d\overline{x}^3 + \mathcal{O}(c^{-4}), \quad (21)$$

$$\overline{Q}^k = \rho^* \dot{\overline{x}}^k + \frac{1}{c^2}\left(\overline{p}^{ks}\dot{\overline{x}}^s - \frac{1}{2}G\rho^*\int_V \rho^*(t,\overline{x}')\dot{\overline{x}}'^s\frac{7\delta^{ks} + \overline{n}^k\overline{n}^s}{|\overline{x} - \overline{x}'|}d\overline{x}'^3\right) + \mathcal{O}(c^{-4}), \quad (22)$$

$$\rho^* = \frac{1}{c^2}(-\overline{g})\overline{T}^{00} = \frac{1}{c^2}(-g)T^{00}, \quad (23)$$

$$\overline{p}^{ij} = P^{ik}P^{js}p^{ks}, \quad (24)$$

$$\overline{n}^i = P^{ij}n^j = \frac{\overline{x}^i - \overline{x}'^i}{|\overline{x} - \overline{x}'|}, \quad (25)$$

$$\overline{C}^{ij}(t) = \int_V \rho^*(\delta^{ij}\overline{x}^s\overline{x}^s - \overline{x}^i\overline{x}^j)d\overline{x}^3 + \frac{1}{c^2}\varepsilon_{iak}\varepsilon_{jbs}\int_V \overline{x}^a\overline{x}^b\overline{p}^{ks}d\overline{x}^3$$
$$- \frac{G}{2c^2}\left(7\delta^{ij}\overline{\alpha}^{ss} - 7\overline{\alpha}^{ij} + \overline{\beta}^{ij}\right) + \mathcal{O}(c^{-4}), \quad (26)$$

$$\overline{\alpha}^{ij}(t) = \int_V\int_V \rho^*(t,\overline{x})\rho^*(t,\overline{x}')\frac{\overline{x}^i\overline{x}'^j}{|\overline{x} - \overline{x}'|}d\overline{x}'^3 d\overline{x}^3, \quad (27)$$

$$\overline{\beta}^{ij}(t) = \int_V\int_V \rho^*(t,\overline{x})\rho^*(t,\overline{x}')\frac{(\overline{x}\times\overline{x}')^i(\overline{x}\times\overline{x}')^j}{|\overline{x} - \overline{x}'|^3}d\overline{x}'^3 d\overline{x}^3, \quad (28)$$

$(\overline{x}\times\overline{x}')^i = \varepsilon_{ijk}\overline{x}^j\overline{x}'^k$, and $\overline{V}$ is the support of the body in the RS+. When deriving (19)–(29) we supposed that (15) is valid in the RS. Let us note that $\overline{Q}^k$ (and, therefore, $\overline{S}^i$) is explicitly proportional to the matter velocity $\dot{\overline{x}}^i$ relative to the RS+. Moreover, $\overline{Q}^k$ and $\overline{S}^i$ have the same functional form as the original spin density Q^k and the spin S^i relating to the RS (see, (12)–(16)). Not pretending to a rigorous physical meaning, we can call $\overline{S}^i$ "post-Newtonian angular momentum of the body relative to the RS+". On the other hand, the matrix $\overline{C}^{ij}$, being symmetric $\overline{C}^{ij} = \overline{C}^{ji}$ can be called "post-Newtonian tensor of inertia in the RS+".

6. Post-Newtonian Tisserand axes and principal axes of inertia

It is to note that although all the quantities entering (19) contain explicit post-Newtonian terms, (19) itself looks formally analogous to the Newtonian rotational equations of motion relative to a rigidly rotating reference system. This allows us to introduce the post-Newtonian Tisserand axes and the principal axes of inertia in a formally Newtonian way as it was done, e.g., in Moritz, Mueller (1987). Thus we can define

- *The post-Newtonian Tisserand reference system* RS_1^+ by imposing the condition

$$\overline{S}^i = 0. \tag{29}$$

- *The post-Newtonian reference system of principal axes of inertia* RS_2^+ by imposing the condition

$$\overline{C}^{ij}(t) = \mathrm{diag}(\mathcal{A}(t), \mathcal{B}(t), \mathcal{C}(t)). \tag{30}$$

Both (29) and (30) should be considered as the definitions of the rotational matrix $P^{ij}(t)$ which relates the RS and the RS^+.

In the RS_1^+ the rotational equations of motion (19) read

$$\frac{d}{dt}\left(\overline{C}^{ij}\overline{\omega}^j\right) + \varepsilon_{ijk}\overline{C}^{ks}\overline{\omega}^j\overline{\omega}^s = P^{ij}F^j + \mathcal{O}(c^{-4}). \tag{31}$$

The condition (29) which fixes the RS_1^+ can be expressed through the coordinates of the RS as

$$\varepsilon_{ijk}\int_V x^j Q^k dx^3 = 0, \tag{32}$$

$$Q^k = \rho^* V^k + \frac{1}{c^2}\left(p^{ks}V^s - \frac{1}{2}G\rho^*\int_V \rho^*(t,\mathbf{x}')V'^s\frac{7\delta^{ks} + n^k n^s}{|\mathbf{x}-\mathbf{x}'|}dx'^3\right), \tag{33}$$

$$V^i = \dot{x}^i - \varepsilon_{ijk}\omega^j x^k, \qquad V'^i = \dot{x}'^i - \varepsilon_{ijk}\omega^j x'^k, \tag{34}$$

where

$$\omega^i = P^{ji}\overline{\omega}^j = \frac{1}{2}\varepsilon_{ijk}P^{mj}\dot{P}^{mk} \tag{35}$$

is the angular velocity of rotation of the RS^+ relative to the RS projected onto the axes of the RS. Then (32)–(34) can be written as a system of linear algebraic equations defining ω^i at each moment of time

$$S^i(t) - C^{ij}(t)\omega^j(t) = 0, \tag{36}$$

$$C^{ij} = P^{ai}P^{bj}\overline{C}^{ab} = \int_V \rho^*(\delta^{ij}x^s x^s - x^i x^j)dx^3 + \frac{1}{c^2}\varepsilon_{iak}\varepsilon_{jbs}\int_V x^a x^b p^{ks}dx^3$$
$$- \frac{G}{2c^2}\left(7\delta^{ij}\alpha^{ss} - 7\alpha^{ij} + \beta^{ij}\right), \tag{37}$$

$$\alpha^{ij} = P^{ai}P^{bj}\overline{\alpha}^{ab} = \int_V \int_V \rho^*(t,\mathbf{x})\rho^*(t,\mathbf{x}')\frac{x^i x'^j}{|\mathbf{x}-\mathbf{x}'|}dx'^3 dx^3, \qquad (38)$$

$$\beta^{ij} = P^{ai}P^{bj}\overline{\beta}^{ab} = \int_V \int_V \rho^*(t,\mathbf{x})\rho^*(t,\mathbf{x}')\frac{(\mathbf{x}\times\mathbf{x}')^i(\mathbf{x}\times\mathbf{x}')^j}{|\mathbf{x}-\mathbf{x}'|^3}dx'^3 dx^3, \qquad (39)$$

where $(\mathbf{x}\times\mathbf{x}')^i = \varepsilon_{ijk}x^j x'^k$, S^i is defined by (12)–(16), and both S^i and $C^{ij} = C^{ji}$ are obviously independent of ω^i.

According to (36) the post-Newtonian spin $S^i = C^{ij}\omega^j$ and, therefore, it can be split into a product of the angular velocity ω^i and the tensor of inertia C^{ij}. Hence the rotational equations of motion (11) can be re-written as

$$\frac{d}{dt}\left(C^{ij}\omega^j\right) = F^i, \qquad (40)$$

ω^i being defined by (36).

It is interesting to note that the post-Newtonian tensor of inertia C^{ij} defined by (37)–(39) formally coincides with the expression for the post-Newtonian tensor of inertia derived in Soffel (1994) in the first post-Newtonian approximation for an isolated, rigidly rotating body with the aid of the approach proposed by Thorne and Gürsel (1983). The approach used in Soffel (1994) accounts only for first-order terms in $\epsilon = \omega R/c$, where R is typical linear size of the body (that is, the body is supposed to rotate slowly enough). On the contrary, we make use of the general post-Newtonian approximation scheme.

7. The post-Newtonian torque and the BD moments

In Voinov (1988), and Damour, Soffel and Xu (1993) it has been shown that the local reference system (GRS) is physically adequate for modelling rotational motion of a body being a member of an N-body system. From now on, we consider the "nonrotating" reference system (t, x^i) used above to coincide with the GRS. In the latter of the two papers cited above it has been shown also that the post-Newtonian torque F^i defined by (17) cannot be expressed through the Blanchet-Damour mass and spin moments of the body under consideration and that it is possible to define the post-Newtonian spin as $\tilde{S}^i = S^i + c^{-2}S'^i$, where S'^i is an additional term vanishing together with external potentials. Then the rotational equations of motion read

$$\frac{d}{dt}\tilde{S}^i = \tilde{F}^i + \mathcal{O}(c^{-4}), \quad \tilde{F}^i = F^i + \frac{1}{c^2}\frac{d}{dt}S'^i, \qquad (41)$$

and S'^i can defined in such a way that

$$\tilde{F}^i = \varepsilon_{iab}\sum_{l=0}^{\infty}\frac{1}{l!}\left[M_{aL}G_{bL} + \frac{1}{c^2}\frac{l+1}{l+2}S_{aL}H_{bL}\right] + \mathcal{O}(c^{-4}), \qquad (42)$$

where $L = i_1 i_2 \ldots i_l$ is the multiindex, M_L and S_L are the BD mass and spin moments, and G_L and H_L are the electric- and magnetic type tidal moments of external gravitational field (see, Damour, Soffel, Xu (1993) for details).

Starting from (41) and applying the same arguments, which allowed us to derive (31) and (40) from (11), we can introduce another version of the post-Newtonian Tisserand reference system RS_3^+. Making use of the RS_3^+, the rotational equations of motion can be written as

$$\frac{d}{dt}\left(C^{ij}\tilde{\omega}^j\right) = \tilde{F}^i + \mathcal{O}(c^{-4}), \tag{43}$$

$$\frac{d}{dt}\left(\overline{C}^{ij}\overline{\tilde{\omega}}^j\right) + \varepsilon_{ijk}\overline{C}^{ks}\overline{\tilde{\omega}}^j\overline{\tilde{\omega}}^s = \tilde{P}^{ij}\tilde{F}^j + \mathcal{O}(c^{-4}). \tag{44}$$

where $\tilde{\omega}^i$ is defined by

$$\tilde{S}^i(t) - C^{ij}(t)\tilde{\omega}^j(t) = 0, \tag{45}$$

and the relations between angular velocities $\tilde{\omega}^i$, $\overline{\tilde{\omega}}^i$ and the corresponding orthogonal matrix $\tilde{P}^{ij}$ are the same as above (see, (20) and (35)).

The definition of angular velocity of a non-rigid body is not unique even in Newtonian physics (as we mentioned above one can introduce the Tisserand axes and the principal axes of inertia which give different definitions of angular velocity). In general relativity there is also another reason of the nonuniqueness: the post-Newtonian spin itself of a body being a member of an N-body system is not unique due to contributions of the gravitational field binding the system (see, also Damour, Soffel, Xu (1993)). Considering this nonuniqueness of the angular velocity, the definitions (36) and (45) can be considered as conventions which make the laws of rotational motion as simple as possible.

8. A note on astronomical applications

We showed above that the post-Newtonian rotational equations of motion can be written in a formally Newtonian way. Therefore, many of the Newtonian results concerning the rotational motion of non-rigid bodies are also valid in the post-Newtonian approximation of general relativity. For example, if we suppose the body to be "dynamically rigid" in the RS^+ $\overline{C}^{ij}$ = const then (31) or (44) are precisely equivalent to the Euler equations. The influence of general relativity results in the non-Newtonian expressions for the torque $\tilde{F}^i$ and for the tensor of inertia C^{ij}. As for the latter, one could try to derive a system of differential equations defining time-dependence of C^{ij} and $\overline{C}^{ij}$ (related to the post-Newtonian hydrodynamical equations). However, since the internal structure of celestial bodies (including the Earth) is not known well enough to obtain the tensor of inertia with sufficient accuracy by evaluating the integrals (37)–(39) or the like, C^{ij} is to be derived from various kinds of astronomical observations. Therefore, probably the most important difference between the post-Newtonian rotational equations of motion derived above and their Newtonian counterparts is the relativistic corrections to the torque. Eq. (31) can be used to derive relativistic effects in the forced polar motion. Eq. (43) can be used to investigate relativistic effects in the forced precession and nutation (see, Bizouard, *et al.* (1992)). More detailed evaluation of the astronomical consequences will be published elsewhere.

Acknowledgement. The author is indebted to Prof. M.Soffel and Dr. A.Voinov for insightful discussions.

References

Bizouard C., Schastok, J., Soffel M.H., Souchay J., (1992) Étude de la rotation de la Terre dans le cadre de la relativité général: premiere approche. In: *Journées 1992*, N. Capitaine (ed.), Observatoire de Paris, pp. 76-84

Brumberg, V.A. (1972) *Relativistic Celestial Mechanics.* Nauka, Moscow. (in Russian)

Brumberg, V.A., Kopejkin, S.M. (1989) Relativistic Theory of Celestial Reference Frames. In *Reference Frames*, edited by J. Kovalevsky, I.I. Mueller and B.Kołaczek. Kluwer Academic Publishers, pp. 115

Damour, T., Soffel, M., Xu, C. (1991) General Relativistic Celestial Mechanics I. Method and definition of reference systems *Phys. Rev. D*, Vol. no. 43, pp. 3273-3307

Damour, T., Soffel, M., Xu, C. (1992) General Relativistic Celestial Mechanics II. Translational Equations of Motion *Phys. Rev. D*, Vol. no. 45, pp. 1017–1044

Damour, T., Soffel, M., Xu, C. (1993) General Relativistic Celestial Mechanics III. Rotation Equations of Motion *Phys. Rev. D*, Vol. no. 47, pp. 3124–3137

Fock, V.A. (1959) *Theory of space, time and gravitation.* Pergamon, Oxford.

Klioner, S.A., Voinov, A.V. (1993) Relativistic Theory of Astronomical Reference Systems in Closed Form. *Phys. Rev. D*, Vol. no. 48, pp. 1451–1461

Kopejkin, S.M. (1988) Celestial Coordinate Reference Systems in Curved Space-Time. *Celestial Mechanics*, Vol. no. 44, pp. 87–115

Landau, L.D., Lifshitz, E.M. (1971) *The Classical Theory of Fields.* Pergamon Press, Oxford.

Moritz, H., Mueller, I.I. (1987) *Earth Rotation: Theory and Observation* Ungar, New York.

Soffel, M. (1994) The problem of rotational motion and rigid bodies in the post-Newtonian framework. *unpublished notes*

Synge, J.L. (1960) *Relativity: the General Theory.* North-Holland Publishing Company, Oxford.

Thorne, K.S., Gürsel, Y. (1983) The free precession of slowly rotating neutron stars: rigid-body motion in general relativity, *Mon. Not. R astr. Soc.*, Vol. no. 205, pp. 809–817

Voinov, A.V. (1988) Motion and rotation of celestial bodies in the post-Newtonian approximation, *Celestial Mechanics*, Vol. no. 41, pp. 293–307

RELATIVISTIC ROTATIONAL EFFECTS: APPLICATION TO THE EARTH-MOON SYSTEM

DAVID VOKROUHLICKÝ
*Observatoire de la Côte d'Azur, Dept. CERGA,
Av N. Copernic, 06130 Grasse, France* [1]

Abstract. Relativistic spin effects involved in the Earth-Moon dynamics are reviewed. They enclose: (i) the coordinate system effects, and (ii) the relativistic physical librations. The geodetic precession is the only relativistic spin phenomenon which has been firmly detected so far. The best candidates of the effects which might be detected in the forthcoming period are the lunar physical librations and coordinate nutations. As for the latter, however, a fine cancellation between the geodetic and the Lense-Thirring coordinate effects results in decreasing their amplitude just below the possibility of the Lunar Laser Ranging technology.

1. Introduction

The relativity theory became a well-established tool for the Solar system studies. Planetary and satellite dynamics, reference frame as well as time scales definition need to account for tiny but observable relativistic effects. Small values of "the compactness parameter" (relating the gravitational radius and the true radius of a given body) and "the slow motion" (expressed in units of the light velocity) indicate that using a first post-Newtonian approximation (1PN) is appropriate in the Solar system. Exception is reserved for a group of proposed high-precision experiments which need to account for the higher order effects. These will not be discussed in this text.

The most exhaustive and ambitious approach to the 1PN dynamics of the system of N extended bodies has been recently presented by Damour

---

[1] On leave from the Institute of Astronomy, Charles University Prague, Švédská 8, 15000 Prague 5, The Czech Republic.

et al. (1991, 1992, 1993, 1994) [hereafter called DSX]. Not only the theory accounts in an elegant way for all details of the multipole structure of the gravity field(s) but also harbors in a natural way the theory of the local and global coordinate systems and time scales. We base our approach on the DSX series of papers.

In what follows, we discuss the relativistic spin effects in the Earth-Moon system with special care for the lunar motion. Thanks to a high precision of the Lunar Laser Ranging (LLR) technology we dispose of a data of superior quality if compared to the orbital motion data of the other solar system bodies[2]. This places the lunar motion among a good candidates for testing the relativistic effects.

Indeed, the lunar motion gives now the best quantitative evidence for the weak equivalence principle [see, for instance, discussion in Damour and Vokrouhlický (1995)]. Also, a high quality constraints on the 1PN parameters β and γ, as well as variation of the gravitational constant $\dot{G}/G$ are available [e.g. Williams et al. (1995)]. Finally, a special interest for our discussion is due to about 1 % verification of the de Sitter (geodetic) precession through the secular rate of the lunar longitude of perigee [Williams et al. (1995)]. Formally, it can be understood as "the relativistic spin" effect related to the local Earth-Moon reference frame definition, although the interpretation considering this phenomenon as a relativistic perturbation of the lunar orbital motion is equally possible [this is the way how historically de Sitter (1916) discovered the effect].

In the next section, we give an outlook for possibilities to measure another relativistic spin effects in the Earth-Moon system in the forthcoming period.

2. 1PN rotational dynamics and its application to the Earth-Moon system

In the third paper of the DSX series the authors succeeded to derive 1PN equations of the rotational motion relating the local-time T_A derivative of the properly defined spin vector S_a^A of body A to the perturbing torques

$$\frac{dS_a^A}{dT_A} = \sum_{l \geq 0} \frac{1}{l!} \epsilon_{abc} \left[M_{bL}^A G_{cL}^A + \frac{1}{c^2} \frac{l+1}{l+2} S_{bL}^A H_{cL}^A \right] + \mathcal{O}(c^{-4}) . \qquad (1)$$

The first term in the bracket describes coupling of the body A mass-multipoles M_L^A with the gravitoelectric multipoles G_L^A of the external grav-

---

[2]The best current LLR performance is 1 to 2 centimeters as a formal error of "the normal points" for the lunar motion. Technological improvements are supposed to suppress this precision to about 3 millimeters in the forthcoming years (C. Veillet, private communication).

itational field measured in the local frame of body A. The latter split into a piece which can be considered as "formally Newtonian" and a 1PN correction. The lowest order 1PN effect related to these terms comes from putting $l = 1$ in Eq. (1) and results from coupling of the mass quadrupole-moment M_{ab}^A [dominated by the flattening parameter J_2^A] with the 1PN piece of the gravitoelectric field G_{ab}^A. Corresponding relativistic "librations" has been recently examined for the Earth [Bizouard et al. (1992)] and for the Moon [Bois and Vokrouhlický (1995)]. In the former case their amplitude is very small, but in the lunar case it reaches the order of 1 milliarcsecond. The latter value might be in the range of detection possibilities of the improved LLR technology in the next years.

The presence of the $1/c^2$ factor in front of the second term in the right hand side of Eq. (1) clearly points out that it is entirely the 1PN order. Of these terms, the most important corresponds to fixing $l = 0$. It can be *formally* interpreted as the local coordinate system precession with the angular velocity given by the appropriate piece of the gravitomagnetic field $H_a(T_A)$. The latter quantity depends on the external gravitational potentials (both the scalar one, resulting in the geodetic precession, and the vectorial one, resulting in the Lense-Thirring precession) measured at the origin of the local frame of a given body A and also on the acceleration of the body A (resulting in the Thomas precession). Such secular relativistic rotational effects called "precession" are also accompanied by the periodic ones called "nutations". In the following, we briefly review recent calculations of these phenomena for the Earth and Moon referentials [see for details Fukushima (1991), Brumberg et al. (1992), Vokrouhlický (1995)].

The effects arising from the interaction of the solar gravity field with the Earth-Moon center-of-mass motion are the same for both the Earth and Moon referentials and, thus, can be attributed to the Earth-Moon center-of-mass referential system. They enclose the (observed) geodetic precession with the rate of 1919.4 mas/cy. Of particular interest is the Lense-Thirring (LT) precession, because this important prediction of the general relativity theory has not been experimentally verified yet. Careful analysis shows, that there are two sorts of LT precession: (i) the first related to the solar gravity field coupling with the Earth-Moon total angular momentum having the rate of about 0.47 mas/cy, and (ii) the second given by the solar angular momentum having the rate of about -0.28 mas/cy. It is interesting to note that some recent works, pioneered by Brumberg et al. (1992), prefer to join the former LT effect to the definition of "the geodetic precession constant" (in the broader sense). The reason is that this piece of the LT precession can be calculated with sufficient precision contrary to the latter effect which is submitted to at most 50 % error due to not precisely known value of the total angular momentum of the Sun. More detailed discussion can be found in Vokrouhlický (1995). Apart from the common terms, the mutual Earth-Moon dynamics results in the individual precession of the lunar and terrestrial reference frames. About 29.8 mas/cy advance of the lunar frame represents a major effect of these terms. The effect is potentially "observable" using the forthcoming year LLR data, but its secular nature troubles it decorrelation from the effects of the tidal origin. On contrary, it might be suggested to subtract this value from the lunar rotation in longitude to achieve a better precision of the tidal parameters adjustment.

The lunar librations associated with these coordinate effects has been studied by Vokrouhlický (1995). As a result of a fine cancellation of the geodetic and LT phenomena their magnitude is unfortunately small (approximately 0.03 mas). Individually, both effects are one order of magnitude greater (but opposite sign), about the same as the principal annual term discovered by Fukushima (1991). A wide spectrum of these relativistic nutations has been computed by Brumberg *et al.* (1992) for the terrestrial referential. Generally their amplitudes are of the order of microarcseconds (or smaller). Necessity of their use will attend the next generation of the astrometric projects.

Acknowledgement. The author worked on this paper when staying at OCA/CERGA (France) thanks to the H. Poincaré research grant.

References

Bizouard, C., Schastok, J., Soffel, M. and Souchay, J. (1992) Etude de la rotation de la Terre dans la cadre de la relativite generale: Premier approche, in: *Journee 1992: Systemes de reference spatio-temporels* ed. N. Capitaine, Paris Observatory, pp. 76-84.

Bois, E. and Vokrouhlický, D. (1995) The relativistic spin effects in the Earth-Moon system, *Astron. Astrophys.*, **300**, p. 559.

Brumberg, V.A., Bretagnon, P. and Francou, G. (1991) Analytical algorithms of relativistic reduction of astronomical observations, in: *Journee 1991: Systemes de reference spatio-temporels* ed. N. Capitaine, Paris Observatory, pp. 141-148.

Damour, T. (1987) chapter 6 of *300 years of gravitation*, eds. S.W. Hawking and W. Israel, Cambridge Univ. Press, Cambridge, pp. 128-198.

Damour, T. and Vokrouhlický, D. (1995), The equivalence principle and the Moon, *Phys. Rev. D*, in press.

Damour, T., Soffel, M. and Xu, C. (1991) *Phys. Rev. D*, **43**, p. 3273; (1992) *Phys. Rev. D*, **45**, p. 1017; (1993) *Phys. Rev. D*, **47**, p. 1017; (1994) *Phys. Rev. D*, **49**, p. 618; see also Damour, T. and Vokrouhlický, D. (1995) *Phys. Rev. D*, **52**, p. 4455.

Dickey, J.O. et al., *Science*, **265**, p. 482.

Fukushima, T. (1991) *Astron. Astrophys.*, **244**, p. L11.

Sitter, W. de (1916) *Mon. Not. R. astr. Soc.*, **76**, p. 699; (1916) *Mon. Not. R. astr. Soc.*, **77**, p. 155.

Vokrouhlický, D. (1995) The relativistic spin effects in the Earth-Moon system, *Phys. Rev. D*, in press.

Williams, J.G., Newhall, X.X. and Dickey, J.O. (1995) Relativity parameters determined from Lunar Laser Ranging, *Phys. Rev. D*, in press.

ON THE ORBITAL MOTION DESCRIPTION WITH A PROPER REFERENCE FRAME IN A COMPLETE SCHWARZSCHILD FIELD

J.M. GAMBI AND P. ZAMORANO
Departamento de Matemáticas. Escuela Politécnica Superior. Universidad Carlos III. Madrid.

AND

P. ROMERO AND M.L. GARCIA DEL PINO
Sección Departamental de Astronomía y Geodesia. Facultad de Ciencias Matemáticas. Universidad Complutense. Madrid.

It is a fact that, none of the diverse coordinate systems unified in the Post-Newtonian formalism used to describe the exterior Schwarzschild field can be regarded as being materialized by a reference frame. Only the polar Gaussian coordinates $(\rho, \vartheta, \varphi, t)$, or their naturally associated Fermi coordinates, can be shown to have this property (Synge, 1960).

As it is known, each one of them has a precise physical meaning and by means of them at least four new results appear.

In effect, by using the general expression for the world function (Synge, 1960) in these coordinates, it is easy to see that for an observer at rest with respect to the Sun, the relativistic distance $|\rho_1 - \rho_2|$, between two particles aligned with him, becomes their Euclidean distance; this is the first result.

Now, starting from the lagrangian L, written again in these coordinates, corresponding to a test particle moving in the exterior field, and using the first integrals in the usual way, the equation of the trajectory $(\vartheta = \pi/2, \vartheta' = 0)$ results to be

$$\left(\frac{du}{d\varphi}\right)^2 = f(u) = \alpha^2(\beta^2 - 1) + 2m\alpha^2\beta^2 u + 4m\alpha^2(\beta^2 - 1)u\log u + \\ + (-1 + 4m^2\alpha^2)u^2 + 6m^2\alpha^2\beta^2 u^2 \log u - 2mu^3 \log u \qquad (1)$$

where $u = 1/\rho$, $\alpha^{-1} = \frac{\partial L}{\partial \varphi'}$ and $-\beta = \frac{\partial L}{\partial t'}$ so that, as can be seen, the right side in (1) is not a cubic as it happens to be when the standard coordinates, for example, are used. This is the second result.

Nevertheless, it can be shown that not even being $f(u)$ a cubic (it can be shown that it has only two roots) the solution of (1) can be given in terms of elliptic functions. This is the third result.

Finally, the fourth result may help to better understand some properties in Orbital Dynamics and Astrometry: as it is known, for coordinates of a static weak gravitational field, for which the metric is written as $\eta_{ij} + h_{ij}$, the equations of motion for a test particle in first approximation contain the terms

$$\frac{1}{2} h_{44,\alpha} \quad \text{and} \quad -\frac{1}{2} h_{\alpha\beta} h_{44,\beta} \qquad (2)$$

(Brumberg, 1991). The second term comes from the unique non-linear term taken in the Christoffel symbols and is shown to play an important role as a source of relativistic perturbations. For example, in standard coordinates, this term appears as the first one in the right hand side of the equations

$$\ddot{\vec{r}} + \frac{m}{r^3}\vec{r} = \frac{m}{r^3}\left[\left(\frac{2m}{r} - 2\dot{\vec{r}}^2 + 3\frac{(\vec{r}\dot{\vec{r}})^2}{r^2}\right)\vec{r} + 2(\vec{r}\dot{\vec{r}})\dot{\vec{r}}\right] \qquad (3)$$

But, if polar Gaussian coordinates are used, we obtain

$$\ddot{\vec{\rho}} + \left(\frac{m}{\rho^3} - \frac{m^2}{\rho^4}(1 - 2\log\rho)\right)\vec{\rho} = \frac{m}{\rho^3}\left[\left(-(1+\log\rho)\dot{\vec{\rho}}^2 - \frac{(1-3\log\rho)}{\rho^2}\right.\right.$$
$$\left.\left. \cdot \left(\vec{\rho}\dot{\vec{\rho}}\right)^2\right)\vec{\rho} + \left(2(2 - \log\rho) - \frac{2m}{\rho}(1 - 2\log\rho)\right)\left(\vec{\rho}\dot{\vec{\rho}}\right)\dot{\vec{\rho}}\right] \qquad (4)$$

so that the analogous to this term in (3) is now the third one in the left hand side of (4) and although, obviously, their effects in the perturbed motion are analogous (but not equal in any way, as can be imagined from the previous results) an important difference between their origins emerges, and this is that, whereas the cited term in (3) comes from the second one in (2) as was said, the cited term in (4) comes from the first one in (2), so that no non-linear term is needed to be introduced in the Christoffel symbols to explain the perturbations when polar Gaussian (or Fermi) coordinates are used. In fact, if the second term in (2) is evaluated in these coordinates the result is that its value is, simply, zero.

References

Synge, J.L., *Relativity: The General Theory*. North Holland. N.Y., 1960.
Brumberg, V.A., *Essential Relativistic Celestial Mechanics*. Adam Hilger, N.Y., 1991.

DYNAMICAL SCREENING OF INTERACTIONS IN GRAVITATING SYSTEMS AND THE EPHEMERIS TIME

A.V. VITYAZEV AND A.G. BASHKIROV
Institute of Planetary Geophysics, RAS
B.Gruzinskaya Str. 10, 123810 Moscow, Russia
e-mail: abas@orig.ipg.msk.su

The concept of the screening of interparticle interactions has its origin in electrolyte and plasma theories. The most known example is the Debye-Hückel screening of the potential of the resting test charge provided with the availability of charges of the opposite sign and the total electroneutrality of the plasma. When this charge is moving, the static Debye screening decreases and an anisotropic dynamic screening develops due to excitation of waves of charge density. As a result, an effective potential of the positive moving ion becomes alternating with a characteristic length of space oscillations of order of the Debye length (Peter, 1990). Such dynamic screening is due to perturbations of charges of both signs by the varying field of moving test charge. This effect does not call for an electroneutrality of the system and is associated with the long-range character of the Coulomb interaction potential only.

As a result of the similar long-range character of the Newtonian gravitational potential, analogous effects are to be expected for a gas of gravitating bodies (Binney and Tremaine, 1987; Saslaw, 1987). According to Saslaw (1987, Ch.15), gravitational screening is caused by waves of the gas density generated by a moving gravitating test body. A gravitational potential of the unperturbed constant density of the homogeneous gas can be neglected (the so-called "Jeans Swindle"). Zones of reduced densities imitate negative mass densities and, as a result, the negative layers screen effectively outer bodies from the test moving body. In the same manner, zones of augmented densities generate additional gravitational interactions which strengthen the action of an effective force field of the test body on outer bodies. So, we get space oscillations of the effective potential of a moving body with a characteristic length of order of the Jeans length.

Here, we discuss briefly a more rigorous analysis (Bashkirov and Vityazev, 1996) of the above scenario and the observable consequences of such effect for Solar planetary System.

The formal analysis is based on the linearized Vlasov equation for the time-dependent perturbation $f_1(\mathbf{r}, \mathbf{v}, t)$ of the distribution function of the gas of gravitating bodies over coordinates $\mathbf{r}$ and velocities $\mathbf{v}$

$$\frac{\partial f_1(\mathbf{r},\mathbf{v},t)}{\partial t} + \mathbf{v}\frac{\partial f_1(\mathbf{r},\mathbf{v},t)}{\partial \mathbf{r}} + \nabla \Phi \frac{\partial f_0(\mathbf{r},\mathbf{v},t)}{\partial \mathbf{v}} = 0 \qquad (1)$$

where $f_0(\mathbf{v}) = \rho_0/(2\pi\tilde{v}^2)^{3/2}\exp\{-v^2/2\tilde{v}^2\}$, ρ_0 and $\tilde{v}$ are the mean mass density and thermal velocity, correspondingly.

The self-consistent gravitational potential taking into account the mass density perturbation $\int d^3v f_1(\mathbf{r}, \mathbf{v}, t)$ and a test gravitating body of mass m_1 inserted into the system at the instant t_0 at the point $\mathbf{r}$, with the velocity $\mathbf{u}$, is determined by the Poisson equation:

$$\nabla^2 \Phi(\mathbf{r},t) = -4\pi G \int d^3v f_1(\mathbf{r},\mathbf{v},t) - 4\pi G m_1 \delta(\mathbf{r} - \mathbf{u}(t - t_0)) \qquad (2)$$

For the sake of simplicity, the masses of all bodies are taken to be identical. Such a set of equations is widely used for the description of relaxation processes and stability analysis of gravitating systems (Binney and Tremaine, 1987; Saslaw, 1987).

Then, we solve these equations and get the Fourier transform of the perturbed gravitational potential in the form (Bashkirov and Vityazev, 1996)

$$\Phi(\mathbf{k},\omega) = \frac{8\pi^2 G m_1}{k^2 \varepsilon_g(\mathbf{k},\omega)} \delta(\omega - \mathbf{k}\mathbf{u}). \qquad (3)$$

Here,

$$\varepsilon_g(\mathbf{k},\omega) = 1 - \frac{k_J^2}{k^2}\left[1 + F\left(\frac{\omega}{\sqrt{2}k\tilde{v}}\right)\right] \qquad (4)$$

where $k_J = (4\pi G \rho_0/\tilde{v}^2)^{1/2}$ is the Jeans wave number and $F(z) = i(\pi)^{1/2} z \times \exp\{-z^2\}\mathrm{erfc}(-iz)$.

The function $\varepsilon_g(\mathbf{k},\omega)$ is the gravitational susceptibility of the system. This function determines the response of a gravitating system to a gravitational perturbation(Binney and Tremaine, 1987; Saslaw, 1987; Kukharenko et al, 1994), and its zeroth value gives the dispersion equation for waves of density of the system. The complex form of the function F accounts for the Landau damping of the collective excitations (Binney and Tremaine, 1987; Saslaw, 1987; Kukharenko et al, 1994).

If the test body is one of the system bodies, the Fourier transform (3) of its interaction potential is to be averaged over all the possible values

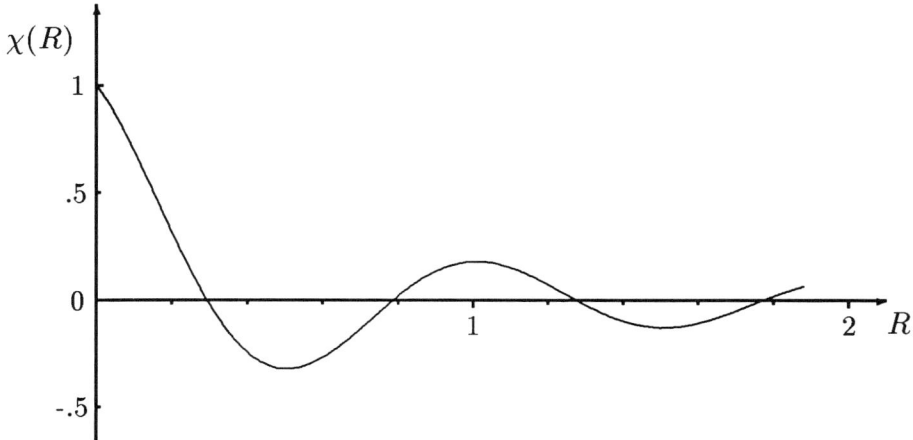

Figure 1. Screening factor $\chi(R)$ of the effective interaction potential $\tilde{\Phi}(r)$

and directions of the velocity **u** with the use of the Maxwellian distribution function $f_0(\mathbf{u})$. Then, as a result of the averaging and reverse Fourier transformation of Eq.(3) we get (Bashkirov and Vityazev, 1996)

$$\tilde{\Phi}(\mathbf{r}) = \frac{Gm}{r}\chi(R) \qquad (5)$$

where $R = r/\lambda_J$, $(\lambda_J = 2\pi/k_J)$. The function $\chi(R)$ has been obtained as a result of analytical and numerical integrations (Bashkirov and Vityazev, 1996). It is represented in the graphical form in Fig. 1.

Thus, we have found that the renormalized (or effective) interaction potential in the Maxwellian gravitating system not only decays faster than Newtonian potential, but it also becomes oscillating.

The final computational results for $\chi(R)$ are represented at Fig.1. The resulted graphic can be approximated as

$$\chi(R) = e^{-\alpha R}\cos(\beta R), \qquad (6)$$

where the constants α, β take the values $\alpha = 1.9$ and $\beta = 6.0$. Notice that such approximation is impaired after some oscillation periods as the periods are scaled down and the decay constants α decrease slightly with the distance R.

The dynamic screening effect is manifested itself in such systems as the Solar planetary system as well. Among other things, the attractive Sun potential's variation from the Newtonian potential is to cause an increase of the Keplerian periods of the planets. In this regard, the Earth is a sensitive detector of the Sun potential's properties, as inhabitants of the Earth have determined to date the Keplerian period of their planet with a high degree of precision.

As the Keplerian period $T \propto 1/\sqrt{\chi}$, the correction for non-newtonian potential is $\Delta T/T = (1 - \sqrt{\chi})/\sqrt{\chi}$.

We estimate now the value χ for the Sun in the system of gravitating stars of the Galaxy. The Jeans length is estimated as

$$\lambda_J = \sqrt{\pi} \left[\frac{\tilde{v}}{1.55 \cdot 10^6}\right] \left[\frac{G}{6.7 \cdot 10^{-8}}\right]^{-1/2} \left[\frac{\rho_0}{1.9 \cdot 10^{-23}}\right]^{-1/2} \text{cm} = 2.5 \cdot 10^{21} \text{ cm} \tag{7}$$

where we use the known values of the thermal velocity $\tilde{v}$ and the average matter density ρ_0 in the vicinity of the Sun.

For the Earth orbit $\Delta T_\oplus/T_\oplus \simeq 5.7 \cdot 10^{-9}$ or $\Delta T_\oplus \simeq 0.18$ sec/year.

The sign and the order of this value and its annual variation correspond to the correction for the ephemeris time of the Earth revolution.

Account for the screening of the Sun potential cause to much greater corrections to Keplerian periods of giant planets. In particular, we get for Jupiter $\Delta T_J \simeq 11$ sec/period, for Saturn $\Delta T_S \simeq 50$ sec/period, for Uranus $\Delta T_U \simeq 290$ sec/period, for Neptune $\Delta T_N \simeq 890$ sec/period and, for Pluto $\Delta T_P \simeq 1700$ sec/period.

References

Bashkirov, A. and Vityazev, A.:1996, "Screening of Gravitational Interactions in Newtonian World". *Phys. Lett. A,*, in press.

Binney, J. and Tremaine, S.: 1987, *Galactic Dynamics.* Princeton Univ. Press, Princeton, NJ.

Kukharenko, Yu., Vityazev, A. and Bashkirov, A.:1994, "Dynamical Screening of Long-Range Interactions in Nonrelativistic Self-Gravitating Systems". *Phys. Lett. A*, **195**, 27-30

Peter, T.: 1990, "Linearized Potential of an Ion Moving through Plasma". *Journ. Plasma Physics,* **44**, 269-280.

Saslaw, W.C.: 1987, *Gravitational Physics of Stellar and Galactic Systems.* Cambridge Univ. Press, Cambridge.

EVOLUTION OF EPHEMERIDES REPRESENTATION AND DIFFUSION

P.K.SEIDELMANN
U.S.Naval Observatory

Abstract. There has been, and continues to be, a close interaction between celestial mechanics used for the generation of ephemerides, mathematical techniques, and computer technology. As the computer capabilities of the ephemerides offices and the users of ephemerides have improved, the methods of determining and the accuracies of ephemerides have changed and the medium and representation of the ephemerides provided to the user have evolved.

Ephemerides have been provided in the form of theories, tables, tabulations, polynomials, graphics, and subroutines by means of the printed page, punched cards, magnetic tape, floppy disks, CD/ROMs, and electronic mail. As mathematical techniques and computer technology continue to develop and the requirements for ephemerides evolve, the methods of representing and diffusing ephemerides will continue to improve.

1. Introduction

The evolution of ephemerides representation and diffusion can be tied to the technological and cultural developments over the years. In the latter years, technological development has been mostly the development of computer technology, including the speed, size, and memory capabilities of the computer and the media used for storage of computer data. In many cases, the changes that have taken place in ephemerides can be tied to people and dates, when significant changes in procedure were introduced. In parallel, the changes of ephemerides can be tied to scientific developments that required changes, or observations that differed from the ephemerides and thus required scientific developments.

The paper will cover the early years of ephemerides, the development period as ephemerides representation and diffusion flourished, and then

TABLE 1. Early Ephemerides

DATES	PREPRINTING
3000 - 2000 BC	Stonehenge (equinoxes, solstices)
700 - 100 BC	Linear zigzag functions (eclipses, lunar crescent visibility)
200 BC	Apollonius–Planetary motion (eccentric and epicyclic motion)
200 BC	Hipparchus–star catalog–precession
	Numerical values in geometric models
200 AD	Ptolemy–"Almagest"–"Handy Tables"
800 AD	Venerable Bede–computers–Easter from lunar solar cycles
900 AD	al-Khworizmi "Zij"–lunar crescent–astrology
900 AD	al-Battani "Zij"–sines replace chords
1000 AD	Toledan Tables
1200 AD	Roger Bacon–Easter and astronomy differences
1252 (1320)	"Alfonsine Tables" (Paris or Castile?)
1474	Johann Muller "Regiomontanus"
1496	Abraham Zactus "Almanach Prepetuum"
1500	Almanacs – configurations, Tables (Phases of Moon, Transits of planets rising & settings, wit and wisdom)
1545	Copernicus "De Revolutionibus" tables
1550	Erasmus Reinhold's "Prutenic Tables"
1600	John Tapp "Seaman's Kalendar" (Sun, Moon, bright stars for seamen)
1627	Kepler "Rudolphine Tables" (elliptic orbits, parameters from Tycho)
1676	Mariners New Calendar

the computer period when diffusion techniques were driven by computer technology. Finally, conclusions and expectations for future development of ephemerides are considered.

In this paper, I will define an ephemeris as a communication of a predicted astronomical event. This definition allows the consideration of oral tradition which does not survive forever, but is sophisticated and enduring. There is also the question of the purpose of ephemerides, and we will touch on some of these different purposes as we trace their history. Cultural requirements, calendars, religion, navigation, and scientific interests have been involved, and also astrology has been a consideration throughout the whole period.

2. Early Years

The earliest representations of ephemerides (Table 1) that survive to this day are the standing stones as exemplified by Stonehenge, which provides information on the equinoxes and solstices, and dates back to 2000 to 3000

TABLE 2. Ephemerides and Scientific Developments

Date	Person	Scientific Development
350 BC	Aristotle	Geocentric universe
200 BC	Apollonius	Epicycle, eccentric motion
200 BC	Hipparchus	Precession
200 AD	Ptolemy	Mathematical astronomy
1545	Copernicus	Sun-centered solar system
1609-19	Kepler	Laws of motion
1687	Newton	Universal law of gravity
1750	Mayer	Lunar Tables
1772	Euler	Lunar Tables
1772	Lagrange	Libration points
1784	Laplace	Solar system stability
1801	Gauss	Orbit determination
1835	Hamilton	Generating functions, quaternions
1846	LeVerrier, Adams	Discovery of Neptune
1860	LeVerrier	Planetary theories
1878	Hill	Lunar equations of motion
1900	Newcomb	Astronomical standards, planetary theories
1919	Einstein	Theory of relativity
1927	DeSitter	Variation in Earth rotation
1930	Brown	Lunar theory
1958	Brouwer, Vinti Kozai, Garfinkel	Artificial satellite theory
1960	Danjon, Clemence	Ephemeris Time

BC. Usually Aristotle is credited with the concept of the geocentric universe, and Apollonius with the introduction of eccentric and epicycle circular motions to provide planetary motions. Based on the construction of a star catalog in the second century BC, Hipparchus was able to detect precession and he was also able to apply numerical values to the parameters in the geometric planetary models. In 200 AD, Ptolemy developed mathematical astronomy in his "Almagest." Then, his "Handy Tables" enabled one to calculate celestial longitude and latitude of the Sun, Moon, and planets and the phenomena for different terrestrial latitude regions. In the eighth century, the Venerable Bede was concerned with the calculation of the date of Easter based on solar/lunar cycles of various degrees of accuracy. In the Middle Ages, Ptolemy was followed by the Islamic astronomers who were interested in the visibility of the lunar crescent for Islamic calendar purposes. They also were, of course, interested in astrology. al-Khwarizmu produced the first major Islamic tables, "Zij," in the ninth century. The Alfonsine Tables were introduced in the thirteenth century. Their origin is uncertain, either being the court of Alphonso X in Castile or in Paris. The Copernicus tables in "De Revolutionibus" of 1543 were difficult to use. But Erasmus Reinhold's "Prutenic Tables" made the methods of Copernicus more accessible to almanac makers until the Rodolphine tables of Kepler, which were based on elliptic orbits and parameters from Tycho.

In about 1500 the printing press became available and almanacs became popular. These almanacs included configurations of the planets, phases of the Moon, transits of the planets, times of risings and settings and, in addition, some wit and wisdom.

The reasons for ephemerides and their diffusion developed over the years: scientific curiosity, the cultural reasons of calendar and religious dating, seasons, agricultural purposes, and, of course, astrological thoughts. Ephemerides became paramount for purposes of navigation and the determination of time.

In the 1600s, the Seaman's Calendar became available containing the positions of the Sun, Moon, and bright stars for sailors. During this period of introduction of the methods of printing and presenting ephemerides, there was a corresponding process of scientific development (Table 2) which included the Sun-centered solar system of Copernicus, the laws of motion of Kepler, the Universal law of gravity of Newton, and lunar tables of Euler.

The situation with regard to ephemerides, computing, and scientific development was addressed in the seventeenth century by Leibniz who wrote "Also the astronomers surely will not have to continue to exercise the patience which is required for computation. It is this that deters them from computing or correcting tables, from the construction of Ephemerides, from working on hypotheses, and from discussions of observations with each other. For it is unworthy of excellent men to lose hours like slaves in the labor of calculation which could safely be relegated to anyone else if machines were used."

The foundations of celestial mechanics include the concepts of libration points by Lagrange, the question of solar system stability by Laplace, the orbit determination method necessitated by discoveries of the minor planets as developed by Gauss, and the generating function and quaternions developed by Hamilton. A primary scientific discovery based on ephemerides was the accurate prediction of Neptune by LeVerrier and Adams. Planetary and lunar theories were developed by LeVerrier and Newcomb which, based on their accuracy levels and those of the observations, led to the requirement for the theory of relativity by Einstein, and the recognition that the rotation of the Earth was variable by de Sitter.

TABLE 3. Examples of National Almanacs (Navigation & Astronomy)

1679	Connaissance des Temps (France)
	(French Academy; Bureau des Longitudes)
1767	Nautical Almanac and Astronomical Ephemeris (UK)
1776	Berliner Astronomisches Jahrbuch (Germany)
	Nautisches Jahrbuch
1791	Efemerides Astronomicas (Spain). Almanaque Nautico
1855	The American Ephemeris and Nautical Almanac (USA)
1923	Annuaire Astronomique (USSR)
1933, 1941	American Air Almanac (USA)
1937	British Air Almanac (UK)
1943	Japanese Ephemeris (Japan)
1958	Indian Ephemeris and Nautical Almanac (India)
1975	Almanac for Computers (USA)
1982	Floppy Almanac
1993	Multiyear Interactive Computer Almanac (MICA)
1994	Redshift

TABLE 4. Ephemerides Accuracy Values

Date	Purpose	Accuracy	
BC	Eclipse prediction	30"	
1750	Navigation	1"	
1900	Scientific investigation (masses)	1"	
1960	Dynamical reference frame	0."1	
1965	Space mission	0."01	Moon 10 km
1990	Dynamical reference frame	0."001	Moon 1 cm

3. Development Period.

In the development of ephemerides, there arose the need for almanacs for navigation and astronomy. Thus, the Connaissance des Temps from France in 1679 became the first of the national almanacs. It predates the Bureau des Longitudes, whose bicentennial we celebrate at this meeting. Other countries followed with astronomical and navigational publications (Table 3). There arose from the national ephemerides and almanacs, the need for international standardization which led to the establishment of a standard meridian, standard time, standard astronomical constants, a single reference frame, and the beginnings of a general method of international cooperation. During the course of this time, there was also a continuing progress in the accuracy of ephemerides (Table 4). This progress is tied to the scientific developments which were either driven by the accuracy requirements or made possible the improvements in accuracies.

TABLE 5. Ephemerides Dissemination Techniques

Date	Method	New Technology	Technique
2000 BC	standing stones	alignments	observations
700 BC	tablets	clay	tables
200 BC	scribes	paper	tables
1500	book	printing press	tables, graphs
1900	books	photographs	theories
1940	punched cards	punched cards	tables
1960	magnetic tapes	computers	tabular
1965	magnetic tapes	computers	theories
1975	printed	hand calculators	Chebychev series
1982	floppy disks	personal computers	programs
1990	CD/ROM	compact disks	graphics, images
1994	electronic	networks	tables, graphics, images, programs

TABLE 6. Computers for Ephemerides

Date	Place	Person	Equipment	Purpose
1623	Tubingen	Schickard	desk calculator	
1642	France	Pascal	adding machine	
1693	Mainz	Leibniz	desk calculator	
1822	Cambridge	Babbage	difference integrate	
1854	Sweden	Seheitz	Babbage design	planetary distances
1863	Sweden	Wiberg	difference	logarithms
	America	Grant	difference	logarithms
1909	Berlin	Hamann	difference	Peters logarithms
1929	UK NAO	Comrie	printing calculator	Almanacs
1933	Columbia U	Eckert	punched cards	orbit computation
1940	USNO	Eckert	punched cards	Air Almanac
		Clemence		Theory of Mars
1947	Cincinnati	Herget	punched cards	minor planets
1948	IBM HQ	Eckert	IBM SSEC	outer planets
1948	Yale U	Brouwer	punched cards	star measuring
1954		Herget	UNIVAC I	Jupiter VIII
1954	Dahlgren VA	Herget	NORC	minor planets

4. Computer Development and Dissemination Techniques.

There has been a close relationship between the advancement of computers and computer technology and the methods in computing and disseminating ephemerides. In the early days, before electronic computers, tabular representations of theories were the most efficient methods for people to calculate ephemerides and to determine daily positions of the bodies (Table 5). With the arrival of punched cards, the application of those tables could be automated and the ephemerides more rapidly and correctly computed (Table 6). The electronic computer made the computation of special perturbations, or numerical integration, a much more attractive means of computing ephemerides, so that the IBM Selective Sequence Electronic Calculator (SSEC) was used to compute by numerical integration the ephemerides of the five outer planets. The general method of computing ephemerides became by means of numerical integration. At the present time, the only means of achieving the accuracy required for lunar laser ranging or radar ranging is by means of numerical integration.

The method of disseminating ephemerides has followed somewhat different routes. Tabular methods of providing data have existed since the technology of clay tablets. With the advent of paper and the printing process, the printing of the theories of the motions of the bodies and tables for computing ephemerides was introduced, both as a means of communicating a scientific accomplishment, and as a means for others to compute the ephemerides themselves. With the introduction of punched cards, both the ephemerides in tabular type form and the theories could be recorded in machine-readable format (1940 - 1965). This was followed by magnetic tapes, which provided a more compact, or more rapidly readable, format for providing the information (1962-1985). On magnetic tapes, data at uniformly spaced time intervals covering long periods of time could be provided so that people could interpolate for whatever time they wanted from the ephemerides.

With the introduction of the hand-held calculator (1974), there was a desire for a method whereby one could make a limited number of entries into the hand-held calculator and interpolate the data to determine celestial positions for specific times. This led to the publication of Chebychev polynomials for specific time limits, to be used with the hand-held calculator. With the advent of the personal computer (1980) came the floppy disk. Now the previously printed data could be stored on a floppy disk and accessed by a program also stored thereon. This technology let to the supplement of the printed book with a floppy disk to provide ephemeris data. The arrival of the CD ROM (1990) permitted much more data storage and led to the ability to include graphics. The availability of networks whereby people can access data from all over the world (1994), has provided a new means of disseminating the data in the forms of tables, graphs, programs, and pictures.

In the early days the data were provided in a tabular form and in graphs, or plots, of the planetary positions with respect to the stars, although I have yet to see an example of such an early plot that still exists. Then the capability for computing and the availability of the theories for people to calculate their own positions became desirable in the early 20th century. With the advent of computers and magnetic tapes, the tabular material was widely used on sequential storage. Now with the advent of the faster computers and more storage capability, programs based on theories, or based on compact Chebychev representation of sequential data, are more desirable than reading long sequential series of data. With the development of computer graphics and imaging capability, we now see graphics coming back as a popular augmentation to the accurate ephemerides.

5. Conclusion

Over the years there has been a close tie between ephemerides and cultural, scientific, and technological developments, computer technology, and individuals. There has also been, and unfortunately continues to be, a close tie between the ephemerides and astrology. It appears that a change in this progression should not be anticipated. It is my expectation that the future should see the distribution of data and information via networks, so the users (astronomers, space scientists, navigators, or the general public) can call up a graphic presentation of selectable scales showing the configurations of the planets, satellites, and stars for any time. In addition, they can call up numerical values to whatever accuracy is desired. In addition, standardized software packages will be available so that computations can be embedded in their own programs, without each individual writing all the software oneself.

References

Duncombe, R.L., Seidelmann, P.K., and Janiczek, P.M. (1974) "Planetary Ephemerides" In *Highlights of Astronomy*, G. Contopoulos, ed. 223-227.
Explanatory Supplement to the Astronomical Almanac, P.K. Seidelmann, ed., University Science Books, Mill Valley, California, 1992.
Gingerich, O. (1993) *The Eye of Heaven Ptolemy, Copernicus, Kepler* American Institute of Physics, New York.
Goldstein, B.R.(1972) "Theory and Observation in Medieval Astronomy," *Isis* **63**, 39-47.
Hawkins, G.S. (1965) *Stonehenge Decoded*, Doubleday and Co., Garden City, NY.
Neugebauer, O. (1957) *The Exact Sciences in Antiquity*, Brown U. Press, Providence.
Pedersen, O. (1993) *Early Physics and Astronomy*, 2nd ed," Cambridge U. Press.
Seidelmann, P.K. (1976) "Celestial Mechanics," In *Encyclopedia of Computer Technology 4.*, K. Belzer, A.G. Holzman and A. Kent, eds., Marcal Dekker Inc., New York.
Seidelmann, P.K. (1979) "The Ephemerides: Past, Present and Future" In *Dynamics of the Solar System*, R.L. Duncombe, ed., 99 - 114.
Seidelmann, P.K. (1993) "Review of Planetary and Satellite Theories," *Celestial Mechanics and Dynamical Astronomy* **56**, 1 - 12.
Seidelmann, P.K., Janiczek, P.M., Haupt, R.F., (1976), "The Almanacs - Yesterday, Today and Tomorrow," *Navigation* **24** 303-312.
van der Waerden, B.L. (1974) *Science Awakening II; the Birth of Astronomy* Oxford U. Press.

DES ÉPHÉMÉRIDES ASTRONOMIQUES ANNUELLES EN PRÉLIMINAIRE À L'ANNUAIRE DU BUREAU DES LONGITUDES

SUZANNE DÉBARBAT
Observatoire de Paris - DANOF/URA 1125
61 avenue de l'Observatoire, F-75014 Paris

La Convention nationale, après avoir entendu le rapport de ses Comités de marine, des finances et d'instruction publique, décrète:
ARTICLE PREMIER
Il sera formé un Bureau des longitudes.
V.

Le Bureau des longitudes est chargé de rédiger *la connoissance des temps*, qui sera imprimée aux frais de la République, de manière que l'on puisse toujours avoir les éditions de plusieurs années à l'avance; il perfectionnera les tables astronomiques, et les méthodes des longitudes, et s'occupera de la publication des observations astronomiques et météorologiques. (Extrait du Rapport sur l'établissement du Bureau des longitudes par Grégoire, Séance du 7 Messidor, l'an 3 de la République une et indivisible; Suivi du décret de la Convention nationale, et Imprimé par son ordre. Le document, de seize pages se termine par la mention: *Adopté*; il est dit: A Paris, De l'Imprimerie nationale, Messidor l'An III)

Si l'on en croit Lalande (1732-1807) et sa *Bibliographie astronomique* (1803), la plus ancienne des éphémérides conservées date de 1150 (*Tabulae astronomicae et ephemerides*, R. Salomon Iarchi, Iarchus). Au 15e siècle, tables et almanachs, calendriers et éphémérides se partagent la vedette et, à propos de Regiomontanus (Johann Muller, 1436-1476), Lalande note que *Ces éphémérides sont les premières qui aient été publiées et, pour ainsi dire, les premières qui aient été faites*. Elles couvrent les années 1475 à 1506. C'est d'ailleurs au siècle suivant que vont être principalement publiées des éphémérides pour plusieurs années successives: 1536-1550 (15 ans); 1552-1562 (11); 1554-1568 (15); 1554-1570 (17); 1557-1575 (19); 1558-1577 (20); 1564-1584 (21); 1576-1600 (25); 1577-1590 (14); 1581- 1615 (35); 1584-1607 (24); 1589-1600 (12); 1590-1610 (21); 1595-1630 (36). La durée couverte est variable, mais toujours longue; le "clou" est une publication (Paris, 1571) in-quarto *Ephemerides, ou almanach du jour et de la nuit, pour cent ans*, et celui de la lune durant le temps de 19 ans; composé et *revu par J. Gosselin de Vize.* Il est vrai que les voyages à la mer, nécessitant des connaissances astronomiques, duraient de nombreuses années... Le 17e siècle, qui permet à Képler (1571-1630) l'exploitation des observations de Tycho Brahe (1546-1601), voit se publier à la fois des éphémérides de longue durée: 1607-1618 (12); 1609-1617 (9); 1621-1640 (20); 1629-1640 (12); 1637-1700 (64); 1641-1660 (20); 1659-1671 (13); 1661-1675 (15); 1672-1681 (10); 1666-1680 (15); 1675-1684 (10), et des éphémérides de durée plus réduite: 1633-1636 (4); 1636-1640 (5); 1638-1642 (5); 1682-1684 (3). Cependant, vont paraître aussi des éphémérides annuelles, pour 1615 par exemple, pour 1617, pour 1650, 1666, 1683, 1685,...

Dans le même temps se crée la *Connoissance des tems ou calendrier et éphémérides... pour l'année 1679, calculées sur Paris...* car, indique Lalande, Les *éphémérides de Hecker couvraient les années 1666 à 1680*. Il est bien connu qu'avec cette *Connoissance des tems* de 1679 commençait une série d'éphémérides annuelles dont l'année 1995 marque le 317e volume. Il est bien connu, aussi, que l'*Adresse Au Roy* de ce premier volume précise qu'il a été établi **après l'avoir épuré de toutes les choses ridicules dont ces sortes d'Ouvrages ont esté remplis jusqu'à présent**. Éphémérides remplies de choses ridicules et aussi éphémérides ridiculisant leurs auteurs cohabitaient puisqu'à propos d'un in-octavo paru, en allemand en 1607, Lalande écrivait qu'il s'agissait de *La grande mère de tous les almanachs, ou almanach universel, qui ne contient que des prédictions ridicules des choses assurées; fait pour se moquer des astrologues*. Le nom de calendrier, d'éphéméride, de table, d'almanach est souvent usité de manière indifférente pour des publications contenant des indications astronomiques. Le terme d'almanach sera d'ailleurs celui retenu en Grande-Bretagne quand paraîtra, créé en 1767, le *Nautical Almanac*.

Malgré les remarques de Lalande, la confusion des contenus continuera au 18e siècle, mêlant d'ailleurs, pour le public, astronomie, astrologie, météorologie ainsi qu'en témoigne un *Almanach astrologique pour l'année 1769 qui marque exactement le tems de chaque jour, excepté les tems d'orage* (Strasbourg, 1769). La CLEF qu'il contient donne sous forme de pictogrammes les conditions météorologiques et l'on constate que celles-ci figurent, effectivement, à chacun des jours de l'année... Un *Calendrier royal pour l'année 1790* est beaucoup plus prudent puisqu'il ignore les prédictions météorologiques tout comme d'autres calendriers lesquels constituent, au 18e siècle, les éphémérides populaires, se limitant à l'énoncé des jours et des saints de l'année (Paris, 1741) même lorsque, plus tard, ils sont présentés à la Nation, la Loi et le Roi, à Paris en 1792.

Quant aux scientifiques du 18e siècle, peu soucieux de laisser prise aux croyances astrologiques, ils publient des volumes d'éphémérides strictement

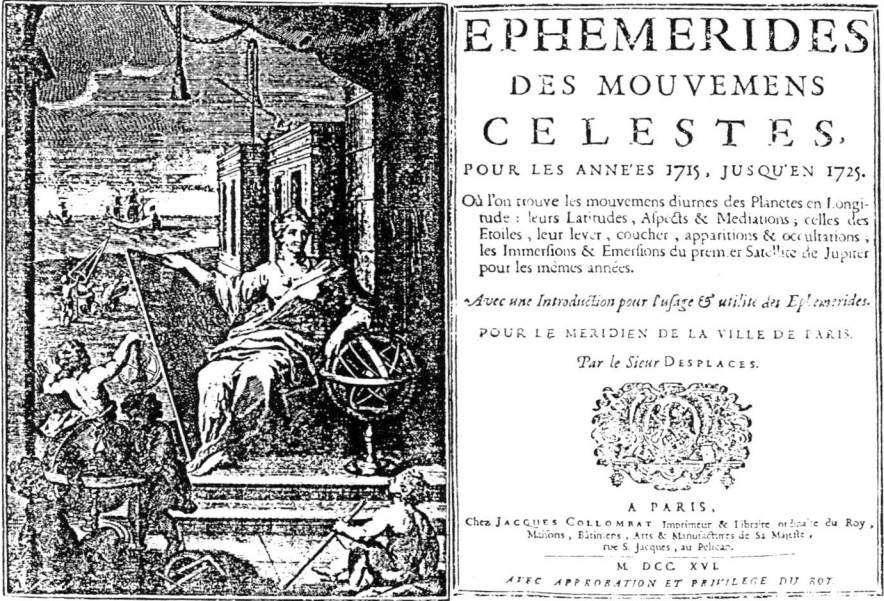

Figure 1. Frontispice et page de titre des éphémérides de Desplaces pour les années 1715 à 1725. *(Collection particulière)*

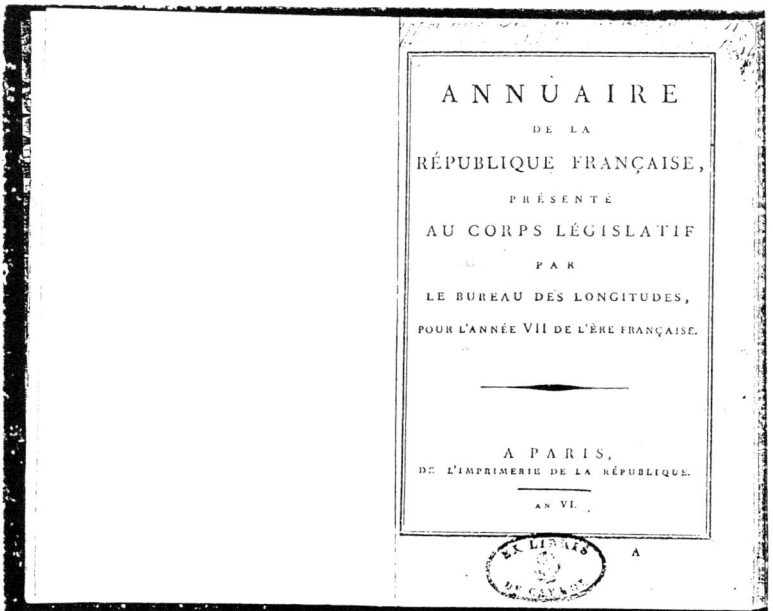

Figure 2. Page de titre de l'Annuaire pour l'année VII. *(Collection du Bureau des longitudes*

Jours du mois.	Vendémiaire.	Anc. style, 1798.	Lever du Soleil H. M.	Couc. du Soleil H. M.	Déclin. du Soleil. Austr. D. M.	T. moyen au midi vrai. H. M. S.	Var. de l'horl. par jour. S.	Âge de la Lune.
1	Primedi.	22 Septembre.	5. 55	6. 4	0. 8B	11. 52. 32		13
2	Duodi.	23	5. 57	6. 2	0. 16A	11. 52. 11	21	14
3	Tridi.	24	5. 59	6. 0	0. 39	11. 51. 50	21	15
4	Quartidi.	25	6. 1	5. 58	1. 2	11. 51. 30	20	16
5	Quintidi.	26	6. 3	5. 56	1. 26	11. 51. 10	20	17
6	Sextidi.	27	6. 4	5. 55	1. 49	11. 50. 50	20	18
7	Septidi.	28	6. 6	5. 53	2. 13	11. 50. 30	20	19
8	Octidi.	29	6. 8	5. 51	2. 36	11. 50. 10	20	20
9	Nonidi.	30	6. 10	5. 49	2. 59	11. 49. 51	19	21
10	Décadi.	1 Octobre.	6. 11	5. 48	3. 23	11. 49. 32	19	22
11	Primedi.	2	6. 13	5. 46	3. 46	11. 49. 13	19	23
12	Duodi.	3	6. 15	5. 44	4. 9	11. 48. 55	18	24
13	Tridi.	4	6. 17	5. 42	4. 32	11. 48. 37	18	25
14	Quartidi.	5	6. 19	5. 40	4. 56	11. 48. 19	18	26
15	Quintidi.	6	6. 21	5. 39	5. 19	11. 48. 2	17	27
16	Sextidi.	7	6. 22	5. 37	5. 42	11. 47. 45	17	28
17	Septidi.	8	6. 24	5. 35	6. 5	11. 47. 29	16	29
18	Octidi.	9	6. 26	5. 33	6. 28	11. 47. 13	16	30
19	Nonidi.	10	6. 28	5. 32	6. 50	11. 46. 57	16	1
20	Décadi.	11	6. 29	5. 30	7. 13	11. 46. 42	15	2
21	Primedi.	12	6. 31	5. 28	7. 36	11. 46. 28	14	3
22	Duodi.	13	6. 33	5. 26	7. 58	11. 46. 14	14	4
23	Tridi.	14	6. 35	5. 24	8. 21	11. 46. 0	14	5
24	Quartidi.	15	6. 36	5. 23	8. 43	11. 45. 47	13	6
25	Quintidi.	16	6. 38	5. 21	9. 5	11. 45. 35	12	7
26	Sextidi.	17	6. 40	5. 19	9. 27	11. 45. 23	12	8
27	Septidi.	18	6. 42	5. 17	9. 49	11. 45. 12	11	9
28	Octidi.	19	6. 43	5. 16	10. 11	11. 45. 1	11	10
29	Nonidi.	20	6. 45	5. 14	10. 32	11. 44. 51	10	11
30	Décadi.	21	6. 47	5. 12	10. 54	11. 44. 42	9	12

Les jours décroissent, pendant ce mois, de 51′ le matin et de 51′ le soir.

Jours du mois.	Lever de la Lune. H. M.	Couch. de la Lune. H. M.	Phases de la Lune.
1	5. 46 Soir	2. 39 Matin.	P. L. le 4
2	6. 2	3. 47	à 2h. 13′
3	6. 17	4. 54	du mat.
4	6. 31	5. 59	D.Q. le 12
5	6. 47	7. 5	à 0h. 40′ du mat.
6	7. 4	8. 14	N.L. le 18
7	7. 22	9. 23	à 3h. 46′ du soir.
8	7. 45	10. 33	P.Q. le 25
9	8. 15	11. 44	à 3h. 26′
10	8. 53	0. 55 S	du soir.
11	9. 44	1. 59 soir.	
12	10. 51	2. 55	Autom.
13	matin.	3. 40	Le 1 le S.
14	0. 6	4. 14	dans la
15	1. 29	4. 43	balance à 8h 6′
16	2. 57	5. 7	du soir
17	4. 24	5. 28	Le 2 lune
18	5. 49	5. 47	apogée.
19	7. 17	6. 9	Le 8 équ. ascend.
20	8. 44	6. 32	de la L.
21	10. 9	7. 0	Le 17 lun. périgée.
22	11. 31	7. 37	Le 20 éq.
23	0. 46 Soir.	8. 21	descend.
24	1. 46	9. 15	de la L.
25	2. 34	10. 18	Le 30 lun. apogée.
26	3. 7	11. 26	
27	3. 30	matin.	
28	3. 59	0. 36	
29	4. 18	1. 44	
30	4. 33	2. 53	

Jours.	Lever. des Planètes H. M.	Couch. des Planètes H. M.	Passage des Planètes au Mérid. H. M.
Mercure. ☿			
1	7. 0 Matin.	6. 2 Soir.	0. 31 S.
11	5. 24	5. 23	11. 22 Mat.
21	4. 46	5. 8	10. 57
Vénus. ♀			
1	3. 39 Matin.	5. 23 Soir.	10. 30 Matin.
11	4. 7	5. 15	10. 40
21	4. 40	5. 0	10. 50
Mars. ♂			
1	5. 39 Matin.	3. 25 Soir.	10. 32 Soir.
11	4. 52	2. 53	9. 57
21	4. 17	2. 22	9. 19
Jupiter. ♃			
1	8. 8 Soir.	11. 12 Matin.	3. 40 Matin.
11	7. 29	10. 29	2. 58
21	6. 44	9. 48	2. 18
Saturne. ♄			
1	11. 54 Soir.	3. 32 Soir.	7. 47 Matin.
11	11. 22	3. 0	7. 15
21	10. 46	2. 24	6. 39
Herschel. ♅			
1	4. 52 Soir.	5. 44 Soir.	11. 22 Matin.
11	4. 22	5. 12	10. 51
21	3. 46	4. 40	10. 16

Diamètre du Soleil. Le 1.er 31′ 58″ ; Le 16 32′ 6″

Figure 3. Pages 8 et 9 de l'Annuaire pour l'année VII. (*Collection du Bureau des longitudes*)

Jours.	Jours complémentaires.	Anc. style, 1799.	Lever du Soleil H. M.	Couc. du Soleil H. M.	Déclin. du soleil. Boréal. D. M.	T. moyen au midi vrai. H. M. S.	Var. de l'horl. par jour. S.	Âge de la Lune.
1	Primedi.	17 Septembre.	5. 46	6. 13	2. 10	11. 54. 21		18
2	Duodi.	18	5. 48	6. 11	1. 47	11. 54. 0	21	19
3	Tridi.	19	5. 50	6. 10	1. 24	11. 53. 39	21	20
4	Quartidi.	20	5. 51	6. 8	1. 0	11. 53. 18	21	21
5	Quintidi.	21	5. 53	6. 6	0. 37	11. 52. 57	21	22
6	Sextidi.	22	5. 55	6. 4	0. 14	11. 52. 36	21	23

Jours.	Lever de la Lune. H. M.	Couch. de la Lune. H. M.	Phases de la Lune.
1	7. 31 Soir.	9. 2 Matin.	D. Q. le 6
2	7. 49	10. 9	à 7h 32′
3	8. 12	11. 20	du mat.
4	8. 41	0. 31 Soir.	
5	9. 25	1. 38	Le 1 lune
6	10. 7	2. 39	apogée.

Jours.	Lever. des Planètes H. M.	Couch. des Planètes H. M.	Passage des Planètes au Mérid. H. M.
Mercure. ☿			
1	4. ≚46	6. ∽34	11. ≚13
Vénus. ♀			
1	9. ≚23	6. ∽43	2. ∽ 3
Mars. ♂			
1	5. ≚39	6. ∽17	0. ∽ 2
Jupiter. ♃			
1	10. ∽16	2. ∽18	6. ≚21
Saturne. ♄			
1	1. ≚16	4. ∽30	8. ≚57
Herschel. ♅			
1	5. ≚45	6. ∽30	11. ≚58

Diamètre du Soleil. Le 1.er 31′ 55″.

Pendant les six jours complémentaires, les jours décroissent de 9′ le matin et de 9′ le soir.

EXPLICATION

DES

PRINCIPAUX ARTICLES

DE L'ANNUAIRE.

Le calendrier et les phénomènes astronomiques dont notre Annuaire présente le tableau, ne supposent pas une connaissance approfondie de l'astronomie, mais des notions élémentaires que nous rappellerons en peu de mots pour le grand nombre des lecteurs.

Les éclipses sont les phénomènes les plus remarquables pour le public. Il y en a quelquefois sept dans un an, quelquefois deux seulement, dans les saisons où le soleil se trouve vers les nœuds de la lune, c'est à dire vers les points où l'orbite de la lune traverse ou coupe l'orbite apparente du soleil, c'est-à-dire, l'écliptique.

Après un espace de 18 ans et 10 jours, les éclipses reviennent à-peu-près dans le même ordre, et dans cet intervalle il y a environ 70 éclipses, 29 de lune et 41 de soleil, visibles en quelque endroit de la terre.

Si l'on prend un lieu déterminé, comme Paris, on ne trouve qu'environ 25 éclipsos de soleil en 56 ans; il n'y en aura pas de visible à Paris avant le 8 août 1802.

Le calendrier républicain, sur lequel notre Annuaire est

Figure 4. Pages 32 et 33 de l'Annuaire pour l'année VII. (*Collection du Bureau des longitudes*)

consacrés aux éléments astronomiques tels que levers et couchers du Soleil et de la Lune, positions des planètes, etc. En France, la **Connaissance des temps** poursuit sa route annuelle depuis son premier volume de 64 pages. Lalande précise, à son propos: *Lefebvre y mit son nom en 1685, Lieutaud en fut chargé en 1702, Godin en 1730, Maraldi en 1734* (Godin va partir pour l'expédition du Pérou). Puis il indique: *J'ai commencé en 1760, Jeaurat en 1776, Méchain en 1788;* **J'ai repris la rédaction de l'an 4 (1795) et des années suivantes.** Ainsi Lalande se trouve en charge de cette éphéméride quand le Bureau des longitudes est créé par la Loi du 7 Messidor an III (25 juin 1795).

Trois mois plus tard, le 4e jour complémentaire An III (20 septembre 1795), est édicté le *Règlement du Bureau des longitudes* dont l'*Article 9* est ainsi rédigé: **Le Bureau présentera chaque année au Corps législatif un Annuaire propre à régler ceux de toute la République.** Tout au long du 18e siècle dans différentes villes d'Europe, telles Berlin, Bologne, Londres, Venise, Leipzig, Stockholm, Edinburgh, Florence, Vienne, Hamburg, Amsterdam, Milan, Lisbonne, Rome,... nombre d'ouvrages d'éphémérides seront publiés. En France, à Rouen, dès 1701 - en in-quarto - paraissent des éphémérides annuelles que commente Lalande: *(Jean de Baulieu, ou Desforges) Le même auteur en donna, en 1703, la suite jusqu'en 1715 que commencèrent les éphémérides de Desplaces, continuées en 1745 par Lacaille, et par moi depuis 1775 jusqu'en 1800.* La figure 1 donne, du volume de la période 1715-1725 (en fait couvrant les années 1715-1724), le frontispice et la page de titre détaillant les éléments figurant chaque année. En fait Desforges (décédé en 1714) avait repris le nom de Beaulieu, mathématicien de Paris qui avait calculé des éphémérides avant lui.

Ces éphémérides ont couvert les périodes successives: 1715-1725; 1725-1735; 1735-1745; 1745-1755; 1755-1765; 1765-1775; 1775-1784; 1784-1793; 1793-1800. Chaque volume comprend essentiellement, pour chaque année, 26 pages dont deux pour chaque mois: à gauche *Mouvemens diurnes des planètes* (dont le Soleil et la Lune), à droite *Aspects des planètes* incluant dans les FESTES les saints du jour avec, en bas de page, les immersions et les émersions du 1er satellite de Jupiter. Le dernier volume de la publication commencée par Desplaces (1659- 1736), poursuivie par Lacaille (1713-1762) puis Lalande, est donc paru quand - en 1795 - ce dernier prend en charge la Connaissance des temps, au titre du Bureau des longitudes, et qu'est décidée, le 20 septembre 1795, la création d'un Annuaire. Dans l'introduction de ce volume (1793-1800), Lalande précise: **La connoissance des Tems que M. Méchain vient de publier pour 1792, contient sur les Mouvemens célestes de plus grands détails, ainsi que les Ephémérides de Londres, de Berlin, de Milan, de Vienne en Autriche; mais les nôtres offrent, plusieurs années d'avance, tout ce qui suffit au Public et même** *à des Astronomes qui, allant dans des Pays éloignés, ne pourroient pas se procurer les ouvrages plus étendus.* Dans ce volume, toujours à deux pages par mois, figurent les immersions et les émersions des quatre satellites de Jupiter.

Le premier volume de l'*Annuaire de la République française* comme il s'appelle alors, ne paraîtra qu'en juillet 1796, *l'an IV de la République française*. Il sera établi *pour l'année V de l'ère française (1797 ancien style)*. La figure 2 montre la page de titre de l'Annuaire pour l'année VII. L'Avertissement précise qu'il est extrait de la *Connaissance des tems* et qu'il doit *parvenir facilement et en nombre suffisant dans toutes les parties de la France*. Les éléments sont évidemment donnés dans le *Calendrier républicain* (figure 3) incluant, néanmoins (à partir, comme il se doit, du début de l'année républicaine, le 22 septembre 1798), les dates *ancien style*. Les éphémérides des levers et couchers de la Lune et du Soleil (figures 3 et 4), qui constituent l'essentiel du volume, se terminent par les 6 jours complémentaires de l'an VII (17 au 22 septembre 1799).

De dimensions fort modestes par rapport aux volumes des éphémérides précédemment établies, par Desplaces et les autres, le nouvel *Annuaire propre à régler ceux de toute la République* se révélait plus maniable pour atteindre les populations (sachant lire) et sans doute aussi moins coûteux pour le budget de l'époque; son prix, pour plusieurs années, sera de 60 centimes. C'est cet Annuaire qui, en 1995, se présente dans un format à peine plus petit que le volume qui, en 1800, terminait la série des *Ephémérides des mouvements célestes* initialisée au début du 18e siècle, mais qui compte - pour chaque année - environ 300 pages. Quant à l'*Almanach des postes*, il suffit - d'après les données du Bureau des longitudes - à fournir, au public, une très large part de ce que contenait l'Annuaire de l'An VII, puisqu'il comprend les levers et couchers de la Lune et du Soleil. Pour ceux, d'ailleurs, auxquels ces éléments ne suffiraient pas, ils peuvent se rapporter à l'*Annuaire du Bureau des longitudes* de l'année en cours, sous-titré *Ephémérides astronomiques* ou au Minitel, puisqu'il existe depuis plusieurs années le *3616 BDL*.

Références

Annuaires du Bureau des longitudes pour les années V, VI, VII.
Connaissances des temps des années 1679, an IV, an V, an VI.
Ephémérides des mouvemens célestes de Desplaces, 1715 à 1800.
Rapport sur l'établissement du Bureau des longitudes par Grégoire, Séance du 7 Messidor, l'an 3 (24 juin 1795) de la République une et indivisible.
Règlement du Bureau des longitudes du 4e jour complémentaire an III (20 septembre 1795).
Lévy, J.: 1976, "La création de la Connaissance des temps" *Vistas in Astronomy* **20**, 75-77.

FAST EVALUATION OF EPHEMERIDES BY POLYNOMIAL APPROXIMATION IN THE CHEBYSHEV NORM

J. C. COMA, M. LARA AND T. J. LÓPEZ MORATALLA
Real Instituto y Observatorio de la Armada
11110 San Fernando. Spain

Normally the planetary and satellite ephemerides are provided in tabular form, where the user interpolates between points in order to obtain the ephemerides. There are other methods of providing ephemerides by means of polynomial representations. The user is supplied with the coefficients of a set of polynomials which allow him a fast ephemerides evaluation.

In relation to Astronomy polynomial approximations are usually computed from a least squares fit. The disadvantage of this method is the absence of an estimate for the error: the magnitude of the error grows from the center of the interval towards the ends.

Another method no so frequently used is the uniform approximation in the Chebyshev norm. With this polynomial approximation, contrary to least squares, we obtain an uniform approximation of the function: the maxima and minima absolute values of the errors derived from this approximation are equal along the complete considered interval.

Fitting analytical functions, the method is well known. For instance, since 1980 the Bureau des Longitudes in their "Connaissance des Temps" computes series expansions in Chebyshev polynomials which, for certain intervals, represents their analytical theories.

The topic of this contribution is different. We use the uniform approximation in the Chebyshev norm in order to fit a data base. For the theory and details on the programming the reader is addressed to the references.

1. Uniform approximation and "Almanaque Náutico"

The "Real Instituto y Observatorio de la Armada en San Fernando" is editing the *Almanaque Náutico* continuously since 1792. Nowadays we use the data base DE200/LE200 of the Jet Propulsion Laboratory to compute our almanac. From basic ephemerides we select the desired data and then

Polynomial degree and error estimation. Period 35 days.

Body	R.A. $\varepsilon < 10^{-5}$ rad.	δ $\varepsilon < 10^{-5}$ rad.	r $\varepsilon < 10^{-6}$ AU	V $\varepsilon < 10^{-2}$
Sun	4	4	4	
Moon	24	21	7	
Venus	6	7		3
Mars	5	4		2
Jupiter	4	3		2
Saturn	3	3		2

RA: Right ascension, δ: declination, r: distance, V: visual magnitude.

TABLE 1. Approximation polynomials for *Almanaque Náutico*.

we use uniform approximation in order to compress the data: taking an interval of 35 days and establishing a bound for the error of 10^{-5} radian, about 0.03 minutes of arc, we compute the planetary ephemerides using low degree polynomials except for the Moon. The necessary degrees are listed in Table 1.

The approximation polynomials provide a fast and simple way to compute the quantities that appear in the *Almanaque Náutico* and this method has been implemented as a part of the navigation system of the Spanish Navy.

2. Acknowledgments

We are indebt with Dr. Deprit who suggested us the application of the minimax method. The first author (J. C. C.) thanks *Grupo de Mecánica Espacial* in the University of Zaragoza where he developed the capital part of this work. This work has been supported in part by the Spanish Ministry of Education (DGICYT Project # PB93-1236-C02-02) and by the Département de Mathématiques Spatiales at the CNES (Toulouse).

References

1. Barrodale, C.P.:1975, ACM Trans. Math. Software, **1**, 264–270.
2. Deprit, A. and Picard, H.:1979, Naval Research Laboratory Report 8280.
3. Remez, E.:1957, *General Computation Methods for Chebyshev Approximation*, Izdat. Akad. Nauk. Ukranisk SSR, Kiev.
4. Schmitt, H.:1971, *Discrete Chebyshev curve fit*, Comm. ACM **14**, 355–357.
5. Stiefel, E.L.:1958, Numerical Methods of Chebyshev Approximation, on *Numerical Approximation*, R. Langer (Ed.), U. Winsconsin, pp. 217–232.
6. Valleé-Poussin, C. de la.:1919, *Leçons sur l'approximation des fonctions d'une variable réelle*, Paris Gauthier-Villars, p. 75.

EXPERIENCE OF NUMERICAL INTEGRATION AND APPROXIMATION WITH APPLYING CHEBYSHEV POLYNOMIALS FOR CONSTRUCTING EPHEMERIDES OF THE SOLAR SYSTEM NATURAL AND ARTIFICIAL BODIES

A.A.TRUBITSINA
Institute of Theoretical Astronomy, RAS
Nab. Kutuzova 10, St. Petersburg, 191187 Russia
e-mail 1108@ita.spb.su

Abstract.

Successful experience of applying the Chebyshev polynomials as a power "mathematical tool" for numerical integration and approximation techniques in celestial mechanics is presented. Detailed analysis of approximation function behavior inside an integration step allows to elaborate a special technique for high accuracy and rapid integration of piece-wise continuous functions, modeling the Earth's shadow effect for artificial satellite orbits. Original software is elaborated for creating the ephemeris file simultaneously with the process of numerical integration. This technique is applied for the construction of ephemerides of natural and artificial celestial bodies as well as for the compact polynomial representation of different geodynamic parameters.

1. Approximation method

Let the motion of a celestial body be described by a system of the differential equations

$$Y'' = F(Y', Y, t), \qquad (1)$$

under initial conditions

$$Y|_{t=0} = Y_0, \quad Y'|_{t=0} = Y_0'. \qquad (2)$$

Here Y, Y' and Y'' denote three-dimensional vectors of the position, velocity and acceleration of a body. In INCH method (Belikov, 1993) of numerical integration each right-hand side of the system (1) is presented as a truncated Chebyshev expansion

$$F(Y', Y, t) = \sum_{m=0}^{M} a_m T_m^* \left(\frac{t}{h}\right), \quad t^{(1)} \leq t \leq t^{(2)}. \qquad (3)$$

Here T_m^* is shifted Chebyshev polynomials, $h = t^{(2)} - t^{(1)}$ is an integration step, and a_m are coefficients which are evaluated by an iteration procedure. The acceleration $Y''(t)$ is presented over an interval $T = t_2 - t_1, (T > h)$ in form of another truncated series

$$F^{(N)} = Y''(t) = \sum_{n=0}^{N} A_n T_n^* \left(\frac{t}{T}\right), \quad t_1 \leq t \leq t_2. \qquad (4)$$

The initial conditions at the left-hand boundary of an approximation interval are

$$Y|_{t=t_1} = y_0, \quad Y'|_{t=t_1} = y_0'. \qquad (5)$$

Thus the problem of the polynomial approximation is reduced to the determination of the coefficients A_n. The latter are presented in form of integrals

$$A_n = \frac{2\delta_n}{\pi} \int_0^1 \frac{Y''(x) T_n^*(x)}{\sqrt{x(1-x)}} dx, \quad \delta_0 = \frac{1}{2}, \quad \delta_i = 1 \, (i > 1). \qquad (6)$$

Here $Y''(x)$ means a function which is defined by formula (3) over the whole interval T. In INCH method the value of function $Y''(x)$ can be determined with a sufficient accuracy at an arbitrary point of the time interval T. Thus for calculating the integral (6) one can use the quadrature formula of the highest accuracy class. The resulting formula is

$$\int_0^1 \frac{Y''(x) T_n^*(x)}{\sqrt{x(1-x)}} dx \approx \alpha Y''(0) T_n^*(0) + \beta Y''(1) T_n^*(1) + \alpha^* \sum_{k=1}^{N} Y''(x_k) T_n^*(x_k), \qquad (7)$$

where

$$x_k = \frac{1 + \cos(\pi k/(N+1))}{2}, \quad k = 1, 2, ... N, \qquad (8)$$

$$\alpha = \beta = \frac{\pi}{2(N+1)}, \quad \alpha^* = \frac{\pi k}{N+1}. \tag{9}$$

By applying the quadrature formulae (7)-(9), one can determine the whole set of the desired coefficients A_n in (4). The value of function $Y''(x_k)$ in (7) is found as follows:

$$Y''(x_k) = \sum_{m=0}^{M} a_m T_m^*(z_k), \quad z_k = x_k T/h, \quad t_1 \leq t_k \leq t_2, \tag{10}$$

where a_m are the above mentioned coefficients computed by integrator INCH. Then, after twofold analytical integrating (4) with consideration of (5), one has:

$$Y'^{(N)} = \sum_{n=0}^{N+1} B_n T_n^*(t/T), \quad Y^{(N)} = \sum_{n=0}^{N+2} C_n T_n^*(t/T). \tag{11}$$

The high accuracy of the representation is proved by Powel's estimates establishing the closeness between the approximations by Chebyshev interpolating polynomial, Chebyshev truncated series and by polynomials of the best uniform approximation (Luke, 1975).

2. Applications of the method

The calculations were performed on PC Dell 486/66 computer with double precision (16 decimal digits of mantissa).

2.1. CONSTRUCTION OF EPHEMERIDES OF THE MOON, SUN AND PLANETS

The gravitational interaction of the Solar System bodies is modeled by Einstein-Infeld-Hofmann's equations, defining the orbital barycentric motion of the Sun, major planets and the Moon as non-rotating masses in the barycentric isotropic coordinate system (Eroshkin, Trubitsina, 1992). Additional perturbations in the motions are caused by the attraction of five most massive asteroids. The total system, consisting of 40 ordinary differential equations of the second order, forms the basis of the given model. 39 of these equations describe the barycentric motion of the Sun, 9 major planets and geocentric motion of the Moon, and remaining four equations describe the Moon's rotation around its own center of the masses. In Table 1 the results of the numerical testes of the problem is shown. The accuracy criterion for AE ephemerides is that an approximation error should be less than 1 millimeter at each point of approximation interval.

Similar tests have been conducted for constructing the specialized geocentric ephemerides of the Moon and Sun (AEMS) (Eroshkin, Taybatorov, Trubitsina, 1994). The ephemerides are optimized by accuracy and compactness for practical requirements of numerical integration for some types of artificial satellites.

TABLE 1. Maximum residuals between the ephemeride AE94 and DE200/LE200 over 50 year span

Object	Residuals (mm)	Approx.interval (days) AE(LE/DE)	Polyn.degree AE(LE/DE)
Mercury	52	8(8)	12(11)
Venus	7	32(32)	11(11)
E-M barycenter	10	16(16)	12(14)
Mars	18	32(32)	8(9)
Jupiter	16	32(32)	8(8)
Saturn	59	32(32)	7(7)
Uranus	31	32(32)	7(7)
Neptune	33	32(32)	5(5)
Pluto	14	32(32)	5(5)
Geocentr.Moon	5	8(4)	14(11)
Sun	0.034	32(32)	12(14)
Earth	9	16	12

2.2. ARTIFICIAL SATELLITE NUMERICAL EPHEMERIDES

The numerical tests were conducted for GPS satellite orbit (Taybatorov, Trubitsina, 1992). The results of both numerical integration over the time interval of 7 days and polynomial approximation for GPS satellites are presented in Table 2. As the reference ("exact") solution in "Integration" part of Table 2 the results of numerical integration are taken, obtained on computer ELBRUS-1CB with a longer mantissa (24 decimal digits). In the part of Table 2, titled as "Polynomial representation", the maximum deviations of the ephemeris position components from the numerical integration results are given, where Q denotes the ephemerides file size, δt is additional computer time for ephemeride file construction procedure, M, N, T, h are defined in (3)-(4).

TABLE 2. Comparison of integration and compact polynomial representation procedures for GPS satellites numerical ephemerides for (7 days arc)

Integration	M	h(days)	ΔR (m)	$\Delta\lambda$(sec)	$\Delta\beta$ (sec)		
	7	0.05	0.35e-3	0.11e-3	0.49e-4		
Polynomial	N	T(day)	ΔR (m)	$\Delta\lambda$ (sec)	$\Delta\beta$ (sec)	δt	Q(Kb)
Approximation	14	0.05	0.11e-7	0.23e-8	0.25e-8	4%	49.3
	14	0.15	0.51e-4	0.42e-6	0.31e-6	2%	16.2

3. Numerical integration of piecewise-continuous functions in satellite dynamics

If we need to take into account the solar pressure in problems of satellite dynamics we deal with piecewise-continuous functions. The characteristics of the numerical integration for such a problem were investigated on a model, simulating the disturbing motion of Lageos with taking into account the direct solar pressure and passing a satellite the Earth's shadow.

Three methods are compared with varying basis parameters: RA(15) (Everhart, 1985), INCH(9), INCHE(9) (Belikov, 1993). All the conclusions are made by comparison of the differences between the forward and backward solutions for one day time interval with four gap points.

The earlier elaborated method INCHE(9) (Belikov, 1993) is tested for the case with discontinuities. In this procedure the bisection principle is used to find the appropriate length of a variable step in the neighborhood of discontinuity. The high accuracy of solution is reached in this procedure at the expense of significant increasing CPU time due to a lot of step subdivisions. It is not convenient for practical application.

In this connection special procedure for an optimal subdivision of an integration step is elaborated at present work, in which the property of Chebyshev approximation to reach the maximum approximation error in the neighborhood of a gap (Gibbs phenomenon) is used for the optimal choice of an integration step. This modified INCHE method shows its essential time reducing, when solving the numerical problems for piecewise-continuous functions with sufficient accuracy. The results are summarized in Table 3, where N_{step} is number of intergation steps, N_{force} is number of calls of force subroutine (computer time measure).

Acknowledgements

The author is grateful to the Russian Foundation of Fundamental Research for financial support of this investigation (grant N95-02-04304-a).

TABLE 3. Comparison of the efficiency of the different numerical integration software in solving the system of differential equations with piecewise continuous functions

Method	$\frac{Y''_{sp}}{Y''_e}$	ΔR (m)	$\Delta \lambda$ (sec)	$\Delta \beta$	N_{step}	N_{force}
RA(15)	0	0.34e-7	0.65e-8	0.34e-8	194	2974
	1.e-9	0.30e-3	0.11e-4	0.28e-4	183	2809
	1.e-5	0.32e+1	0.12e+0	0.34e+0	190	2914
INCHE(9)	0	0.23e-7	0.93e-8	0.59e-8	149	2886
	1.e-9	0.56e-6	0.33e-9	0.58e-7	1060	22769
	1.e-5	0.46e-6	0.18e-7	0.36e-7	1060	22769
INCHE(9) modified	1.e-9	0.33e-7	0.11e-7	0.58e-8	231	5664
	1.e-5	0.33e-7	0.12e-7	0.25e-7	238	5570

References

Belikov M.V. (1993) Methods of numerical integration with uniform and mean square approximation for solving problems of ephemeris astronomy and satellite geodesy, *Manus. Geod.*, **Vol. 15**, N 4, pp. 182–200.

Eroshkin G.I., Trubitsina A.A. (1992) A new results on constructing of Sun, Moon and planets ephemeris AE92, *Abstracts of the Conference "Organizing of observation program for high-orbital Earth's artificial satellite and Solar system celestial bodies" (St.-Petersburg, 21-26 Spt., 1992)*, St.-Petersburg: ITA of the RAS, pp. 49-50 (in Russian).

Eroshkin G.I., Taybatorov K.A. and Trubitsina A.A. (1994) Construction of the specialised numerical ephemerides of the Moon and the Sun for solving Earth artificial satellite dynamics problems, *Preprint ITA RAS, St.-Petersburg*, N 4, 33 p. (in Russian).

Everhart E. (1985) An efficient integrator that uses Gauss-Radau spacings, *In Dynam. of Comets: Their Origin and Evolution*, Eds. A.Carusi, G.B.Valsecchi, Reidel Publ. Co., pp. 185-202.

Luke Y.L. (1975) *Mathematical functions and their approximations*. Academic Press Inc., New York, San Francisco, London.

Taybatorov K.A., Trubitsina A.A. (1993) Effective method for constructing artificial satellite polynimial ephemerides, *Proceedings of the Conference on Astrometry and Celestial Mechanics, Poznan, Poland, September 13-17, 1993*, pp. 245-250.

CONTROLLING THE OBSERVATIONAL DATA ON MINOR PLANETS WITH THE "CERES" SOFTWARE PACKAGE

YU.CHERNETENKO, V.L'VOV, V.SHOR, R.SMEKHACHEVA
AND S.TSEKMEJSTER
Institute of Theoretical Astronomy, RAS
Nab. Kutuzova 10, St. Petersburg, 191187 Russia

CERES is a powerful software package for calculating the ephemerides of the major planets, minor planets and comets and for executing related tasks.

At the present time the principal features of the package are as follows:

1. To store in its integrated database the elements and other characteristics of all the numbered minor planets, as well as to browse and to update the data.

2. To calculate ephemerides of various types for members of the solar system. It is possible to obtain spherical or rectangular coordinates (geometric positions) of any specified object referred to the center of the Sun, the Moon or a major planet. For the Earth the geocentric or topocentric astrometric coordinates are also available, as are the apparent coordinates with their first and second derivatives. The coordinates can be referred to the planes of equator, ecliptic or horizon. It is also possible to get a set of heliocentric osculating elements. The ephemerides can be calculated with different accuracy and be represented in various forms. One can obtain the coordinates of a minor planet, either from simple Keplerian motion or by numerically integrating the perturbed motion, taking into account the perturbations by any or all of the major planets, the Moon and the four most massive minor planets.

3. To model visually a dynamical picture of the motions of several selected minor planets, the Sun, the Earth, Mars and Jupiter simultaneously. The point from which the picture is to be viewed can be specified arbitrarily, and the scale and speed of action are also adjustable.

4. To store and to browse the coordinates of more than 500 stations (MPC list) for which topocentric positions of minor planets can be computed.

5. To visualize the tracks of several minor planets in a selected area of the sky and to show the motion of any minor planet against the stellar background. The package contains data on about 200 000 stars in the PPM catalog within a zone 80 degrees wide centered on the ecliptic. There are special utilities for the user to create his own star catalog.

6. To get information on different astronomical notions.

CERES runs on IBM PC compatibles under MS-DOS version 3.0 or later and requires 640 Kb of RAM, about 6.5 Mb of hard-disk storage and an EGA/VGA adapter. A math coprocessor is recommended. CERES (version 2.3) is distributed on five 3.5-inch 1.44-Mb or six 5.25-inch 1.2-Mb diskettes. These diskettes contain both executable files and data files, including the star catalog and a numerical ephemeris of the Sun, the Moon and the major planets for the years 1950–2020.

The actual and new options of the CERES software are now under developing. In addition to the augmented catalogs of numbered minor planets and observing stations the catalogs of selected unnumbered minor planets and periodic comets are introduced. It is now possible to integrate the motion of any object for an interval of at least of two centuries (1850–2049). Objects in planetocentric orbits and that have parabolic or hyperbolic orbits can also be processed by CERES. Perturbed orbital motion can be modelled on the screen. New options for processing the observations of minor planets are to be added, including comparison of available observations with computed positions and orbit determination. A search for all the objects observable within a selected area of the sky at a given moment is very useful for an observer, as is an image of the sky on the screen and/or printer. The problem of identifying objects (and therefore discovering new ones) at short notice is facilitated.

It is supposed that CERES will provide ephemeris support for observations of occultation phenomena associated with minor planets, namely, occultations of stars by objects and of objects by the Moon. The task can be carried out, both in terms of predicting the local circumstances for any observatory and of constructing charts of the Earth's surface showing the limits where the phenomena will be visible.

These new facilities are designed to help observers in planning their photographic, CCD or photoelectric occultation observations, in controlling the accuracy of observations before publishing them and in searching for new objects. In other words, by using CERES software one can test the quality of observations and perhaps obtain a list of unknown objects. This will make the CERES software package useful for monitoring all the members of the solar system, including the near-Earth asteroids.

EPHEMERIDES SOFTWARE OF NATURAL SATELLITES

N.V.EMELIANOV
Sternberg State Astronomical Institute
Universiretskij Prospect, 13, 119899 Moscow, Russia

To make a photograph of a satellite, we have to know how it is located relative to the planet. After processing the plates, the identification of satellites is to be made. Precise coordinates of natural satellites need to be known for this purpose.

The accuracy of determination of relative coordinates of satellites increases with the decreasing of their apparent angular distance. If the CCD-matrix is used, the field of view is small. It is necessary to precalculate the moment when the images of satellites are close to each other.

In some rare moments, mutual phenomena occur: the image of one satellite overlaps with another or one satellite enters the shadow of the other. In this case, the relative coordinates of satellites are obtained with maximal accuracy than it can be achieved from the Earth. These phenomena are very rare and short-term. For their detection, it is necessary to calculate satellite coordinates with highest accuracy.

In every case of satellite observations the condition of planet visibility for a given observatory, at a given time moment, is required. It is necessary to know definitely whether it is possible to observe or not.

The problems listed above lead to the necessity of calculations of ephemerides of natural satellites. The printed almanacs and the existing tables for this purpose are not convenient enough in modern conditions. In some cases, they are unsuitable.

Therefore, we have considered the problem of creating an ephemeris means which best satisfies all modern requirements. Some version of ephemerides software of natural satellites is suggested in the present report.

The main idea consists in the refusal of using the tables with beforehand calculated coordinates of a satellite, for a number of moments of time. The coordinates will be calculated for any moment and only when they will be required to inform an observer how the satellites are located in the telescope field of view and what are the conditions of visibility of a planet on the observatory. Therefore, the main ephemerides tool is a screen of a computer on which a planet and satellites are seen how they can be viewed through the telescope. A field of sight is displayed on a screen in a kind of window and around it the comments and menus are placed to control the "telescope". Pressing the keys, you can turn the model telescope to the left and to the right, upwards or downwards, as well as quickly change its magnification. Inscriptions beside the window show where the "telescope" is directed and what is the magnification. On the screen are given also: the time moment, elevation of a planet over the horizon and angular "depth" of the sun under the horizon. Position and phase of the moon are also given. Satellite coordinates can instantly be issued on the screen in digital form. A copy of the window can be printed together with the digital table of all calculated data. All this can be done for any given moment of time both in the past and in the future. The mutual phenomena of satellites are reproduced as they could be viewed through the telescope with the resolution of 0.005 seconds of arc per pixel. You can see as the disk of one satellite occults or eclipses the disk of the other.

The constructed tool comprises a program product ensuring the handling problems for natural satellites ephemerides. It includes the theories of planets motion and their satellites, as well as a facility of control and visualization of ephemeris data.

This idea of natural satellite ephemerides was already realized earlier in Bureau des Longitudes (the program SATELL11). We have known about this program after the first version of our program had been made. We have endeavored to make our software more advanced and convenient.

The created means permits to predict apparent approaches of the satellites at any given angular distance. Mutual occultations and eclipses of satellites are also automatically registered. The moments of the phenomena are recorded in the table. Subsequently, the program allows to take a moment of the phenomena from this table and reproduce the phenomenon. Eclipses can be demonstrated as they were viewed from the Sun.

At present, the moments of all mutual phenomena of the major satellites of Saturn in 1995 – 1996 and all mutual phenomena of Galilean satellites of Jupiter in 1997 – 1998 are calculated and saved in the files.

In this version, the motion of those satellites for which there exist good analytical theories is calculated with maximal accuracy. For the other satellites, simplified models of movement are accepted.

The existing version of the program is adapted for personal computers IBM PC and compatible. The program is transmitted free of charge to all astronomical institutions by inquiry to the author. It can be copied through the computer network Internet from our FTP-server with anonymous access. The designation of the software is MONS-EPH.

PRODUCTION D'ÉLÉMENTS ORBITAUX DE COMÈTES SUR PC

P. ROCHER ET C. CAVELIER
Bureau des longitudes
77, avenue Denfert-Rochereau
75014, Paris FRANCE

1. Introduction

La puissance de calcul des ordinateurs PC actuels permet leur utilisation pour des travaux qui étaient, jusqu'à ces dernières années, effectués exclusivement sur de gros ordinateurs. Le mode interactif utilisé pour la gestion et la saisie des données, facilement mis en œuvre sur les PC, permet un gain de temps important par rapport au travail classique en traitement par lot (batch). Nous présentons ici un logiciel PC, permettant le calcul et l'amélioration des éléments orbitaux des comètes. Une bonne connaissance de ces éléments est indispensable pour le calcul des éphémérides, notamment pour la redécouverte des comètes périodiques.

2. Description du logiciel

Le programme est écrit en Pascal 7.0 (Borland Pascal ®) et utilise un logiciel de gestion d'écrans et de menus déroulants Hyper-Screen 5.5 (PC Soft ®). Il fonctionne sur des ordinateurs PC (486 recommandé) ayant au moins 4Mo de RAM. Il occupe environ 15Mo sur le disque dur. La totalité du programme fonctionne en mode interactif, en utilisant des menus déroulants et des écrans de saisies et d'affichages, le tout étant géré à l'aide de la souris.

Ce programme permet donc le calcul et l'amélioration des éléments orbitaux des comètes à l'aide des observations. On peut distinguer trois phases importantes : la gestion des différentes bases de données liées au problème, le calcul proprement dit, et la sortie et la publication des résultats.

3. Gestion des bases de données

Ce programme utilise de nombreuses bases de données.

3.1. POUR LES OBSERVATIONS

On dispose de trois bases de données, une pour les observations des comètes périodiques (environ 28000 observations), une pour les comètes non périodiques (environ 18000 observations) et une pour les lieux d'observations (environ 530 sites). Ces différentes bases sont mises à jour régulièrement à l'aide des données publiées dans les *Minor Planet Circulars*. Le programme permet la gestion de ces bases, notamment le tri et la mise à jour des fichiers d'observations.

3.2. POUR LES ÉLÉMENTS ORBITAUX

Les éléments orbitaux sont sauvegardés sous deux formes distinctes : soit d'une part des fichiers ASCII contenant les éléments sous la forme de conditions initiales (vecteurs position et vitesse), soit d'autre part des bases de données « éléments cométaires » dans des fichiers séquentiels indexés utilisant la méthode des arbres B. La première forme est utilisée uniquement par ce programme alors que la seconde est utilisée par d'autres programmes, notamment pour le calcul et l'édition des éphémérides de comètes.

3.3. POUR LES ÉPHÉMÉRIDES DES PLANÈTES PERTURBATRICES

Pour le calcul des orbites, on tient compte de l'ensemble des perturbations planétaires et lunaire. Les éphémérides de ces corps sont issues des théories élaborées au Bureau des longitudes, VSOP82 (P. Bretagnon 1982) et TOP82 (J.L. Simon, 1983) pour les planètes et ELP-2000/82 (M. Chapront-Touzé et J. Chapront, 1983) pour la Lune. Les positions sont calculées à partir de représentations en séries de polynômes de Tchebycheff des éphémérides issues de ces théories (G. Francou *et al.*, 1983). Les éphémérides que nous utilisons couvrent un intervalle de temps de deux siècles ayant de 1850 à 2050.

4. Calcul et méthodes

4.1. ÉQUATIONS DU MOUVEMENT

Les équations du mouvement sont écrites dans le repère cartésien équatorial héliocentrique J2000. Pour le calcul de la trajectoire, on tient compte de l'ensemble des perturbations planétaires et lunaire, ainsi que des effets des forces de type non gravitationnel (forces dues au dégazage du noyau au voisinage du Soleil) lorsque cela est nécessaire.

4.2. INTÉGRATION NUMÉRIQUE ET COMPARAISON AUX OBSERVATIONS

Les observations sont des coordonnées équatoriales astrographiques topocentriques J2000. Les instants d'observations sont en Temps universel. On calcule pour chaque observation la position et la vitesse de la comète à l'aide d'une intégration numérique. On utilise pour cela la méthode d'intégration de Gragg-Bulirsh-Stoer (R. Bulirsh. et J. Stœr, 1980), il s'agit d'une méthode de type multi prédicteurs-correcteurs, la prédiction se fait par la méthode du point milieu et la correction par extrapolation en fractions rationnelles. En plus de la position et de la vitesse de la comète, on intègre également un système d'équations différentielles permettant de calculer en chaque point d'observation les dérivées des variables par rapport aux paramètres de l'intégration (conditions initiales et coefficients des forces non gravitationnelles). Ces dérivées sont utilisées par la suite pour l'amélioration de l'orbite.

4.3. AMÉLIORATION DES PARAMÈTRES

Selon que l'on utilise où non des forces non gravitationnelles, le nombre de paramètres à ajuster peut être de 6, 8 ou 9. On ajuste alors les positions et les vitesses initiales, ainsi que les coefficients des forces non gravitationnelles. Cet ajustement se fait par la méthode des moindres carrés, à partir des valeurs des O-C (différence entre les valeurs observées et les valeurs calculées). Deux drapeaux (flags) sont associés à chaque observation, un par coordonnée, ils peuvent être mis à bon ou mauvais suivant que l'observation est considérée comme bonne ou mauvaise en ascension droite ou en déclinaison. Ces drapeaux peuvent être modifiés manuellement à l'aide d'un menu déroulant, ou automatiquement à l'aide du test du χ^2. En même temps qu'il ajuste ces paramètres, le programme fournit les écarts types sur les ajustements à effectuer. L'amélioration de l'orbite se fait donc par une succession d'intégrations et d'ajustements, et l'on arrête lorsque les améliorations des paramètres deviennent négligeables devant leurs écarts types. Le programme permet également de calculer par la méthode des moindres carrés les constantes à utiliser dans le calcul des magnitudes (magnitude du noyau et magnitude totale).

5. Résultats

Après l'ajustement des éléments orbitaux, le programme permet de calculer, par intégration, de nouveaux éléments pour les époques correspondant aux passages au périhélie précédents et suivant. Ces éléments pourront être sauvegardés dans les bases de données « éléments cométaires » et être réutilisés par la suite pour le calcul d'éphémérides.

Les résultats, c'est-à-dire les éléments ajustés, la précision sur ces éléments, les écarts types et les moyennes des O-C, ainsi que les listes d'observations et d'O-C sont publiés et diffusés dans les *Notes Orbites Cométaires du Bureau des longitudes*. Le programme génère automatiquement l'édition de ces notes sur la forme de fichiers TeX.

Les notes sont accessibles sous la forme de fichiers Postscript sur le serveur Internet du Bureau des longitudes (ftp.bdl.fr ou www.bdl.fr), dans le répertoire /ftp/pub/ephem/comets/elements/french pour la version française et dans le répertoire /ftp/pub/ephem/comets/elements/english pour la version anglaise. Les éléments orbitaux calculés à l'aide de ce programme, sont utilisés pour la production des éphémérides de comètes publiées dans l'*Annuaire du Bureau des longitudes* et dans les *Notes scientifiques et techniques du Bureau des longitudes*, ils sont également utilisés dans le cadre du projet spatial ISO, pour produire les éphémérides des comètes retenues par ce projet.

Références

P. Bretagnon. (1982) Théorie du mouvement de l'ensemble des planètes. Solution VSOP82, *Astron. Astrophys.*, **Vol. no. 114**, p. 278–288.

R. Bulirsh. et J. Stœr. (1980), *Introduction to Numerical Analysis*, **Chap. no. 7**, New-York : Springer-Verlag

M. Chapront-Touzé et J. Chapront. (1983) The lunar ephemeris ELP2000, *Astron. Astrophys.*, **Vol. no. 124**, p. 50–62.

G. Francou. (1983) Nouvelles éphémérides du Soleil, de la Lune et des planètes, *Astron. Astrophys.*, **Vol. no. 128**, p. 124–139.

J.L. Simon. (1983) Théorie des quatre grosses planètes. Solution TOP82, *Astron. Astrophys.*, **Vol. no. 120**, p. 197–202.

LES DIAMÈTRES DU SOLEIL DANS LA CONNAISSANCE DES TEMPS DEPUIS 1795

MICHEL TOULMONDE
Observatoire de Paris (DANOF)

Pour calculer les éphémérides nécessaires à l'observation du Soleil, publiées dans la *Connaissance des Temps* (CdT), les astronomes du Bureau des longitudes ont utilisé depuis 200 ans différentes valeurs de son diamètre: à l'apogée ou à la distance moyenne de la Terre. Ces données, provenant des meilleures mesures les plus récentes, ont évolué avec la précision des techniques instrumentales.

1. Le Soleil dans la "Connaissance des Temps" au 19e siècle

En France, le 19e siècle commence pendant l'an IX de la République *"une et indivisible"*: à la fin du 18e siècle, le calendrier républicain devint d'usage officiel pendant un peu plus de 12 ans, du 6 octobre 1793 (15 vendémiaire an II) au 31 décembre 1805 (10 nivôse an XIV). L'Académie des Sciences est supprimée par la Convention (août 1793) et c'est la Commission temporaire des Poids et Mesures qui publie la *CdT pour 1795, "Connoissance des Temps, à l'usage des Navigateurs et des Astronomes pour l'année 1795. Du 12 Nivose de l'an 3 au 10 nivose de l'an 4 de l'Ère Républicaine."*

Le demi-diamètre solaire y est calculé à partir de la valeur de référence provenant de la mesure du diamètre apogée (31'31") par Lalande en 1764 [1]. Lalande, chargé des calculs, l'indique de deux façons: dans les pages mensuelles d'abord, tous les 6 jours (les 1, 7, 13, 19 et 25 du mois), puis à la fin de ces tables mensuelles, les 1 et 16 de chaque mois républicain. Les tableaux mensuels d'éphémérides donnent également la demi-durée de passage du Soleil au méridien (à 0,1 s près). Comme les mois républicains ont tous la même durée (30 jours), c'est la première fois que ces informations sont indiquées dans des tables à des *dates équidistantes* sur une année.

En 1795, la Convention crée le *Bureau des longitudes* (7 messidor an 3 ou 25 juin 1795 *"vieux style"*), lequel est alors chargé de rédiger la CdT "*plusieurs années à l'avance*" (comme le fait le Board of Longitudes anglais) ainsi qu'un "*Annuaire propre à régler ceux de toute la République*". Le premier volume de la CdT publié par le Bureau des longitudes est celui pour l'an IV (1795-1796), édité en septembre 1795. Concernant le diamètre solaire, il est indiqué (p. 178):

"*Le diamètre du Soleil est calculé de 6 en 6 jours sur les tables de Lalande à partir de la valeur apogée de 31'31" observée en 1764 avec un héliomètre de 18 pieds. A cause de l'irradiation, il faut diminuer de 6" les diamètres que l'on trouve ici dans les tables.*"

Le premier volume de l'Annuaire est celui de l'an V (1796-1797): "*Annuaire de la République française présenté au Corps Législatif par le Bureau des longitudes, pour l'année V de l'Ère française.*"

Ce petit livre de 40 pages (8 cm × 10 cm), version pratique et très simplifiée de la CdT, donne entre autres le diamètre du Soleil uniquement le 1 et le 16 de chaque mois républicain; les valeurs proviennent également de la mesure du diamètre apogée (31'31") par Lalande en 1764.

L'*Annuaire pour l'an IX* (1800-1801) donne le diamètre du Soleil "*en tems*" le 1 et le 16 de chaque mois. Les valeurs des durées de passage du Soleil au méridien indiquées dans l'*Annuaire pour l'an XII* (1803-1804) sont manifestement incorrectes (ce qui n'est pas le cas dans la CdT): elles décrivent une évolution sinusoïdale annuelle de cette durée ! Les valeurs pour l'an XIII, identiques à celles pour l'an XII, présentent le même défaut, corrigé dans l'*Annuaire pour l'an XIV* (1805-1806). Celui *pour 1808* donne les demi-diamètres du Soleil "*en tems*", le 1 et le 16 de chaque mois *grégorien*, le calendrier républicain ayant cessé d'être officiel le 31 décembre 1805.

La présentation et le contenu de la CdT vont continuer d'évoluer suivant les rédacteurs et les éditions. En ce qui concerne le Soleil, les éphémérides du diamètre (ou parfois du demi-diamètre) sont regroupées sur deux pages, à partir de l'édition pour 1838, et données de façon continue tous les 5 jours, depuis le 0 janvier, ou le 1 janvier selon les années. La valeur de référence reste le diamètre apogée mesuré par Lalande en 1764, correspondant à un demi-diamètre à la distance moyenne $R_1 = 961,4"$. Parfois, il est indiqué que cette valeur provient des tables de 1806 calculées par Delambre [2] lequel avait repris la valeur mesurée par Lalande.

En 1876, la valeur de référence devient $16'1,82"$ (ou $R_1 = 961,82"$), provenant d'observations effectuées à Greenwich de 1836 à 1847 sous la direction de G.-B. Airy [3]. Cette valeur est conservée jusqu'à l'édition pour 1895 (publiée en août 1892) dont l'avertissement indique (page III): "*Le demi-diamètre du Soleil à sa distance moyenne de la Terre a été adoptée = 16'1",82 ainsi qu'il résulte des observations faites à l'Observatoire de Greenwich, de 1836 à 1847, et sa parallaxe = 8",86. Pour le calcul des éclipses et des occultations, nous adoptons pour valeur du demi-diamètre moyen du Soleil 15′59″,63 (Auwers Astron. Nachrichten, n° 3068)."*

Voici donc 100 ans, cette année, qu'apparaît pour la première fois dans la CdT la valeur établie par Auwers [4] en 1891. En ce qui concerne le diamètre solaire, la CdT pour 1895 donne jour par jour à midi moyen la durée de passage du demi-diamètre (en temps sidéral et en temps moyen, en 0,01s), le demi-diamètre (en 0,01″), et le logarithme du rayon vecteur avec 7 décimales. Les diamètres sont calculés à partir de la valeur $R_1 = 961,82″$ en excès de 2,19″ sur la valeur d'Auwers (959,63″).

2. Le Soleil dans la "Connaissance des Temps" au 20e siècle

En 1911 se tient à l'Observatoire de Paris le *Congrès International des Éphémérides astronomiques* (23-26 octobre 1911). Concernant les dimensions du Soleil, la deuxième résolution indique: *"De même pour [unifier] le calcul des éclipses, on conservera la valeur 15′59″,63 (Auwers) du demi-diamètre apparent du Soleil qui est actuellement employée par toutes les éphémérides".*

Cette résolution est appliquée dès la publication en mars 1912 de la *CdT pour 1914*; on y lit (p. 12): *"Le demi-diamètre vaut 16′1″,18 (d'après Auwers) et 15′59″,63 pour le calcul des éclipses."*

La *CdT pour 1915* précise: *"Le demi-diamètre du Soleil est s = 15′59″,63 (Auwers), valeur adoptée pour le calcul des éclipses par la Conférence internationale des Éphémérides astronomiques. Pour le calcul des éphémérides, la Connaissance des Temps prend, d'après Auwers en tenant compte de l'irradiation s = 16′1″,18.*

Depuis, ces valeurs n'ont pas été modifiées et, pour chaque année jusqu'à 1979, la CdT donne, jour par jour à midi vrai de Greenwich, la durée de passage du demi-diamètre (en temps sidéral) et le demi-diamètre (en ′ et ″).

Mais depuis 1980, la CdT ne donne plus de table du diamètre du Soleil : la "Nouvelle série" fournit des coefficients pour calculer les coordonnées sous forme de développements en polynômes de Tchebychev. La distance du Soleil est ainsi calculable avec une précision inférieure à 1×10^{-6} UA.

Le système des constantes conventionnellement utilisées dans le calcul des éphémérides astronomiques, modifié en 1976 par la 16e assemblée générale de l'UAI (Grenoble), est en vigueur depuis 1984. Parmi les valeurs *"recommandées"* figure le rayon équatorial du Soleil qui doit être pris égal à 696000 km. Le Supplément à la CdT pour les *"Satellites galiléens de Jupiter, phénomènes et configurations"* indique cependant dans chacune de ses éditions annuelles: *"Le Soleil est une sphère de rayon 695980 km."*

L'Annuaire du Bureau des longitudes, utilise les mêmes valeurs de référence. Pour la table des diamètres solaires, jusqu'en 1911, la référence est

961,82″ (Airy 1855). Il est indiqué que la distance moyenne du Soleil est de 149 501 milliers de kilomètres. Dans l'édition *pour 1912*, il est écrit (p. 246): *"Le diamètre du Soleil est de 32′02″,36 [R_1 = 961, 18″] mais cette valeur est un peu trop grande à cause de l'irradiation, et le diamètre réel doit être réduit à 31′59″,26 [R_1 = 959, 63″].*"

L'information reste la même de 1912 à 1951, et dans l'édition *pour 1952*, seule la valeur 31′59,26″ est citée, et ce jusqu'en 1974. Depuis 1975, le diamètre de référence n'est plus indiqué dans l'*Annuaire*.

Pour la table des diamètres du Soleil dans l'*Annuaire*, les valeurs sont indiquées avec une périodicité de 15 jours (de 1796 à 1967), puis de 10 jours de 1968 à 1976, et enfin de 8 jours depuis 1977. La table donne le demi-diamètre (D/2) à 12 h UT et la distance du Soleil (d) à 0 h UT. On ne peut pas vérifier directement la constance du produit $K_1 = d.D/2$, les données n'étant pas établies pour le même instant; une interpolation est nécessaire, et on constate effectivement que K_1 vaut 961, 18″ correspondant à la somme 959, 63″ + 1, 55″ en tenant compte de l'irradiation pour les observations du disque solaire.

Il est à noter que cette valeur 1, 55″ utilisée pour l'irradiation est toujours référencée comme étant due à Auwers; toutefois, son article [4] indique l'écart 3, 14″ (double de 1, 57″) entre la valeur du diamètre apparent utilisée par le Berliner Jahrbuch et celle qu'il déduit des mesures qu'il a effectuées. L'écart cité n'est donc pas de 3, 10″ (double de 1, 55″). Peut-être s'agit-il d'un arrondi qui se perpétue ?

Deux valeurs du demi-diamètre solaire sont utilisées simultanément depuis un siècle: l'une pour le calcul des éclipses (959, 63″), l'autre pour les observations physiques du Soleil (961, 18″). L'évolution de ces valeurs de référence traduit l'amélioration des techniques d'observation, ainsi que les grandes difficultés à déterminer rigoureusement le bord solaire, dont les limites sont souvent vues de façons différentes selon les observateurs [5].

Références

Lalande J.J.(de), (1764), *Astronomie*.
Bureau des longitudes, 1806, *Tables astronomiques*.
Airy G.B., 1855, *Observations Made at the Royal Observatory Greenwich in the year 1853*, London.
Auwers A., 1891, "Der Sonnendurchmesser und der Venusdurchmesser nach den Beobachtungen an den Heliometern der deutschen Venus-Expeditionen.", *Astron. Nachrichten*, **3068**(128), 361.
Toulmonde M., 1995, *Étude comparative de diamètres solaires observés à partir d'instruments astrométriques*, Thèse de doctorat, Observatoire de Paris.

RADAR ASTROMETRY OF ASTEROIDS, COMETS AND PLANETARY SATELLITES

STEVEN J. OSTRO
Jet Propulsion Laboratory,
California Institute of Technology

Groundbased radar has considerable astrometric potential for asteroids, comets and natural satellites that enter the detectability windows of available telescopes. For very closely approaching asteroids and comets, measurements of the distribution of echo power in time delay (range) and Doppler frequency (radial velocity) can achieve a fractional precision between 10^{-5} and 10^{-9}, and consequently are invaluable for refining orbits and prediction ephemerides. Even for mainbelt asteroids and the more readily detectable planetary satellites whose orbits are very well known, radar can collapse range uncertainties from a few target radii to a few percent of a target radius, with direct implications for the navigation of spacecraft on flyby or rendezvous trajectories.

Radar refinement of orbits is tightly coupled to radar determination of physical properties. For this reason, almost every radar measurement that produces new information about a target's size, shape, rotation, or surface properties also provides an astrometrically useful measurement. The relative utility of the different kinds of information provided depends on the caliber of the radar data and the available prior information, as well as on the sophistication of analysis techniques.

The tally of radar-detected targets includes the Moon, Venus, Mercury, Mars, Saturn's rings, the Galilean satellites, Phobos, Titan, five comets, 37 mainbelt asteroids (MBAs), and 34 near-Earth asteroids (NEAs, almost all of which are Earth-crossers). The Arecibo and Goldstone radars are responsible for almost all this work. Ostro (1993) offers a review of planetary radar astronomy, including an outline of techniques and observational highlights. The terrestrial bodies have been subjected to delay-Doppler radar observations for decades. Whereas lunar radar time-delay measurements ("ranging") have become less useful than laser measurements, radar ranging to the inner planets still helps to maintain the accuracy of ephemerides

TABLE 1 – Residuals for optical (O) and optical+radar (OR) predictions of the sky positions of four Earth-crossing asteroids during the post-discovery apparition when they were recovered optically. The last column in the table demonstrates that radar astrometry produced a several-order-of-magnitude reduction in the sky area that one would have to search to achieve any given probability of recovering one of these objects. Each object's designation starts with the year of discovery.

Object	Recovery Date	Residuals O	OR	O/OR	(O/OR)^2
1989 PB	May 1990	24"	0.4"	60	3.6 ×10^3
1991 AQ	Sep 1994	57°	0.15°	380	1.4 ×10^5
1986 DA	Oct 1994	56"	0.9"	60	3.6 ×10^3
1991 JX	Mar 1995	3600"	4.6"	780	6.1 ×10^5

and in the case of Mercury is motivated also by long-term tests of gravitation theories. The first radar ranging to Ganymede and Callisto was reported by Harmon et al. (1994). Those authors' interpretation of the ranges used antiquated values for the satellite radii and has been redone by E. M. Standish (pers. comm.), who assess the consistency between the radar data and the most recent planetary and satellite ephemerides. Radar ranging has been carried out on 3 MBAs and 14 NEAs (in two cases during two different apparitions), but all observations of comets, Io, Phobos and Titan have been Doppler-only. The bulk of asteroid/comet radar astrometry was reported by Ostro et al. (1991), and has been incorporated in orbit estimates by Yeomans et al. (1992).

Radar astrometry of a near-Earth asteroid (NEA) during its discovery apparition can ensure its optical recovery (Yeomans et al. 1987). Observational experience has demonstrated that radar astrometry commonly improves upon the accuracy of optical-only ephemerides of newly discovered NEAs by one to five orders of magnitude (e.g., Table 1).

Radar can image NEAs if the echoes are strong enough, and if the orientational coverage of the images is adequate, inversion of the data by the methods introduced by R. S. Hudson (1993) can yield a three-dimensional model and an estimate of the delay-Doppler trajectory of the center of mass with very fine precision: a few decameters for work reported so far (Hudson and Ostro 1994, 1995) and a few meters for results of observations and/or analyses anticipated for the next few years. Radar aperture-synthesis observations of asteroids (currently possible only by transmitting from Goldstone and receiving with the Very Large Array) can measure absolute angular positions in the quasar reference frame at a level approaching 0.01 arc-

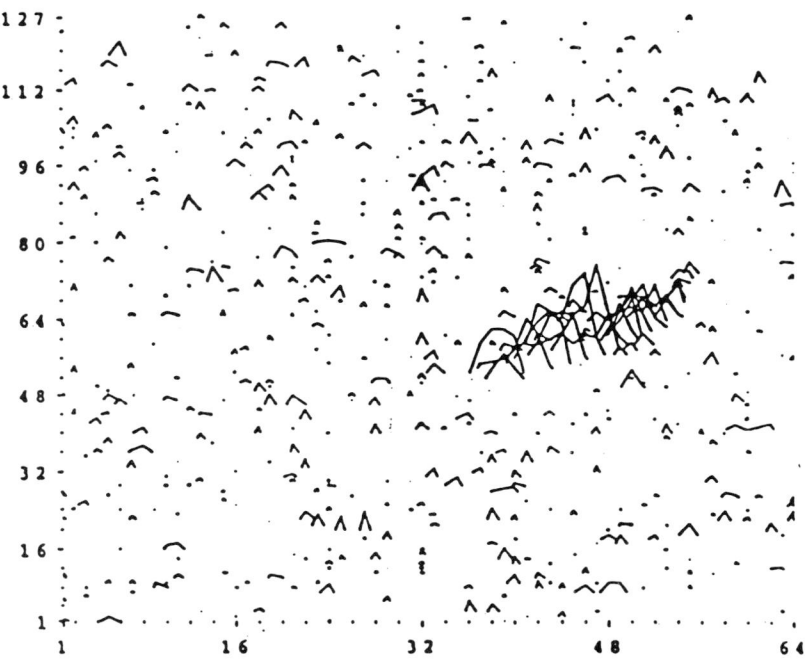

FIGURE 1 – Real-time display of a Goldstone radar image of asteroid 1620 Geographos. Power above half the peak value is plotted for 64 Doppler-frequency cells and 127 time-delay cells. The delay-Doppler resolution (1 μs by 2.9 Hz) corresponds to spatial resolution of 150 m by 151 m; prior knowledge of the spin vector from optical lightcurves indicated that our view was equatorial and set the Hz-to-m conversion. Range increases toward the top of this figure and frequency (or radial velocity) increases toward the right. The asteroid, rotating clockwise, is illuminated from below. The "leading edge" of the asteroid is seen prominently in this display. The asteroid's orientation is between end-on and broadside, and the spatial extent of the echo is ∼ 4.7 km. The total integration time is 46 s. The data were obtained less than one hour after the beginning of observations on 1994 Aug. 28, the first day of a one-week experiment. Analysis of radar movies from Aug. 28-29 indicates that at the beginning of the Aug. 28 imaging the correction to the range-prediction ephemeris was -10.9 ± 0.3 km and was becoming more positive at a rate of about 0.16 km/h. (Here "range" is the distance from the asteroid's center of mass to a reference point on the Goldstone 70-m antenna.) See Ostro et al. (1996) and Ostro, Rosema et al. (1995).

TABLE 2 – Geographos radar astrometry reported by Ostro, Rosema et al. (1995). Goldstone (8510 MHz) time delays and Doppler frequencies correspond to hypothetical echoes from Geographos' center of mass received at the intersection of the azimuthal and elevation axes of the 70-m antenna, DSS-14. Postfit residuals are with respect to an orbit calculated by D. K. Yeomans and J.D.Giorgini and reported in the same paper.

UTC epoch of echo reception (hh:mm)	Time delay (UTC μs) Estimate	Residual	Doppler frequency (Hz) Estimate	Residual
1994 08 28 07:10	38936537.06±1.76	-0.91	-364880.6±0.9	-0.7
1994 08 29 04:40	42596456.40±1.13	0.33	-427851.0±0.2	-0.1
1994 09 02 07:20			-615553.2±0.5	0.1

sec, complementing delay-Doppler measurements (P. Palmer, unpublished results).

With adequate radar support, it would be possible for a spacecraft lacking onboard optical navigation to be guided into orbit around, or collision course with, an asteroid. For example, consider how Goldstone observations would have shrunk the positional error ellipsoid of Geographos just prior to a Clementine flyby of that target on Aug. 31, 1994. Before the Goldstone observations, the error ellipsoid's typical overall dimension was ~11 km. Ranging on Aug. 28-29 (e.g., Fig. 1) and a preliminary shape reconstruction collapsed the ellipsoid's size along the line of sight to several hundred meters, so its projection toward Clementine on its inbound leg would have been 11 × 2 km. Goldstone-VLA radar aperture synthesis could have shrunk the error ellipsoid's longest dimension to about 1 km, about half of Geographos's shortest overall dimension. Table 2 gives radar astrometry from analysis of a low-resolution subset of the Geographos data (Ostro et al. 1996).

The imaging of 4179 Toutatis by Goldstone and Arecibo in 1992 was most elaborate asteroid radar experiment reported so far (Ostro, Hudson et al. 1995; Ostro and Hudson 1995). Inversion of a subset of the images yielded center-of-mass astrometry for which the estimation uncertainties and the postfit residuals in the orbit solution are typically a few mm/s in radial velocity and a few decameters in range. As noted by Yeomans and Chodas (1994), Toutatis will pass 0.010360AU (4 lunar distances) from Earth on 2004 Sep 29.56711 UTC, the closest approach predicted for any asteroid or comet between now and 2060. Inclusion of the recently reported radar astrometry reduces the uncertainty in the miss distance to 40 km.

In an asteroid radar imaging experiment, an error in the Doppler prediction ephemeris causes motion of the asteroid with respect to the predicted delay trajectory, so images will be smeared in delay. Hence the achievable delay resolution is limited by the accuracy of the Doppler prediction ephemeris. A standard strategy is to first obtain echoes using a coarse delay resolution, fold the resultant astrometry into a new orbit solution, generate a refined prediction ephemeris, and then image the object with a finer delay resolution. This iterative process continues until the motion of the target through the delay-prediction ephemeris is negligible, i.e., until it does not compromise the finest delay resolution achievable in light of the available echo strength (typically ~100 nanoseconds for modern radar systems). The cycle time for updating delay-Doppler prediction ephemerides has recently been shrunk dramatically by D. K. Yeomans and J. D. Giorgini's JPL On-site Orbit Determination Program (OSOD). Installed at Goldstone in 1994, OSOD contributed significantly to the most recent NEA imaging experiment, on 1991 JX in June 1995. We were able to cycle through three generations of orbit solutions before the object's closest approach, settling on an ephemeris that turned out to be accurate to 0.3 mm/s in radial velocity (0.02 Hz at the Goldstone frequency of 8510 MHz).

NEA radar opportunities will expand significantly upon completion of upgrades in the Arecibo telescope, which by 1996 should be producing thousand-pixel images of several NEAs annually. A dedicated optical search program (the so-called Spaceguard Survey) could discover some 100,000 NEAs, most of which could, in principle, be studied with groundbased radar at least once every few decades. However, Arecibo and Goldstone are already heavily oversubscribed, so observation of more than a small fraction of the objects discoverable in proposed optical surveys will require dedicated radar telescopes.

Acknowledgements

This research was conducted at the Jet Propulsion Laboratory, California Institute of Technology, under contract with the National Aeronautics and Space Administration.

References

de Pater, I., Palmer, P., Mitchell, D. L., Ostro, S.J., Yeomans, D. K. and Snyder, L. E. (1994), "Radar Aperture Synthesis Observations of Asteroids". *Icarus* **111**, 489-502.

Harmon, J. K., Ostro, S. J., Chandler, J. F. and Hudson, R. S.(1994), "Radar Ranging to Ganymede and Callisto". *Astron. J.* **107**, 1175-1181.

Hudson, R. S. (1993), "Three-dimensional reconstruction of asteroids from radar observations". *Remote Sensing Revs* **8**, 195-203.

Hudson, R. S. and Ostro, S. J. (1994), "Shape of Asteroid 4769 Castalia (1989 PB) from Inversion of Radar Images". *Science* **263**, 940-943.

Hudson, R. S. and Ostro, S. J. (1995), "Shape and Non-Principal-Axis Spin State of Asteroid 4179 Toutatis". *Science* **270**, 84-86.

Ostro, S. J., Campbell, D. B., Chandler, J. F., Shapiro, I. I., Hine, A. A., Velez, R., Jurgens, R. F., Rosema, K. D., Winkler, R. and Yeomans, D. K. (1991), "Asteroid Radar Astrometry". *Astron. J.* **102**,1490-1502.

Ostro, S. J. (1993), "Planetary Radar Astronomy". *Reviews of Modern Physics* **65**, 1235-1279.

Ostro, S. J. (1994), "The Role of Groundbased Radar in Near-Earth Object Hazard Identification and Mitigation" In *Hazards Due to Comets and Asteroids* (T. Gehrels and M. S. Matthews, eds.), Univ.of Arizona Press, pp. 259-282.

Ostro, S. J., Rosema, K. D., Hudson, R. S., Jurgens, R. F., Giorgini, J. D., Winkler, R., Yeomans, D. K., Choate, D., Rose, R., Slade, M. A., Howard, S. D. and Mitchell, D. L. (1995), "Extreme Elongation of Asteroid 1620 Geographos from Radar Images". *Nature* **375**, 474-477.

Ostro, S. J., Hudson, R. S., Jurgens, R. F., Rosema, K. D., Winkler, R., Howard, D., Rose, R., Slade, M. A., Yeomans, D. K., Giorgini, J. D., Campbell, D. B., Perillat, P., Chandler, J. F. and Shapiro, I. I. (1995), "Radar Images of Asteroid 4179 Toutatis". *Science* **270**, 80-83.

Ostro, S. J., Jurgens, R. F., Rosema, K. D., Hudson, R. S., Giorgini, J. D., Winkler, R., Yeomans, D. K., Choate, D., Rose, R., Slade, M. A., Howard, S. D., Scheeres, D. J. and Mitchell, D. L. (1996), "Radar Observations of Geographos". *Icarus*, in press.

Yeomans, D. K., Ostro, S. J. and Chodas, P. W. (1987), "Radar Astrometry of Near-Earth Asteroids". *Astron. J.* **94**, 189-200.

Yeomans, D. K., Chodas, P. W., Keesey, M. S., Ostro, S. J., Chandler, J. F. and Shapiro, I. I. (1992), "Asteroid and Comet Orbits using Radar Data". *Astron. J.* **103**, 303-317.

Yeomans, D. K. and Chodas, P. W. (1994), "Predicting Close Approaches of Asteroids and Comets to Earth". In *Hazards Due to Comets and Asteroids* (T.Gehrels and M. S. Matthews, eds.), Univ.of Arizona Press, pp. 241-258.

Appendix

History of Asteroid Radar Detections (Table)

Observatory abbreviations (and wavelengths unless otherwise indicated)

[A]   Arecibo (13cm);
[H]   Haystack (3.8cm);
[G]   Goldstone (3.5cm);
[GV]   Goldstone-VLA (3.5cm);
[EF]   Evpatoria-Effelsberg (6cm);
[GE]   Goldstone-Evpatoria (3.5cm);
[GK]   Goldstone-Kashima (3.5cm).

(For multi-observatory experiments, the order in which observatories detected echoes is indicated).

<u>Underlines</u> identify single-polarization experiments.

^{2,3,4,5,6} means second, third, etc. apparition yielding radar detection.

History of Asteroid Radar Detections

Year	Mainbelt	Ref.	Near-Earth	Ref.
1968			1566 Icarus	1 [G], 2 [H]
1972			1685 Toro	3 [G], λ13]
1975			433 Eros	4 [G, λ3.5, λ13], 5 [A, λ70]
1976			1580 Betulia	6 [A]
1977	1 Ceres	7 [A]		
1979	4 Vesta	8 [A]		
1980	7 Iris	9,26 [A]	1685 Toro^2	10 [A]
	16 Psyche	9 [A]	1862 Apollo	20 [A], 12 [G]
1981	97 Klotho	9 [A]	1915 Quetzalcoatl	20 [A]
	4 Vesta^2	9 [A]	2100 Ra-Shalom	13 [A]
	8 Flora	9 [A]		
1982	2 Pallas	9 [A]		
	12 Victoria	9,26 [A]		
	19 Fortuna	9 [A]		
	46 Hestia	9 [A]		
1983	5 Astraea	9 [A]	1620 Geographos	20 [A]
	139 Juewa	9 [A]	2201 Oljato	20 [A]
	356 Liguria	9 [A]		
	80 Sappho	9 [A]		
	694 Ekard	9 [A]		
1984	9 Metis	9,26 [A]	2101 Adonis	20 [A]
	554 Peraga	9 [A]	2100 Ra-Shalom^2	20 [A]
	144 Vibilia	9 [A]		
	1 Ceres^2	9 [A]		
	7 Iris^2	14,26 [A]		
1985	6 Hebe	9 [A]	1627 Ivar	18 [A]
	41 Daphne	9 [A]	1036 Ganymed	20 [A]
	21 Lutetia	15 [A]	1866 Sisyphus	20 [A]
	33 Polymymnia	15 [A]		
	84 Klio	15 [A]		
	192 Nausikaa	15 [A]		
	230 Athamantis	15 [A]		
	216 Kleopatra	15,26 [A]		
	18 Melpomene	15 [A]		
	16 Psyche^2	* [A]		
1986	1 Ceres^3	* [A]	6178 (1986 DA)	11 [A]
	393 Lampetia	15 [A]	1986 JK	16 [G]
	27 Euterpe	14 [A]	3103 Eger	20 [A]
	19 Fortuna^2	14 [A]	3199 Nefertiti	20 [A]
	9 Metis^2	14,26 [A]		

Year	Mainbelt	Ref.	Near-Earth	Ref.
1987	5 Astraea[2]	14 [A]	1981 Midas	20 [G]
	532 Herculina	14 [A]	3757 (1982 XB)	20 [A]
	2 Pallas[2]	* [A]		
	20 Massalia	14 [A]		
1988	654 Zelinda	20,26 [A]	1685 Toro[3]	20 [A]
	4 Vesta[3]	14 [A]	3908 (1980 PA)	20 [A,G]
	105 Artemis	20 [A]	433 Eros[2]	20 [A]
1989	12 Victoria[2]	26 [A]	4034 1986 PA	20 [A]
			1580 Betulia[2]	20 [G,A]
			1989 JA	20 [A,G]
			4769 Castalia	19 [A,G]
			1917 Cuyo	20 [A,G]
1990	78 Diana	* [A]	1990 MF	20 [A,G]
	1 Ceres[4]	* [A]	1990 OS	20 [G]
	194 Prokne	24 [G]	4544 Xanthus	20 [A]
1991	2 Pallas[3]	* [A]	1991 AQ	21 [A]
	7 Iris[3]	26[G],17[GV]	6489 1991 JX	21 [A,G]
	324 Bamberga	*[A,G],17[GV]	3103 Eger[2]	24 [G]
	796 Sarita	* [A]	1991 EE	*[A],17[GV]
1992	4 Vesta[4]	*[A],24[G]	1981 Midas[2]	24 [G]
			5189 (1990 UQ)	24 [G]
			4179 Toutatis	22[G,A],23[VE], 17 [GV]
1994	97 Klotho[2]	24 [G]	4953 (1990 MU)	24 [G]
			1620 Geographos[2]	* [G]
1995	1 Ceres[5]	** [G]	2062 Aten	* [G]
	18 Melpomene[2]	** [G]	6489 1991 JX[2]	*[G,GV], ***[GE],****[GK]
	7 Iris[4]	** [G]		
Totals:	37 Mainbelt	+	34 Near-Earth	= 71 total

References [1] Goldstein (1969), *Icarus* **10**, 430; Goldstein (1969), *Science* **162**, 903. [2] Pettengill et al (1969), *Icarus* **10**, 432. [3] Goldstein et al (1973), *Astron. J.* **78**, 508. [4] Jurgens and Goldstein (1976), *Icarus* **28**, 1. [5] Campbell et al (1976), *Icarus* **28**, 17. [6] Pettengill et al (1979), *Icarus* **40**, 350. [7] Ostro et al (1979), *Icarus* **40**, 355. [8] Ostro et al (1980), *Icarus* **43**, 169. [9] Ostro et al (1985), *Science* **224**, 442. [10] Ostro et al (1983), *Astron. J.* **88**, 565. [11] Ostro et al (1991), *Science* **252**, 1399. [12] Goldstein et al (1981), *Icarus* **48**, 59. [13] Ostro et al (1984), *Icarus* **60**, 391. [14] Ostro et al (1988), *Bull. Amer. Astron. Soc.* **20**, 863 (abstract). [15] Ostro et al (1986), *Bull. Amer. Astron. Soc.* **18**, 796 (abstract). [16] Ostro et al (1989), *Icarus* **78**, 382. [17] de Pater et al (1994), *Icarus* **111**, 489. [18] Ostro et al (1990), *Astron. J.* **99**, 2012. [19] Ostro et al (1990), *Science* **248**, 1523. [20] Ostro et al (1991), *Astron. J.* **102**, 1490. [21] Ostro et al (1991), *Bull. Amer. Astron. Soc.* **23**, 1144 (abstract). [22] Ostro et al (1995), *Science* **270**, 80. [23] Zaitsev et al (1993), *Radiotekhnika Elektronika* **38**, 1842-1850. [24] Ostro et al (1994), *Bull. Amer. Astron. Soc.* **26**, 1165 (abstract). [25] Ostro et al (1995), *Nature* **375**, 474. [26] Mitchell et al (1995), *Icarus* **118**, 105-131. * Ostro et al, unpublished. ** Mitchell et al, unpublished. *** Zaitsev et al, unpublished. **** Koyama et al, unpublished.

LONG-FOCUS CCD ASTROMETRY OF PLANETARY SATELLITES

D.PASCU
U.S.Naval Observatory

Abstract.
CCD detectors are rapidly replacing the photographic plate and photomultiplier in satellite observations used for orbital improvement. This includes both phenomena timings as well as tangent plane astrometry. In most cases this change has been for the better, but in some areas there has been no gain – even a loss. We will review this change in terms of the recent history of satellite observations. The impact of the CCD will be discussed in terms of its applications, and the increase in precision it affords. Finally, a few things will be said about future directions, especially about spin-off applications.

1. RECENT HISTORY

The modern period of astrometric satellite observing began about 1965, anticipating the spacecraft reconnaissance of the outer planets, but apparently not motivated by it. Rather, it resulted directly from the need to improve the badly out-of-date ephemerides published in the national almanacs. Photographic observing programs were begun, principally at the Naval Observatory in the United States, and at the Pulkova Observatory in the Soviet Union, with an emphasis on the Martian satellites (because of an interest in the secular acceleration of Phobos) and the Galilean moons of Jupiter.

It became clear to the planners of NASA's Grand Tour of the outer planetary systems that the accuracy of the existing satellite ephemerides was incompatible with the more severe requirements of space reconnaissance. The observational effort was greatly expanded, driven largely by

NASA funding (Seidelmann 1977, 1979). This period of expansion lasted throughout the Voyager years, 1973 - 1989. While photographic techniques predominated (Pascu 1977, 1979), this period saw the introduction of several new observational methods – in particular, mutual phenomena techniques (Millis 1974; Aksnes and Franklin 1975), space techniques (Born and Duxbury 1975; Duxbury and Callahan 1988, 1989), and CCD techniques (Pascu et al. 1983, 1987).

At present, space reconnaissance is still the major driver of satellite observation, but there is also a broader component to this demand. Much of it is fueled by an interest in the physical study of the satellites using the Hubble Space Telescope (HST), and more accurate ephemerides are required for such observations. An added complication is that many of these satellites were discovered by the Voyager spacecrafts and cannot be observed from the ground. Consequently, the astrometry of these satellites must also be done with HST. CCD techniques are now in general use, but a number of promising techniques have appeared in the past few years. HST has already been applied to the astrometry of the faint satellites of Pluto (Null et al. 1993), Uranus (Zellner et al. 1994; Currie et al. 1994; Pascu et al. 1995), and Saturn (Bosh and Rivkin 1995; Showalter et al. 1995). Infrared techniques have proved useful for ground-based observations of the inner systems of Jupiter and Saturn (Nicholson and Matthews 1991; Nicholson et al. 1992), and may have some limited application to the brightest of the inner satellites of Uranus and Neptune. The radar technique (Campbell et al. 1978) and the Very Large Array technique (Muhleman et al. 1986) are beyond the scope of this review, but are mentioned because they produce highly accurate positions.

2. OBSERVATION TYPES

CCDs are applied to two observation types – astrometric and phenomena. Astrometric observation refers to the measurement of relative coordinates, rectangular or polar, in the tangent plane. The measurements can be intersatellite (one satellite referred to another) or with respect to the planet. The simple observations themselves have generally been reported, rather than compound observations such as normal points. This is the most common observation type since it can be applied to all satellites, at all apparitions. It has the additional advantage that a full complement of orbital parameters can be determined, in particular, the semi- major axis from which the system mass is obtained. A recent example of this is the Pluto-Charon mass (Tholen and Buie 1995). The major drawback to this observation type is its need for large instruments, both in aperture and focal length.

Phenomena observation refers to the measurement of the epoch of an event, such as an eclipse or occultation of a satellite. This event can be due to the planet or it can be due to another satellite (a mutual event). The event can be measured photometrically or geometrically (astrometrically) in the case of a mutual occultation. The "epoch" refers to some identifiable part of the event, such as the instant of minimum light, or central eclipse for a photometric event, or the instant of minimum separation in the case of an astrometrically measured mutual occultation. This epoch is sometimes referred to as the "mid-event time." Photometric phenomena observations have the advantage that they can be made with relatively small telescopes (Mallama 1992a,b) and are independent of instrumental parameters. It is for these reasons that relatively precise phenomena observations of the Galilean satellites, due to Jupiter, have been made for more than three centuries. Another notable success of the phenomena technique was the determination of the orbital motions in the Pluto-Charon system from observations of their mutual phenomena during the late 1980s. For the Galilean moons however, phenomena observations due to Jupiter are not of the positional accuracy of the mutual phenomena or the astrometric observations (Lieske 1995b). The principal drawback of the mutual phenomena technique is the rarity of events for the great majority of satellites. In addition, there is a limitation on the precision attainable for the orbital orientation parameters, and the orbital scale - the semi-major axis - can not be determined at all. While the mutual phenomena technique provided the means for analyzing the motions in the Pluto-Charon system, it was the astrometric technique, applied with the HST, which produced an accurate scale for the system, and thus an accurate system mass (Tholen and Buie 1995).

3. IMPACT OF CCDs

The CCD has had a favorable impact on both astrometric and phenomena observations. The high quantum efficiency (70 percent) benefitted the astrometric technique the most. Now, even twentieth magnitude satellites can be imaged without differential guiding, and a higher signal-to-noise ratio means higher astrometric precision for the faint moons. For the mutual phenomena technique, also, the high quantum efficiency means that the technique can be applied to fainter satellites, as in the Saturnian system.

Linear response and area photometry are the two features which make the CCD especially applicable to the observation of the photometric mutual phenomena; both phenomena satellites and reference objects can be recorded simultaneously. And, for both phenomena observation and astrometry, one can account for the scattered light from the primary more precisely.

But, the CCD also has several limitations: The small size of the CCD chip makes it difficult to calibrate scale, and especially the orientation. Even in small standard fields (cluster fields), relative star positions cannot be measured with enough precision to compete with a field which is linearly an order of magnitude greater (on a photographic plate). The reason that orientation is more difficult to calibrate, is that the orientation of the chip must be determined at the point of observation, or else the polar axis alignment must be calibrated.

The small size of the chip also makes it unlikely that enough reference stars will appear on the frame to provide spherical equatorial coordinates (R.A., Dec.) for objects in the field. For some applications, the size problem can be solved with a mosaic of CCDs, but other problems, which require a single chip, must await the arrival of the large 8-cm chips which are expected in a few years (Harris 1995).

Since the CCD chip is not only the detector but the measuring machine as well, the astrometric integrity of the chip is a major consideration. The astrometric integrity of the chip has at least four components, three structural and the fourth electronic. Its flatness, the orthogonality of its rows and columns, and the thickness and spacing of the rows and columns are the structural components. Many of the early chips were crinkled or bowed, while others were rhomboidal (Harris 1995). The electronic component has to do with image form factors. If the image is distorted by blocked columns, or charge transfer inefficiency, for example, the centroid will not represent its position well. Because of charge transfer inefficiency, the Mark IV TI 800 X 800 CCD, prototype of the HST CCDs (now in use at the USNO), is used well below full well for the highest astrometric precision (Dahn 1994). If a CCD is to be used for astrometry, it should be constructed and calibrated in the same manner as were the metrics of the modern plate measuring machines. While most of these problems can be solved, they need to be addressed in the specifications and construction of the CCD.

The photometric integrity of the CCD is also a consideration. While it also has several components, the most important for photometric observations is the linearity. Few CCDs are linear over their entire dynamic range and need to be calibrated.

Brightness attenuation and magnitude compensation are problematic with the CCD. The inaccessibility of the focal plane (chip) makes it difficult to reduce the brightness of the planet without reimaging. An alternative is a prefocal stop, which limits the distance from the planet at which a satellite can be observed. Another is placing the attenuator inside the dewar, either directly on the CCD chip or on a pellicle placed immediately before the chip.

4. APPLICATIONS AND TECHNIQUES

4.1. PHENOMENA

Observations of phenomena due to the planet refer to eclipses of the Galilean satellites by Jupiter (the eclipse/occultation events of Pluto and Charon are more properly classified as mutual events). As mentioned above, positional information derived from the Jupiter events are no longer competitive in accuracy with that from mutual phenomena or astrometric observations. This is not a calibrational problem, but due to an inability to suitably model the event. Although Mallama (1991, 1992b) is making some progress in this area, he suggests that the future value of such observations appears to be in deriving information about the atmosphere of Jupiter itself (Mallama 1991, 1995).

Observations of the photometric mutual phenomena, however, are an entirely different matter. Since their first modern observation in 1973 it was recognized that this observation type would yield highly accurate positional data (Brinkmann and Millis 1973). The mutual events of the Galileans have been observed photometrically every six years, beginning in 1973. Since 1979 the Bureau des Longitudes has organized these observations as international campaigns involving many observatories (Arlot et al. 1982; Arlot and Thuillot 1988; Arlot et al. 1990). The next series of events will occur in 1996, and plans are already underway for a campaign to observe those (Arlot 1995). With the widespread use of CCDs at those events, it will be possible to measure the events astrometrically as well as photometrically. Astrometric phenomena were first mentioned by Arlot (1982) and attempted by Arlot et al. (1982) at the 1979 mutual events with promising results.

Likewise, the serendipitous discovery of Charon by Christy (1978), and the fortuitous recognition by Andersson (1978) that mutual phenomena were imminent, led to six years of observations of Pluto- Charon (photometric) events which were invaluable to the determination of physical parameters of Pluto and Charon, and to the study of the motions in that system. The Pluto-Charon events ended in 1990 and will not occur again for more than a century. Buratti et al. (1995) describe the CCD observations of these events made with the Palomar 60-inch telescope, and give references for most of the others.

During 1995, activity will be focussed on the mutual events of the Saturnian satellites which occur every 15 years. These were last observed at the ring-plane crossing in 1980 (Dourneau 1982; Aksnes et al. 1984). The Bureau des Longitudes organized an international campaign to observe these events in 1995 (Arlot and Thuillot 1993). These phenomena are more difficult to observe than the Galilean events, but the introduction of CCD detectors should ensure their success. While CCDs were used for the Pluto-Charon events as early as 1985 (see Buratti et al. 1995), they were not applied to the Galilean satellites until the 1990-91 mutual events (Colas 1991; Mallama 1992a; Le Campion et al. 1992).

Calibration of the CCD, for photometric use, involves bias subtraction (zero point of pixel counters), flatfielding (correction for variation in pixel sensitivity) and determination of linearity (finding the range of digital counts in the pixels in which these counts are linearly proportional to the incident light). Dark current is not significant for cooled CCDs and short exposures, but for the infrared devices, dark current must also be taken into account. Bias subtraction is fairly straightforward, but flatfielding is not, and its effects on phenomena observations has not been investigated. For the bright Galilean satellites events, this is probably not a problem, but it may be for the total mutual eclipses of the Saturnian events. Because of the large dynamic range of the total eclipses in the Saturnian system, calibration of the linearity of the CCD will also be important. For small dynamic range events, this is not a major issue.

4.2. ASTROMETRY

Astrometric CCD observation of the planetary satellites is widespread, with major programs in the United States (Pascu et al. 1983, 1987, 1992b, 1995; Pascu 1994b; Monet and Monet 1992, A. Monet 1993; Nicholson and Matthews 1991; Nicholson et al. 1992; Rohde and Pascu 1993, 1994), France (Colas and Arlot 1991; Colas 1992, 1994, 1995; Arlot et al. 1989, 1994; Rapaport 1994; Viateau and Rapaport 1995; Le Floch 1994), and Great Britain (Beurle et al. 1993; Harper 1994, 1995; Jones 1995), and smaller programs in Brazil (Vieira Martins and Veiga 1995), Spain (Lopez-Garcia 1994), and Russia (Zamarashkin et al. 1994). Since CCD astrometry is now applied to all of the satellites, the observational techniques of detection and calibration are more varied than for the phenomena.

The most difficult problem of detection is that of a faint satellite near a bright primary. A Lyot-type coronagraph appears to be the best solution to the problem (Baum et al. 1981; Pascu et al. 1983) because it eliminates the diffracted light and makes it possible to attenuate the light of the primary as well. While apodizing diaphragms can be designed to remove the diffraction spikes caused by the secondary supports, the diffraction caused by the entrance pupil is often increased. If prefocal stops are used to attenuate the light of the primary, they must be placed very close to the CCD chip or they will limit the distance from the planet that a satellite can be observed astrometrically (Vieira Martins and Veiga 1995). The most heroic detection was that of Proteus by Colas and Buil (1992). The technique is promising but needs much work to make it astrometrically viable. Special techniques are also needed for satellites which are too bright. Astrometric CCD observations of the Galilean satellites made at the U.S. Naval Observatory use dark filters to lengthen the exposure times to a few seconds to help average the effects of seeing. Additional averaging is accomplished by combining numerous frames into normal points (Monet and Monet 1992; A. Monet 1993; Owen 1995).

Assuming that one has an astrometric grade CCD, or one that is adequately mapped and calibrated, the three remaining calibrational problems are scale, orientation and coordinate origin determination (Pascu 1977). Since CCDs are small and star catalogs have a too low density, plate constants are unlikely until the large chips and denser catalogs become available. This means that some version of the trail/scale method must be used – although the diminutive size of the CCD is less than ideal for accurate results. The focal plane scale has been determined in several ways. The best general way is to use a small scale field in a cluster. Astrometric sequences in M15, M92 and Pleiades are most commonly used. The problem is that astrometric standard fields are not available for all parts of the sky. Veiga and Vieira Martins (1994) used the motion of Uranus to determine both scale and orientation. If one is interested primarily in faint, inner satellites, for which the CCD is best adapted, then one may use the brighter satellites of the system to calibrate the CCD frames. The best example of this is the technique of Veillet and Ratier (1980) who used the ephemeris positions of the four bright moons of Uranus to implement a plate constants solution for scale, orientation and coordinate origin to obtain positions for Miranda. The same method was applied successfully with a CCD by Pascu et al. (1987). While the configuration around Uranus is unique, both the Jovian and Saturnian systems have at least two well-spaced bright satellites whose relative ephemeris positions are accurate to 30 mas or better. Such configurations will give scale, orientation and coordinate origin. The use of wide double star pairs is not suitable for accurate scale or orientation calibrations.

Calibration of CCD orientation is more problematic, as explained in Section 3. The most general method for calibration is to record star trails at the same declination as the planet. While it is understood that the short lengths of the trails will fail to produce the accuracy of the longer trails on photographic plates, this disadvantage is reduced by the sensitivity of the CCD, which permits one to record many trails by simply stopping the drive. Some observers have used the scale field clusters to calibrate both scale and orientation. This will work only if the cluster and planet are at the same declination. Otherwise, a systematic error will result due to polar axis misalignment. Jones (1995) has attempted to calibrate this misalignment so that orientation may be calibrated in scale fields as well. As for scale calibration, the brighter moons can also be used to calibrate orientation, and coordinate origin if only the fainter moons are required.

Coordinate origin (reference object) determination is in most cases simple, but in a few instances it is quite involved. For the great majority of the satellites, only intersatellite observations are feasible. And for most satellites, there is at least one other satellite of comparable brightness, or at least within the spatial or dynamic range of the CCD (about 6 magnitudes), to provide a reference zero-point. The exceptions are: the outer moons of Jupiter and Saturn, the Tethys and Dione Lagrange librators, Triton and Nereid. The solution for the outer satellites is to image them only when three or more HST Guide Star Catalog (GSC) stars are in the field (about ten arcmin across). The R.A. and Dec. obtained in this way can be compared to the ephemeris position of the planet. While the GSC star positions are not precise enough to calibrate the scale and orientation, the positions derived in this way are as good or better than the photographic because the images are not trailed in the shorter CCD exposures. For the remainder of the problem satellites, a neutral density filter can solve the problem if a coronagraph is used in the observations. With or without a coronagraph, the method suggested by Pascu et al. (1983) can be used successfully. In that method, shorter exposures of the brighter satellites are taken alternately with the longer exposures of the faint satellites. The motions on the CCD frames (in pixel units) of the bright satellites, as a function of time, are determined from these short exposures. The positions of these bright moons can then be interpolated from these functions for the mean times of the long exposures. This gives, finally, the positions of the faint satellites relative to the bright ones on the long exposure frames.

5. ACCURACY

There is no issue in satellite astrometry more contentious than the claims of observational accuracy among the various techniques. Much of this is due to the mistaken comparison of dissimilar quantities, such as the error of a single observation with that of a normal point or other compound quantity. Another invalid comparison is the internal precision of one method with the external precision of another. A third problem is the failure to take into consideration other technique-independent sources of astrometric error, such as the increase in error with the angular separation of the measured bodies (Pascu et al. 1991; Pascu 1994a). Only the external error (precision) of one observation (i.e. the rms residual from a definitive orbit) should be used in these comparisons, and proper consideration should be given to independent parameters, such as separation of the two satellites and the integration time. Comparisons of this type would not only indicate directions to improvement, but would show that most modern observational techniques are competitive.

The mutual phenomenon "mid-event time," the quantity used in the orbital adjustment, is not, strictly speaking an observation. It is a quantity derived from numerous photometric or astrometric observations by a least-squares procedure which could rival that of the orbital adjustment itself. It has been referred to here as a compound observation akin to a normal point, although technically, only one observation is necessary if only the mid-event time is determined. In that case, the mid-event time would be comparable to a single observation. Recent estimates of the accuracy of the photometric mutual phenomena (compound observations) of the Galilean moons were made by Morando and Descamps (1994), and by Lieske (1995b). Morando and Descamps reported preliminary errors of 28 mas (100 km) for data uncorrected for the deviation of center of light from center of figure, and 4 mas (15 km) for corrected data. Lieske reports a rms residual of 30 mas (100 km) for the corrected data with "wide differences for the same event by different observers." These results suggest systematic errors in the observations and/or non- standard procedures for reduction of these complex data; and there is good evidence for the latter (Lieske 1995a). In any case, it appears that with the introduction of CCD techniques and improved reduction procedures, the best observations of the photometric mutual events of the Galilean satellites will produce data with an external precision in the range of 10 to 30 mas.

Sources of systematic error in the astrometric observations are due to uncalibrated scale and orientation errors and to a non-astrometric CCD chip. In the first case, the residuals increase with separation of the satellites measured. While these can be significant for CCD observations because of the difficulty in calibrating a diminutive focal plane, relative observations of satellites of small angular separation will be negligibly affected (Pascu et al. 1991; Pascu 1994a). A non-astrometric CCD chip is more problematic and should be mapped or avoided.

The external accidental rms error ranges from 20 to 120 mas for a single ground-based observation. This value usually includes the systematic error, if any. It also varies with separation of the measured satellites, and with the integration time and signal-to-noise ratio as well (as affecting the centroiding precision). Centroiding precision is about 5 - 20 mas, leaving the remainder to some limiting physical factor related to the separation and integration time. This limiting factor is believed to be image motion caused by atmospheric seeing. Lindegren (1980) relates the mean error, m.e. (arcsec), in the measured separation, S (radians), of two images, and integration time, T (sec) by

$$\text{m.e.} = 1.3 S^{0.25} T^{-0.5}$$

This formulation is a somewhat optimistic representation of the mean errors of CCD as well as photographic observations. It also demonstrates why any gain in precision due to the CCD would be primarily for the fainter satellites; the CCD is so sensitive that the bright satellite images saturate before the seeing excursions have time to average out. One must resort to attenuation and normal points, as described earlier.

For Hubble Space Telescope observations, the accuracy is limited by the measuring precision. Separation measurements of images near full well are expected to be accurate to 2 mas in the Planetary Camera. The accuracy for faint images is about 5 - 20 mas (Pascu et al. 1995). However, in the final analysis, for both the phenomena and astrometric observations, the astrometric precision of the observation depends on how accurately one can correct for the difference between the center of light, which is measured, and the center of figure (assumed to be at the center of gravity). Descamps (1992) and Mallama (1991) have had considerable success in developing algorithms for doing just that. They used Voyager images, observed photometric models and light scattering theory to compute corrections to center-of-light observations of the Galilean satellites. The ultimate precision for observations of the Galilean moons can be estimated from the rms residual given by Lieske (1995b) for the Voyager observations which were rigorously center-of-figure. He reports 45 km (15 mas). This suggests that the greatest astrometric precision possible will be for the smallest satellites!

6. NEAR FUTURE

CCD use in satellite observation can only increase in the near future. In the next few years, much activity will center about the mutual events of the moons of Saturn and the Galilean moons of Jupiter. There will also be considerable activity with HST observations of the faintest satellites not visible from the ground. To date only the inner satellite system of Neptune has not been imaged with HST and that is expected to change in two years.

In about five years, 9k x 9k CCD chips are expected to be available. It will then be possible to make CCD observations of the larger satellite systems with an accuracy comparable to the photographic ones, because scale calibration can then be done with standard scale fields, and orientation determined from longer trails. In fact, when the Tycho or faint star catalogs become available, it will be possible to implement a plate constants solution − calibrating both parameters simultaneously.

The ring-skimming satellites of Jupiter − Metis and Adrastea − are best observed from the ground at 2-microns (Nicholson and Matthews 1991), and periodic, if not regular, observations should be made. The IR technique should also be developed and extended to Puck and Proteus.

Some of the more interesting possibilities for future CCD observations lie in spin-off applications. When it is possible to implement a plate constant solution using the 9k x 9k CCDs and Tycho stars, it will then be feasible to obtain a five-fold increase in the positional accuracy of the planets (R.A. and Dec.) from relative observations of their satellites (Pascu and Schmidt 1990). These would be on the Hipparcos Catalog system and have an estimated precision of 30 mas. While this could be done better photographically for Jupiter (because of attenuation problems) and possibly for Saturn, certainly the advantage would go to the CCD for Uranus, Neptune and Pluto.

Another spin-off of the differential astrometry is differential photometry. The same observational strategy and data, used to obtain the astrometric results, can be used to obtain magnitudes, colors, albedos and light curves of the faint satellites from accurately known photometric parameters of the bright satellites (Pascu et al. 1992a, 1993, 1994a, 1995; Pascu and Rohde 1993). These data in turn will reveal information on the rotation of the satellites and their surface composition, as well as information on the severe conditions and processes in the strong magnetospheres of their primaries.

An ingenious spin-off of the mutual event observations was the monitoring and mapping of the volcanos and hot spots on Io from infrared photometry of the mutual occultations of Io. Goguen et al. (1988) applied this technique first at the 1985 mutual events using a photometer. Descamps et al. (1992) improved the technique at the 1991 events by using a CCD and combining simultaneous infrared and blue observations. The blue observations were used to derive corrections to the ephemerides of Io and the occulting moon, resulting in a more precise mapping of the volcanos.

7. SUMMARY AND CONCLUSIONS

The demand for high quality observations of the planetary satellites remains high, motivated by the needs of space reconnaissance and other astrometric projects. The principal advance in astrometric observation in the past 15 years was the introduction of CCD detectors. The quantum efficiency and linear response of the CCD detector have been largely responsible for its success, both for tangent plane astrometry and for the observation of the mutual phenomena of the bright moons of Jupiter and Saturn. At present, the CCD has been applied to all satellites detectable from the ground or with HST, except the inner system of Neptune, and there are substantial observing efforts in several countries. However, caution should be used when applying CCDs to large scale systems because the small size and complexity of the CCD chip make it difficult to calibrate. In its present configuration, the CCD is best suited to the observation of faint satellites near their primaries. This caveat notwithstanding, it is now possible to obtain relative positions of the satellites with a precision in the range 2 to 30 mas for HST observations, while a precision of 10 to 30 mas can be reached for ground-based observations of mutual phenomena or astrometric observations made when two satellites are at small separation (<50 arcsec). The limitation on precision appears to depend on how well the positions of the centers of light are known relative to their centers of figure. CCDs also offer several spin-off applications to the satellite astrometry. The volcanos of Io have been monitored and mapped with observations of the photometric mutual phenomena, while color photometry of the faint satellites relative to the bright ones result from the astrometric measurements.

References

Aksnes, K., and Franklin, F.A. (1975). "Mutual Phenomena of the Galilean satellites in 1973. I. Total and near-total occultations of Europa by Io," *Astron. J.* **80**, 56.

Aksnes, K., Franklin, F., Millis, R., Birch, P., Blanco, C., Catalano, S., and Pirronen, J. (1984). "Mutual phenomena of the Galilean and Saturnian satellites in 1973 and 1979/80," *Astron. J.* **89**, 280.

Andersson, L.E. (1978). "Eclipse phenomena of Pluto and its satellite," *Bull. Amer. Astron. Soc.* **10**, 586.

Arlot, J.-E. (1982). "Amélioration des Éphémérides des Satellites Galiléens de Jupiter par l'Analyse des Observations," Thèse de Doctorat d'Etat, Bureau des Longitudes.

Arlot, J.-E. (1995). personal comm.

Arlot, J.-E., Bernard, A., Bouchet, P., Daguillon, J., Dourneau, G., Figer, A., Helmer, G., Lecacheux, J., Merlin, Ph., Meyer, C., Mianes, P., Morando, B., Naves, D., Rousseau, J., Soulie, G., Terzan, A., Thuillot, W., Vapillon, L., and Wlerick, G. (1982). "Les resultats de la campagne d'observation PHEMU79 des phenomenes mutuels des satellites galileens de Jupiter en 1979," *Astron. Astrophys.* **111**, 151.

Arlot, J.-E., Bouchet, P., Gouiffes, Ch., Schmider, F.X., Thuillot, W. (1989). "Mutual Events of the Galilean Satellites: An Analysis of the Observations Made in 1985 at ESO," *Astron. J.* **98**, 1890.

Arlot, J.-E., Colas, F., Thuillot, W. and Vu, D.T. (1994). "CCD observations at the Bureau des Longitudes: analysis of the positions of satellites," in *Galactic and Solar System Optical Astrometry* (L.V. Morrison and G.F. Gilmore, eds.) Cambridge University Press, Cambridge, p. 297.

Arlot, J.-E., and Thuillot, W. (1988)."The coordination of the PHEMU85 international campaign," in *Coordination of Observational projects in Astronomy* (C. Jascheck and C. Sterken, eds.) Cambridge University Press, Cambridge, p. 171.

Arlot, J.-E., and Thuillot, W. (1993). "Eclipses and Mutual Events of the First Eight Saturnian Satellites during the 1993-1996 Period," *Icarus* **105**, 427.

Arlot, J.-E., Thuillot, W., Sevre, F., Vu, D.T., and Descamps, P. (1990). "A new campaign of observation of the mutual events of the Galilean satellites in 1991," *Astron. Astrophys.* **236**, L19.

Baum, W.A., Kreidl, T., Westphal, J.A., Danielson, G.E., Seidelmann, P.K., Pascu, D. and Currie, D.G. (1981). "Saturn's E Ring I. CCD Observations of March 1980," *Icarus* **47**, 84.

Beurle, K., Harper, D., Jones, D.H.P., Murray, C.D., Taylor, D.B. and Williams, I.P. (1993). "Preliminary analysis of CCD observations of Saturn's Satellites," *Astron. Astrophys.* **269**, 564.

Born, G.H., and Duxbury, T.C. (1975). "Phobos and Deimos ephemerides from Mariner 9 TV data," *Cel. Mech.* **12**, 77.

Bosh, A.S., and Rivkin, A.S. (1995). "Saturn ring-plane crossing, May 1995," *Bull. Amer. Astron. Soc.* **27**, 1131.

Brinkmann, R.T., and Millis, R.L. (1973). "Mutual Phenomena of Jupiter's Satellites in 1973-74," *Sky and Telescope* **45**, 93.

Buratti, B.J., Dunbar, R.S., Tedesco, E.F., Gibson, J., Marcialis, R.L., Wong, F., Bennett, S., and Dobrovolskis, A. (1995). "Modeling Pluto-Charon Mutual Events II. CCD Observations with the 60-inch Telescope of Palomar Mountain," *Astron. J.* **110**, 1405.

Campbell, D.B., Chandler, J.F., Ostro, S.J., Pettengill, G.H., and Shapiro, I.I. (1978). "Galilean satellites: 1976 radar results," *Icarus* **34**, 254.

Christy, J. (1978). "1978 P 1," *IAUC* No.3241.

Colas, J.F. (1991)."L'observation des phenomenes mutuels avec une cible CCD," Note technique No.18, Bureau des Longitudes.

Colas, F. (1992). "Astrometric observations of Phobos and Deimos during the 1988 opposition of Mars," *Astron. Astrophys. Supp. Ser.* **96**, 485.

Colas, F. (1994). "CCD observations at Pic du Midi Observatory: the techniques to be used," presented at PHESAT95 Workshop, 19-21 September 1994, Bucharest, Romania.

Colas, F. (1995). "Bureau des Longitudes and modern observations," presented at IAU Symposium 172, 3-8 July 1995, Paris, France.

Colas, F., and Arlot, J.-E. (1991). "Comparisons of observations of the Martian satellites made in 1988 with ephemerides," *Astron. Astrophys.* **252**, 402.

Colas, J.F., and Buil, C. (1992). "First Earth-based observations of Neptune's satellite Proteus," *Astron. Astrophys.* , **262**, L13.

Currie, D., Dowling, D., Seidelmann, P.K., Pascu, D., Rohde, J.R., Wells, E., Storrs, A., and Zellner, B. (1994). "HST Images of the Planet Uranus: Satellites and Ring System," *Bull. Amer. Astron. Soc.* , **26**, 1376.

Dahn, C.C. (1994). personal comm.

Descamps, P. (1992). "Etude des effets de surface sur la réduction astrométrique des observations de phénomènes des satellites Galiléens de Jupiter," Thèse de Doctorat, Observatoire de Paris.

Descamps, P., Arlot, J.-E., Thuillot, W., Colas, J.F., Vu, D.T., Bouchet, P., and Hainaut, O. (1992). "Observations of the Volcanos of Io, Loki and Pele, made in 1991 at ESO During an Occultation by Europa," *Icarus* **100**, 235.

Dourneau, G. (1982). "Observation of 2 mutual events involving the satellites of Saturn in April 1980," *Astron. Astrophys.* **112**, 73.

Duxbury, T.C., and Callahan, J.D. (1988). "Phobos and Deimos astrometric observations from Viking," *Astron. Astrophys.* **201**, 169.

Duxbury, T.C., and Callahan, J.D. (1989). "Phobos and Deimos astrometric observations from Mariner 9," *Astron. Astrophys.* **216**, 284.

Goguen, J.D., Sinton, W.M., Matson, D.L., Howell, R.R., Dyck, H.M., Johnson, T.V., Brown, R.H., Veeder, G.J., Lane, A.L., Nelson, R.M., and McLaren, R.A. (1988). "Io Hot Spots: Infrared Photometry of Satellite Occultations," *Icarus* **76**, 465.

Harper, D. (1994). "CCD observations at La Palma: results and projects," presented at PHESAT95 Workshop, 19-21 September 1994, Bucharest, Romania.

Harper, D. (1995). "CCD astrometry of Saturn's satellites in 1990-1994 from La Palma," presented at IAU Symposium 172, 3-8 July 1995, Paris, France.

Harris, F. (1995). personal comm.

Jones, D.H.P. (1995). "Limitations on the accuracy possible in astrometric observations of satellites of the major planets," presented at IAU Symposium 172, 3-8 July 1995, Paris, France.

Le Campion, J.F., Montignac, G., Chauvet, F., Colin, J., Desbats, J.M., Dourneau, G., and Rapaport, M. (1992). "First CCD observations of mutual phenomena of the Galilean satellites," *Astron. Astrophys.* **266**, 568.

Le Floch, J.C. (1994). "L'experience de l'Observatoire de Bordeaux dans les observations et le traitement des donnees sur les satellites de Saturne," presented at PHESAT95 Workshop, 19-21 September 1994, Bucharest, Romania.

Lieske, J. (1995a). "Making sense out of the 1985 and 1991 mutual events," *Bull. Amer. Astron. Soc.* **27**, 1197.

Lieske, J.H. (1995b). "Galilean satellites," presented at IAU Symposium 172, 3-8 July 1995, Paris, France.

Lindegren, L. (1980). "Atmospheric Limitations of Narrow-field Optical Astrometry," *Astron. Astrophys.* **89**, 41.

Lopez-Garcia, A. (1994). "CCD observations of major planet satellites," presented at PHESAT95 Workshop, 19-21 September 1994, Bucharest, Romania.

Mallama, A. (1991). "Light Curve Model for the Galilean Satellites during Jovian Eclipse," *Icarus* **92**, 324.

Mallama, A. (1992a). "Astrometry of the Galilean Satellites from Mutual Eclipses and Occultations," *Icarus* **95**, 309.

Mallama, A. (1992b). "CCD Photometry for Jovian Eclipses of the Galilean Satellites," *Icarus* **97**, 298.

Mallama, A. (1995). "Detection of Very High Altitude Fall-out from the Comet Shoemaker-Levy 9 Explosions in Jupiter's Atmosphere," *J. Geophys. Res.-* Planets (submitted).

Millis, R.L. (1974). "Mutual Occultations and Eclipses of the Galilean Satellites," *Bull. Amer. Astron. Soc.* **6**, 382.

Monet, A.K.B. (1993). "Ground-based Astrometry for the Galileo Mission," *Bull. Amer. Astron. Soc.* **25**, 1234.

Monet, D.G., and Monet, A.K.B. (1992). "Galilean Satellite Astrometry," *inter-office memo*, 22 Oct. 1992. U.S. Naval Observatory, Flagstaff Station.

Morando, B., and Descamps, P. (1994). "Some results obtained during the 1991 campaign of the observation of the mutual events of the Galilean satellites," in *Galactic and Solar System Optical Astrometry* (L.V. Morrison and G.F. Gilmore, eds.) Cambridge University Press, Cambridge, p. 329.

Muhleman, D.O., Berge, G.L., Rudy, D., and Niell, A.E. (1986). "Precise VLA Positions and Flux-Density Measurements of the Jupiter System," *Astron. J.* **92**, 1428.

Nicholson, P.D., Hamilton, D.P., Matthews, K., and Yoder, C.F. (1992). "New Observations of Saturn's Coorbital Satellites," *Icarus* **100**, 464.

Nicholson, P.D., and Matthews, K. (1991). "Near-Infrared Observations of the Jovian Ring and Small Satellites," *Icarus* **93**, 331.

Null, G.W., Owen, W.M., and Synnott, S.P. (1993). "Masses and densities of Pluto and Charon," *Astron. J.* **105**, 2319.

Owen, W.M. (1995). personal comm.

Pascu, D. (1977). "Astrometric Techniques for the Observation of Planetary Satellites," in *Planetary Satellites*, (J.A. Burns, ed.) University of Arizona Press, Tucson, p. 63.

Pascu, D. (1979). "The Naval Observatory Program for the Astrometric Observation of Planetary Satellites," in *Natural and Artificial Satellite Motion*, (P.E. Nacozy and S. Ferraz-Mello, eds.) University of Texas Press, Austin, p. 17.

Pascu, D. (1994a). "An appraisal of the USNO program for photographic astrometry of bright planetary satellites," in *Galactic and Solar System Optical Astrometry*, (L.V. Morrison and G.F. Gilmore, eds.) Cambridge University Press, Cambridge, p. 304.

Pascu, D. (1994b). "CCD observations of planetary satellites at the U.S. Naval Observatory," presented at PHESAT95 Workshop, 19-21 September 1994, Bucharest, Romania.

Pascu, D., Adler, C.A. and Bloomfield, J.F. (1991). "An Analysis of Photographic Astrometric Observations of the Galilean Moons," *Bull. Amer. Astron. Soc.* **23**, 1255

Pascu, D., Colas, J.F., and Rohde, J.R. (1992b). "Astrometric CCD Observations of Thebe (JXIV)," *Bull. Amer. Astron. Soc.* **24**, 1059.

Pascu, D., Panossian, S.P., Schmidt, R.E., Seidelmann, P.K., and Hershey, J.L. (1992a). "B, V Photometry of Thebe (JXIV)," *Icarus* **98**, 38.

Pascu, D., and Rohde, J.R. (1993). "BVRI Photometry of Helene (SXII), Telesto (SXIII), and Calypso (SXIV)," *Bull. Amer. Astron. Soc.* **25**, 1302.

Pascu, D., Rohde, J.R., and Colas, J.F. (1993). "Relative BVRI Photometry of Thebe (JXIV) and Amalthea (JV)," *Bull. Amer. Astron. Soc.* **25**, 1114.

Pascu, D., Rohde, J.R., and Foechterle (1994a). "B, V Photometry of Helene (SXII)," *Bull. Amer. Astron. Soc.* **26**, 1163.

Pascu, D., Rohde, J.R., Seidelmann, P.K., Currie, D.G., Dowling, D.M., Wells, E., Kowal, C., Zellner, B., and Storrs, A. (1995). "HST Astrometry of the Uranian Inner Satellite System," *Bull. Amer. Astron. Soc.* **27**, 829.

Pascu, D., and Schmidt, R.E. (1990). "Photographic Positional Observations of Saturn," *Astron. J.* **99**, 1974.

Pascu, D., Seidelmann, P.K., Baum, W.A. and Schmidt, R.E. (1983). "Observations of Faint Planetary Satellites with a Charge- Coupled Device," in *The Motion of Planets and Natural and Artificial Satellites*, (S. Ferraz-Mello and P.E. Nacozy, eds.) Universidade de São Paulo, Brasil, p. 253.

Pascu, D., Seidelmann, P.K., Schmidt, R.E., Santoro, E.J. and Hershey, J.L. (1987). "Astrometric CCD Observations of Miranda: 1981-1985," *Astron. J.* **93**, 963.

Rapaport, M. (1994). "Observations of Saturn's satellites made with the CCD camera of Bordeaux Observatory; comparison with the theories," presented at the PHESAT95 Workshop, 19-21 September 1994, Bucharest, Romania.

Rohde, J.R., and Pascu, D. (1993). "Astrometric Observations of Helene (SXII), Telesto (SXIII), and Calypso (SXIV)," *Bull. Amer. Astron. Soc.* **25**, 1235.

Rohde, J.R., and Pascu, D. (1994). "CCD Astrometry of Helene (SXII), Telesto (SXIII), and Calypso (SXIV): 1993 Observations," *Bull. Amer. Astron. Soc.* **26**, 1024.

Seidelmann, P.K. (1977). "Tabulations of Satellite Positional Observations and their Discussion," in *Planetary Satellites* (J.A. Burns, ed.), University of Arizona Press, Tucson, p. 533.

Seidelmann, P.K. (1979). "Planetary Satellites, a Review of the Past and Assessment of the Future," in *Natural and Artificial Satellite Motion* (P.E. Nacozy and S. Ferraz-Mello, eds.), University of Texas Press, Austin, p. 3.

Showalter, M.R., Nicholson, P.D., Danielson, G.E., Dones, L., French, R.G., Larson, S., Lissauer, J., McGhee, C., Seitzer, P., and Sicardy, B. (1995). "HST Observations of Saturn during the August 1995 Ring Plane Crossing," *Bull. Amer. Astron. Soc.* **27**, 1131.

Tholen, D.J. and Buie, M.W. (1995). "The orbit of Charon from Hubble Space Telescope observations," presented at IAU Symposium 172, 3-8 July 1995, Paris, France.

Veiga, C.H. and Vieira Martins, R. (1994). "A method to define a reference system for the reduction of astrometric positions of natural satellites," presented at PHESAT95 Workshop, 19-21 September 1994, Bucharest, Romania.

Veillet, C., and Ratier, G. (1980). "Astrometric Study of the Uranus Satellite Miranda," *Astron. Astrophys.* **89**, 342.

Viateau, B., and Rapaport, M. (1995). "Observations astrometriques a l'Observatoire de Bordeaux," presented at IAU Symposium 172, 3-8 July 1995, Paris, France.

Vieira Martins, R., and Veiga, C.H. (1995). "Astrometric observations of faint satellites," presented at IAU Symposium 172, 3-8 July 1995, Paris, France.

Zamarashkin, K.N., Kirsanov, N.O., Kisseleva, T.P., Kisselev, A.A. and Batrakov, J.V. (1994). "CCD astrometric observations of the Saturnian satellites system on the 26-inch refractor at Pulkovo to be made in 1995-1996 campaign and their comparison with parallel photographic observations," presented at PHESAT95 Workshop, 19-21 September 1994, Bucharest, Romania.

Zellner, B, Seidelmann, P.K., Pascu, D., Kowal, C., Wells, E., and Currie, D.G. (1994). "Recovery of Inner Satellites of Uranus," *Bull. Amer. Astron. Soc.* **26**, 1163.

OBSERVATION OF SMALL SOLAR SYSTEM OBJECTS WITH SPACEWATCH

JAMES V. SCOTTI AND ROBERT JEDICKE
Lunar and Planetary Laboratory
The University of Arizona
Tucson, AZ, 85721 USA

Abstract. Since beginning an automated full-time survey for Near-Earth objects in September 1990, the Spacewatch project has discovered 95 new Near-Earth asteroids (NEAs), 3 new comets, and 3 new Centaur asteroids. Spacewatch typically identifies about 2000 main-belt asteroids each lunation while covering about 150 square degrees to a limiting magnitude of $V_{lim} \sim 20.9$. We report automatically measured astrometric asteroid detections to the Minor Planet Center where known and multiply detected objects are identified. NEAs and other interesting objects are identified by their angular rates of motion near opposition at the time of discovery and are scheduled for astrometric follow-up on subsequent nights. Objects with exceptionally high rates of motion, called very fast moving objects, have been detected in near real-time by the observer and followed for several hours to several days. These objects are the smallest yet detected outside the Earth's atmosphere. Careful analysis of their discovery rates and orbits have indicated an enhancement of their magnitude-frequency distribution over that anticipated before the Spacewatch survey began – of about a factor of 40 for objects near absolute magnitude $H \sim 29$ (Rabinowitz, 1993; 1994). A subset of these small objects which have almost circular orbits and perihelia near the orbit of Earth have been recognized as having significantly different orbits from those of the previously known NEAs (Rabinowitz et al., 1993). Their origin is still under debate, with possible sources including Earth or Lunar impact ejecta, Earth–Sun Trojans, or more complicated secular resonance interactions of NEA orbits with the giant planets combined with stochastic perturbational encounters with the inner planets (Bottke, 1994; N.W. Harris, 1995, personal communication). The large volume of asteroid detections allows magnitude–frequency studies of the detected main-belt asteroids and Jupiter Trojans. New discoveries of Comets and Centaur asteroids (whose orbits cross those of the outer planets) may allow studies of their magnitude–frequency distributions as well.

1. Introduction

The Spacewatch project has been developing the capability and techniques to detect solar system objects using electronic detectors since 1981. At that time, the decision was made to make use of longer focus telescopes and state-of-the-art electronic detectors rather than wide field Schmidt telescopes and photographic techniques for new object discovery. By 1984, an RCA 320 × 512 pixel CCD was being used at the $f/5$ Newtonian focus of the 0.91-meter Spacewatch Telescope to observe asteroids. The techniques of astrometry and asteroid detection were developed and improved between 1984 and 1988. The prototype for our present automated detection software was first implemented in January 1985 (McMillan et al., 1986). The small format CCD, however, did not allow enough sky coverage to successfully discover NEAs.

A Tektronix TK 2048 CCD with 2048 × 2048 pixels was purchased in 1988 and first used at the telescope in 1989. This CCD suffered from low quantum efficiency and low charge transfer efficiency, but allowed enough sky coverage to faint enough limiting magnitudes to become an effective detector for discovery of NEAs. Our present detection software, called the "Moving Object Detection Program" or MODP, was first tested in early 1990 and has been used in an operational mode since September 1990. A thinned Tektronix TK 2048E CCD with higher quantum efficiency and improved charge transfer efficiency was installed in September 1992 and immediately doubled our discovery rates.

2. Observation and Data Reduction

A Tektronix TK2048E CCD is used at the $f/5$ Newtonian focus of the 0.91-m Spacewatch Telescope of the University of Arizona on Kitt Peak (Gehrels, 1991; Scotti et al., 1992; Scotti, 1994). The CCD is operated in slow scanning mode in which the sky is made to drift across the focal plane, usually by turning off the diurnal tracking. The accumulating electronic charges in the CCD are transferred in sync with the drifting images and read out and transferred into our data processing computer. MODP recieves the data and is used to detect moving objects either by their consistent motion or their trailed appearance (Rabinowitz, 1991). Briefly summarizing the operation of MODP, the program provides a real–time observer interface while carrying out 3 modes of object detection. The first mode of detection is the automatic detection of slower moving objects in which three consecutive images of the same location on the sky are obtained usually at half-hour intervals. Each image in the three "passes" are identified and their location and brightnesses are measured. Their relative positions are compared as soon as the third image is available and objects are identified by their consistent motion. The images are flagged in the real–time display and their rates of motion are reported to the observer. The observer can then make use of the rates of motion to identify likely NEAs. The second, and least effective detection method, is automatic streak detection in which trails that are bright enough are identified by MODP. The third detection method is the visual identification of faint trailed images by the observer. The real–time display and online tools in MODP allow the observer to schedule follow–up for these objects in real–time.

MODP also identifies the locations of Hubble Space Telescope Guide Star Catalog (GSC) stars in the image data so that automated astrometric solutions can be computed off-line. Combined with automatically measured positions of confirmed moving objects, another program determines the astrometric solutions and calculates the positions for every confirmed asteroid detection.

3. Astrometry

The technique of drift scan astrometry was first tried and reported by Gehrels et al., 1986. The process required manual measurement of bright (saturated) SAO and later AGK3 reference stars. The availability of the GSC with stars as faint as about $V = 16$ has allowed the process to be automated even though the catalog suffers serious uncertainties in the positions it gives for stars and does not include proper motion. Drift scanning improves the astrometric solutions by sampling a long strip of sky typically about 7 degrees long in Right Ascension and 0.5 degrees wide in Declination so that we sample through the boundaries of the original plates used in the generation of the GSC. The scan to scan consistency of the residuals of a given GSC star is better than 0.2 arcseconds while the typical standard deviation of the full set of several hundred GSC stars in a single scan is typically about 0.7 arcseconds, indicating that our ability to provide astrometric positions is catalog limited. Improved catalogs are still difficult to use with CCDs since the exposures required to obtain images of faint asteroids or comets results in saturation of the much brighter reference stars.

Comparison of the residuals of automatically derived astrometric measurements of asteroids with orbits determined from the automatic measurements and from measurements at other observatories confirms the relative consistency of observations on a given night and the larger night–to–night differences of the measurements. Table I from Scotti (1994) demonstrates this point by reproducing two examples from the Minor Planet Circulars.

4. Magnitude-Frequency Studies

Studies of the magnitude-frequency relationship of any population of asteroids detected by Spacewatch requires understanding the photometric accuracy and detection efficiency of the system.

The agreement between the magnitudes reported by Spacewatch and that expected from measurements of the same object obtained elsewhere have been surprisingly good since there has been relatively little effort to ensure that our magnitudes are calibrated. A comparison of the calculated magnitude for the same object in the three passes over the same section of sky indicates that we are accurate to better than 0.5 magnitudes. This measurement is complicated by the fact that an asteroid's brightness may change markedly during the 1.5 hour observation time due to its rotation. There is also a small tail of observations with very large differences between the brightest and faintest magnitude reported for the same object. These instances are understood as being due to the effect of cirrus and the chance that an asteroid may appear on or near a brighter star which the automated photometry routine mistakenly measures as the asteroid itself.

Knowledge of the detection efficiency and scanning history of the system is essential in order to debias both frequency and orbital distributions. Two important considerations are the efficiency as a function of asteroid magnitude and rate of motion. These two parameters are not independent since the faster an asteroid moves the smaller the signal will be in the peak pixel which is used for object identification. Complicating matters even further, it is not simple to measure the detection efficiency in a consistent and unbiased manner.

Several methods have been tested for determining the efficiency yet it appears that the best technique is to predict the location of known asteroids with good orbits in the scans and ask if the software detects the object. Twenty-two lunation's of data were searched for numbered asteroids yielding 1328 objects which should have appeared in the scans and 668 were detected by the system. Figures 1a and 1b show the detection efficiency as a function of magnitude and rate of motion respectively. The number of asteroids found in this study does not justify a two-dimensional determination of the efficiency in both parameters. Since most of the numbered asteroids are in the main belt, the efficiency in magnitude is dominated by these objects which have characteristic rates of motion near 0.2deg/day.

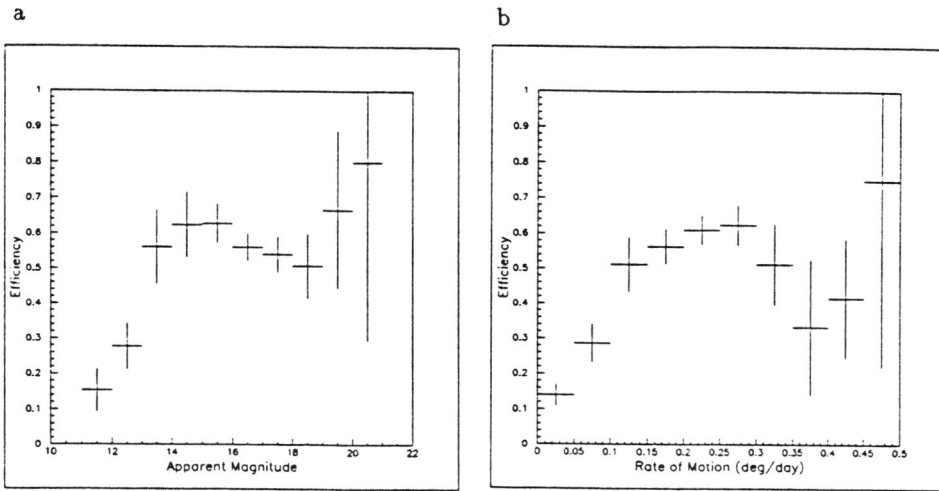

Figure 1. Figure 1a on the left shows the detection efficiency of MODP as a function of object magnitude. Figure 1b on the right shows the detection efficiency of MODP as a function of object motion

The efficiency is low for bright objects because they are saturated in the CCD image which makes centroiding difficult. Where the asteroid images are not saturated the efficiency appears to be almost constant. One of the problems with this technique is that the number of known objects which are very faint in our scans is quite small - this explains the large errors in the efficiency determination near the system's detection threshold. The efficiency is rate dependent due to the combined requirement that an asteroid must move at least a certain distance between scans in order to be detected and must not have moved too far in the same time. Spacewatch observers are now using scans of a fixed length (about 30 minutes and chosen to maximize the discovery rate of NEAs) in order to minimize changes in the detection efficiency due to this factor. Figure 1b shows that the efficiency is low for fast and slow objects and at a maximum for objects moving with main belt rates.

If the detection efficiency were independent of rate or magnitude it would be possible to measure the magnitude-frequency or orbital distribution of the detected objects in a relative sense. Since the rates of motion and the orbital characteristics of objects are well correlated near opposition a study of the orbit distribution of Spacewatch objects must take into account the detection efficiency as a function of their rate of motion.

Work is continuing on measuring our efficiency and it is hoped that by including 16 extra months of data, and un-numbered asteroids with good orbits, we can significantly reduce the errors on the measurement. This will allow a determination of the magnitude-frequency (or limits thereon) and orbital distributions of various classes of asteroids discovered by Spacewatch.

5. Near-Earth asteroid discoveries

The rate of discovery of NEAs by Spacewatch has been steadily improved since our first automated detection of an NEA in 1990 September. Our early rate of discovery of about 12 per year was doubled at the end of 1992 when we upgraded our CCD detector from a thick Tektronix CCD to a thinned Tektronix CCD with about twice the detective quantum efficiency. Figure 2 shows the number of NEAs discovered per year by all observers and by Spacewatch. Spacewatch presently is discovering about 2/3 of the NEAs found worldwide. Of the 95 new NEAs detected through 1995 June 6, 23 are brighter than $H = 18.3$ (~ 1 km diameter), 39 have $18.3 < H < 23.3$ (between 1 and 0.1 km diameter), and 33 have $23.3 < H$ (smaller than about 0.1 km diameter). NEAs are discriminated from the rest of the detected asteroids by their rates of motion. Near the opposition point where the search is concentrated, the angular ecliptic rates of motion are diagnostic of the source orbit and distance of the object (Bowell et al., 1990; Scotti et al., 1992; Scotti 1994). Jedicke (1995) has developed a technique which provides the probability that an object is a NEA based upon its rates of motion and location with respect to opposition. This measure will be incorporated into our discrimination criteria in the future.

The dynamical and physical characteristics of the smallest NEAs discovered by Spacewatch have been studied by Rabinowitz et al. (1993) and by Rabinowitz (1994). These investigations have found that there is an overabundance of objects smaller than about 50 meters diameter which increases to about a factor of 40 at a diameter of about 5–10 meters ($H \sim 29$) when compared to a power law extapolation of the numbers of larger NEAs discovered by Spacewatch and elsewhere. A possibly significant number of objects with nearly circular orbits and perihelia near the orbit of the Earth has also been recognized. Several of these small NEAs have been observed spectrophotometrically when the opportunity has arisen of having an object discovered before its closest approach to Earth. Although the data are individually of low signal-to-noise, taken as a whole, their colors can be compared with those of the main-belt asteroids. Collectively they are significantly different in color from the main-belt asteroids.

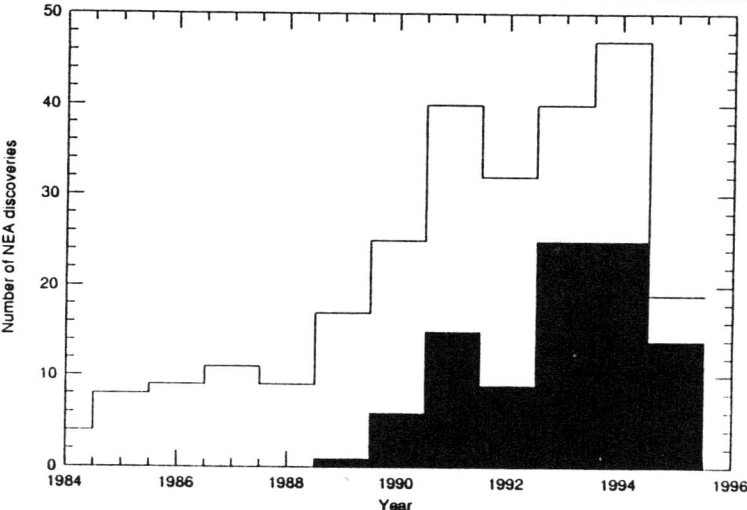

Figure 2. The number of NEA discoveries per year between 1984 and 1995. This histogram shows the number of Spacewatch NEAs as black.

6. Main-belt asteroid discoveries

A byproduct of the search for NEAs with Spacewatch is the discovery of several thousand Main-belt asteroids each lunation. These objects are not followed for a better orbit determination but we report automatically generated astrometric positions for each of the detections. They are regularly employed by the Minor Planet Center to update the orbits of known asteroids reported in our "incidental astrometry" as well as identifying objects which we have accidentally detected on more than one occasion, sometimes during different lunations or even during different oppositions.

The availability of these observations allows study of their magnitude-frequency relationships. In the past two and a half years we have obtained positions, motion vectors, and magnitudes for about 50,000 mainbelt asteroids. We are exploring the possiblity of using circular orbits in order to estimate the distance to each asteroid and thereby determine its absolute magnitude. The error in the apparent magnitude and the distance to the object (by assuming it is on a circular orbit) combine to give an error on the absolute magnitude of about 0.75 - we believe that this is sufficient for a statistical treatment of the magnitude-frequency of the main belt asteroids. Since Spacewatch can detect 1 km diameter objects at the inner edge of the belt, this measurement will have useful application to many main belt evolutionary studies.

A side affect of any future large scale survey for NEAs conducted using techniques similar to those employed by Spacewatch will include a similar (but correspondingly larger) collection of main-belt asteroid detections and ultimately will provide a survey of the main-belt complete to near the limiting magnitude used in the NEA survey. New families and possibly new orbit types amongst the smaller and more complete sample of asteroids may become known.

7. Centaur Asteroid Studies

There now exist over two dozen detected Kuiper Belt Objects yet only six known Centaurs – this is presumably due to their relatively short dynamical lifetime. Three of the Centaurs were discovered by Spacewatch and a fourth was accidentally re-discovered by Spacewatch. We are currently working on converting our observations into a limit on their number. This study depends critically on the efficiency of the system as a function of both magnitude and rates of motion as described above.

The efficiency for detecting Centaurs was determined using a reasonable model population to represent their orbit distribution. With only six known members it is impossible to draw any conclusions about the actual orbit distribution. A series of similar studies will be performed to determine the systematic effect of assuming a specific orbit distribution. In this preliminary study, the generated distribution in semi-major axis was flat in the range $10 \leq a \leq 29$ AU in order to span the range of the known Centaurs. Since their orbits are chaotic and characterized by high eccentricity the distribution in e was generated normally with a mean of 0.4 and width of 0.3. Even though the work of Holman & Wisdom (1993) suggests that circular orbits are not stable between the gas giants, low eccentricity orbits were included in this preliminary study and the effect of excluding them and altering the distribution will be studied in detail in the future. Finally, the inclination distribution was motivated by that of the main belt asteroids and was generated according to a probability distribution of the form $P(i) \propto e^{62 \cos i}$. The distribution in the other three orbital elements and the absolute magnitude were flat in the range $(0, 2\pi)$ and $(5, 10)$ respectively.

Simulations such as this one must use *generated* orbit and magnitude distributions which mimic the actual population as closely as possible in order to properly interpret the results. In this particular case, since only the magnitude-frequency relation is being studied, the generated distribution in absolute magnitude may have any shape since the efficiency is being determined as a function of this parameter.

The model Centaur population of 10^5 objects was subjected to the Spacewatch set of scans, and the efficiency as a function of rate and magnitude as determined above was used to determine if we might have discovered the object. For this particular orbit distribution, the efficiency for detecting simulated Centaurs drops from about 9% to 2% in the range $H = 5 - 10$.

The implied 90% confidence limit on the total number of Centaurs with orbits similar to the generated distribution and brighter than $H = 10$ is about 250. The number of Centaurs is clearly much less than the number of main belt or Kuiper Belt objects.

8. Comet discoveries

Spacewatch has discovered 3 new comets and re-discovered one long-lost comet. Several other comets have been accidentally re-discovered during the survey, however, based on the rate of discovery reported by Gehrels (1981) during the course of a photographic Schmidt survey for faint comets, we would have expected to find approximately 6 new comets each year! This discrepancy is at first surprising. Several factors are apparently conspiring against our finding faint comets. First, the faint comets we should see may not be active enough to show an obvious coma. Second, since we have observed during our first 5 years of our survey with an un-coma-corrected $f/5$ Newtonian, a marginally diffuse object might be masked by the substantial coma present in the images. Third, a recent estimate suggests that about 80% of the comets that we would detect will have rates of motion similar to the main-belt asteroids which are also detected. Fourth, MODP has been designed to minimize false detections as much as possible and it detects many fictitious objects around bright stars with large regions of scattered light, as well as comets with substantial coma and tails. A software filter was installed in the software to specifically guard against the numerous false alarms caused by bright stars, and has the undesireable side affect of removing active comets. Presumably, the observer will notice the comet – if it looks like a comet – and flag the object manually, but for fainter diffuse comets, or those which do not have a distinctive cometary appearance, this may not happen. We have recently installed a coma corrector on the Spacewatch Telescope and a coma corrector is to be part of the design of the 1.8-meter telescope which is under construction, so that the second point regarding badly comatic images should be eliminated. Another possibility which might explain our lack of comet discoveries would be a cutoff in the size distribution of small comets. Alternatively, small comets may not be active enough to be detected or may disintegrate more rapidly than the larger comets.

9. Conclusions and Future Research

Spacewatch has demonstrated the automated discovery and astrometry of small bodies in the Solar System and particularly of the population of NEAs. We are presently constructing a 1.8-meter telescope which will continue to improve our ability to discover and follow-up comets and asteroids.

Continued improvements in our equipment and techniques allow us to study the different populations in the solar system and their interrelationships.

References

Bottke, W.F. (1994) Provenance of the Spacewatch Small Earth-Approaching Asteroids, to be published in *Proceedings of the Astronomical Society of the Pacific: Inventory of the Solar System Conference*, in press.

Bowell, E.L.G., Skiff, B.A., Wasserman, L.H., and Russell, K.S. (1990) Orbital information from Asteroid motion vectors, in *Asteroids, Comets, Meteors III*, (C.-I. Lagerkvist, H. Rickman, B.A. Lindblad, and M. Lindgren, eds.), pp. 19–24.

Gehrels, T. (1981) Faint Comet Searching, *Icarus*, **47**, pp. 518–522.

Gehrels, T. (1991) Scanning with charge-coupled devices, *Space Sci. Rev.*, **58**, pp. 347–375.

Gehrels, T., Marsden, B.G., McMillan, R.S., and Scotti, J.V. (1986) Astrometry with a Scanning CCD, *Astron. J.*, **91**, pp. 1242–1243.

Holman, M.J., and Wisdom, J. (1993) Dynamical stability in the outer Solar System and the delivery of short period comets. *Astron. J.* **105**, pp. 1987–1998.

Jedicke, R. (1995) Detection of Near Earth Asteroids based upon their rates of Motion, submitted to *Astron. J.*

McMillan, R.S., Scotti, J.V., Frecker, J.E., Gehrels, T., and Perry, M.L. (1986) Use of a scanning CCD to discriminate Asteroid images moving in a field of stars. In *Instrumentation in Astronomy VI* (D.L. Crawford, Ed.), *Proceedings of the SPIE*, **627**, pp. 151–154.

Rabinowitz, D.L. (1991) Detection of Earth-Approaching Asteroids in near real time." *Astron. J.* **101**, pp. 1518–1529.

Rabinowitz, D.L. (1993) The size distribution of the Earth-Approaching Asteroids." *Astrophys. J.* **407**, pp. 412–427.

Rabinowitz, D.L. (1994) The size and shape of the Near-Earth Asteroid Belt." *Icarus* **111**, pp. 364–377.

Rabinowitz, D.L., Gehrels, T., Scotti, J.V., McMillan, R.S., Perry, M.L., Wisniewski, W., Larson, S.M., Howell, E.S., & Mueller, B.E.A. (1993) Evidence for a Near-Earth Asteroid Belt." *Nature* **363**, pp. 704–706.

Scotti, J.V. (1994) Computer aided Near Earth Object detection." In *Asteroids, Comets, Meteors 1993* (A. Milani, M. DiMartino, and A. Cellino, Eds.), pp. 17–30. Kluwer.

Scotti, J.V., Rabinowitz, D.L., and Gehrels, T. (1992) Automated detection of Asteroids in real-time with the Spacewatch Telescope." In *Asteroids, Comets, Meteors 1991* (E. Bowell and A. Harris, Eds.), pp. 541–544. Lunar and Planetary Institute, Houston.

CARLSBERG OPTICAL ASTROMETRY OF THE OUTER SOLAR SYSTEM

L.V. MORRISON AND M.E. BUONTEMPO
Royal Greenwich Observatory
Madingley Road
Cambridge CB3 0EZ, UK

Abstract. The Carlsberg astrometric telescope has made about 17 000 observations of outer Solar System objects since it began operation in 1984. The observed positions of the major planets are compared with JPL DE200 and DE403. The agreement with DE403 is good in general, but unresolved discrepancies of the order 0''.1 are found in Jupiter and Saturn. The run-off between the observations and DE200 which was fitted to observations before 1980 emphasize the need to continue optical observations of the outer planets.

1. Carlsberg astrometric telescope

The Carlsberg astrometric telescope (formerly known as the Carlsberg Automatic Meridian Circle) has been operating almost continuously since 1984 on the island of La Palma at the international observatory *Roque de los Muchachos* of the Instituto Astrofísica de Canarias. It is situated at a latitude of 28.7° north and an altitude of 2400 m and is operated jointly by Copenhagen University Observatory, the Royal Greenwich Observatory and the Real Instituto y Observatorio de la Armada, San Fernando. The operating procedure is described in Helmer & Morrison (1985) and a description of the scanning-slit micrometer and photoelectric detector system can be found in Helmer *et al.* (1991). The positions of the Solar System objects are measured once nightly as they cross the prime meridian. Only one satellite in each planetary system is observed each night. About 2% of the observing time is spent on Solar System objects.

2. Observations

The observations discussed in this paper are published in an annual series of catalogues, Carlsberg Meridian Catalogues Numbers 1-8 (1985-1994). The number of observations of Solar System objects in each catalogue is listed in Table 1. The accuracy of the positions is a function of zenith distance

TABLE 1. Carlsberg observations of outer Solar System objects

Catalogue	Year	Ganymede	Callisto	Rhea	Titan	Hyperion	Iapetus
CMC1	1984	–	–	–	–	–	–
CMC2	1985	–	–	–	–	–	–
CMC3	1986	–	67	–	–	–	–
CMC4	1987	–	26	–	70	–	–
CMC5	1988-89	–	44	–	89	–	–
CMC6	1990	–	24	–	40	–	30
CMC7	1991-92	24	76	7	63	4	59
CMC8	1992-93	35	28	5	39	17	36

Catalogue	Year	Uranus	Oberon	Neptune	Pluto	Minor planets($\sim$60)
CMC1	1984	54	–	78	–	667
CMC2	1985	64	–	47	–	1632
CMC3	1986	101	–	103	–	2501
CMC4	1987	65	–	64	–	2030
CMC5	1988-89	105	–	110	32	2467
CMC6	1990	76	–	105	11	1106
CMC7	1991-92	148	8	184	107	2108
CMC8	1992-93	57	21	72	26	1793

and magnitude (see Table 2). The best accuracy of $\pm 0''\!.12$ is obtained in the zenith (Dec $\sim +30°$). The diminution in the errors with time is a consequence of improvements in instrumentation and processing of the raw data. The positions are referred to a smoothed FK5 system, as described by Morrison et al. (1990).

3. Comparison with JPL DE200 and DE403

The comparison of the observations for the years 1984-1995 with JPL DE200 and DE403 are shown in a series of plots. They show the residuals in RA and Dec in units of arcseconds for individual observations which are centred on the date of opposition each year and extend over about five

TABLE 2. Accuracy of Carlsberg observations

Catalogue	Year	RA		Dec	
		$\delta=+30$	$\delta=-30$	$\delta=+30$	$\delta=-30$
CC1	1984	0.″19	0.″28	0.″18	0.″34
CC2	1985	″	″	″	″
CC3	1986	″	″	″	″
CC4	1987	″	″	″	″
CC5	1988-89	0.″15	0.″22	0.″15	0.″27
CC6	1990	″	″	″	″
CC7	1991-92	0.″12	0.″19	0.″12	0.″26
CC8	1992-93	″	″	″	″

months. Where observations of satellites were made, an ephemeris based on the theory of their orbital motion was used to reduce the observed positions to the barycentre of the system. So the plots of the residuals for Ganymede and Callisto, for example, effectively show the comparison of Jupiter with DE200 and DE403. The DE200 comparison is the zero line in each plot, and the DE403 comparison is the continuous wavy line.

The following points are noted from these plots.

3.1. JUPITER (GANYMEDE AND CALLISTO)

The ephemeris in the *Connaissance des Temps* (CdT), which is based on the G-5 theory (Arlot, 1982), was used to reduce the satellite positions to the barycentre of the Jovian system. The residuals in Dec show systematic deviations with respect to DE403. The mean offsets in 1992 and 1993 for both Ganymede and Callisto are nearly −0.″1, yet in 1994 the offset appears to be about +0.″05. The scatter is greater in 1994 because the observations for that year are still provisional. Further treatment of the observations will reduce the scatter but will not alter them systematically.

The explanation of these offsets in Dec lies in the construction of DE403 which attempts to reconcile the inconsistencies between the long series of optical data, which are referred to the FK5 frame, and the fairly recent high-precision radio data which are referred to the extagalactic VLBI frame. The latter frame is now to replace the FK5 as the International Celestial Reference Frame and DE403 is referred to this new frame. This problem is addressed by Standish (1995) elsewhere in the proceedings of this Symposium. The problem seems to be common to the optical data from other telescopes.

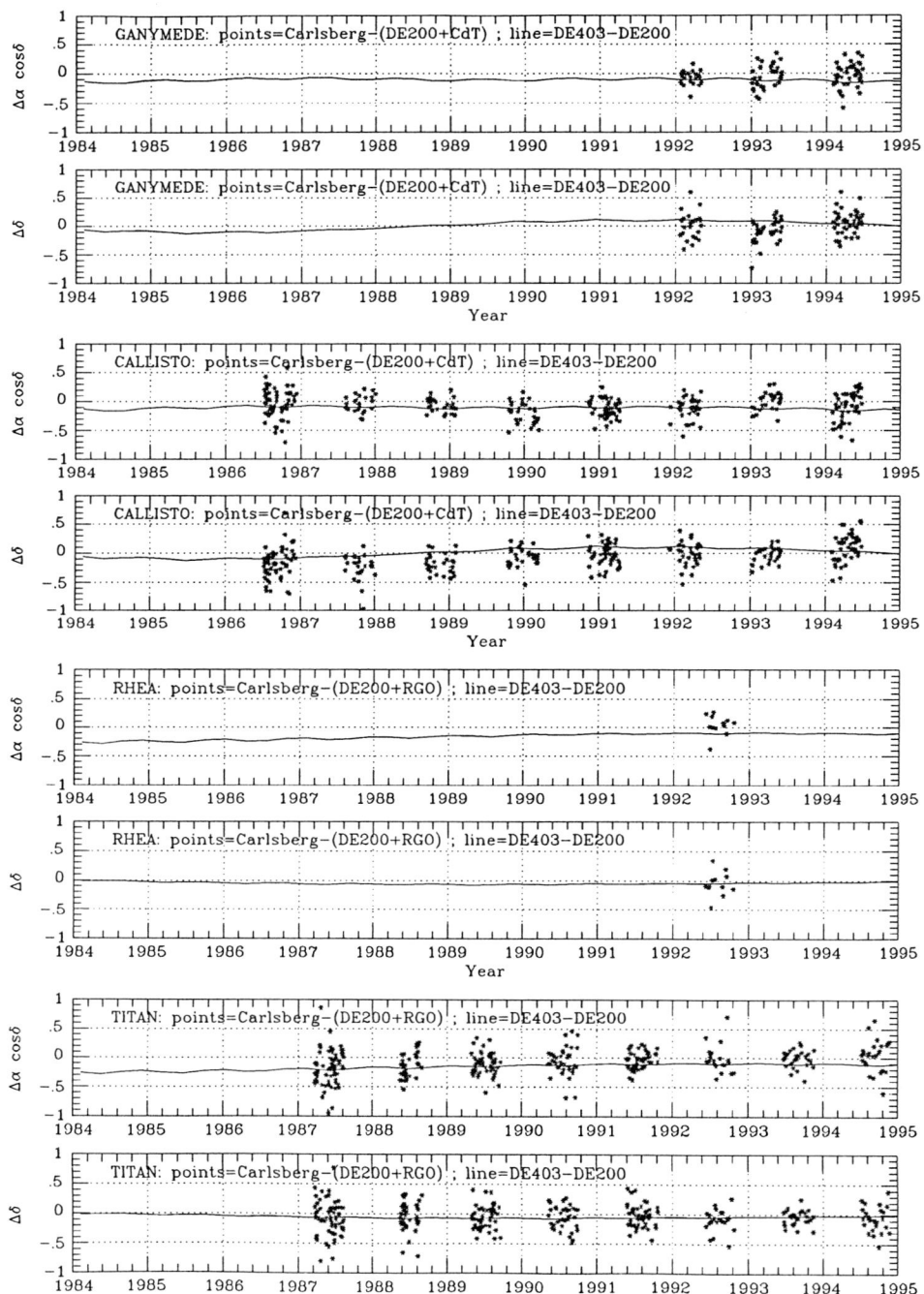

ASTROMETRY OF THE OUTER SOLAR SYSTEM

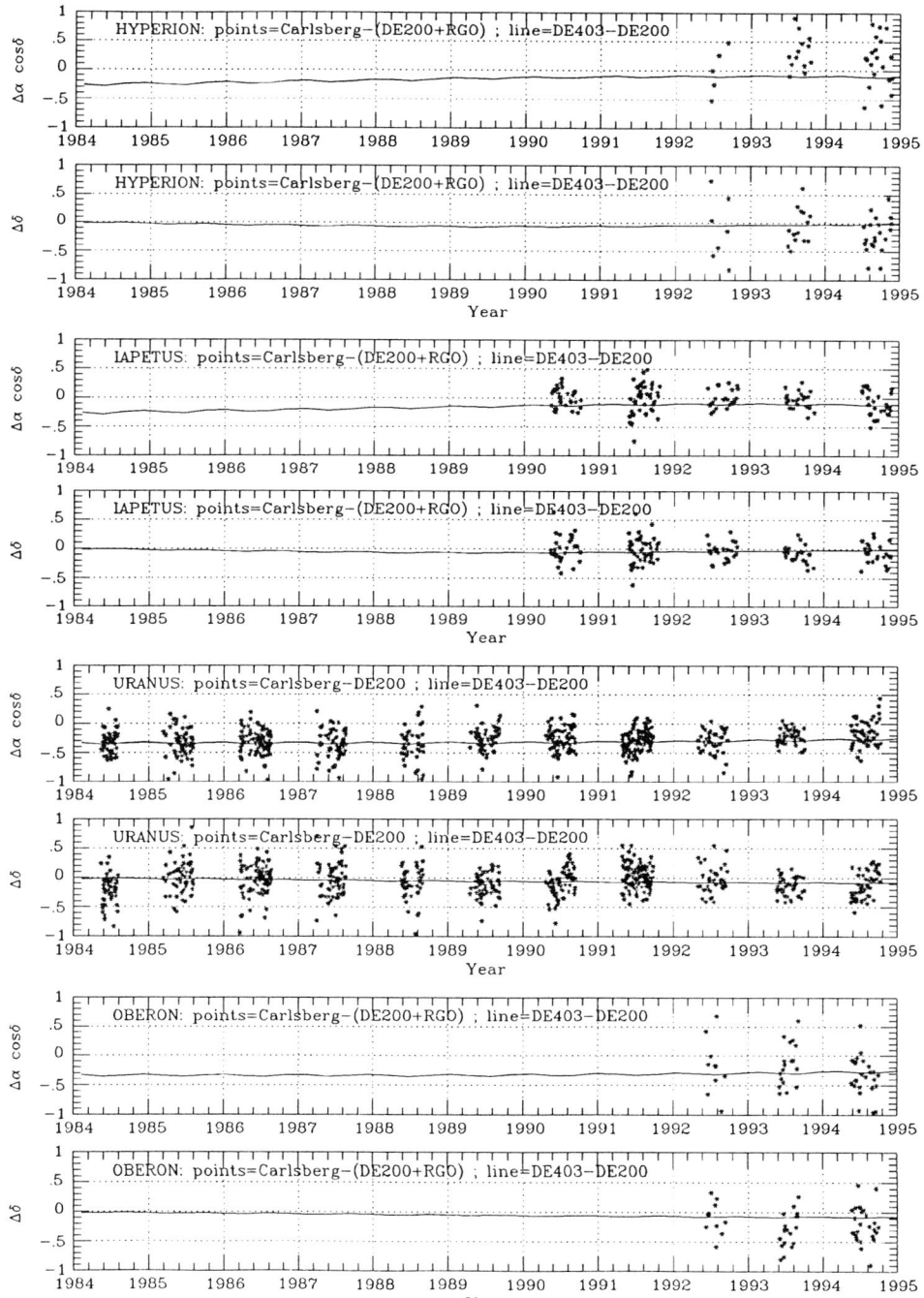

The reconciliation of the two data types requires a rotation of 0″.2 about the x-axis (in the plane of the equator and directed to zero RA) between the FK5 and extragalactic frames. However, the results from other work of comparing the optical and radio positions of galaxies (as yet unpublished), produces a rotation between the reference frames of less than half this amount. A rotation of 0″.2 about the x-axis produces a wave of 0″.2 sin(RA) in declination. This wave appears with the ~12-yr orbital period of Jupiter in the comparison of the optical observations with the ephemeris. Such a wave was found to be incompatible with the optical observations. So, in the production of DE403 the rotations about the x (and y-axis) were reduced to less than 0″.1. This reduced the discrepancy between the optical observations and DE403, but did not entirely remove it, as can be seen from the plot.

It is known (Lindegren et al., 1995) that in the zodiacal belt of the FK5 there is a large-scale systematic warp in declination. This warp is in the sense that the optical declinations which are referred to the FK5 should be increased by 0″.06. This is in the right direction to reduce the offset in the plots between Carlsberg and DE403, but is not the complete answer, particularly in 1994. When *Galileo* reaches Jupiter at the end of 1995 we may have the answer to this conundrum.

3.2. SATURN (RHEA, TITAN, IAPETUS, HYPERION)

DE403 is in good overall agreement with the observations in Dec, but there appears to be an offset of about 0″.1 in RA in both Titan and Iapetus from 1990 onwards. There are no observations of Saturn made with respect to the extragalactic VLBI frame, so DE403 is basically dependent on the optical data, and therefore the conflict between the reference frames which arose with Jupiter does not arise here. The cause of the 0″.1 bias is unknown. The residuals for Hyperion show a greater scatter on account of its faintness and the weakness of the theory of its orbit used in reducing the observations.

3.3. URANUS AND OBERON

The observations show the run-off in RA from DE200 which reaches 0″.35. DE200 was fitted to observations before 1980; whereas DE403 was fitted to observations up till 1994.

3.4. NEPTUNE

The continuing run-off in RA of observations from DE200 can be seen clearly in the plot.

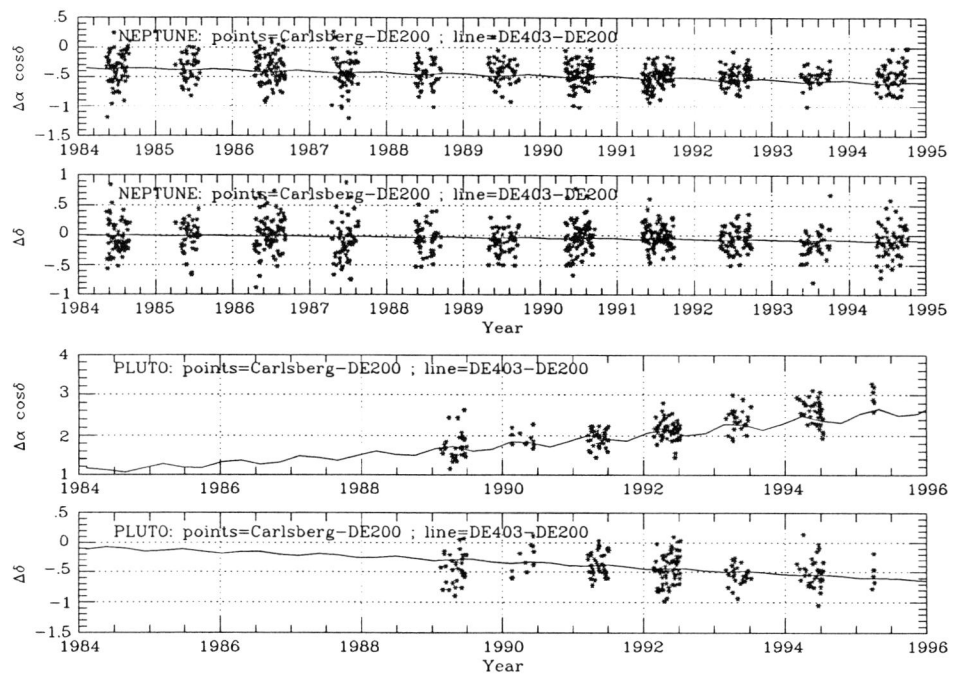

3.5. PLUTO

Pluto has run-off by 3".0 in RA from DE200. Even DE403 does not quite keep up with the observations in 1994. A few preliminary observations in 1995 appear to confirm this.

4. Future developments

Improvements in the accuracy of the optical observations will follow soon from the availability of the HIPPARCOS catalogue, the axes of which will be accurately aligned with the extragalactic frame. The use of CCD detectors on the Carlsberg Telescope in some form or another (as reported by Stone at this Symposium) will enable planetary positions to be measured relative to the HIPPARCOS frame. This will reduce both the systematic and accidental errors of optical positions to well below 0".1, provided that the satellites rather than the planetary discs are observed. This proviso does not apply to Pluto. The magnitude limit will be extended to V=16 which will enable more satellites and asteroids to be included in the observational programme.

5. Conclusion

The considerable run-offs from DE200 in the past 15 years and the absence of high-precision VLBI data beyond Jupiter, emphasize the need to continue monitoring the positions of the planets in the outer Solar System.

References

Arlot, J.-E. (1982) New constants for Sampson-Lieske theory of the Galilean satellites of Jupiter, *AA*, **107**, 305-310

Carlsberg Meridian Catalogues Numbers 1-8 (1985-1994) Copenhagen University Observatory, Royal Greenwich Observatory, Real Instituto y Observatorio de la Armada en San Fernando

Helmer, L., and Morrison, L.V. (1985) Carlsberg Automatic Meridian Circle, *Vista Astron.*, **28**, 505-518

Helmer, L., Fabricius, C. and Morrison, L.V. (1991) The micrometers of the Carlsberg Automatic Meridian Circle, *Experimental Astr.*, **2**, 85-99

Lindegren, L., Röser, S., Schrijver, H., Lattanzi, M.G., van Leeuwen, F., Perryman, M.A.C., Bernacca, P.L., Falin, J.L., Froeschlé, Kovalevsky, K., Lenhardt, H., Mignard F. (1995) A comparison of ground-based stellar positions and proper motions with provisional Hipparcos results, *AA*, in press.

Morrison, L.V., Argyle, R.W., Réquième, Y., Helmer, L., Fabricius, C., Einicke, O.H., Buontempo, M.E., Muiños, J.L. and Rapaport, M. (1990) Comparison of FK5 with Bordeaux and Carlsberg Meridian Circle Observations, *AA*, **240**, 173–177

Standish, M.E. (1995) New accuracy levels for Solar System ephemerides, *IAU Symp.172*, Kluwer Acad. Publ., Dordrecht

ACCURACY ESTIMATION OF NEW SETS OF THE SUN AND PLANETS OBSERVATIONS

A. S. KHARIN
Main Astronomical Observatory of the Academy of Sciences,
Goloseevo, Kiev-22,252650, Ukraine

YU. B. KOLESNIK
Institute for Astronomy of the Russian Academy of Sciences,
109017,48 Piatnitskaya St.,Moscow, Russia

AND

O. E. CHUICHENKO
Astronomical Observatory Kharkov University,
310022,35 Sumskaya St.,Kharkov, Ukraine

The PLANETS database has been conceived and firstly compiled at the Golosiiv Observatory in 1984, Kiev (Kharin et al. 1987). Since 1988, it is updated, maintained and analysed in cooperation with the Institute of Astronomy, Moscow. By this time the database comprises most of the published optical observations of the Sun and 7 major planets made from 1960 onwards with 21 meridian instruments, 15 astrographs and 11 astrolabes at 29 observatories.

The method of random accuracy estimation applied to the actual data covering the interval from 1960 to 1994 is presented in details in Kolesnik (1995). These estimates (the weighted mean standard deviations with weights being proportional to a number of observations in a series $\bar{\sigma}_i$) with respect to the method of observation and object are given in Table 1 (see next page).

References

Kharin, A. S., Voronkevich, V. L., Minyailo, N. F.: 1987, in *Modern Astrometry – Proc. of the 23 Astrom. Conf. USSR*, Polozentzev, D. D. (ed.),Leningrad, p.306.
Kolesnik, Yu.B.: 1995, *Astron. Astrophys.* **294**, 874.

TABLE 1 – Number of series (NS), total number of observations (NO) and the weighted mean standard deviation $\bar{\sigma}_i$ attributed to meridian instruments (MI), astrographs (APH) and astrolabes (ASTR) as they are represented in the PLANETS database

	Right ascension			Declination		
	MI	APH	ASTR	MI	APH	ASTR
Sun						
NS	12		4	8		
NO	8088		972	8178		
$\bar{\sigma}_i$	0.052^s		0.040^s	$0.64''$		
Mercury						
NS	12			10		
NO	1980			2014		
$\bar{\sigma}_i$	0.044			0.47		
Venus						
NS	12	8		10	8	
NO	6504	562		6302	562	
$\bar{\sigma}_i$	0.054	0.041^s		0.62	$0.43''$	
Mars						
NS	15	10	5	13	10	5
NO	1621	673	114	2254	673	114
$\bar{\sigma}_i$	0.029	0.024	0.021	0.43	0.30	$0.26''$
Jupiter						
NS	10	6	4	10	5	5
NO	1562	423	80	1600	423	80
$\bar{\sigma}_i$	0.025	0.020	0.040	0.46	0.21	0.38
Saturn						
NS	10	9	7	10	9	7
NO	1557	524	175	1514	524	175
$\bar{\sigma}_i$	0.025	0.016	0.036	0.46	0.20	0.32
Uranus						
NS	11	2	4	9	2	4
NO	1771	124	187	1663	124	187
$\bar{\sigma}_i$	0.016	0.014	0.022	0.27	0.23	0.33
Neptune						
NS	9	3		10	3	
NO	1639	100		1604	100	
$\bar{\sigma}_i$	0.016	0.013		0.26	0.16	

Acknowledgements

The financial support from BDL and Russian Foundation for Fundamental Research is gratefully acknowleged

McDONALD OBSERVATORY LUNAR LASER RANGING:

BEGINNING THE SECOND 25 YEARS

P.J. SHELUS, R.L. RICKLEFS, J.G. RIES, A.L. WHIPPLE AND
J.R. WIANT
McDonald Observatory and Department of Astronomy
University of Texas at Austin
Austin, Texas 78712-1083 USA

1. Introduction

Lunar laser ranging (LLR) (Dickey *et al.*, 1994) consists of measuring changes in the round-trip travel time for a laser pulse traveling between a transmitter on the Earth and a reflector on the Moon. The lunar surface reflectors are still operating normally after almost three decades of use. The ranging data exhibit a rich spectrum of change due to many effects.

During the early days of the experiment, McDonald Observatory was the only facility that could routinely range to the Moon (Abbot *et al.*, 1973, Mulholland *et al.*, 1975, Shelus *et al.*, 1975). It used as its fundamental component the 2.7-m telescope (Silverberg, 1974). This system was decommissioned in 1985 to be superseded by a dedicated system, the McDonald Laser Ranging Station (MLRS) (Shelus, 1985, Shelus, *et al.*, 1993b), that can range to both artificial satellites as well as the Moon. Due primarily to distance, it is more than a trillion times more difficult to range to the Moon than it is to range to an artificial satellite.

The new station, initially placed in the saddle between Mt. Locke and Mt. Fowlkes at McDonald Observatory, became operational in 1983. Wind tunneling effects at the saddle site produced serious problems with seeing and the MLRS was moved to the top of Mt. Fowlkes in early 1988. Although the MLRS observing emphasis has always been placed upon ranging to artificial satellites (see Figure 1), the Moon continues to be a vital part of the operation (Shelus, 1987). At this point in time, it is the only station in the US that is capable of ranging to the Moon.

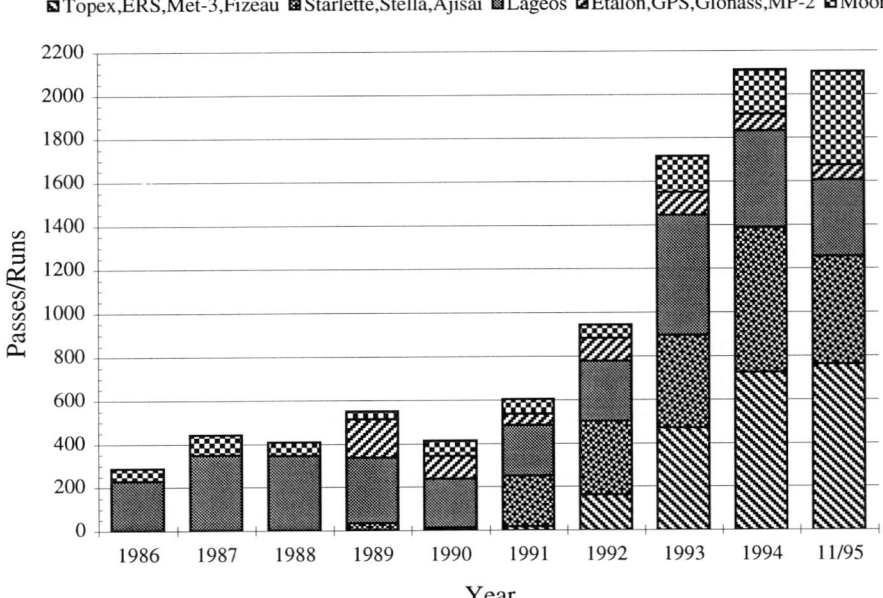

Figure 1. MLRS Laser Ranging Activity.

2. MLRS LLR Upgrade History

The transition of LLR operations at McDonald Observatory from the original 2.7-m system to the 0.76-m MLRS caused a marked decrease in the volume of data, mainly because of the reduction in aperture. The number of normal points obtained annually with the MLRS in the late 1980's was well under 100, while those obtained with the 2.7-m system had been close to 400. Data accuracy and precision were improved, however, due to a decrease in laser pulse length. In the early 1990's a program to improve MLRS LLR data volume, and improve precision and accuracy, was begun.

The first upgrade was an x-y offset guiding stage (Shelus et al., 1993a). Such a device allows an observer to guide on a sun-lit, off-axis feature, or a star, while the reflector is on the shadow side of the lunar terminator. Not only does this provide for a greater number of observing opportunities during a lunation, ranging to a reflector in the dark produces virtually noise free data. Installation was completed and operation began in the spring 1993 with a dramatic increase in the amount of LLR data obtained.

An MLRS laser pulse contains 3×10^{17} photons. In lunar mode, only a very few photons per minute make it back through the receive path. This mandates single photon detection and as large a system efficiency as possible. The spectral filter must eliminate as many noise photons as possible, but must transmit as many signal photons as possible. Also, spectral filter requirements change with lunar phase and sky conditions. A third spectral filter, of intermediate specification between those already in hand, was purchased in spring 1993 and immediately placed into service. Further, in August 1994, because of aging and laser induced damage, the telescope's # 3 dichroic mirror was replaced. The additional scheduling flexibility and extra energy throughput that these two changes provided resulted in a significant increase in the amount of LLR data.

With the MLRS, intensive manual guiding is required to keep the telescope on target. In spring 1993 we specified the MLRS Auto-Guiding and Imaging System (AGIS), an integrated hardware/software system that accepts real-time video signals, i. e., a highly magnified image of the lunar surface or a stellar or artificial satellite image. It performs real-time image processing and allows the user to select among various levels and types of image enhancement. It provides tracking error signals to the control computer for guiding control. The AGIS was received and installed in February 1995. Integration of the error signals into the pointing control loop is in progress. A correlation tracking board for the AGIS is under development.

The German laser ranging group at Wettzell designs and builds avalanche photo-diode (APD) devices. An APD detector in use at the French LLR site exhibits a significant increase in sensitivity and an improved accuracy and precision. Other APD's are used for artificial satellite laser ranging elsewhere. In spring 1993 the Wettzell group built an APD for use at the MLRS for LLR operations. Although this APD was received at the MLRS in early 1994, early difficulties were encountered. A new device was received in mid-summer 1995. Improvements in sensitivity and jitter were immediately noted. Normal APD use at the MLRS will begin shortly.

The MLRS station clock is a cesium-beam device. These are expensive and must undergo considerable preventive maintenance on a time-critical basis. A cheaper, easier to maintain steered oscillator can provide short-term stability and a GPS receiver, used as a fly-wheel, can provide long-term stability. With such a combination of devices, a significant cost and time savings in operation and maintenance could result, with an improvement in timing accuracy and precision. We have taken delivery of a Totally Accurate Clock (TAC), designed by Dr. Thomas A. Clark. This may form the basis for a new timing system to replace the current Cesium device.

We have just completed the replacement of the 15 year old Data General NOVA control computer with a LynxOS based, X-windows, real-time UNIX system running on PC hardware. This was coordinated with upgrades at other NASA laser ranging stations. It has created a system with compatible hardware and software architecture in an open-systems environment. It allows a maximum amount of software portability and sharing.

3. The LLR Data Set

Some parameters in LLR analysis separate on relatively short time scales. Others separate on the 18.6 year period of the lunar nodal regression, or longer. When analyzing any data set, it is important to consider the distribution of observations over the relevant parameter space. The histograms in Figure 2 show the distribution of LLR data with respect to the fundamental arguments of the lunar theory. After more than 25 years of observations, the data distribution with respect to the lunar mean anomaly, l, the solar mean anomaly, l', and the argument of the lunar latitude, F, is reasonably flat. The fact that Ω, the mean longitude of the ascending node of the lunar orbit, covers significantly more than a full period is extremely important. New and full moon effects are seen in the distribution of observations with respect to the mean elongation of the Moon from the Sun, D. This is a consequence of scheduling lunar operations at 1st and 3rd quarter phases, when data is easiest to obtain.

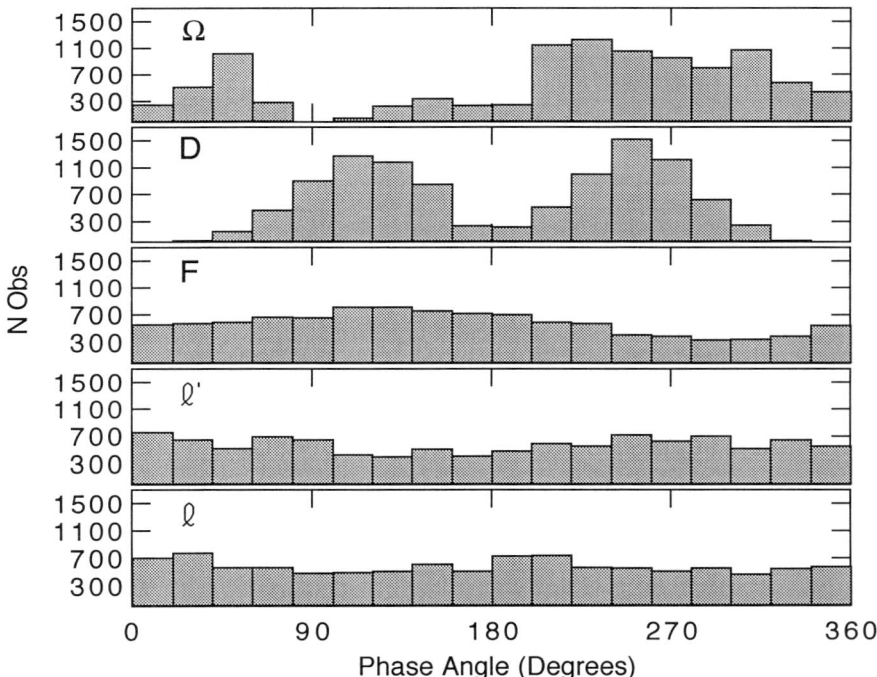

Figure 2. Distribution of LLR observations with respect to the fundamental arguments of the lunar orbital motion. The sampling interval is 20°.

Illustrating the long span of McDonald LLR data, data volume, and formal uncertainty is Figure 3. We show yearly totals of normal points and the weighted root-mean-square of post-fit residuals. As a result of the recent MLRS up-grades, not only are the accuracy and precision of its LLR data significantly better than that of the 2.7-m system, the MLRS data volume matches it as well. There is also a significant drop in the rms of MLRS LLR data during 1995. Although not shown here, the number of photons per normal point is now double what it had been prior to the beginning of the MLRS upgrade. Table 1 details this increased MLRS data volume from 1992 to the present and compares it to the French LLR effort.

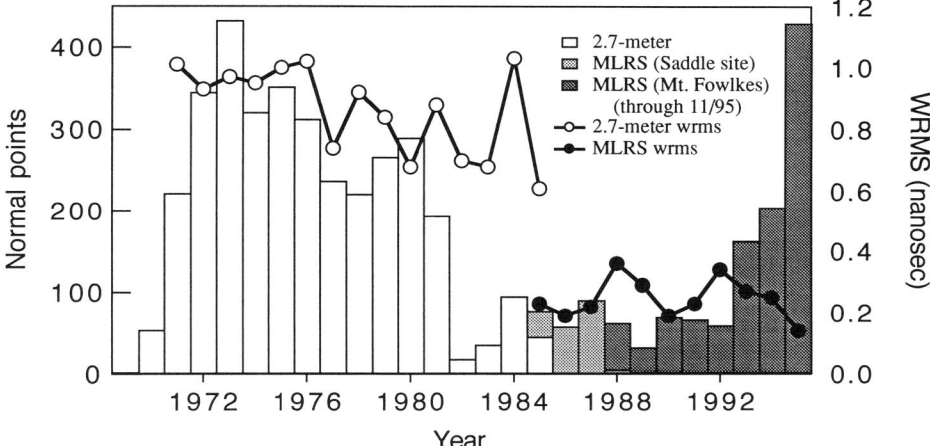

Figure 3. The number of normal points and the weighted RMS of the post-fit residuals from our analysis of the McDonald LLR observations. To insure large enough samples to be statistically significant, the RMS values were calculated from ranges to the Apollo 15 retroreflector only.

Table 1. MLRS/CERGA Comparison (1992-11/95)

		MLRS				CERGA			MLRS/CERGA		
		1992	1993	1994	11/95	1993	1994	11/95	1993	1994	11/95
LLR	Ap 11	-	3	25	34	53	55	56	6%	45%	61%
Normal	Ap 14	-	8	17	49	53	44	46	15%	39%	107%
Points	Ap 15	58	151	160	341	433	499	446	35%	32%	76%
	Lnk 2	-	1	3	6	12	17	10	8%	18%	60%
	Total	58	163	205	440	551	615	558	30%	33%	77%
Nights	Ap 11	-	2	16	21	27	23	23	7%	70%	91%
LLR Data	Ap 14	-	5	12	27	26	17	23	19%	71%	117%
Taken	Ap 15	23	56	61	95	75	71	63	75%	86%	151%
	Lnk 2	-	1	3	4	11	9	7	9%	33%	57%
UT0 pts.			5	26	33	54	49	44	60	53%	70%	90%

4. Conclusions

McDonald Observatory laser ranging operations has pursued a significant and substantial LLR up-grade at the MLRS. That effort has been a remarkable success. We have just a short time ago celebrated the 25th anniversary of the first Apollo manned placement of a retroreflector package on the lunar surface. The lunar laser ranging experiment remains the only active Apollo experiment and it is still marching at the forefront of science. Concerning the availability of the LLR data type, the Crustal Dynamics Data Information System (CDDIS), maintained at Goddard Space Flight Center, archives all lunar laser ranging data. Normal points are available through an on-line data base; filtered photon returns are archived on magnetic tape. Further information about the LLR data set and the CDDIS may be obtained via the Internet from pjs@astro.as.utexas.edu or from noll@cddis.gsfc.nasa.gov.

References

Abbot, R. I., P. J. Shelus, J. D. Mulholland and E. C. Silverberg, Laser observations of the Moon: Identification and construction of normal points for 1969-1971, *Astron. Jour.*, **78**, 784-793, 1973.

Dickey, J. O., P. L. Bender, J. E. Faller, X X Newhall, R. L. Ricklefs, J. G. Ries, P. J. Shelus, C. Veillet, A. L. Whipple, J. R. Wiant, J. G. Williams and C. F. Yoder, Lunar laser ranging: a continuing legacy of the Apollo Program, *Science*, **265**, 482-490, 1994.

Mulholland, J. D., P. J. Shelus and E. C. Silverberg, Laser observations of the Moon: normal points for 1973, *Astron. Jour.*, **80**, 1087-1093, 1975.

Shelus, P. J., J. D. Mulholland and E. C. Silverberg, Laser Observations of the moon: Normal points for 1972, *Astron. Jour.*, **80**, 154-161, 1975.

Shelus, P. J., MLRS: A Lunar/Artificial Satellite Laser Ranging Facility at the McDonald Observatory, *IEEE Trans. on Geosci. and Rem. Sens.*, **GE-234**, 385-390, 1985.

Shelus, P. J., To the Moon ... and Back, *Discovery: Research and Scholarship at the University of Texas at Austin*, **10**, No. 4, 33-37, 1987.

Shelus, P. J., A. L. Whipple, J. R. Wiant, R. L. Ricklefs, and F. M. Melsheimer, A Computer-Controlled x-y Offset Guiding Stage for the MLRS, *NASA Conf. Pub. 3214*, 101-105, 1993a.

Shelus, P. J., R. L. Ricklefs, A. L. Whipple, and J. R. Wiant, Lunar Laser Ranging at McDonald Observatory: 1969 to the present, *AGU Geodynamics Series*, ed. J. J. Degnan, Vol. 25, 183-187, 1993b.

Silverberg, E. C., Operation and Performance of a Lunar Laser Ranging Station, *Appl. Opt.*, **13**, 565-574, 1974.

OBSERVATION OF SOLAR SYSTEM BODIES ON BOARD THE FUTURE RUSSIAN ASTROMETRIC SATELLITE

V.N.YERSHOV
Pulkovo Observatory,
196140 St Petersburg, Russia
e-mail: yersh@gao.spb.su

Abstract. A future Russian astrometric satellite (*Struve*) is aimed at an extension of the fundamental coordinate system down to 18-th magnitude stars. The positional accuracy of observations is planned to be about 0.6 mas. Many of the Solar System objects will be observed by the Struve satellite. Problems of registration and reduction of these observations are discussed.

1. Introduction

The future Russian astrometric satellite *Struve* (Yershov, Kanayev 1994) will be able to detect many of asteroids, comets, planets and their satellites as well as more than 4 million program stars. The limiting magnitude of two on-board telescopes is estimated to be $V = 18$ and positional accuracy 0.6 *mas* of observations is expected. A 40-cm catadioptric Schmidt telescope scheme (2.5 m focal length) is chosen for the Struve satellite. Charge coupled devices are considered as the basic detectors for the focal plain assembly. Microchannel plates are also under consideration which are more suitable for detection of moving objects such as close asteroids or space debris.

2. Scheme of observations

The Struve satellite will produce observations in a scanning mode inertially rotating with a frequency of about 10 turns per day. Four viewing directions are organized by the use of two beam-combiners placed in front of the

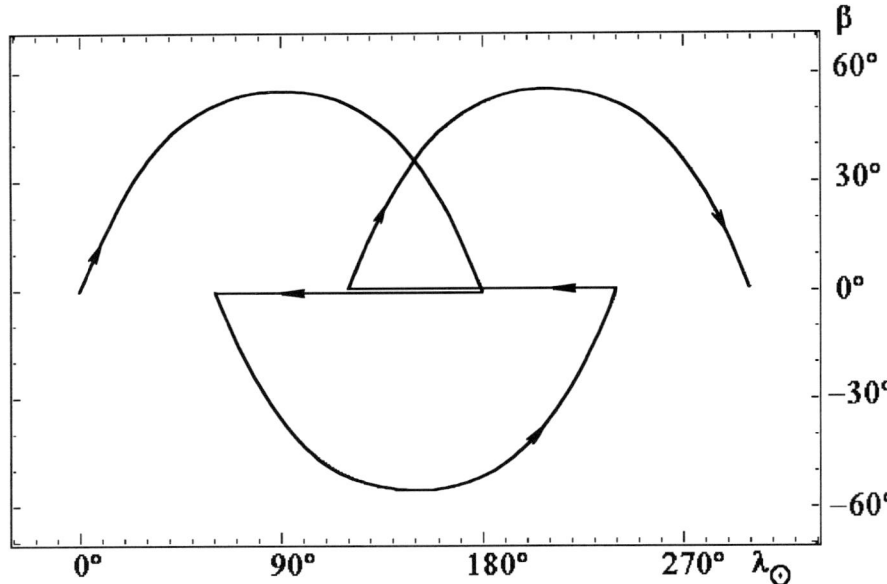

Figure 1. The Vlasov scanning law. After completing half of the current great circle the spin axis is to be switched back to the previous trajectory great circle. The aspect angle varies from 55° to 60°, and the spin axis passes three great circles each year.

entrance pupils of two on-board telescopes. 62° and 74° basic angles have been chosen for the measurements. Two scanning laws are under consideration: the Hipparcos-like rotational scanning with 55° aspect angle and the Vlasov scanning mode, when the spin axis moves along three intersected orthogonal great circles from one intersection point to another (Fig.1).

Solar System objects at their oppositions are not available for observations but all planets except Mercury are visible within elongations ±35° − 145°.

As most Solar System objects concentrate close to the ecliptic, it is interesting to analyze the distribution of scans there. Fig.2 and Fig.3 show this distribution for the Vlasov and the Hipparcos-like scanning modes, respectively. The Vlasov scanning mode provides more uniform coverage of the ecliptic zone.

3. Detection

Computer simulations of CCD detection of bright planets show relatively poor accuracy (for example, a single transit of Mars gives 50 − 80 mas).

OBSERVATION OF SOLAR SYSTEM BODIES 417

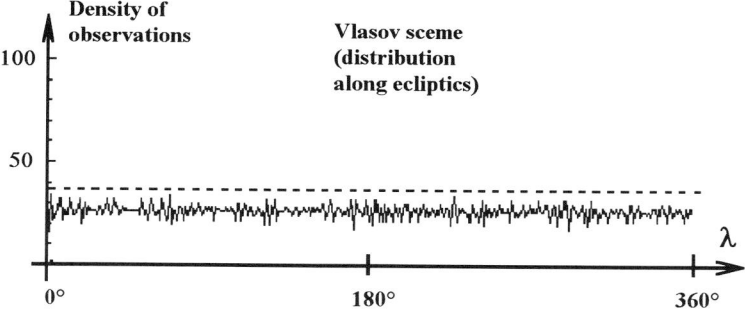

Figure 2. Distribution of density of observations along the ecliptic for the Vlasov scanning law (one year of observations).

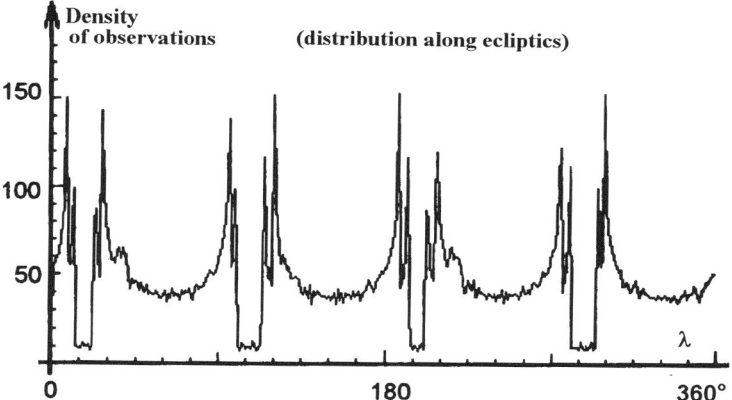

Figure 3. Distribution of density of observations along the ecliptic for the rotational scanning law (one year of observations).

Much better accuracy one may achieve by observing natural satellites of the planets. Practically, all main satellites of major planets are available for observation (except Charon, Pluto's satellite). Table 1 shows the accuracy of a single observation (in milliarcseconds) for different Solar System objects.

4. Identification of objects

Four observations during one scan are separated by time intervals 29.6^m, 42.4^m, 24.8^m, and 47.2^m (corresponding to the viewing directions). Each object is available for observations within three consecutive scans.

The Solar System objects move noticeably even within a single tran-

TABLE 1. Accuracy of observations of Solar System objects (a single observation, 5^S integration time).

Object (magnitude, V)	σ_1, mas
Pluto (15^m)	11
10-km asteroid (16^m)	13
Triton (17^m)	41
1-km asteroid near Mars (18^m)	95

sit time (24 seconds, one viewing direction). Pluto, for example, moves on 20 mas. Completing the scan, one may calculate the image velocity and acceleration. Then one may pre-calculate positions of the object for future scans. Using the Pulkovo Observatory technique of "apparent motion parameters" (Kisselev, Bykov 1976) one may estimate a preliminary orbit of the object and then identify the object 2 – 4 months later.

A simulation of Struve observations of two minor planets (451 *Patientia* and 1993 *MF*) shows rather big errors of $(O - C)$ values (up to $1°$) for 4 months prediction time interval. But derivatives of the coordinates are in good correspondence with those calculated from the object's real orbit. This is very important for identification of unknown asteroids.

5. Conclusions

The Struve astrometric satellite will be an efficient instrument for investigation of the Solar System objects. The use of the Vlasov scheme of observations gives a uniform coverage of the ecliptic zone of the sky. A great number of new asteroids is expected to be discovered. The total number of moving objects to be detected by the Struve satellite is estimated as 50 to 80 thousands after 3 years of observations.

References

Kisselev, A.A., Bykov O.P. (1976) Determination of the elliptical orbit of a satellite with the use of parameters of its apparent motion, *Astron.J.*, **53**(4), 879–888.
Yershov, V.N., Kanayev, I.I. (1994) Development of the Russian Space Astrometric Satellite, *Third International Workshop on Positional Astronomy and Celestial Mechanics*, ed. A. López García, Cuenca.

ASTROMETRIC OBSERVATIONS OF FAINT SATELLITES

R. VIEIRA MARTINS, C. H. VEIGA
Observatório Nacional
Rua General José Cristino 77, CEP 20921-030 Rio de Janeiro,
RJ - Brazil. E-mail: rvm@on.br or cave@on.br

AND

M. ASSAFIN
Observatório do Valongo/UFRJ
Ladeira Pedro Antônio 43, CEP 20080-090 Rio de Janeiro,
RJ - Brazil. E-mail: massaf@vms1.nce.ufrj.br

Abstract. We present a method to obtain the reference system for isolated observations of faint satellites made with CCD[1]. The method consists in the construction of a secondary catalogue of faint stars using 'The Digitized Sky Survey' and 'The Guide Star Catalogue' corrected by modern astrometric catalogues.

1. Introduction

In 1982, we started a systematic program of astrometric observations of faint satellites. From 1982 to 1988, all observations were made using photographic plates and, since then, CCD devices have been used.

For the reduction of our observations, some methods were developed. The center positions for the images were determined using special photometric methods (Veiga and Vieira Martins,1995) and the reference system was obtained through the method described by Veiga and Vieira Martins (1994) which uses the apparent motion of the planet.

Nevertheless, the method of the motion of the planet cannot be used for all observations. This happens if, between the first and the last frame

---

[1] Based on observations made at Laboratório Nacional de Astrofísica/CNPq/MCT-Itajubá-Brazil.

of a observations mission, the planet do not cover an arc longer than the distance between the satellite-planet or satellite-reference satellite and also for isolated observations.

So, we have now developed a method for isolated observations of faint satellites with small CCD fields. This method is similar to the classical one which consists in the construction of secondary catalogues. However, in our method, we make use of 'The Digitized Sky Survey of the Space Telescope Science Institute' in the place of the astrographic plate.

Next, we present some data about our observations, the general idea of the method of astrometric calibration and, as an example, an application to observations of Nereid (magnitude 19).

2. Observations

The faint satellites observed in our program are: the Martian satellite Deimos; the Jovians Amalthea, Thebe and the eight outer satellites; Helene, Calypso, Telesto and Phoebe of Saturn; the satellites Miranda, Ariel Umbriel, Titania and Oberon of Uranus; Triton and Nereid of Neptune.

The observations were made at the Cassegrain focus of the 1.6 m reflector of the Laboratório Nacional de Astrofísica, Brazil ($\phi \approx -23°$). This telescope has a focal distance equal to 15.8 m which gives the scale of $13''.0$/mm at focal plane (Veiga et al.,1987).

The CCD used was a EEVP 88231 of 770 × 1152 square pixels with 22μm. No filter was used. As the image of the planet is always saturated, the diffraction spikes were avoided by placing a mask with 8 circular apertures between the secondary mirror vanes. To avoid the saturation due to the light of the planets (Mars, Jupiter and Saturn), circular masks with appropriate diameter are placed on the CCD window. The distance from the window to the chip is 10mm and so we can observe faint satellites when the distance from the edge of the planet is greater than $13''$.

So, we have images of the satellites surrounded by a small field ($5'30'' \times 3'40''$) of stars whose magnitudes are smaller than or equal to the satellites magnitudes.

3. Astrometric Calibration

The usual method to obtain positions of satellites in an equatorial reference system is the same used in astrometric reductions of faint objects. It consists in the construction of a secondary astrometric catalogue of faint stars in the neighborhood of the satellites using high quality astrometric plates taken by telescopes with small focal distance. This procedure presents two problems: first, it is necessary to have an astrometric plate of the field and second, we must measure this plate carefully.

To avoid these problems we developed a variation of this method using facilities available in almost all observatories. The method consists in the following steps:

1. We take a field with about 4° × 4° centered in the satellite position using a good astrometric catalogue and the Guide Star Catalogue (GSC) (Russel et al.,1990). We correct the GSC positions using a third degree plate model. So, we arrive to a local version of the GSC corrected by a selected astrometric catalog.

2. Using the Digitized Sky Survey (DSS), we determine the center of GSC stars in a field with 30' × 30' surrounding the CCD frame. We determine also the center of some stars on the CCD.

3. With the corrected GSC positions and using a first degree polynomial, we obtain a catalog for the CCD stars.

4. Using the CCD images, we measure the center of the stars and the satellites.

5. With the catalogue constructed for the CCD stars, we calculated the positions of the satellite using a polynomial of degree one or two. The degree of the polynomial depends on the relative positions of the satellites in the CCD matrix.

The astrometric details of the method which was used for many extragalactic radio sources are given in Assafin and Vieira Martins (1995).

In order to present an example we give the results of the application of the method for 7 observations of the system Triton-Nereid. The observations were made in two nights and Triton and Nereid are near opposite edges of the CCD matrix (their distance was about 310").

The astrometric catalogue used was the Carlsberg Automatic Meridian Circle Catalogue-4 (CAMC-4) (1990). For the first step, the GSC correction was made with 23 CAMC-4 stars in a field 4° × 4° with a third degree polynomial and the adjustment yielded $\sigma_x = 0''.22$, $\sigma_y = 0''.20$. For the step 2, the error in the center process were typically $0''.04$. The step 3 was done with 11 stars from a field 12' × 12' with a first order plate model and the standard errors were $0''.11$ for x and $0''.09$ for y.

The center of the CCD images (step 4) were reduced using the program ASTROL (Colas and Serrau,1993) and the errors were typically $0''.03$ for the two directions. The catalog for CCD stars (step 5) was calculated with a second order polynomial with 14 stars and the typical errors were $0''.11$ for x and y. Finally, the O-C for the relative positions Triton-Nereid are $\bar{x} = 0''.06$ ($\sigma_x = 0''.03$) and $\bar{y} = -0''.01$ ($\sigma_y = 0''.09$) for 7 images taken in two nights. The theoretical positions of Triton and Nereid were calculated following Veiga et al. (1995).

4. Conclusion

We presented a method to obtain the reference system for isolated observations of faint satellites. It presents some advantages:
- large field plates are not required;
- it can be used for small CCD fields;
- as the method provides the equatorial coordinates of satellites, it works well to obtain positions of satellites far from the planets.

Acknowledgements

This work was concluded while the authors (Roberto Vieira Martins and Carlos Henrique Veiga) were visiting scientists at Bureau des Longitudes. This work was partially supported by CNPq-Brazil.

References

Assafin, M., Vieira Martins, R.: 1995, in preparation.
Carlsberg Meridian Catalogue No.4, 1990, Copenhagen University Observatory, Royal Greenwich Observatory and Real Instituto y Observatório de la Armada en San Fernando.
Colas, F., Serrau, M.: 1993, *ASTROL et INTERPOL*, version 3.10, Edition du Bureau des Longitudes. Paris.
Russel, J.L., Lasker, B.M, Macleau, B.J., Sturch, C.R., Jenkner, H.: 1990, *Astron. J.* **99** 6, 2059.
Veiga C.H., Vieira Martins, R., Veillet C., Lazzaro D.: 1987, *Astron. Astrophys. Suppl. Ser.* **70**, 325.
Veiga, C.H., Vieira Martins, R.: 1994, *Astron. Astrophys. Suppl. Ser.*, **107**, 551.
Veiga, C.H., Vieira Martins, R.: 1995, *Astron. Astrophys. Suppl. Ser.* **111**, 387.
Veiga, C.H., Vieira Martins, R., Le Guyader, Cl., Assafin, M.: 1995. *Astron. Astrophys. Supp. Ser.* in press.

THE PHESAT95 CAMPAIGN OF OBSERVATIONS OF THE PHENOMENA OF THE SATURNIAN SATELLITES

J.-E. ARLOT, W. THUILLOT, F. COLAS, P. DESCAMPS,
J. BERTHIER, B. MORANDO, C.H. VEIGA, D.T. VU AND
CH. RUATTI
Bureau des longitudes
Unité de Recherche Associée au CNRS URA 707
77 avenue Denfert-Rochereau, 75014 PARIS, FRANCE

J. LECACHEUX
Observatoire de Meudon
5 Place J. Janssen, 92195 MEUDON, FRANCE

AND

P. LAQUES
Observatoire Midi Pyrénées
Observatoire du Pic du Midi
BAGNÈRES DE BIGORRE, FRANCE

Abstract. This paper reports on the international PHESAT95 campaign of observations of the Saturnian events coordinated by Bureau des longitudes. Thanks to CCD or photometric receptors, accurate astrometric data can be get from the observation of the eclipses by Saturn and mutual events of the Saturnian satellites. These events occur from 1994 to 1996 and we give our first results.

1. Introduction

Every 15 years, the Earth and the Sun cross the equatorial plane of Saturn. Since the rings become less bright, various observations may be performed close to these events: research of new satellites, study of the faint ring structures, astrometry of the faint satellites. Furthermore during about three years around the plane crossing events, eclipses of the satellites by Saturn and mutual occultations and eclipses are observable. In advance to the exploration of the Saturnian system by the CASSINI space probe, these events are the occurrence to get accurate astrometric measurements of the Saturnian satellites. They are also a rare opportunity to get information on some physical parameters related to the surface parameters of these satellites.

2. The PHESAT95 Campaign

We have organized an international campaign to observe the eclipses and mutual events of the Saturnian satellites. Observers using photometric or CCD receptors have been called to join this campaign. A workshop was organized jointly by Bureau des longitudes and the Bucharest Institute of Astronomy and was held in Bucharest in September 1994. Problems related to acquisition and treatment of data were discussed. The proceedings are under publication but they are already available on the Web Server of Bureau des longitudes (http://www.bdl.fr/PHESAT95/phesat95.html). Predictions of the most observable phenomena have already been published (Arlot and Thuillot, 1993). Aksnes and Dourneau (1995) have also published their own predictions. More recently a full list of phenomena, including grazing events, have been made available on the ftp anonymous server ftp://ftp.bdl.fr/pub/ephem/satel/phesat95, or on floppy disk on request to the authors. A Web page at the address http://www.bdl.fr/phemu gives access to similar informations using the HTML facilities.

3. First observations

According to the predictions (Arlot and Thuillot, 1993),(Thuillot and Arlot, 1995), the main period of observations of eclipses and mutual events of the Saturnian satellites begins in summer 1995 and ends early in 1996. But some eclipses by Saturn were observed in 1994. These events are generaly difficult to observe because of the close distance to Saturn. Nevertheless several stations allow this type of observation. This is the case of Pic du Midi Observatory were we success to observe several such events. For example, figure 1 shows the reappearance of Rhea from eclipse by Saturn on November 1994. This event occurred at about 5 arcsec. from the edge of the planet and was observed with a CCD camera (THX7863) on the 1 meter telescope of Pic du Midi Observatory.

More recently mutual events have been observed in several sites. In France, at Haute-Provence Observatory, on June 17 1995, we success to observe the eclipse of Tethys by Enceladus at the 0.80 meter telescope (fig.2), using a CCD camera (THX7863). Such events do not involve the atmosphere of Saturn, contrarily to the eclipses of the satellites by the planet, so we expect to get a high accuracy in the astrometric measurement inferred from this photometric observation. According to previous similar

THE PHESAT95 CAMPAIGN OF OBSERVATION OF SATURNIAN EVENTS

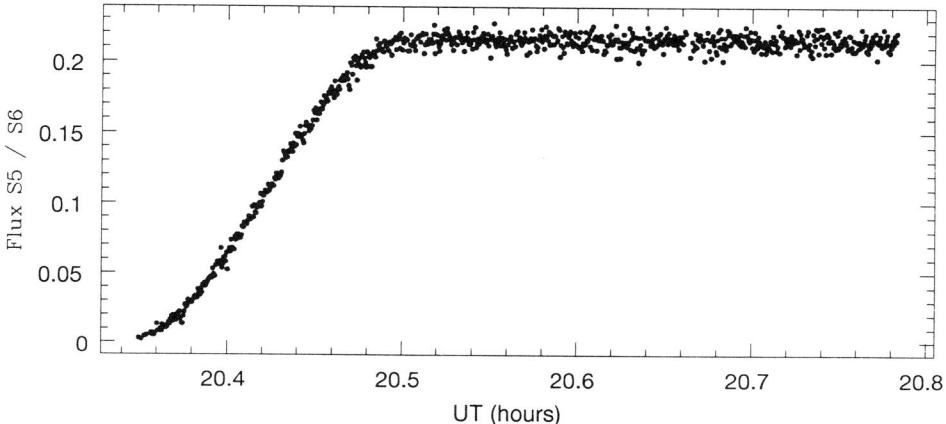

Figure 1. Reappearance of Rhea after an eclipse by Saturn observed on November 1994 at the 1m telescope of Pic du Midi Observatory

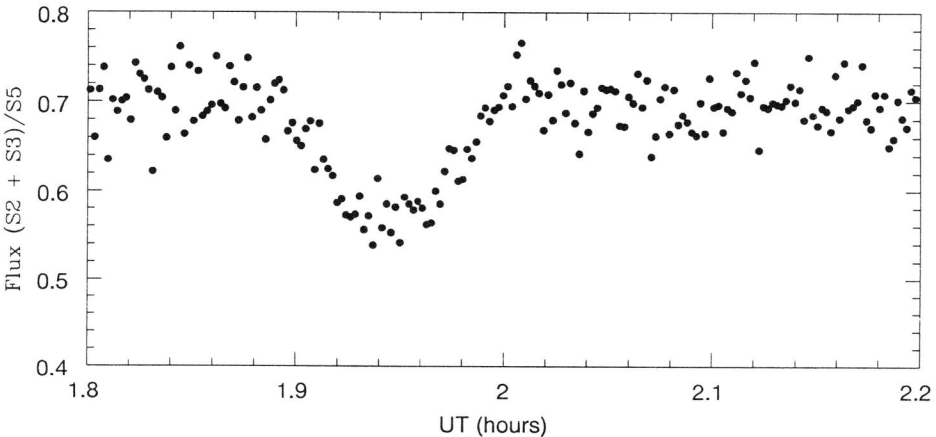

Figure 2. Occultation of Tethys by Enceladus observed on June 17 1995 at the 0.8m telescope of Haute-Provence Observatory

observations we made concerning the Galilean system of Jupiter, a precision better than 200km (0.03 arcsec.) can be expected. Furthermore some surface parameters of the satellites could be deduced from the modelling of these events.

4. Conclusion

The period of phenomena of the Saturnian satellites, which mainly occur from 1994 to 1996, is a great opportunity to get accurate astrometric measurements of these satellites. The PHESAT95 campaign of observation that we organize will get such data. Subsequently, they could be used in adjustment of the dynamical model of the motion of the Saturnian satellites.

References

Aksnes, K., Dourneau, G. (1994) *Icarus*, **Vol. no. 112**, p. 545-548
Arlot, J.-E., Thuillot, W. (1993) *Icarus*, **Vol. no. 105**, p. 427-440
Thuillot, W., Arlot, J.-E. (1995) Proceedings of the Workshop *CCD et récepteurs photométriques appliqués à l'observation des satellites de Saturne durant la période favorable de 1994-1996* , held in Bucharest, Bureau des longitudes and Institute of Astronomy, (also available on the Web Server http://www.bdl.fr/PHESAT95/phesat95.html)

THE RESULTS OF PHOTOGRAPHIC OBSERVATIONS OF GALILEAN SATELLITES WITH 26-INCH REFRACTOR AT PULKOVO.

T.P.KISSELEVA
The Main Astronomical Observatory
Russian Academy of Science,
Pulkovskoe shosse, 65-1,
St.Petersburg, 196140, Russia.

Abstract.
Photographic observations of the Galilean satellites of Jupiter have been made in Pulkovo with 26-inch refractor during two periods 1976-1981 and 1986-1993. The internal mean square errors of jovicentric coordinates and mutual distances is equal to 0.10 arcsec. The external errors are equal to 0.08 and 0.17 arcsec in AR and Decl. The systematic errors were analysed. The errors of the theory G-5 of J.E. Arlot do not exceed 0.1 arcsec.

Systematic observations of the Galilean satellites of Jupiter were carried out at Pulkovo using the 26-inch long-focus refractor in two periods: 1976-81 and 1986-93. 289 photographic plates with nearly 2500 exposures were obtained. Each plate contains two diurnal trails and ten images of Jupiter and its satellites. The high-contrast ORWO WO-3 plates were used. Exposure times of about 30-60 sec were chosen. No filters were used to reduce the planet brightness. The measurements were made with the help of a measuring machine Ascorecord. The scale-trail technique was used for astrometric reduction (Kisselev, 1989, Pascu, 1980).

In this paper, we consider the results of observations made in 1976-1981. The jovicentric coordinates of the satellites and their relative (mutual) coordinates were analysed by comparison with corresponding theoretical coordinates derived from the modern theory of Galilean satellites G-5, developed by J.E.Arlot (1982). The values (O-C) for jovicentric and mutual coordi-

TABLE 1. The mean (O-C) and m.s.e. in jovicentric coordinates (arcseconds)

S/J	$(O-C)_x$	$(O-C)_y$	E_x	E_y	IE_x	IE_y	N
1	-0.029	-0.062	0.100	0.116	0.090	0.090	102
2	-0.021	-0.039	0.107	0.166	0.090	0.100	113
3	-0.029	-0.044	0.116	0.187	0.090	0.100	114
4	-0.012	0.000	0.149	0.229	0.100	0.110	92

TABLE 2. The mean (O-C) and m.s.e. in mutual coordinates (arcseconds)

S/S	$(O-C)_x$	$(O-C)_y$	E_x	E_y	N
1-2	-0.011	-0.026	0.075	0.131	95
1-3	+0.008	-0.014	0.074	0.121	93
1-4	-0.029	-0.081	0.087	0.192	73
2-3	+0.008	-0.012	0.076	0.188	102
2-4	-0.002	-0.041	0.098	0.164	82
3-4	-0.022	-0.035	0.088	0.224	81

nates of satellites were used for analysis of the precision of observations, theory of motion, the phase effect of Jupiter and orbital phase effect of satellites, the errors of scale and orientation and the remaining refraction effects.

1. The precision of observations.

The mean square error (m.s.e.) of one observation was calculated from the deviation of the observed and theoretical relative coordinates in the whole 6-years period. The mean square error of one observation and the mean values (O-C) are presented in tables 1 and 2. The internal errors do not exceed 0.1 arcsec, but the external errors reach 0.2 arcsec for the 3rd and 4th satellites in Y-coordinate and depend on the distances between the satellites (table 3). Table 3 shows that the scale-trail method provide a high precision of relative coordinates (0.05 arcsec) if the X-coordinates do not exceed 100 arcsec.

The m.s.e. in the relative coordinates of each satellite with respect to the "mean satellite" are 0.062 and 0.107 arcsec (in X and Y).

TABLE 3. The dependence of m.s.e.
on X-coordinates (arcseconds)

X	E_x	E_y	N
0" - 20"	0.053	0.046	27
20 - 100"	0.050	0.049	118
100 - 1000"	0.083	0.170	425

2. The systematic errors of observations and reduction

The phase correction reduction model for Jupiter was computed on the basis of orthotropic law of light reflection with the coefficient 0.5 in X and Y-coordinates. This coefficient has been obtained by the analysis of the (O-C) near the satellite elongations.

The geometrical scale of 26-inch refractor was found to be 19.8078 ± 0.0006 arcsec/1 mm. Hence, the error of coordinates depending on the scale, for mutual distances up to 1000 arcsec, is not over 0.05 arcsec. This conclusion is confirmed by the high precision of coordinate X for all satellites in tables 1-2. The most considerable error in scale-trail method is the inaccuracy of orientation and effect of remaining refraction. These errors were investigated in detail using two trails on the photographs. The value of systematic orientation error in our case attained 0.018 degree leading to errors in relative Y-coordinate of about ± 0.22 arcsec for X-distances up to 600 arcsec (maximum for the 4th satellite). Thus, the errors of Y-coordinate increase with the distances of X-coordinates. The analysis of observational errors depending on zenith distances, hour angles and seasons have demonstrated small influence of these conditions. The total effect does not exceed ± 0.1 arcsec.

3. The precision of the theory of motion of Galilean satellites

The (O-C) of the satellite positions in their jovicentric orbits made possible to estimate the precision of the theory of motion of the satellites. The distribution of (O-C) on the jovicentric longitudes does not show any systematic deviations for the 1st and 2nd satellites. There is a small trend in (O-C)$_y$ for the 3rd satellite and appreciable wave in (O-C)$_x$ for the 4th satellite with maximum amplitude 0.1arcsec near the conjunctions. This wave may be conditioned by the error of longitude of the 4th satellite and also orbital phase effects of satellite. These two errors cannot be determined separately in this investigation. The error of longitude of the 4th satellite may be obtained from the solution of the system of equations

$$V_x T + XM = (O-C)_x \qquad (1)$$

where T is the unknown correction of longitude (the time argument), M is the unknown correction of scale, V_x is the X-velocity of the motion of the 4th satellite. From (1) we have obtained:

$$T = 0.354 \pm 0.141 \text{ min}$$

$$M = 0.00005 \pm 0.00004 \qquad (2)$$

$$E_o = \pm 0.146 \text{ arcsec}$$

Thus we may estimate the errors of theory of motion, because the other errors are known from our investigations:

$$E_{cx} = 0.082, \qquad E_{cy} = 0.132 \text{ arcsec.} \qquad (3)$$

Thus the precision of the theory G-5, as derived by their comparison with the observations of 26-inch refractor, is close to 0.1 arcsec; in other words, its accuracy is within observational errors.

4. The determination of coordinates of Jupiter by the measurements of satellites and stars of FK5 catalogue.

A by-product of this work was the determination of the positions of Jupiter by the satellites and fundamental stars which were measured on the photographs with Galilean satellites. In this case, the images of the planets were not measured. The positions of the FK5 stars and theoretical distances of satellites to Jupiter and the method scale-trail were used. The precision of this technique is higher than usual photographic one. The error of one position of the planet is 0.15 arcsec. In conclusion, the results of this work may be used in future observations of satellites of planets with CCD and 26-inch refractor, this year, in Pulkovo. The author thanks Dr.N.I.Glebova (Institute of Theoretical Astronomy) for computing theoretical coordinates of Galilean Satellites.

References

Pascu, D. (1980). "Methods of astrometric observations of natural satellites." In *Planetary satellites* (J.A.Burns, ed.) Mir, Moscow, p.71-105.

Kisselev, A.A. (1989). *The theoretical foundations of the photographic- astrometry* Nauka, Moscow.

Arlot, J.-E. (1982). "New constants for Sampson-Lieske Theory of the Galilean Satellites of Jupiter." *Astron. and Astrophys.* **107**, 305.

LIMITATIONS ON THE ACCURACY POSSIBLE IN ASTROMETRIC OBSERVATIONS OF THE SATELLITES OF THE MAJOR PLANETS

D.H.P. JONES
Royal Greenwich Observatory and
Queen Mary and Westfield College
Madingley Road, Cambridge CB3 0EZ
United Kingdom

Astrometric accuracy has two components; the accuracy with which an image can be centred on a CCD or photographic plate and the accuracy with which two image centres can be mapped on to standard co-ordinates. Photographic plates cover a sufficient area of sky that several standard stars can be measured with the satellite images and the mapping to standard co-ordinates can be done in one operation. CCDs have much smaller areas and the number of stars with sufficiently accurate positions may not be sufficient. The surface densities of the best available catalogues are shown in Table I.

TABLE 1. Available Catalogues

Catalogue	Mean Position Error (arcsec)	Number of Stars	Stars/Square Degree
GSC	0.5 (>1.0)	1.9×10^7	461
PPM (North)	0.27	1.8×10^5	8.7
PPM (South)	0.11	2.0×10^5	9.7

The accuracy with which the scale can be measured is shown in Table II using the 1-metre Jacobus Kapteyn Telescope (JKT) on La Palma as an example.

Centering accuracy depends on whether the image is under- or over-sampled, the signal-to-noise and the centering algorithm used. Experience with an ongoing trigonometric parallax programme shows that repeated CCD exposures of the same field may be mapped on to each other with an

TABLE 2. Characteristics of 1-metre JKT

Focal Ratio	f/15	f/8
Detector	CCD	Photography
Field Diameter (arcsec)	400	5400
Field Area (square Deg)	0.010	1.77
Expected Number of Stars		
GSC	4.6	816
PPMN	0.1	15
PPMS	0.1	17
Error of Scale ($\times 10^4$)		
GSC	12	0.9
PPMN	7	0.5
PPMS	3	0.2

error of 0.05 pixels. As our CCDs are typically 1000*1000 pixels we have to establish the scale with a fractional error ($\times 10^4$) of 0.5 and the rotation within 0.003 degrees. Table II shows that only the PPM catalogue used with the photographic camera can yield sufficient accuracy and only the GSC has sufficient stars to calibrate a typical CCD exposure. The best policy is to set up fiducial fields by photography using the PPM as primary standards to find the positions of a sample of stars within one CCD field.

Three photographic plates of the Pleiades taken with the JKT. There are approximately sixty PPM (North) stars on each plate and the mean error from the PDS measures is ± 0.30 arcsec in one co-ordinate, showing that the PPMN accuracy has been degraded by errors of plate measurement. Within the CCD fiducial field the inter-agreement of the measures was ± 0.18 arcsec in one co-ordinate. Six CCD images of the fiducial field which contains nine stars of roughly the fourteenth magnitude, were measured and the agreement with the photographic positions was ± 0.05 arcsec in one co-ordinate on one image. The mean of the six CCD images was used to correct the photographic star positions; after correction the error improved to ± 0.008 arcsec which is 0.03 pixels. Thus we have more than achieved the desired accuracy in centering the images but the accuracy of the scale and orientation of the chip are still limited by errors in the star positions.

Thirty stars were measured in the field of M92 and treated in the same way as the Pleiades. The five images were taken on different nights. Thirteen stars were measured in NGC 6823 and reduced relative to the GSC; four stars were measured in the field of Saturn on 1993 August 28/29 and also

LIMITATIONS ON THE POSSIBLE ACCURACY

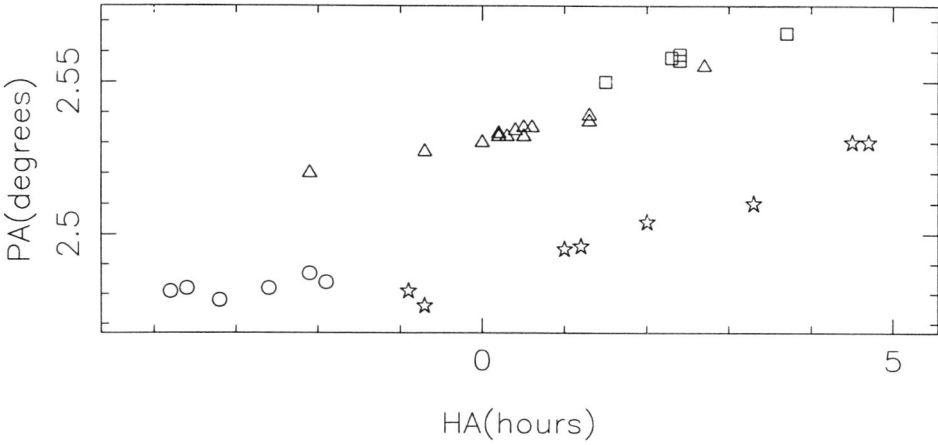

Figure 1. Rotation of fields at different declinations with Hour Angle, Squares M92, declination 43.1, Circles Pleiades 24.5, Stars NGC 6823 23.2, Triangles Field of Saturn -14.2; the errors for M92 and the Pleiades are ~ 0.001 degrees and the other fields ~ 0.01

measured relative to the GSC. Fig. 1 shows the angle in degrees between the columns of the chip and north plotted as a function of hour angle. The scatter after correction for hour angle in each field is commensurate with the errors which implies that the telescope is moving elastically without significant hysteresis or plasticity.

TABLE 3. Coefficients of Pointing Errors

Parameter	Hour Angle	Declination	Position Angle
Altitude of Pole	$\tan\delta\sin\tau$	$\cos\tau$	$\sec\delta\sin\tau$
Azimuth of Pole	$\tan\delta\cos\tau$	$\sin\tau$	$\sec\delta\cos\tau$
Non-perpendicularity of axes	$\tan\delta$	.	$\sec\delta$
Collimation	$\sec\delta$	.	$\tan\delta$
Flexure of Declination Axis	$\sec\delta(\sin\phi\sin\delta + \cos\phi\cos\delta\cos\tau)$	.	$\sin\phi\sec\delta$
Flexure of Tube	$\sec\delta\cos\phi\sin\tau$	$\sin\phi\cos\delta - \cos\phi\sin\delta\cos\tau$	.

Some of the rotation in Fig.1 is caused by imperfections of the equatorial mounting. The correction formulae for hour angle, declination and field rotation were developed by Bessel (1841), Arend (1951). The corrections

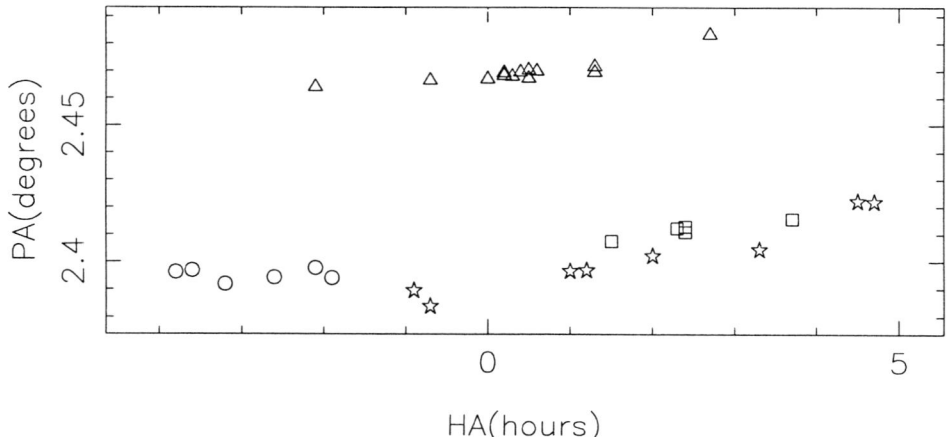

Figure 2. Rotation after correction for errors of the equatorial mounting, symbols as in Fig.1

are regularly determined so that the telescope can be accurately pointed. The effect of applying these corrections to Fig.1 is shown in Fig.2. The discrepancies are reduced but still significant.

Alternatively the measured position angles may be used to find the pointing coefficients of the equatorial mounting. If this is done the residuals show no correlation with position on the sky and have a standard deviation of 0.003 degrees. However the pointing coefficients are then significantly different from those used to point the telescope. These discrepancies may arise from errors in the star positions or from Bessel's model of the equatorial telescope being incorrect in this case e.g. the assumption that the telescope tube and declination axis flex proportionately to the sine of the zenith distance may be false.

It is possible to calibrate CCD astrometry adequately and absolutely from calibration fields. However these fields should have star positions more accurate than those currently available and should lie close in the sky to the target field.

References

Bessel, F.W. (1841) Theorie eines, mit einem Heliometer versehen Aequatoreal-Instruments, *Astronomische Untersuchungen*, **Erster Band**, pp. 1–54

Arend, S. (1951) Théorie de l'équatorial visuel et de l'équatorial photographique., *Monographies Observatoire Royal de Belgique*, **2.**, pp. 51–57

CCD ASTROMETRY ORIENTED ANALYSIS OF DIGITIZED MULTIPLE IMAGES (PLANETS, SATELLITES, ETC.) OBTAINED WITH THE PULKOVO PVC

GEORGE A. GONCHAROV
Pulkovo observatory
196140, Saint-Petersburg, Russia

The Charge Coupled Device (CCD) has become the detector of choice in astrometry. The mechanical or electronic (drift scanning) tracking of objects by the telescope leads among observing strategies developed for CCD observations (Stone and Dahn, 1995).

Another strategy can be adopted when several bright objects are available in one frame, as in differential observations of the Solar system bodies (brighter than 12^m) with respect to the Hipparcos/Tycho stars. Such observations will be the program of the Pulkovo photographic vertical circle (PVC) in the near future when the photographic camera will be exchanged with a CCD. In this case the PVC's field of 30×30 arcmin (focal scale is 0.1 arcsec/micron) will contain roughly 5 reference stars and a Solar system body. In this method the telescope is kept stationary and the objects move across the field (at the diurnal rate) providing trails. A precise vane is used to recognize simultaneously observed parts of the trails of different objects. Finally the relative coordinates of the simultaneous parts are measured and averaged. A similar method was considered in (Schildknecht, Hugentobler, Verdun, 1994) as applied to fast moving objects.

In the methods with the tracking the position of accumulated image is determined with accuracy $\sigma(T) = 0.33" \cdot (T + 0.65^s)^{-1/4}$, where T is exposure (Høg, 1968). In the method of trails the exposure T is divided into n independent exposures of t ($T = n \cdot t$), and the accuracy is higher because $\frac{\sigma(t)}{\sqrt{n}} < \sigma(n \cdot t)$. Both differential methods greatly reduce the effects of refraction and atmospheric turbulence, but the method of trails takes into account a co-turbulence of the simultaneous images.

To test the method, we used plates photographed by the PVC in 1987 – 1995 which contain trails of simultaneously observed objects: double stars,

Jupiter with Galilean satellites and so on (a by-product of usual meridian observations). Several tens of the frames were digitized with a pixel size of about 10 microns and 256 levels of gray scale. It was proved that photographic effects are negligible (the grain size is 3 microns).

The analysis of the frames indicates that the atmospheric turbulence is of crucial importance in the shift and distortion of the images. It has a periodicity of 1 − 10 seconds and mean amplitude of ±1 arcsec. The turbulent oscillations of the trails in one frame are correlated so that a scatter of distances between trails is 0.1 to 0.5 of the scatter of a trail about its parallel. This correlation is significant everywhere over the windowed telescope's field of 6 arcmin.

The analysis demonstrates the advantage of the method of trails. The length of the trail's parts used in the treatment for unit relative coordinates was optimized for the amplitude and periodicity of the atmospheric turbulence for every night apart. In most cases we treated the trails exposed within 40 seconds dividing them into the parts corresponding to about 0.2 seconds. The centres of the parts were determined with accuracy of ±0.5 pixel because of poor signal/noise relation (this is the drawback of the method). By this means, the differential coordinates of similar-in-magnitude objects were obtained with a mean internal error of ±0.04 arcsec.

A method using informative (mainly peripheral) parts of bright images was applied for different-in-magnitude objects in one field: Jupiter (-2.5^m) with Galilean satellites (4.6^m to 5.6^m), Polar star (2.0^m) with its satellite (9.0^m) and so on. Their differential coordinates were obtained with the mean internal error of ±0.08 arcsec.

Acknowledgements

The calculations were performed with a computer provided by the Russian foundation of fundamental investigations (grant # 93-02-3056).

References

Høg, E. (1968) Refraction anomalies: the mean power spectrum of star image motion, *Zeitschrift für Astrophisik*, **Vol. 69**, pp. 313–325

Schildknecht, T., Hugentobler, U., Verdun, A. (1994) CCD astrometry of artificial satellites, In *Dynamics and Astrometry of Natural and Artificial Celestial Bodies*, (K.Kurzyńska et al., Eds), Poznań, pp. 103–108

Stone, R.C., and Dahn, C.C. (1995) CCD astrometry, In *Proc. of the IAU Symp. No. 166, Astronomical and Astrophysical Objectives of Sub-Milliarcsecond Optical Astrometry* (E.Høg and P.K.Seidelmann, Eds), Kluwer Acad. Publ., pp. 3–8

SOLAR SYSTEM OBJECTS OBSERVED BY HIPPARCOS.

D. HESTROFFER AND B. MORANDO
Bureau des Longitudes
77 Av. Denfert Rochereau
F-75014 Paris

Abstract. The ESA satellite Hipparcos provides valuable photometric and astrometric results on minor planets and natural satellites. Observations of 48 asteroids, J II-Europa, S VI-Titan and S VIII-Iapetus were made from November 1989 to March 1993. A twenty seconds averaged normal place is constructed, providing thus positions accurate to a few hundredths of arcseconds and relative to a very precise and homogeneous reference frame.

1. Introduction

The systematic scanning of the sky by Hipparcos from November 1989 to March 1993 yields accurate astrometric parameters for about 120 000 stars. It will provide an homogeneous optical counterpart of the ICRS's reference frame (ICRF). Because the Hipparcos sphere has no orientation, it is tied to the reference frame of extra-galactic objects. It can also be tied to the dynamical reference frame defined by the motion of solar system objects. Three natural satellites and 48 minor planets have been added to the list of stars to be observed by the Hipparcos satellite (they are listed on Table 1 and 2). These are small and bright objects not larger than $\approx 1"$, and with magnitude lower than 12.5.

2. Observations

The payload design of the instrument and the global reduction of the FAST consortium are described in (Perryman et al., 1992) and (Kovalevsky et al., 1992); the reduction procedure of the astrometric data for minor planets is described in (Hestroffer et al., 1995).

TABLE 1. Solar system objects observed by Hipparcos
Asteroids

N°	Name	Nbr	N°	Name	Nbr	N°	Name	Nbr
1	Ceres	67	18	Melpomene	103	63	Ausonia	13
2	Juno	62	19	Fortuna	31	88	Thisbe	36
3	Pallas	63	20	Massalia	61	115	Thyra	32
4	Vesta	51	22	Kalliope	64	129	Antigone	43
5	Astraea	80	23	Thalia	66	192	Nausikaa	32
6	Hebe	93	27	Euterpe	36	196	Philomela	16
7	Iris	66	28	Bellona	36	216	Kleopatra	20
8	Flora	57	29	Amphitrite	63	230	Athamantis	35
9	Metis	47	30	Urania	48	324	Bamberga	66
10	Hygiea	47	31	Euphrosyne	15	349	Dembowska	96
11	Parthenope	67	37	Fides	33	354	Eleonora	97
12	Victoria	24	39	Laetitia	103	451	Patientia	29
13	Egeria	36	40	Harmonia	105	471	Papagena	108
14	Irene	47	42	Isys	50	511	Davida	65
15	Eunomia	87	44	Nysa	53	532	Herculina	38
16	Psyche	46	51	Nemausa	15	704	Interamnia	72

TABLE 2. Solar system objects observed by Hipparcos
Satellites

Jupiter			Saturn		
N°	Name	Number	N°	Name	Number
J II	Europa	88	S VIII	Titan	111
			S VIII	Iapetus	65

Due to the particular scanning law of the Hipparcos satellite, the amount of observations varies between the different objects (columns 'Nbr' and 'Number' of Tables 1 and 2). Moreover the observation of a solar system body occurs in a 86° wide zone around the quadratures, which can yield a photocentre offset of a few mas for the largest planetoids.

Because the fundamental measure is a photon count, Hipparcos not only provides positions of the observed objects, but also magnitudes with a precision of a few 0.01 mag (Mignard et al., 1992), (Morando and Mignard, 1993). So, reduced light curves for a few objects and phase curves between 15° and 25° can be constructed.

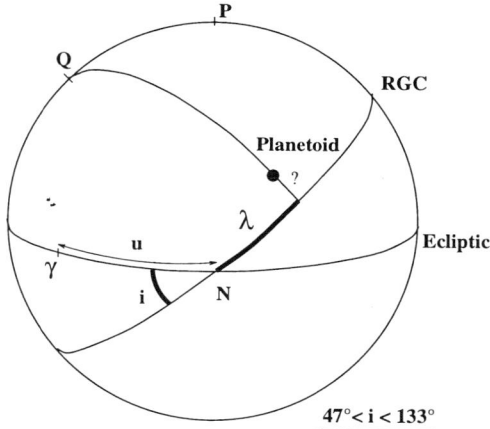

Figure 1. Position of a planetoid given by Hipparcos

3. Astrometry

As the measures are made through a modulating grid with parallel slits, Hipparcos observations only provide a one-dimensional abscissa λ on a reference great circle (RGC) as shown on Figure 1. This precessing great circle oscillates, with a period of 57 days, around a plane perpendicular to the ecliptic, yielding thus more information on the planetoids latitude than on their ecliptic longitudes.

The reduction procedure takes into account effects of the order of 1 mas. Care has also been taken to avoid modelisation errors of such an amplitude so that no correction of the photocenter offset has been added. The given abscissa corresponds then to the astrometric direction of the planet photocentre for each transit in the field of view. In general, this normal point is distant to the center of mass by only a few 0.1 mas. The accuracy of the position obtained depends on the body's magnitude and size, it is of the order of 0″.015.

The O–C's obtained, for the satellite S VI-Titan and for all the minor planets, from the comparison to the Hipparcos observations are shown on Figure 2. These are differences on the great circles; the dispersion present on these graphs arises from the measurement noise, the variation over successive transits on a same great circle, and, on a larger time scale, from the geometry of the scanning direction relative to the planets' orbit.

The calculated places for the minor planets were obtained by numerical integration with initial conditions taken from the "Ephemerides of minor planets, 1992". The O–C's depend on a rotation of the Hipparcos sphere

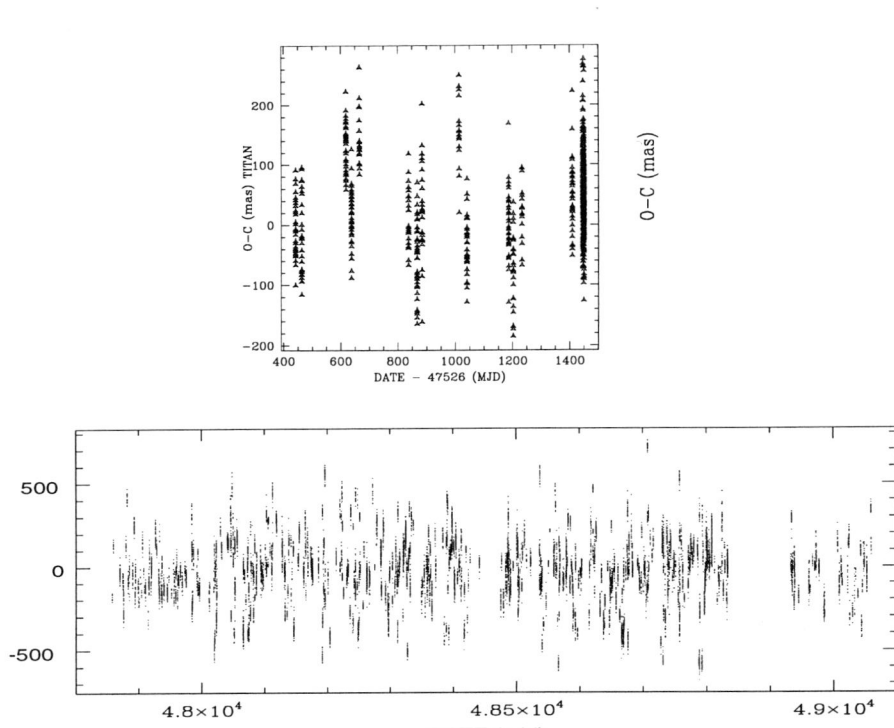

Figure 2. O–C's for S VI-Titan (*top*) and for the 48 minor planets (*bottom*).

relative to the dynamical reference frame, on the systematic errors due to phase effect, on errors of the Earth's ephemerides; but mostly on the initial conditions from the minor planets' ephemerides. For Titan the ephemerides were taken from the theory of Dourneau (1987) and the ephemerides of DE200 for Saturn, the O–C's are then in the range $\pm 0\rlap{.}''2$; next step will be to compare these observations to the more recent theories TASS (Duriez and Vienne, 1991) or of Dourneau (1993), with improved ephemerides of Saturn.

4. Reference Frame

The link of the Hipparcos sphere to the dynamical frame – defined by the minor planets' motions – is given by an infinitesimal time dependant rotation: $\mathbf{W} = \mathbf{W}_o + (t - t_o) \cdot \mathbf{W}_1$. Other parameters outlined in Section 3 are the initial conditions of the minor planet ephemerides. With 3 years of observations it is not possible to determine the whole set of initial conditions. Parameters such as eccentricity, semi-major axis or mean longitude are not or are poorly determined; nevertheless, the orientation of the osculating plane can be provided to an accuracy of a few mas.

Correction to the orientation of the Earth's osculating plane can not be determined simultaneously with those of the minor planets and the orientation of the sphere (Hestroffer et al., 1995). The orientation $\mathbf{W}_o$ and the rotation rate $\mathbf{W}_1$, given with respect to the DE 200 system, are provided with an accuracy of 1–2 mas and 1–3 mas/year respectively; where the orientation of the ecliptic is better determined than the position of the equinox (Hestroffer, 1994).

5. Conclusion

Hipparcos observations of minor planets and natural satellites, made over a period of 3 years, yield magnitude determination with a precision of a few 0.01 mag and astrometric positions accurate to about 0".015. From the minor planets' astrometric data, it is possible to obtain a partial improvement of their ephemerides. These observations can however profitably complete the ground-based observations made on a wider time span. It is also possible to determine the rotation of the Hipparcos sphere relative to a dynamical reference frame with a precision of 3 mas and 3 mas/year.

Hipparcos observations of solar system objects are of great value for orbital improvement or frames linking, but they are spread over a short time span when compared to the minor planets' sidereal periods. These observations will be advantageously completed by high precision observations from ground (CCD observations of occultations or close approaches) or by astrometry from space (HST, Struve, GAIA).

References

Dourneau G. (1987) "Observations et étude du mouvement des huit premiers satellites de Saturne", *Thèse*, Bordeaux.
Dourneau G. (1993) "Orbital elements of the eight major satellites of Saturn determined from a fit of their theories of motion to observations from 1886 to 1985", *A&A*, **167**, 292–299.
Duriez L. and Vienne A. (1991) "A general theory of motion for the eight major satellites of Saturn", *A&A*, **243**, 263–275.
Ephemerides of minor planets for 1992, Institute of Theoretical Astronomy, Leningrad.
Hestroffer D. (1994) "Astrométrie et photométrie des astéroïdes observés par le satellite Hipparcos. Apport à l'élaboration d'un système de référence dynamique", *Thèse*, Observatoire de Paris.
Hestroffer D., Morando B., Mignard F., Bec-Borsenberger A. (1995) "Astrometry of minor planets with Hipparcos", *accepted by A&A*.
Kovalevsky J., Falin J.L., Pieplu J.L., Bernacca P.L., Donati F., Froeschlé M., Galligani I., Mignard F., Morando B., Perryman M.A.C, Schrijver H., van Daalen D.T.,

van der Marel H., Villenave M., Walter A.G., Badiali M., Borriello L., Brouw W.N., Canuto E., Guerry A., Hering R., Huc Cl., Iorio-Fili D., Lacroute P., Lattanzi M., Le Poole R.S., Murgolo F.P., Preston R.A., Röser S., Sansò F., Wielen R., Belforte P., Bernstein H.H., Bucciarelli B., Cardini D., Emanuele A., Fassino B., Lenhart H., Lestrade J.F., Prezioso G. and Tommasini Montanari T. (1992) "The FAST Hipparcos Data Reduction Consortium: overview of the reduction software", A&A.,258, 7-17.

Mignard F., Froeschlé M. and Falin J.L. (1992) "Hipparcos photometry: FAST main mission reduction", A&A.,258, 142-148.

Morando B. and Mignard F. (1993) In:*Developments in Astrometry and their impacts on Astrophysics and Geodynamics*, I.I. Mueller and B. Kolaczek (eds.), p 2.

Perryman M.A.C, Høg E., Kovalevsky J., Lindegren L., Turon C., Bernacca P.L., Crézé M., Donati F., Grenon M., Grewing M., van Leeuwen F., van der Marel H., Murray C.A., Le Poole R.S. and Schrijver H. (1992) "In-orbit performance of the Hipparcos astrometry satellite", A&A.,258, 1-6.

ASTROMETRY BY ASTEROID OCCULTATIONS

L. KAHL KRISTENSEN
Institute of Physics and Astronomy
University of Aarhus, DK-8000 Aarhus, Denmark

Abstract. For the determination of the dynamical equinox in fundamental star catalogues observed by astrometric satellites it is proposed to observe occultations of stars by selected asteroids. Being a long term programme now is the time for planning. Quadrature observations and photometric back-up are essential.

1. Introduction

The Hipparcos satellite provides a high accuracy fundamental catalogue at epoch 1991.5. However, the equator and ecliptic, which by tradition fix the coordinate axes, require separate determinations. The equator and precession must be determined by instruments taking part in the rotation of the Earth, as for instance VLBI radio observations. The determination of the ecliptic requires observations of planets. Direct observations by Hipparcos are affected by systematic errors due to the deviation of the photocentres at the large phase angles $15\text{-}25^0$ caused by the scanning mode. This displacement is also indicated in high accuracy photographic positions (Kristensen 1980). In the JPL series of ephemerides, partly based on radar and radio observations, coordinates of the Earth are available with superior accuracy. The problem is, however, to determine the rotation between the two systems. Here we shall investigate the possibility of using asteroid occultations as a direct link between the stars and the planets.

2. Accuracy of occultation observations

Asteroids have been proposed for fundamental astrometry by several authors (Dyson 1929, Numerov 1932, Brouwer 1935, Strömgren 1950) because their positions may be obtained relative to the stars with high systematic accuracy. The most accurate positions of asteroids relative to stars are obtained from occultations. Formally an event timed to ±0.2 s gives, with the typical daily motion 0.25^0, a position error of order ±0.002''; corresponding to 2-3 km on the surface. Ultimately the errors will depend on how well the shape can be modelled and on surface irregularities. A mean error $\sigma = \pm 5$ km in the relative positions on the sky-plane does not seem unrealistic. The rotation periods and poles must be determined from lightcurves and the occultations themselves will then contribute to the determination of shape. This work can be made separately and requires access to and combination of all the original observations. The results give the simple input to the astrometric programme as **four** numbers for each event: time, relative coordinates, asteroid minus star, and a star identification label. For many years the unknown shape will be of minor importance compared to the present ± 0.15 ''/century error in proper motions. Full benefit of occultations must await a new astrometric satellite!

3. Determination of rotation to dynamical frame

A simplified model illustrates the essential aspects in the determination of the rotation. Nearly circular orbits in the (X,Y)-plane are assumed for the Earth and the asteroid. The Earth has orbital radius a and the coordinates taken from the ephemeris are rotated through the angles (p, q, r) around the X, Y and Z axes. The elements of the asteroid are supposedly determined by more than three occultations. We assume that the eccentricity, perihelion and semi-major axis a' are well separated from the rotations (p', q', r') by a uniform distribution of the observations. The perfect situation would be if these three quantities were determined in the system of the inner planets by transmitters or reflectors on the surface of the asteroids. This would give the additional condition (p',q',r')=(p,q,r) and make the adjustment trivial. Let λ and λ' be the longitudes of the Earth and the asteroid, E the elongation and β the phase angle. The left hand side of the equations of condition in the Z-direction on the sky plane is:

$$a'p' \sin \lambda' - a'q' \cos \lambda' - ap \sin \lambda + aq \cos \lambda \tag{1}$$

and in the direction of motion:

$$a'r' \cos \beta + ar \cos E \tag{2}$$

These equations will have the same weights because their mean errors are equal and of order $\sigma = \pm 5$ km.

The unknowns really wanted are simply (p, q, r), the corresponding elements for the planet being dummies. We note immediately that if observations are concentrated near opposition ($\lambda' = \lambda$) it is not possible to separate a·(p, q, r) from a'·(p', q', r'). Classical transit or photographic observations tend to concentrate at opposition and occultations have a great advantage in this respect. To separate the variables we must have different coefficients in (1) and (2) so the difference $\lambda' - \lambda = 180^0$-E- β must be as large as possible. The phase angle $\beta(E)$ is a function of the elongation E. Other thing being equal, distant asteroids are preferable by increasing the differences between the coefficients i (1) and (2). Equation (2) shows that quadrature observations separate r from r' (cos E = 0, sin $\beta = a/a'$). The separation of the node and inclination variables p and q is also mainly determined by the distribution of the observations in elongation. We solve (1) by least squares and assume symmetry with respect to opposition and average the normal equations over λ. Define the average $< \lambda' - \lambda >_k$ of N_k observations for asteroid k by:

$$N_k \cos < \lambda' - \lambda >_k = -\sum_{j=1}^{N_k} \cos(E_{jk} + \beta_{jk}) . \qquad (3)$$

The mean error ϵ in the determination of p and q is then estimated by

$$\epsilon^2 = 2\sigma^2/a^2 \sum_k N_k \sin^2 < \lambda' - \lambda >_k . \qquad (4)$$

The optimal accuracy is thus obtained if $\lambda' - \lambda$ is uniformly distributed around $\pm 90^0$ or elongation E_k given by $\tan E_k = a_k/a$.

4. The two stages of an occultation programme

Lunar occultations have proved their importance in fundamental astrometry and Newcomb (1878) demonstrated their accuracy and increasing value with time. A main point is that it is a "dependency" method where re-reductions are easily made and advantage can be taken of improved star places. High accuracy photographic observations must be stated by 3-4 dependencies to keep their value but for occultations this is simplified to a single star.

Work on asteroid occultations was pioneered by G.E.Taylor and is today well organized by D.W.Dunham and others. The main difficulty till now is the uncertain predictions. Today star errors in large surveys are at best $\pm 0.3''$ (PPM North). Orbits are based on observations with residuals $\pm 1.0''$ mainly due to plate measurements. Ephemerides average hundreds of such positions and many are at present more accurate than the stars. The star errors reduce to $\sigma^* = \pm 0.040''$ when the Tycho Catalogue becomes available and the predictions will immediately be improved. In a second stage the orbit can be determined by three occultations and the accuracy of the ephemeris will be comparable to that of the stars. To have 68% probability (1σ limit) that an event will actually occur on the central-line of a predicted track, the diameter D must satisfy:

$$2\sigma^* \Delta \leq D \qquad (5)$$

When occultation observations at this second stage become more a routine than happy chance, the necessary number of observations for each object may be secured. The number of chords may not be large though. Examples are known where only two well- placed chords give good accuracy (Kristensen 1981). Equation (5) has important consequences for the brightness of the stars. For a distant (a' = 3.30) object D > 134 km and assuming a low albedo $p_V = 0.04$ the planet magnitude is V(a,0) = 12.9 mag. The star must be at least 0.5 mag brighter in order to have a 1 mag drop and V* < 12.4 mag. The table gives the faintest V* for albedoes 0.04, 0.10 and 0.16. At quadrature phase effects may correct V* by + 0.7 mag, especially for dark (C-type) objects.

a'	D km	0.04	0.10	0.16	80°	60°	Yearly	
2.20	70	11.5	10.5	10.0	2.0	4.5	1.4	3.1
2.75	102	12.0	11.0	10.5	1.6	2.4	1.3	1.9
3.30	134	12.4	11.4	10.9	1.0	1.5	0.8	1.3

In the case of V*=10.0 a search in the Hipparcos catalogue is nearly complete but the effective density of stars is a little reduced by unusable fainter stars. In the opposite situation V*=12.4 some faint events are lost in the catalogue but all 24 stars per square degree are usable. The areas within the parallax 8.80″/Δ from the opposition loops to elongations 80° and 60° are multiplied by 24 stars/sq.deg. to give the mean number of predictions per opposition. The last two columns give the number of events per year.

References

Brouwer D.: 1935, *Astron. Journ.* **44**, 57-63.
Dyson F.: 1929, *Trans. I.A.U.* **3**, 227.
Kristensen L.K.: 1980, *Mitt. d. Astron. Ges.* **48**, 50-55.
Kristensen L.K.: 1981, *Astron.Astrophys.Suppl.Ser.* **44**, 375.
Newcomb S.: 1878, *Motion of the Moon. Washington Observations for 1875*. Washington, Appendix II.
Numerov M.B.: 1932, *Bull. l'Inst. Astron.* (Leningrad) **32** 139-147.
Strömgren B.: 1950, *Trans. I.A.U.* **7**, 220.

HIPPARCOS MINOR PLANETS:

TOWARDS AN IMPROVEMENT OF THE MODEL ANALYSIS BY DETECTING INFLUENCE FACTORS.

M. L. BOUGEARD
Observatoire de Paris, DANOF/IERS, 75014 Paris and Université Lyon 1, Dep. Math., 69622 Villeurbanne, France

J.-F. BANGE
Bureau des Longitudes, URA CNRS 707
77 Avenue Denfert-Rochereau, Paris, France

M. MAHFOUZ
Université Lyon 1, Dep. Math. Statistique

AND

A. BEC-BORSENBERGER
Bureau des Longitudes, URA CNRS 707

Abstract. In order to evaluate a possible rotation between the Hipparcos and the dynamical reference frames, Hipparcos minor planets preliminary data are analysed. The resolution of the problem is very sensitive to correlations induced by the short length of the interval of observation. Several statistical methods are performed to appreciate the factors of bad conditioning. A procedure for variable selection and model building is given.

1. Hipparcos Minor Planets (HMP) model

For each of the 48 minor planets observed by the astrometric satellite Hipparcos, we denote by 'O' the observed abscissae on a Reference Great Circle (RGC) related to a Hipparcos satellite revolution (see Hestroffer et al, 1995, for the reduction procedure). Subtracting the corresponding position 'C' calculated through numerical integration of the motion equations leads to the following equation of condition, written with sufficient approximation as:

$$\mathbf{O} - \mathbf{C} = \mathbf{A}\Delta\overrightarrow{u^0} + \mathbf{B}\Delta\overrightarrow{u_T^0} + \mathbf{D}\overrightarrow{\theta} \qquad \text{(HMP)}$$

where $\mathbf{A}$ is a matrix depending on the minor planet's coordinates partial derivatives with respect to its initial elements; $\mathbf{B}$, a matrix depending on the earth's coordinates partial derivatives with respect to its initial elements and **D** a matrix linking the Hipparcos reference frame and the dynamical one. The unknowns are the 6-component vectors Δu^0 (corrections to the minor planet orbital elements), Δu_T^0 (corrections to the Earth elements) and θ (link between the dynamical and Hipparcos reference frame). The initial epoch is t_0=JD 2447800.5 .

2. Statistical exploration of the HMP-Model

Here we focus our attention to the estimation of the 6×1 vector θ by using preliminary FAST data covering 37 months of the mission. The statistical approach is illustrated on a subset of 3 minor planets: 6-Hebe (91 observations on 20 RGC), 9-Metis (43 observations on 14 RGC) and 42-Isis (52 observations on 15 RGC). Performing a Least Squares regression by pooling the 186 equations of condition (parameters Δu_T^0 and θ are common) leads to a very large instability of the results that are not even meaningful! As a measure of possible multicollinearity phenomena, the **VIF** (Variance Inflation Factor) has been performed for each predictor in two cases: model 1 with full design matrix (A, B, D) and model 2 without the B-factors. We report some values for model 2 (Table 1).

Although both models fit the data quite well (coefficient of determination R2=0.97 and R2 adjusted=0.96) the very large **VIF** values assert that many factors are involved in the multicollinearity (Tassi, 1992). This concerns A-factors as well as D-factors: the overall model 1 including B-factors being worst (**VIF** values are larger). At this stage, the LS regression through any numerical approach (singular values or QR decomposition) is statistically misleading especially in terms of tests and error variance estimates (Bougeard and Michelot, 1991).

TABLE 1. Variance inflation factors (model 2)

	D_1	D_2	D_3	D_4	D_5	D_6	A_1	A_2	A_3	A_4	A_5	A_6
VIF	18	152	401	16	62	508	2997	6	23	203	5451	457

3. Procedures for multicollinearity detection

Consider the 42-Isis HMP data. Due to the short interval-length of observation several factors are measuring nearly similar physical phenomena. Although the correlation matrix is of full rank, three eigenvalues $\lambda_6 \leq \lambda_5 \leq \lambda_4$ appear as negligeable (Table 2).
Let T be the 6×3 matrix with columns V_6, V_5, V_4 the related eigenvectors. To detect the factors that may produce multicollinearity, the norm of each 3×3 sub-matrix inverse of T is computed. Those with minimal norm give indication on the influential factors (Table 3).

In conclusion, for obtaining a model meaningful from a statistical viewpoint, the variables/predictors to be dropped are: first A_1 and A_5, second A_4 or A_2 or A_3.

TABLE 2. 42-Isis data correlation matrix

	\multicolumn{6}{c}{Correlation matrix}	Eigenvalues	Cumulative proportion
	A_1	A_2	A_3	A_4	A_5	A_6		
A_1	1	0.04	-0.03	-0.98	-0.99	0.84	$\lambda_1=3.135$	0.619
A_2		1	-0.99	-0.12	0.03	-0.17	$\lambda_2=2.0415$	0.959
A_3			1	0.11	-0.04	0.18	$\lambda_3=0.2316$	0.998
A_4				1	0.95	-0.74	$\lambda_4=0.0102$	0.999
A_5					1	-0.91	$\lambda_5=0.0030$	1.0
A_6						1	$\lambda_6=0.0001$	1.0

4. Statistical building of the HMP model and Detection of influential observations

The above variable selection procedure was extended to 6-Hebe and 9-Metis predictors $A_1...A_6$ but also on predictors $B_1...B_6$ associated to the correction of the Earth orbit and $D_1...D_6$ associated to the link between reference frames (recall that large **VIF** values were associated to all these variables). Results obtained after a least squares regression on the selected model are given in Table 4. VIF and condition numbers are then acceptable. By reducing the effects of multicollinearity, estimates of θ with smaller mean squared error were produced. At this time, statistical tests to detect 'atypical' observations can be used. Here, we resort to the use of the Belsey-Kuh Welsch influence diagnostic statistics (already applied in Bougeard, 1987). Influential observations are given in table 5.

TABLE 3. Sub-matrices of rank 3

norm	2.301	2.305	2.498	2.500	3.004	3.005	3.357	3.361	1006.71
lines of T	1,2,5	1,3,5	2,4,5	3,4,5	1,2,4	1,3,4	1,2,6	1,3,6	2,3,6

5. Conclusion

In this paper, a statistical methodology for HMP model building was tested on a subset of three minor planets. Detection of multicollinearity was performed, factors of influence analysed on 42-Isis data (see also Bec-Borsenberger et al, 1994), a model built that provides estimate with smaller mean squared error. Then statistical influence tests have allowed the detection of influential observations.

TABLE 4. Results and collinearity diagnostics (arcsecond or milliarcsecond.JJ^{-1}: D_5, D_6)

predictor	parameter estimate	standard error	variance inflation	n°	eigenvalue	condition index
D_2	-0.513	0.006	17.7	1	4.00	1.0
D_3	-0.005	0.012	34.1	2	2.37	1.3
D_5	-0.04	0.01	12.1	3	2.26	1.3
D_6	-0.08	0.02	29.1	4	1.84	1.5
...	...	...	...	...	...	...
$42 - A_2$	0.077	0.005	4.8	15	0.07	7.7
$42 - A_3$	-0.019	0.006	7.8	16	0.03	10.9
$42 - A_6$	-0.007	0.005	4.0	17	0.01	19.4

TABLE 5. Belsley-Kuh Welsch influence statistics (*: observation potentially suspicious)

atypical obs. ref.	minor planet	RGC	numb. of obs on this RGC	Rstudent	Dfitts	HATDIAG	comment
34	6	450	8	*			?
61	"	1404	3	*	*	*	suspicious
78	"	2003	2	*	*	*	suspicious
91	"	2641	2	*	*	*	suspicious
94	9	771	4	*	*		suspicious
102	"	947	4	*	*		suspicious
120	"	1414	1	*	*		suspicious
130	"	1533	1	*	*		suspicious
135	42	258	2	*	*		suspicious
140	"	411	4	*	*		suspicious
148	"	549	6	*			?

References

Bange, J-F.: 1993, *Mémoire de DEA*, Observatoire de Paris.
Bec-Borsenberger, A., Bange, J.-F., Bougeard, M. L.: 1995, *Astron. Astrophys.* **304** 176.
Bougeard, M. L.: 1987, *Astron. Astrophys.* **173** 191.
Bougeard, M. L., Michelot, C., 1991: *Journées Systèmes de référence Spatio-temporels*, Observatoire de Paris, p. 86.
Hestroffer, D. *et al* : 1995, *Astron. Astrophys.* **304** 168.
Tassi, P.: 1992, Méthodes Statistiques, Economie.

CCD-OBSERVATIONS OF ASTEROIDS: ACCURACY AND THE NEAREST PERSPECTIVES FOR APPLICATION OF THE LAPLACIAN ORBIT DETERMINATION METHOD

O.P.BYKOV
Pulkovo observatory, Russia
E-mail: bykov@venvi.usr.pu.ru

The modern crowded CCD-observations of the Solar system small bodies are very effective and accurate. They easily provide close positions on a short topocentric arc for any celestial body – from several hours during one night to several successive nights at a single observatory. It is obvious, now, that the informational value of such short arc observations is very high. Statistical treatment of these accurate near positions gives an opportunity to calculate the first ($\dot{\alpha}$, $\dot{\delta}$) and second derivatives ($\ddot{\alpha}$, $\ddot{\delta}$) of spherical asteroid coordinates or an angular topocentric velocity and its positional angle, a topocentric angular acceleration and a curvature of visible trajectory of observed celestial body.

Pulkovo observatory has a long and successful experience of processing short arc observations of Artificial Earth Satellites, Double Stars, Minor Planets and Comets. The classical Laplacian method and the new Apparent Motion Parameters Method developed by Dr.A.A. Kiselev together with colleagues are applied for the orbits determinations.

Practically, the Laplacian method was usually at a disadvantage if compared to the Gaussian initial orbit determination method, in spite of efforts of known scientists – from A.Leuschner to J.Kovalevsky and F.Barlier. The author explains this situation by difficulties of obtaining the first and second derivatives of spherical coordinates from positional observations of celestial body: traditionally these derivatives were calculated from three positions only (it was the common practice in the case of Gaussian orbit determination method). But, at Pulkovo observatory, special parameters of celestial bodies motion were introduced into consideration and special procedures were created for their calculation with the use of crowded accurate sets of positional observations.

The author continues this tradition of Pulkovo investigations. First of all, it is necessary to compare precisions of new CCD and stable photo positional asteroid observations. Up to day, we have no own CCD-device at Pulkovo and the author was compelled to use the observations published in the Minor Planet Circulars. The CERES software package, created at Institute of Theoretical Astronomy, was applied to calculate residuals $(O-C)$ for near 136 numbered minor planets observed irregularly and quasi-simultaneously in 1993 by CCD, as well as by photo techniques, at 25 observatories (ESO, CERGA, Kitt Peak, Oak Ridge etc.). The accuracy of observations was estimated by means of standard error of average $(O-C)$ for each type of observations obtained by each telescope. As a rule, the CCD-observations of the numbered minor planets are considerably more exact than the photo ones [1]. This conclusion was confirmed later by analysis of an accuracy of the first and the second derivatives obtained from similar sets of CCD and photo observations of unnumbered asteroids.

The high accuracy of the CCD-observations and their efficiency allow to solve the problem of orbit determination for any celestial object moving in the field of reference stars. The Laplacian orbital elements are, of course, preliminary, but they are very close to the real asteroids orbits. The author's practice of orbital calculations with the use of the Laplacian method for the Near Earth Objects, the Main Belt and Kuiper Belt asteroids observed by CCD technique allows to affirm that the Laplacian method can give good results in a direct processing of CCD observations immediately, during their execution on a telescope (for a circular orbit), or several days after (for an elliptical one). The Laplacian method can be successfully applied to solve the problem of faint asteroid identification and detailed study of asteroids population in the Main and Kuiper Belts. For any observer, this method can provide an independent ephemeris service on the basis of the own observations only. The CERES software [2] and LAPLACE software, made at the Pulkovo observatory, are the first steps to realize this purpose in the nearest future.

The algorithms of the Laplacian orbit determination method are published in [3].

References

1. Bykov O.P.: 1994 "On the accuracy of CCD and photo observations of asteroids and their current orbit determinations", *Proceedings of IAU Symp. 167*, p.351, 1994.
2. Babaev I.O., Chernetenko Yu.A., L'vov V.N. et al.: 1994 *CERES - Minor Planets Ephemeris Software Package for IBM PC and Compatibles. User's Guide*,ITA, St.-Petersburg, p.24.
3. Bykov O.P.: 1989 "Determination of celestial bodies orbits by the direct methods", *Problemy postroenia koordinatnykh sistem v astronomii*, GAO, Leningrad, pp. 328-356.

THE "BLACK DROP" PHENOMENON AND REDUCTION OF THE MERCURY TRANSITS OBSERVATIONS

A.M. SVESHNIKOV[1] AND M.L. SVESHNIKOV[2]
[1] *Physical Faculty, State St.Petersburg University, Stary Peterhof, St.Petersburg, 198904 Russia*
[2] *Institute of Theoretical Astronomy, RAS Nab. Kutuzova 10, St. Petersburg, 191187 Russia*

Mercury transits are important for the investigation of long-term variations in the Earth rotation. They have been observed for more than 300 years. The basic array of this set is the visual observation of contacts (as a rule the second and third ones). The detection of the instant corresponding to the geometrical contact of the solar and Mercury limbs is a difficult task, since during 10^s to 60^s, the set of phases is observed transforming continuously from one to another. One of the encumbering factors, is the so-called "black drop" phenomenon, i.e., the dark cross-bar formed between the limbs (Struve, 1882; Kuhl, 1929; Wittman, 1974; Morrison et al., 1975). The influence of this factor could be negleted if the observations were distributed uniformly with regard to contacts, observers and observational conditions. In fact, it is far from that and the necessity of reduction corrections arises.

In forming the photometric profile of the border of some object, the following factors have an influence:

1) THE LIMB DARKENING (for the Sun).

2) IMAGE BLURRING including the atmospheric turbulence; the diffraction can be represented by two-dimensional isotropic Gaussian distribution in which the parameter σ depend on the amplitude of blurring.

3) The excitement of the elements of retina is not proportional to the illumination, but to its linear combination with the so-called CONTRAST FUNCTION, which takes into consideration nonlinear variations of illumination upon the retina.

The process of the formation of bar was investigated by numerical simulation (Sveshnikov et al., 1995). It was found that the main factor of the drop formation is image blurring. In good seeing ($\sigma \simeq 0.5''$), the black drop phenomenon is absent. For $\sigma \simeq 1.5 - 2''$ (mean or bad seeing), it is distinctly observed. Increasing the parameter σ up to these values, the filament, i.e. ,a dark narrow strip having the width about 1/3 of Mercury radius, appears for a distance between limbs from $\sigma/2$ to σ. Then, the filament is transformed into a drop (dark wide bar having a width of about Mercury radius R). In addition to that, the irradiation diminishes the apparent Mercury radius and the apparent moment of contact with the drop does not coincide with that of geometrical contact. For $\sigma \simeq R$, the observation of contacts is unrealizable practically. The half-sum of the moment of apparent contact with black drop and that of the drop breaking equals, approximately, the moment of the geometrical contact. The presence of contrast leads to the variation of apparent density of drop and the appearance of fictitious contacts.

The displacement of moment of apparent contact caused by the drop effect can be estimated by:

$$\Delta T^s \simeq \frac{d}{V \sqrt{1 - (c_m/R_s)^2}},$$

where d is the distance between the geometrical limbs in the moment of apparent contact with the drop, in arcsec (d is determined by simulation; its mean value is $\simeq 1.4''$); V is the velocity of Mercury along its apparent trajectory (V equals 1/10 or 1/15 ''/s for November and May transits); c_m is the minimal distance between centers; R_s is the solar radius.

The preliminary comparison of the obtained numerical values of the corrections for reduction of observations to moments of geometrical contacts have shown a good agreement with the real observations for the transits of 1878(II), 1878(III), 1907(III), 1914 (II), 1914 (III), 1953 (II), 1953(III), 1957 (II). In some cases, a significant diminution of the error of averaged moment of contact was recorded (about in 2 times). It should be noted that the photographic method of multiple observations of the position of planet across the solar surface excludes the errors connected with black drop, but, in the other hand, includes errors due to scale factor and oscillation of refraction. At present, at ITA RAS, the archive of Mercury transit observations is being revised. The above method of reduction permits to decrease the error in the calculation of contact's normal points.

References

von Kuhl, A.: 1929. *Phys. Zeitschr.*, Leipzig, Nr.1, pp. 1-34.
Morrison, L.V. et al.: 1975, *Month. Not. Roy. Soc.* **173**(1), 183-206.
Struve, H.: 1882. *Mem. Acad. Sci. St.-Peterburgh*, T.30, pp.1-104.
Sveshnikov, A.M. et al.: 1995. *Preprint ITA RAS* No. 43 (in russian).
Wittmann, A.: 1974. *Astron. and Astrophys.* **31**, 239-243.

REFERENCE FRAMES

JEAN KOVALEVSKY
Observatoire de la Côte d'Azur/CERGA,
Av.Copernic 06130 Grasse, France

AND

MARTINE FEISSEL
Observatoire de Paris/IERS,
61, Av. de l'Observatoire 75 014 Paris, France

Abstract. After a short introduction on the definition and the construction of a reference frame, the presentation concentrates on the actual construction of the new IAU conventional reference frame. The 1991 IAU resolutions on reference systems is reviewed and the present state of the implementation of these resolutions is discussed. The IERS extragalactic reference frame is described and we show that it fulfills the IAU requirements. The Hipparcos catalogue will be linked to this frame before the end of this year. The state of advancement of this job is presented.

All motions observed from the Earth as well as the motions of the Earth itself can be described only if they are referred to a system of coordinates which are supposed to be fixed or having a well known absolute time variations, that is to say with respect to something else that is fixed. But how can one decide whether there exists such a fixed system of coordinates and, moreover, that this property applies to the natural system which defines it? One can overcome this difficulty only by some assumption based on theoretical arguments which lead to consider that a certain physical structure has no rotation in the Universe. Even then, the question may arise whether the Universe is rotating as a whole. Theories predict that in any case, this rotation is very slow, far beyond the present observing capabilities.

This is the assumption that was taken by the International Astronomical Union in 1991 when it adopted a set of nine recommendations which include statements that the new celestial reference system will be based upon the positions of a number of distant extragalactic radio-sources such as quasars (Bergeron, 1992).

Another approach could have been made, following in this the classical precedent FK series which define the coordinate triad by imposing that the solution of equations of motions in the solar system have no centrifugal or linear acceleration. This was fine in Newtonian space, but in General Relativity theory this can have a meaning only in the tangential Euclidean space and provided that the gravitation field is small. The theory of dynamical reference systems in general relativity in presence of masses is a very complex problem, to be treated with respect to the metric tensor of space-time taking into account the motions of the masses themselves. The choice of the IAU has the advantage that it is only based upon the path of light and not on gravitating - and actually also rotating - objects.

The important point in the IAU resolution is the fact that it explicitly introduces General Relativity as the background for all theoretical and observational problems related to time and space. It actually imposed a reduced form of the metric which includes only second order terms in $1/c$

$$ds^2 = -c^2 d\tau^2 =$$
$$= -\left(1 - \frac{2U}{c^2}\right)\left(dx^0\right)^2 + \left(1 + \frac{2U}{c^2}\right)\left[\left(dx^1\right)^2 + \left(dx^2\right)^2 + \left(dx^3\right)^2\right] + ...$$

where $x^0 = ct, x^1, x^2$ and x^3 are the four space-time coordinates, τ is the proper time and U is the sum of the potential of the ensemble of masses of the system and of the tidal potential of external masses.

So from now on, any reference system should be devised so that comply with this description of space-time. This is the case of the celestial extragalactic system which is being designed, but must be also the case of any future dynamical system. The existing ones do not follow exactly the terms of the IAU resolutions, in particular in what concerns the barycentric and geocentric coordinate time scales: the newly defined TCB and TCG do not coincide with the previously used TDB and TDG. For instance, this applies to the dynamical system to which the series of DE200 ephemerides are referred.

The adoption of IAU resolutions incited JPL to modify its policy concerning the reference system. Starting with DE402, planetary ephemerides are rigidly constrained onto the IERS reference system (Standish, 1995). This system is based upon a number of positions of extragalactic objects such as quasars which are assumed to represent globally a fixed structure. Its tie with the preceding dynamic reference system used by JPL was determined with an accuracy of ±3 mas for the z axis and ±1 mas for x and y axes using VLBI and Lunar laser ranging observations (Folkner et al.,

1994). The newest DE403 ephemerides, which have included some corrections to DE402, are of course also in this extragalactic system.

The tie did not include any time-dependent rotation. The evaluation of such a rotation would be a very interesting and important result. It would be obtained from a comparison of two reference systems built strictly within the framework of the IAU resolutions but based strictly upon each of the approaches (kinematic/extragalactic and dynamical/solar system).

However, reference systems, whatever the basics on which they are designed, are not accessible *per se* and one has to materialize them by assigning coordinates to some astronomical objects in the triad and the corresponding coordinate time selected, in other terms to implement a reference frame. How this is done has been described in many instances, and we shall not deal with this topic here. Let us only state that some new treatment of observations used in the DE ephemerides will be necessary. It is also expected that the 3000 accurate minor planet observations realized by Hipparcos should be an additional asset to such an undertaking.

Since it is the choice made by the IAU, and since it is in a good advancement stage, we shall now concentrate on the state of the art concerning what should become the next official IAU celestial reference frame and its optical extension.

The 1991 IAU resolution has set a certain number of conditions to the realization of such a reference frame.

1 - The space coordinate grids with origins at the solar system barycenter and at the centre of mass of the Earth show no global rotation with respect to a set of extragalactic objects.

2 - A list of candidates of such objects be established for 1994 General Assembly of the Union.

3 - The principal plane of the new conventional celestial reference system be as near as possible to the mean equator at J2000.0 and the origin in this principal place be as near as possible to the dynamical equinox of J2000.0.

4 - The positions of the extragalactic objects selected in accordance with point 2 and representing the reference frame be computed initially for the equator and equinox J2000.0 using the best available values of the celestial pole offset with respect to the IAU expressions for precession and nutation.

5 - The celestial reference system should be accessible to astrometry in visual as well as in radio wavelengths. It was in addition specified that as long as the relationship between the stellar optical and the extragalactic radio frames is not sufficiently accurately determined, the FK5 catalogue shall be considered as a provisional realization of the celestial reference system in optical wavelengths.

Let us see now to what extent these five conditions are or will be fulfilled in a foreseeable future. Since 1988, the International Earth Rotation Service (IERS) receives and analyses catalogues of extragalactic radio-sources obtained by several VLBI networks and compiled by different groups (GSFC/NASA, JPL, NGS/NOAA, USNO). It compares them, rotates the catalogues to a single reference system and computes a weighted mean catalogue. During the following years, the IERS sequence of catalogues rapidly converged in the sense that while the number of sources, and particularly the number of primary sources increased, the mean uncertainties of the positions has steadily and significantly decreased.

The rotation between these successive catalogues became also smaller and smaller and reached an rms of 0.002 mas between the 1994 and 1995 versions. They all were realization of a reference system called until 1994 ICRS (IERS celestial reference system at J2000.0, Arias et al., 1995). In the construction of its catalogues, IERS used the IAU conventional precession and nutation respectively defined in 1976 and 1980. Both are now known, from VLBI and lunar laser to be in error (3 mas per year for precession). Computations were performed showing that correct values of these quantities would change the position of the pole of the system by less than 20 mas. This figure is to be compared with the basic uncertainty of the FK5 pole which is estimated to be of the order of 50 mas. Similarly it was found that its offset from the mean equinox of epoch 2000.0 is 78 ± 10 mas. This is also compatible with the error of the FK5 equinox which is of the order of 100 mas as shown by comparisons with meridian and Hipparcos observations. From this, it results that ICRS complies with the conditions 3 and 4 within the uncertainties inherent in the FK5 system realization. Finally, this system is also barycentric and the IERS catalogues indeed show no global rotation of the sources included.

Considering this, the IAU working group on reference frames which met in February 1995 decided that the ICRS should be adopted as the new IAU celestial reference system under the name of ICRS (International Celestial Reference System). Actually, there were no other serious contender to ICRS and, in addition, since seven years at least, all the results concerning the Earth rotation were computed and published in this system. Because of the many implications of Earth rotation parameters in various aspects of astrometry and the Earth's dynamics, any other choice would have brought only unnecessary complications and artificial discontinuities at the time of the official adoption of the new system. Note that this decision implies that the DE400 series of JPL ephemerides are in this new international celestial reference system ICRS.

In the meantime, the work preparing the fundamental reference system progresses. As requested in the second item, a list of about 600 extragalactic sources was prepared and adopted in 1994 by the IAU. Now that the system is defined, the precise positions of these sources are being finalized and will be made available in October 1995 so that, at that time, all IAU conditions concerning the reference system and its realization in radio wavelengths will be met.

The only remaining condition concerns the extension to the stars in visual wavelengths. The best - and by far the best - catalogue of stellar positions, proper motions and parallaxes will be the Hipparcos catalogue including more than 115 000 stars with a median internal precision for each of these quantities in the range 1 to 1.5 mas or mas per year. If the reference system of this catalogue can be identified with ICRS, the last IAU reservation will be overcome. This work is in progress and should be finished by the end of 1995, date at which the Hipparcos catalogue should be completed and be entirely in the new ICRS (Lindegren and Kovalevsky, 1995). Seven methods are used to realize the link and compute the elements of two rotation matrices: the matrix R which rotates the arbitrary Hipparcos coordinate grid on ICRS and the matrix R' which is a linear function of time and which stops the rotation with respect to the extragalactic sources as a whole. These are:

1 - Comparison between positions and proper motions of radio-stars from Hipparcos catalogue and determined by VLBI. About 12 stars are used in this method and each element of R and R' is determined with an rms of 0.5 mas and 0.5 mas per year.

2 - The same method using observation by VLA and MERLIN radio interferometers.

3 - Computation of the elements of R' using the Lick, Yale and Kiev catalogues of proper motions determined with respect to galaxies from photographic observations. More than 8000 Hipparcos stars are in common with these catalogues and the uncertainty on R' seems at present to be significantly below 0.5 mas per year.

4 - Photography by a Schmidt telescope of radio sources of the primary catalogue. More than 100 plates involving approximately the same number of quasars are being analyzed for R and R'.

5 - Two or more epoch observations of specially selected areas of the sky involving Hipparcos stars and quasars. The epoch differences may reach 90 years giving conditions for R'.

6 - Hubble Space Telescope Fine Guidance Sensor observations of some Hipparcos stars with respect to quasars of known positions give information to determine R. Unfortunately, the observations started too late (April 1993) to allow precise determinations of R'.

7 - Comparing the observations of the Earth rotation with Hipparcos catalogue give information on R'. However the necessity to introduce the terrestrial reference frame as an intermediary induces possible systematic effects which are difficult to estimate.

Nine different teams in various countries are working to determine R and/or R' by one of these methods. It will be then necessary to adopt some relative weights for them before attempting a synthesis of the results.

In any case, this link will be realized (actually with the data available at present, an accuracy of 0.5 mas per year is already possible) so that at the end of 1995, the ICRS, the corresponding radio reference frame and the rotated Hipparcos catalogue will be ready, allowing the 1997 IAU general assembly to adopt them as the fundamental celestial reference system and frames.

References

Arias, E. F. , Charlot, P. , Feissel, M. , Lestrade, J. F.: 1995, "The extragalactic reference system of the International Earth Rotation Service, ICRS", *Astron. Astrophys.* in press.

Bergeron, J. (ed.): 1992, "Proceedings of the 21-st General Assembly", *Transaction of the IAU* **21 B**, 41-63.

Folkner, W. M. , Charlot, P. , Finger, M. H. et al.: 1994, "Determination of the extragalactic-planetary frame tie from joint analysis of radio interferometric and lunar laser ranging measurements", *Astron. Astrophys.* **287**, 279-289.

Lindegren, L. and Kovalevsky, J.: 1995, "Linking the Hipparcos Catalogue to the extragalactic reference system" *Astron. Astrophys.* in press.

Standish, E. M. , 1995, "The dynamical reference frame", in *Astronomical and Astrophysical Objectives of Sub-milliarcsecond Optical Astrometry, IAU Coll. 166.*, E. Hoeg and P. K. Seidelmann (eds.), Kluwer Acad. Publ. , Dordrecht, pp. 109-116.

IAU STANDARDS – ITS FUTURE

TOSHIO FUKUSHIMA
National Astronomical Observatory
2-21-1, Ohsawa, Mitaka, Tokyo 181, Japan
(Internet) toshio@spacetime.mtk.nao.ac.jp

Abstract. The ongoing movement of standardization in Fundamental Astronomy was reviewed. Its history was briefly presented with an emphasis on the problems which triggered its creation. The achievements of the first term of the IAU WG on Astronomical Standards were given. The major goals of a second term were presented with the author's view to resolve them.

1. Past

1.1. —1989

The present movement of the standardization in Fundamental Astronomy was initiated in 1989 when the Sub-Group on Astronomical Constants of the IAU WG on Reference Systems (WGRS/SGAC) was formed to give a report on the possible update of the IAU (1976) system of astronomical constants. Refer to Fukushima (1991) and other related papers which appeared in the proceedings of IAU Colloquium 127. At that time, more than a decade after the adoption of the current IAU system of constants, various questions had come up with the system itself and the philosophy implicitly embedded in it. The apparent problems could be listed as:

1. There had been some confusion on the determination of constants mainly based on the difference in the interpretations of their definitions within the general relativistic framework. See Fukushima et al. (1986).
2. Some planetary masses were obsolete, especially that of Pluto and, as well, those of some outer planets. See Fukushima (1991). Note that the IAU (1976) system adopted the masses before the Voyager observations.

3. The discrepancy of the adopted precession constant was already clear. See Fukushima (1991). This discovery was greatly owe to the precision Earth rotation observations such as conducted by the IERS.

Also, in the author's personal sense, there had been some opinions about our mechanism to authorize the system of constants and other general rules under the name of the IAU. Some unspoken ones might be:

1. We were so slow in keeping up with the cutting-edge information both observational and theoretical.
2. We were so drastic in introducing the changes in creating reference works such as the compilation of nautical almanacs and star catalogs.
3. Sometimes, we recommended systems which are inconsistent themselves or incompatible with those authorized by other organizations.

1.2. 1989-1991

During the discussions within the SGAC, we noticed that the problem was not limited within the list of constants to be updated. Imagine to replace the precession constant. A mere revision of this constant never means the way to calculate the precession is updated. We need not only the revised constant of general precession, such as given in the IAU (1976) system, but also a formula to compute the precession matrix as a function of time as provided by Lieske et al. (1977). Then, we reached a conclusion that the scope of revision should be enlarged to cover the actual computational procedure. This matched with the increasing requirements of established routines for basic computations in Fundamental Astronomy. Also, we can not deny that the success of the MERIT standards and the following IERS Standards had spurred us to this direction. Meanwhile, the issues on the general relativistic considerations were the major items of the other two Sub-Groups: Reference Frame and Time. Therefore, the SGAC did not provide a recommended list of constants that time and asked the IAU to extend its activity to study the possibilities to create and maintain the IAU version of the IERS Standards, which we call roughly the *IAU Standards*.

1.3. 1991-1994

The Buenos Aires General Assembly in 1991 permitted us to reform the SGAC into a multi-commissions supported working group: the WG on Astronomical Standards (WGAS). The activity of the WGAS in this period was fully reported in the proceedings of the Joint Discussion 14 of the last IAU General Assembly (IAU 1995a). Also, some of its conclusions were adopted as the IAU (1994) Resolutions B11, C6, and C7 (IAU 1995b). Please refer to them for the details. In summary, the WG conducted four

sub-groups: Numerical Standards, Standard Procedures, Electronical Distribution, and Issues on Time. The major conclusions are:
1. To establish the two-tier mechanism for constants: those maintained in a long term for reference works and those frequently revised for up-to-date researches:
2. To make efforts to create the IAU-authorized standard procedures named the SOFA (=Standards Of Fundamental Astronomy).
3. To continue the WGAS to do these tasks.

As a first step of the first item, we recommended the usage of the IAU (1976) System for reference works and provided the IAU (1994) File of Current Best Estimates of Astronomical Quantities for research use, which is published in the proceedings of JD 14 mentioned above. Here, we reproduce it in a compact form.

Also, the consideration on general relativistic definitions of astronomical units and constants was given as its homework for next three years. We should mention that a new fundamental constant L_C has been introduced: the scale factor or the conversion factor among the newly introduced multiple timescales based on the general relativistic considerations. See details in Seidelmann and Fukushima (1992) and Fukushima (1995).

2. Present

2.1. CONSTRUCTION AND POLICY

At the Hague General Assembly in 1994, a new layer was introduced into the IAU structure: Divisions. Since all the Commissions supporting the WGAS (Commissions 4, 8, 19, 24, and 31) belong to the Division 1, the WG has become one of WGs under the Division 1, automatically. Based on the consultation with Presidents/Vice-Presidents of these Commissions, the Division 1 President has nominated the author as the Chair of the WGAS for its second term, namely for 1994-1997.

To comply with the given missions, we have reorganized the WGAS into three sub-functions: the Maintenance Committee (of Numerical Standards) headed by Dr. D.D. McCarthy, the Reviewing Board (of SOFA) chaired by Mr. Patrick T. Wallace and the Sub-Committee on General Relativistic Issues (on Units and Astronomical Constants) lead by Prof. Victor A. Brumberg. Together with more than 30 Members, who substantially do the jobs, we will continue to invite opinions from the wide communities of Fundamental Astronomy, Earth rotation studies, space geodesy and related sciences, through an electronically published newsletter named IAU/WGAS/Circulars. The activity of the WG will be mainly kept by E-mail exchanges among Members and informal and open discussions held on the Circulars, as we did in the last term.

TABLE I – IAU (1994) File of Current Best Estimates of Astronomical Quantities

Name	Value	Units	Ref
Defining Constants			
k	0.01720209895	$[au^3/day^2]^{1/2}$	
c	299792458.	m/s	
Primary Constants			
L_C	$1.4808268452(1) \times 10^{-8}$		[10]
τ_A	499.00478642(7)	$SI : s$	[3]
	499.00478384(7)	$TDB : s$	
p	5028.83(4)	$"/cy$	[13]
	5028.83(3)		[14]
ϵ	84381.412(5)	$"$	[3]
$\mathcal{M}_{Sun}/\mathcal{M}_1$	6023600.(250.)		[1]
$\mathcal{M}_{Sun}/\mathcal{M}_2$	408523.71(6)		[2]
$\mathcal{M}_{Sun}/\mathcal{M}_B$	328900.56(2)		[3]
$\mathcal{M}_{Sun}/\mathcal{M}_4$	3098708.(9.)		[4]
$\mathcal{M}_{Sun}/\mathcal{M}_5$	1047.3486(8)		[5]
$\mathcal{M}_{Sun}/\mathcal{M}_6$	3497.898(18)		[6]
$\mathcal{M}_{Sun}/\mathcal{M}_7$	22902.98(3)		[7]
$\mathcal{M}_{Sun}/\mathcal{M}_8$	19412.24(4)		[8]
$\mathcal{M}_{Sun}/\mathcal{M}_9$	$1.35(7) \times 10^8$		[9]
$\mathcal{M}_{Moon}/\mathcal{M}_{Earth}$	0.012300034(3)		[3]
G	$6.67259(30) \times 10^{-8}$	$m^3/(gs^2)$	[15,16]
$G\mathcal{M}_{Earth}$	$398600.4415(8) \times 10^9$	$SI : m^3/s^2$	[12]
	$398600.4356(8) \times 10^9$	$TDB : m^3/s^2$	
a_E	6378136.55(1)	m	[15,16]
W_0	62636857.5(1.0)	m^2/s^2	[17]
	62636856.26(1.0)		
J_2	$1082.6269(6) \times 10^{-6}$		[15,16]
$1/f$	1/298.257(1)		[15,16]
ω	$7292115. \times 10^{-11}$	rad/s	[15,16]
Derived Constants			
L_B	$1.550519747(3) \times 10^{-8}$		[10]
$c\tau_A$	149597871475.(30.)	$SI : m$	[3]
	149597870700.(30.)	$TDB : m$	
$\mathcal{M}_{Earth}/\mathcal{M}_{Moon}$	81.30059(1)		[3]
$G\mathcal{M}_{Sun}$	$1.32712440042 \times 10^{20}$	m^3/s^2	[3]
$\mathcal{M}_{Sun}/\mathcal{M}_{Earth}$	332946.05(2)		[3]

Table I (cont.) - References

1. Anderson,J.D., Colombo,G., Esposito,P.B., Lau,E.L. and Trager,G.B.: 1987, "The Mass, Gravity Field, and Ephemeris of Mercury", *Icarus*, **71**, 337-349.
2. Sjogren,W.L., Trager,G.B. and Roldan,G.R.: 1990, "Venus: A Total Mass Estimate", *Geophys. Res. Let.*, **17**(10), 1485-1488.
3. JPL Planetary and Lunar Ephemeris DE245.
4. Null,G.W.: 1969, "A Solution for the Mass and Dynamical Oblateness of Mars Using Mariner-IV Doppler Data", *Bull. Am. Astr. Soc.*, **1**(4), 356.
5. Campbell,J.K. and Synnott,S.P.: 1985, "Gravity Field of the Jovian System from Pioneer and Voyager Tracking Data", *Astron. J.*, **90**(2), 364-372.
6. Campbell,J.K. and Anderson,J.D.: 1989, "Gravity Field of the Saturnian System from Pioneer and Voyager Tracking Data", *Astron. J.*, **97**(5), 1485-1495.
7. Jacobson,R.A., Campbell,J.K., Taylor,A.H. and Synnott,S.P.: 1992, "The Masses of Uranus and its Major Satellites from Voyager Tracking Data and Earth-based Uranian Satellite Data", *Astron J.*, **103**(6), 2068-2078.
8. Jacobson,R.A., Riedel,J.E. and Taylor,A.H.: 1991, "The Orbits of Triton and Nereid from Spacecraft and Earth-based Observations", *Astron. Astrophys.*, **247**, 565-575.
9. Tholen,D.J. and Buie,M.W.: 1988, "Circumstances for Pluto-Charon Mutual Events in 1989", *Astron.J.*, **96**(6), 1977-1982.
10. Fukushima,T.: 1995, "Time Ephemeris", *Astron. Astrophys.*, **294**, 895-906.
11. Bursa,M., Sima,Z., Kostelecky,J.: 1992, "Determination of the Geopotential Scale Factor from Satellite Altimetry", *Studia geoph. et geod.*, **36**, 101-114.
12. Ries,J.C., Eanes,R.J., Shum,C.K. and Watkins,M.M.: 1992,"Progress in the Determination of the Gravitational Coefficient of the Earth", *GRL*, **19**(6), 529-531.
13. Williams,J.G., Newhall,X X and Dickey,J.O.: 1991,"Luni-Solar Precession: Determination from Lunar Laser Ranges", *Astron. Astrophys.*, **241**, L9-L12.
14. Miyamoto,M. & Soma,M.: 1993, "Is the Vorticity Vector of the Galaxy Perpendicular to the Galactic Plane? I. Precessional Correction and Equinoctial Motion Correction to the FK5 System", *Astron J.*, **105**, 691.
15. Bursa,M.: 1992, "Parameters of Common Relevance of Astronomy, Geodesy and Geodynamics", *Bull. Geod.*, **66**(2), 193-197.
16. Moritz,H.: 1992, "Geodetic Reference System", *Bull. Geod.*, **66**(2), 187-192.
17. Rapp: 1994, private communication via E. Groten.

2.2. TOOLS

At the time of writing this summary, we are glad to report that some tools have been prepared already. Among them, we should refer the completion of new JPL DE403 planetary ephemeris which is fully compatible with the IAU (1994) Best Estimates *and* the IERS reference frame. Refer the report by Dr. Standish in the same proceedings. Another work was done by the Bureau des Longitudes: new formulas on the precession calculation. It is remarkable that they contain partial derivatives with fundamental constants such as the planetary masses so that they can keep up with the future change of the constants. Apart from astronomical works, the recent spread of Internet and World Wide Web (WWW) has drastically changed the way to retrieve massive and up-to-date information. For example, thanks to Prof. R.B. Langley of Univ. of New Brunswick, Canada, all the WG's electronic newsletters, the IAU/WGAS/Circulars, are accessible via WWW, under the Canadian Space Forum archive whose URL is

http://www.unb.ca/Geodesy/CANSPACE.html

See IAU/WGAS/Circular No.97 for the details. This and the development of inexpensive and capable computers will reduce the time and labor which will be required to create, maintain and distribute the IAU Standards. As for the non-networked distribution, two will remain as major ways: the 3.5" Floppy disks for small data such as the list of constants and softwares and CD-ROMs for large data such as the ephemerides and/or star catalogs. Fortunately, the recent technological development makes it easy to press CD-ROMs much less expensively.

3. Future

Since the future trend is difficult to predict and the activity of the WGAS in the second term has just begun, here I would like to present very personal views of mine. By the time of this publication, the situation will be different. Please keep in touch with us through subscribing our electronic Newsletter, the IAU/WGAS/Circular.

3.1. MEANING OF STANDARDS

So far, the IAU System of Constants have seemed to be more compulsory than it was intended. In proceeding the new two-tier mechanism on astronomical system of constants, we would like to confirm that the major purpose of these systems is to serve a reference. Also, this can be said with the procedures. In this sense, the word *Standards* is more suitable. No one has to follow these standards. They are there to be used as a scale. Everyone can express his/her parameters, model or method by noting the (small?) differences from the standards, which finally makes it easier to compare with others.

3.2. PRECESSION

Now that the planetary masses were effectively revised in the IAU (1994) Best Estimates and realized by the JPL DE403 planetary/lunar ephemeris and that the schedule of the present and the near future space explorations within the solar system makes us feel that these Best Estimates will be not so drastically changed in the possible near future, the most important and urgent item remained can be the precession constant. Unfortunately the situation is not so clear-cut, since this issue is tightly connected with the problem of the current IAU nutation theory, especially its long-periodic terms. This issue is being discussed by the WG lead by Dr. Dehant. Anyway, it is clear that the present adopted value needs a correction of around 3.0"/cy. I feel that the amount of revision is not so controversial. Rather the timing of introduction could be. If a consensus on the next generation nutation theory will be formed within a few years, I personally think that to recommend the introduction of a new precession constant from J2000.0 would be most convenient.

3.3. OPEN POLICY ON SOFA CREATION/MAINTENANCE

Another issue is how to construct the SOFA actually. For this purpose, I have much expectation on the anonymous public. Though it might sound too revolutionary, we are seriously considering to invite the ideas and contribution both on the submission of software and on their reviewing procedure 1) from the astronomers in other fields via announcing our activities through ADASS and other conferences on the general astronomical/astrophysical data archiving and software and 2) from the really general public through already established computer-based electronic forums such as the `sci.astro.research` USENET newsgroup. I hope that this will bring us great merits: to save huge labors and time of the SOFA center, to advertise the IAU's activity and to introduce the new and fresh energy into our fields.

3.4. COLLABORATION WITH THE IERS AND IAG

Apart from the collaboration with the other IAU WGs such as those for Non-Rigid Earth Nutation Theory and Reference Frame, which is our obligation, we should seek the way to have substantial cooperation with two outside organizations: IERS and IAG. The IERS has its famous Standards Committee, which has continuously published the IERS Standards (McCarthy 1989, 1992, 1996). Also, the IAG has its own dedicated Special Committee, SC on Fundamental Constants, under the Section V. To enhance these collaborations, we have done a few things. First, we have adopted the two-tier system on maintaining the system of constants, which was originally introduced in geodesy. Thus, the two closely-related fields, astronomy and geodesy, have had a similar mechanism to refer standard values of constants. Also, we are very glad to welcome Dr. McCarthy as the Head of Maintenance Committee of our WG. This will be a great step toward the fusion of similar activities, which are now conducted in separately in the IAU, IERS and IAG.

3.5. KYOTO, 1997

The next General Assembly will be held in Kyoto, Japan, during the last two weeks of August, 1997. We are anticipating to hold a Joint Discussion similar to the JD 14 we had in the Hague. Though it would be much earlier, we are also seeking a possibility to propose a satellite IAU Symposium together with other WGs in the same field; say together with the Non-Rigid Earth Nutation Theory WG chaired by Dr. Veronique Dehant and the WG on Reference Frame headed by Dr. Leslie V. Morrison.

4. Conclusion

We presented a quick summary of the activities of WGAS and its predecessor, WGRS/SGAC since 1988. Also the view on its future movement is given. However, please understand that this view is just one of the possible choices, and the actual policy of the WG will be and should be decided through a diverse discussion among the astronomical and other related communities. Not the Members themselves but YOU, who are reading this note, determine the future of this movement! To catch up with the current trend and to reflect your opinions to the activity, please start free subscription of the WGAS's electric newsletter, IAU/WGAS/Circular. To do this, you only have to send a request to the author

toshio@spacetime.mtk.nao.ac.jp

Let's join to discuss the most essential and stimulating issues in Fundamental Astronomy: the *IAU Standards*.

References

Fukushima, T. (1991), Activity report of the IAU Working Group on Reference Systems: Subgroup on Astronomical Constants, *Proc. IAU Colloquium No.127 Reference Systems*, Hughes J.A., Smith C.A., Kaplan G.H. (eds), USNO, Washington D.C., 27-35.
Fukushima T., Fujimoto M.-K., Kinoshita H., and Aoki S. (1986), *Cele. Mech.*, **36**, 215-230.
Fukushima T. (1995), *Astron. Astrophys.*, **294**, 895-906.
IAU (1995a), *Highlights in Astronomy*, Kluwer Acad. Publ., Dordrecht.
IAU (1995b), *Trans. IAU, XXII*, Kluwer Acad. Publ., Dordrecht.
Lieske J.H., Lederle T., Fricke W., and Morando B. (1977), *Astron. Astrophys.*, **58**, 1-16.
McCarthy D.D. (ed.) (1989). IERS Standards (1989), *IERS Tech. Note*, **3**.
McCarthy D.D. (ed.) (1992) IERS Standards (1992), *IERS Tech. Note*, **13**.
McCarthy D.D. (ed.) (1996) IERS Standards (1995), *IERS Tech. Note* (in press).
Seidelmann P.K., and Fukushima T. (1992) *Astron. Astrophys.*, **265**, 833-838.

THE FK5 EQUINOX AND EQUATOR FROM COMBINED RADAR AND OPTICAL DATA OF THE NEAR-EARTH ASTEROIDS

N.V.SHUYGINA AND E.I.YAGUDINA
Institute of Applied Astronomy
of Russian Academy of Sciences
8 Zhdanovskaya st., St.Petersburg, 197042 Russia
e-mail eiya@ipa.rssi.ru

1. Introduction

Minor planets optical observations have long been used for the purpose of establishing a Celestial reference frame. Being in existence since the early 1960s modern high-accuracy radar measurements of the so-called near-Earth asteroids (NEAs) have been widely extended to the orbit determination process and predicting of the next apparition of the asteroid. Even few radar measurements, when added to optical ones, significantly improve asteroid's ephemeris and reduce standard deviations of the orbital elements (Yeomans *et al.*, 1987). The idea to connect optical and radar data in the problem of the catalogue zero-point determination has been stated by several scientists (Boiko, 1975). And even the first attempt of the authors (Krivova *et al.*, 1994) with actual optical and radar observations of two NEAs: (4179) Toutatis and (1862) Apollo appears to have considerable promise. It was demonstrated the possibility of obtaining standard deviations of catalogue orientation parameters 1.5 − 2 times better with radar data included.

In order to choose more suitable near-Earth asteroids for the purpose of the equinox and equator determination a numerical simulation has been developed. Using the results of numerical investigation and taking into account the first attempt with actual observations, we have selected the following near-Earth asteroids: (1620) Geographos, (3908) 1980 PA, (4769) 1989 PB, (1627) Ivar, along with the (1862) Apollo, and (4179) Toutatis.

The processing of both radar and optical observations of these minor planets makes it possible to determine not only a precise orbit for each asteroid, but tiny catalogue zero-point corrections as well. The results lend support to the validuty of the method used.

2. Numerical simulation: the uncertainty analysis

A numerical investigation was undertaken to choose more suitable near-Earth asteroids for the purpose of the equinox and equator determination. This simulating analysis is based on a well-known least-squares method. The three types of measurements presented in this study are optical right ascension and declination, radar time delay, and radar Doppler frequency shift. The standard deviation chosen for the optical observational noise is $1''$ in each coordinate. As for radar measurements, the standard deviation assumed to be about 20 μs in range and 5 Hz in Doppler case.

TABLE 1. 1991 JX type

i (deg)	0	10	20	30
Only optical data				
dA ($''$)	± 2.09	± 0.89	± 1.45	± 3.08
dD ($''$)	± 0.53	± 0.14	± 0.50	± 1.92
dE ($''$)	± 1.84	± 0.43	± 0.67	± 2.45
corr. coeff.(%)	95	91	97	92
normalized rms($''$)	1.029	0.986	1.010	0.960
number of obs.	362	364	366	366
Combined data (radar and optical)				
dA($''$)	± 0.67	± 0.46	± 0.51	± 0.39
dD($''$)	± 0.15	± 0.08	± 0.16	± 0.11
dE($''$)	± 0.24	± 0.14	± 0.21	± 0.15
corr. coeff.(%)	90	89	97	95
normalized rms($''$)	0.958	0.916	0.959	0.889
number of obs.	424	426	430	430

The primary purpose of this simulation was to examine the relationship between the length of optical observational history, the quantity of hypothetical radar measurements and uncertainties of the catalogue corrections. Of special interest is the question of how radar observation can decrease these uncertainties. That is why our strategy is to obtain standard deviations of the equator and equinox corrections as well as of corrections to the orbital elements derived from the processing of (1) optical observations, (2) combined optical and radar data covered several approaches to the Earth.

We have investigated two different types of asteroids. The first one is an asteroid with a short optical history (from half a year to 4 years) and with radar observations made during three of its approaches to the Earth: at the beginning, in the middle and at the end of the interval – we called it "1991 JX type". Another type is a near-Earth asteroid with a long (of more than 25-40 years) optical history and with radar data available during several approaches to the Earth – "Geographos type". In the "Geographos" case, there were simulated 5138 optical observations uniformly covering a time interval of about 35 years. As regards radar measurements, they were distributed according to the following schedule: each asteroid's approach to the Earth lasted for 2 months and repeated approximately every 2nd year. Each observation (both optical and radar) is considered to be made every 5th day.

TABLE 2. Geographos type

i (deg)	0	10	20	30
Only optical data				
dA (″)	± 0.11	± 0.11	± 0.12	± 0.12
dD (″)	± 0.03	± 0.03	± 0.03	± 0.03
dE (″)	± 0.04	± 0.04	± 0.04	± 0.05
corr. coeff.(%)	87	89	90	90
normalized rms(″)	1.001	0.990	0.986	0.984
number of obs.	5138	5138	5138	5138
Combined data (radar and optical)				
dA (″)	± 0.10	± 0.10	± 0.10	±0.09
dD (″)	± 0.02	± 0.02	± 0.02	± 0.02
dE (″)	± 0.03	± 0.03	± 0.03	± 0.03
corr. coeff.(%)	91	91	88	86
normalized rms(″)	0.951	0.941	0.937	0.936
number of obs.	5719	5719	5719	5719

In each case standard deviations for six orbital elements (derived from combined data) along with the uncertainties for parameters: dA – the FK5 equinox correction, dȦ- its secular variation, dD – the FK5 equator correction, and dE – the correction to the mean longitude of the Earth (obtained from the optical data only) have been estimated. The numerical simulation was fulfilled within the ERA (Ephemerides Research in Astronomy) applied program package (Krasinsky et al., 1995). All calculations were performed for each asteroid individually and for the combination of asteroids (the so-called global solution). Tables 1 and 2 demonstrate standard deviations of the parameters under consideration, maximum correlation coefficient, normalized root-mean squares residuals, and number of observations adopted, depending on the different values of the asteroid's inclination to the ecliptic plane.

As evident from the tables, parameters uncertainties do not depend to a large extent on the inclination of the asteroid's orbit. The influence of including radar data was expected to be significant for catalogue orientation parameters determination for asteroids with very short optical history (5–6 times), and only modest in the case when rich optical data exists (1.5–2 times). Obviously, asteroids with a long history of optical measurements and several approaches to the Earth covered with radar observations are preferable, but minor planets with a short optical and rich radar set, can be used too.

3. The FK5 equinox and equator from the processing of actual near-Earth asteroids

Using combined radar and optical observations for the set of NEAs, mentioned in the Introduction, several solutions for the zero-point corrections to the FK5 catalogue were obtained.

The orbits of asteroids are computed by numerical integration of the relativistic equations of motion of NEAs taking into account the perturbations from all major planets and Schwarzschild's terms due to the Sun. The Everhart 15th order method of integration was used. For calculations of the coordinates of perturbing planets and the Moon the DE200/LE200 ephemerides were used. The comparison of the measured values with the computed ones for optical and radar observations was made in barycentric coordinate system J2000.0. The O-C differences for radar observations were combined with those for optical ones in a linearized weighted least squares procedure to produce estimated corrections to non-singular orbital elements for a standard epoch and zero-point corrections (dA, dȦ, dD and dE). This technique was applied to each NEA and for the combination of asteroids. To obtain these parameters the complete set of all available optical (they were mainly taken from the Minor Planet Catalogue) and radar data (Yeomans et al., 1991) were used. Optical observations (which were taken not from the MPC) were transferred to the FK5 system using standard procedure adopted by IAU. It can be seen that we have chosen asteroids with various optical and radar observational histories. Some of NEAs have a long set of optical measurements but only 2 radar observations (Geographos, Ivar);

Toutatis has also the long optical but sufficiently large radar history of about 55 observations; for 1980 PA there exist only short optical history and quite good set of radar data.

The processing of actual data provided support for the conclusion derived from the numerical simulation that standard deviations of the zero-point corrections are approximately two times better with radar data included. Table 5 demonstrates the preliminary results for the FK5 catalogue corrections and their standard deviations obtained from optical observations separately and from combined data for five (except Apollo) asteroids.

TABLE 3. The FK5 catalogue preliminary corrections

Only optical observations				Combined data			
dA(″)	dȦ(″/cy)	dD(″)	dE(″)	dA(″)	dȦ(″/cy)	dD(″)	dE(″)
0.568 ±0.078				0.294 ±0.052			
0.558 ±0.116	− 0.10 ±0.40			0.242 ±0.063	− 0.40 ±0.30		
0.525 ±0.119	− 0.10 ±0.40	0.109 ±0.045		0.242 ±0.060	− 0.35 ±0.30	0.057 ±0.042	
0.677 ±0.181	− 0.10 ±0.40	0.108 ±0.046	− 0.091 ±0.070	0.666 ±0.141	−0.35 ±0.30	0.062 ±0.041	− 0.166 ±0.049

Normalized rms for optical data : 1.035

Normalized rms for combined data : 1.039

Max. correlation coefficient between dA and dE for optical data: 89%

Max. correlation coefficient between dA and dE for combined data: 90%

4. Conclusion

1. The results confirm that the set of combined data of near-Earth asteroids is feasible for zero-point correction determination. The accuracy of zero-point corrections is about two times better in the case of processing combined photographic and radar data than in that of optical observations only. The preliminary corrections to equinox and equator of the FK5 were obtained. They are probable in comparison with other determinations (Branham et al., 1994).

2. As it could be expected, the equator correction is determined with confidence in every case.

3. Considering that the coefficient of correlation between the equinox correction and the correction to the longitude of the Earth is about 90-95%, one could obtain the linear combination of these parameters only. This correlation coefficient is reduced slightly (to 88-86%) for NEAs with a long optical history and radar observations available during several approaches to the Earth. An example of such an asteroid is just (1620) Geographos which was observed by radar technique second time.

4. It is desirable to observe NEA not only during its approach to the Earth, but before and after it. It should be recommended to observe NEAs with higher accuracy, as for asteroids from the main belt for example, "Selected minor planets", at least no more than 0.5".

5. Now there exist more than 60 asteroids with available radar data (Ostro et al., 1993). The inclusion of these observations will provide more conclusive and stable results.

References

Boiko,V.N., (1975) The improvement of FK System from minor planets observations, *Russian Astron. J.*, **52**, 47-55.

Branham,R.L., Sanguin,J.G., (1994) The FK5 equinox and equator, *Proceedings of the Third Workshop on Positional Astronomy and Celestial mechanics, in Cuenca, October 17-21, 1994, Spain*, Editor A.Lopez Garcia et al., Valencia, (in press).

Krasinsky,G.A., Vasilyev,M.V., and Pitjeva,E.V, (1995) *ERA – Ephemeris Research in Astronomy, manual*, Institute of Applied Astronomy, St.Petersburg, Russia.

Krivova,N.V., Yagudina,E.I., (1994) Catalogue corrections determined from combined radar and optical data of minor planets, *Proceedings of the Conference on Astrometry and Celestial mechanics held in Poznan, September 13-17, 1993, Poland*, Editor K.Kurzynska et al., pp. 79-82.

Ostro,S.J., (1993) Planetary radar astrometry, *Reviews of Modern Physics*, **65**, 1235-1280.

Yeomans,D.K., Chodas,P.W., Keesey,M.S., Ostro,S.J., Chandler,J.I., and Shapiro,I.I., (1991) Asteroid and comet orbits using radar data, *Astron.J.*, **103**(10), 303-317.

Yeomans,D.K., Ostro,S.J. and Chodas,P.W., (1987) Radar astrometry of near-Earth asteroids, *Astron.J.*, **94**(1), 189-200.

EFFECTS OF PRECESSION UNCERTAINTIES ON PLANETARY EPHEMERIDES

H.G. WALTER
Astronomisches Rechen-Institut
Mönchhofstr. 12-14, D-69120 Heidelberg, Germany

1. Introduction

Several methods of data analysis applied recently to precise astrometric observations give evidence for a correction of the luni-solar precession (ψ) of the order of $\Delta\psi = -3$ milliarcseconds per year (Williams et al., 1994). When studying the motion of bodies of the planetary system from photographic observations, an inertial stellar reference frame is required. Thus, any imperfection of the luni-solar precession would have repercussions on the determination of the orbit dynamics of the celestial body under investigation.

We calculate the change in the determination of the semimajor axis of orbits of planets and asteroids as a function of the luni-solar precession, and compare the estimated shift of the semimajor axis with its precision obtained from radar ranging measurements.

2. The Basic Equations

Relations between orbital elements and angular observations (λ, β) referring to the ecliptic coordinate system can be derived from the equation of elliptical motion. We employ the customary notations of the orbital elements $(a, e, i, \Omega, \omega, M)$. Let $\mathbf{r}$ be a barycentric radius vector to a planet. Expressed by the orbital elements referring to the triad $(\mathbf{i}, \mathbf{j}, \mathbf{k})$ defining the ecliptic coordinate system, three unit vectors are introduced:

$$\begin{aligned}
\mathbf{e}_1 &= \cos\Omega\,\mathbf{i} + \sin\Omega\,\mathbf{j} \\
\mathbf{e}_2 &= -(\sin\Omega\cos i)\,\mathbf{i} + (\cos\Omega\cos i)\,\mathbf{j} + \sin i\,\mathbf{k} \\
\mathbf{e}_3 &= (\sin\Omega\sin i)\,\mathbf{i} - (\cos\Omega\sin i)\,\mathbf{j} + \cos i\,\mathbf{k}.
\end{aligned}$$

The semimajor axis is expressed by (e.g. McCuskey, 1963)

$$a = \frac{r + B_1(\mathbf{r} \cdot \mathbf{e}_1) + B_2(\mathbf{r} \cdot \mathbf{e}_2)}{1 - e^2} , \qquad (1)$$

where $B_1 = e\cos\omega$ and $B_2 = e\sin\omega$.

On forming the differential of eq. (1) with respect to the precession one gets for the variation of the semimajor axis

$$\Delta a = B_1/(1-e^2)(\frac{\partial \mathbf{r}}{\partial \psi} \cdot \mathbf{e}_1)\Delta\psi + B_2/(1-e^2)(\frac{\partial \mathbf{r}}{\partial \psi} \cdot \mathbf{e}_2)\Delta\psi \qquad (2)$$

comprising the terms contributed by the precession.

Note that Δa depends on a through $\mathbf{r}$, and that the total error in the determination of the semimajor axis contains Δa as a systematic part.

3. Numerical Results and Conclusions

On applying eq. (2) to Jupiter ($e \approx 0.05$), Mars ($e \approx 0.10$), and to an asteroid ($e = 0.20$, $a = 4$ AU) we find $\Delta a = 4.0$ km, 2.3 km and 15.4 km, respectively, 10 years after the epoch of reference.

The ratio of the radar measurement to its uncertainty is about 10^8 for asteroids (Ostro et al., 1991) and slightly larger for planets. Accordingly, the precision of radar ranging to a planet at Jupiter distance lies between 1 to 10 km, and amounts to some 10 km for asteroids. Thus, in conclusion, we infer from the above estimates of Δa caused by the imperfections of the luni-solar precession, that estimation of the semimajor axis of a planetary orbit derived from angular measurements contains a systematic error approximately as large as the precision of planetary radar ranging. This fact deserves consideration when semimajor axes are compared after being determined from angular observations and radar ranging.

4. References

McCuskey, S.W., 1963, Introduction to Celestial Mechanics, Addison-Wesley Publ. Comp., Reading, Massachusetts, p. 70

Ostro, S.J., Campbell, D.B., Chandler, J.F., Shapiro, I.I., Hine, A.A., Velez, R., Jürgens, R.F., Rosema, K.D., Winkler, R., Yoemans, D.K., 1991, Astron. J. 102, 1490

Williams, J.G., Newhall, X X, Dickey, J.O., 1994, Determination of Precession and Nutation from Lunar Laser Ranging Analysis, Proceedings of JD 19 of the IAU XXIInd General Assembly in The Hague, IAU/WGAS Circular No. 99, Article No. 99.1.2

RESIDUAL ROTATION OF THE FK5 FROM OPTICAL OBSERVATIONS OF THE PLANETS 1960-1994

YU. B. KOLESNIK
*Institute for Astronomy of the Russian Academy of Sciences,
109017,48 Piatnitskaya St.,Moscow, Russia*

1. Observational basis

In the interval covering last three decades optical observations of the Sun and major planets have been produced with an unprecedented intensity. Most of the published world-wide observations of the Sun, Mercury, Venus and Mars made from 1960 to 1994 with transit circles, astrographs and astrolabes are incorporated here for investigation of the residual rotation of the stellar system with respect to the dynamical reference frame.

The transformation procedure of observations to the FK5-based stellar system presented in (Kolesnik 1995), henceforth Paper I, was applied to observations of the Sun and five major planets to convert them from the system based on standards IAU 1964 to that defined by the standards IAU 1976. The principal correction includes the transition from Newcomb's to Fricke's value of precession and elimination of the non-precessional rotation of the FK4. Residuals $(O-C)_\alpha$ and $(O-C)_\delta$ have been formed by comparison with DE200 ephemeris.

An independent normal system in the equatorial half of the sky (N70E) has been compiled on the basis of 37 modern catalogues made between 1958 and 1993 (Kolesnik 1996). FK3, FK4, GC and N30 catalogues were incorporated in addition to refine the system of proper motions. The mean formal accuracy of the N70E system in both α and δ is estimated to be around 4.5 mas for positions at the epoch 1970, and 15 mas/cy for proper motions. The overall orientation of N70E positions and proper motions is coincident with FK5 being consistent with the adopted equator and equinox J2000.0. The systematic differences N70E-FK5 have been applied to residuals α and δ of the Sun and planets.

2. Rotation of the FK5-based stellar system with respect to dynamical reference frame

Smoothed right ascension residuals $(O-C)_\alpha$ derived from 15 transit circle and astrolabe instrumental series for the Sun, 12 transit circle series for Mercury, 13 transit circle series for Venus and 30 transit circle, photographic and astrolabe series for Mars are shown in Fig.1a-d. Spectacular $1''$/cy positive secular drift of the Sun residuals is clearly manifested by the whole ensemble instruments. The similar drift can be noticed for Mercury and Venus as well, despite the clear phase effect. On the other hand, the positive trend is absent in the case of Mars. Solutions for the equinox correction ΔE and its secular motion $\Delta \dot E$ are given in Table 1. Conditional equations of Newcomb, for the Sun, and of Sveshnikov (1972), for Mercury Venus and Mars, have been used. The terms of secular variation of origin of the Earth's and planet's mean longitude were omitted in solution for Mercury, Venus and Mars in view of their strong linear correlation with this of equinox motion. Residual rotation of the stellar system and mean motions of the Earth and planets cannot be effectively separated in a LS solution with actual accuracy of optical observations, time interval covered by observations and conditional equations used. The derived quantity $\Delta \dot E$ might be interpreted, therefore, as the combined effect of the equinox drift and inaccurate inertial mean motions of the Earth and planets in the comparison ephemerides. The following hypotheses are analysed here to explain

TABLE 1. The equinox correction ΔE for the epoch 1975 (arcsec) and its secular variation $\Delta \dot E$ (arcsec/cy) derived from optical observations of the Sun, Mercury, Venus and Mars

		Sun	Mercury	Venus	Mars
Number	RA	10482	1988	6524	3232
of observ	DEC	8177	2017	6313	3892
ΔE		+0.016 ± 0.008	+0.056 ± 0.052	+0.010 ± 0.034	+0.045 ± 0.066
$\Delta \dot E$		−1.21 ± 0.08	−0.92 ± 0.22	−0.96 ± 0.14	+0.56 ± 0.12

the $-1''$/cy correction : 1) inaccurate mean motion of the Earth in DE200; 2) inaccurate mean motions of Mercury and Venus in DE200; 3) constant error in FK5 proper motions caused by error in adopted value of luni-solar precession; 4) inadequacy of the procedure applied to transform the FK4-based observations to the FK5 system; 5) constant error in the FK5 proper motions of non-precessional origin appeared due to overestimated Fricke's correction $+1.27'' \pm 0.11''$/cy applied in the FK5; 6) unknown systematic

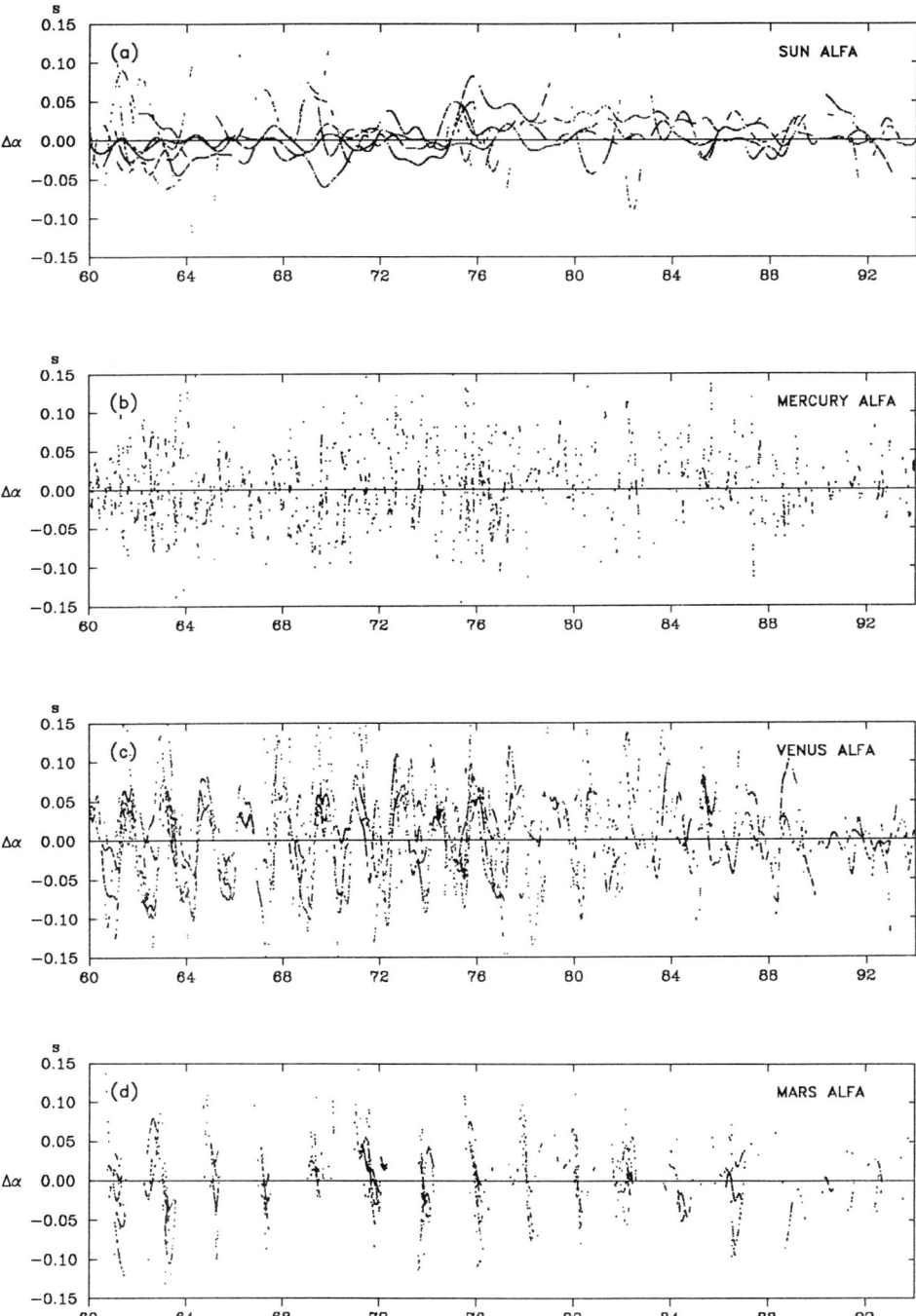

Figure 1. $(O-C)_\alpha$ residuals for the Sun, Mercury, Venus and Mars formed on the basis of modern optical obervations from 1960 to 1994

error in observations of daytime objects varying with time; 7) spurious effect of accidental combination of different instrumental series in a combined solution.

1) The mean motion of the Earth in DE200 cannot be erroneous by $1''/cy$ taking into consideration the investigation of Standish & Williams (1990).
2) To test how the inaccurate mean motions of planets with omitted terms in the conditional equations can affect results of the LS solution, a $1''/cy$ secular error in the origin of planet's longitude has been simulated and applied to the actual observations. Solutions for $\Delta\dot{E}$ with (O-C) computed in this way are $-1.08'' \pm 0.22''$ for Mercury, $-0.98'' \pm 0.14''$ for Venus, $-1.03'' \pm 0.12''$ for Mars, which are to be compared with the respective results in Table 1. It is seen that for Mercury, and especially for Venus, inaccurate mean motions of these planets (at least on the level of $1''/cy$) have a marginal effect on the solutions and cannot essentially affect estimates of given in Table 1. For the outer planets by contrast, in view of the significantly longer periods of revolution, the mean motions are dominant factors affecting the final results.
3) Determination of precession by lunar laser ranging and interferometric technique has shown that Fricke's value cannot have a $1''/cy$ error: see Fukushima (1991). The investigation of Schwan (1988) does not reveal any significant error in FK5 proper motions due to precession.
4) The same procedure has been applied to observations of the inner and outer planets. The latter ones do not manifest the positive trend in right ascension. Hence the $1''/cy$ drift cannot be attributed to an inadequacy of the transformation procedure.

Conclusion: the items 5), 6), 7) are recognized here to be the most probable explanation of the positive secular-like drift in right ascension residuals of modern observations of the Sun, Mercury and Venus.

Acknowledgements

The financial support from BDL and Russian Foundation for Fundamental Research is gratefully acknowleged

References

Fukushima T., 1991, in Hugues J.A., Smith C.A., Kaplan G.H (eds) Reference Systems, USNO, Washington, p.27
Kolesnik Yu.B., 1995, A & A 294, 876
Kolesnik Yu.B., 1996, submitted to MNRAS
Schwan H., 1988, A & A 198, 116
Standish E.M., Williams J.G., 1990, in Lieske J.H. and Abalakin V.K. (eds.) Inertial Coordinate System on the Sky ,Kluwer, Dordrecht, p.173
Sveshnikov M.L., 1972, Bull.ITA 13,131 (in Russian)

AN UPDATED GSC AS THE ASTROMETRIC REFERENCE FOR MINOR PLANET OBSERVATIONS

SIEGFRIED RÖSER
Astronomisches Rechen-Institut
Mönchhofstr. 12-14, 69120 Heidelberg, Germany

Abstract. Minor planet observers generally use the Guide Star Catalog (GSC) as the astrometric reference for the reductions of their CCD frames. Two problems are related with GSC. First, it contains no proper motions, and second, it shows severe systematic errors near the plate edges.

I report on the status of a new reduction of GSC and present a first version of STARNET, a catalogue of 4.3 million positions and proper motions using GSC as second, and a new reduction of the Astrographic Catalogue (AC) as first epoch. In STARNET the average rms-accuracy of the proper motions is 5 milliarcsec/year, that of the present-day positions 0.3".

1. New GSC reduction

A comparison between GSC 1.0 (Lasker et al. 1990) and PPM shows large differences with an rms up to 0.7" especially on the southern hemisphere (Röser et al. 1994). Starting from GSC 1.0 a new plate-by-plate reduction onto the system of PPM has been carried out. This was done by applying a numerical filter with a fixed number (25) of stars (Röser et al. 1994). As a result of this procedure the rms residuals between GSC and PPM coordinates are now better than 0.3". The reason for this improvement is a removal of systematic errors on scales smaller than the size of the plates. These were present in GSC 1.0, and were caused by the fact that SAOC and CPC were used as reference catalogues. This improvement was not found by Lopez Garcia and Yagudin (1995), because they only corrected the zero-point of each plate with respect to PPM.

This re-reduction still showed the well-known (Taff et al. 1990) distortions at the plate edges, when residual differences between the newly reduced GSC and PPM as functions of the plate coordinates averaged over many plates are formed. From GSC plate-overlaps a radial magnitude equation has been detected. This is confirmed by Morrison, Röser et al. (1995) in a comparison with the Lick Northern Proper Motion Catalog (Klemola et al. 1987). The edge-distortions and the magnitude effect will be removed from GSC within the next few months.

2. STARNET - the determination of proper motions

GSC is an observational catalogue. It is *not* a reference catalogue, because it does not contain proper motions. Galactic rotation and solar motion lead to systematic errors, the larger the more one deviates from the epoch of observation. In some regions of the southern hemisphere, after 20 years of epoch difference, these systematic errors amount to 0.2". Referencing observations to GSC means referencing to a coordinate system which rotates at a rate of 5 mas/year.

For about a quarter of the GSC objects, the measurements of the Astrographic Catalogue (AC) can be used as first-epoch observations. These published measurements were put into machine-readable form by Nesterov et al. (1990). Identifying AC with GSC has yielded positions and proper motions for 4.3 million stars. This catalogue of reference stars is called STARNET.

STARNET has the following characteristics: an average star density of 100 stars per square degree, a median magnitude of $B = 12.0^m$ on the southern hemisphere and $V = 11.5^m$ on the northern hemisphere. The present-day rms-accuracy of the positions in STARNET is 0.3", that of the proper motions 5 mas/year. Comparison of STARNET with observations made with the Bordeaux meridian circle (Requième 1995) confirmed these values, but also showed the systematic errors mentioned in the previous section. STARNET is available at Astronomisches Rechen-Institut. A detailed publication on its construction and its properties is in preparation.

References

Klemola, A.R., Jones, B.F. and Hanson, R.B., 1987, *AJ*, **94**, 501.
Lasker, B.M., Sturch, C.R., McLean, B.J., Russell, J.L., Jenkner, H. and Shara, M.M., 1990, *AJ*, **99**, 2019.
Lopez Garcia, A. and Yagudin, L.I., 1995, (this volume).
Morrison, J.E., Röser, S., Lasker, B.M., Smart, R.L. and Taff, L.G., 1995, submitted to *AJ*.
Nesterov V.V., Kislyuk V.S., Potter Kh. I., 1990, In IAU Symposium 141, J.H. Lieske and V.K. Abalakin eds., Kluwer Academic Publishers, Dordrecht, 482.
Röser S., Bastian U., Kuzmin A.N., 1994, in IAU Coll. 148, in press.
Requième, Y., 1995, private communication.
Taff L.G., Lattanzi M.G., Bucciarelli B. et al., 1990, *ApJ*, **353**, L45.

CORRECTED GSC: REFERENCE CATALOGUE FOR CCD OBSERVATIONS. NOW

A. LOPEZ GARCIA
Valencia University Observatory. Dept. Astronomy and Astrophysics, Valencia, Spain.

AND

L.I.YAGUDIN
Pulkovo Observatory, Pulkovo, St.Petersburg, Russia.

Abstract. Astrometrical properties of the Hubble Space Telescope Guide Star Catalogue (GSC) were recently investigated by the same authors through its comparison with the PPM catalogue. In this paper, a new systematic plate-based magnitude dependent error produced by telescope optics and by the incompleteness of plate reduction model has been found after applying a simplified block adjustment procedure. New subroutine for correction of all systematic errors is developed and GSC can be used now as a dense reference catalogue on about 0.4" accuracy level.

1. Introduction

Starting with the GSC authors (Russel at al., 1990) many astronomers were occupied by the GSC astrometric properties, trying to estimate the more evident systematic errors (Lasker at al., 1990; Taff at al., 1990; Bucciarelli at al., 1993; Roeser et al., 1994; Roeser et al., 1995), but nobody gave, up to now, a solution which could be used in astrometric practice immediately.

One year ago, we begun our investigations of the GSC properties with the aim of developing simple software which, connected with the Space Telescope Science Institute GSC CD ROM, could convert the GSC into a reference catalogue for narrow field instruments, representing the FK5 J2000 coordinate system (Lopez et al., 1994). Two systematic errors were found in the comparison of GSC with PPM catalogue: a plate correction due to the transformation into the FK5 system and a correction to be applied to each star according to its position on the GSC photographic plates (Lopez et al., 1995).

2. GSC quasi-block-adjustment analysis

GSC plate field corrections apply to GSC stars for the PPM magnitude range and it should be expected that GSC errors depend also on magnitudes, as faint stars errors have the same origin and nature as the bright stars ones, but on a different scale. Because of faint PPM stars absence, there is only one way to estimate the faint stars systematic errors, a significantly simplified block adjustment procedure applied to the overlapping parts of the GSC plates and based on the following assumptions:

1. Faint stars systematic errors are the same for all stars and all plates in narrow magnitude and declination zone intervals.
2. Faint stars systematic errors after applying the previous corrections have a very simple nearly radial form.

These assumptions are only approximated, but they simplify the procedure and give satisfactory results. To estimate the magnitude-dependent plate-based systematic errors, we have to solve by least squares method about 200 equation systems (for different declination zones and for different magnitude ranges) with a few number of unknown parameters in every system.

We choose the following nearly radial model for magnitude-dependent plate-based errors fitting:

$$\delta_x = A*x*r + B*x*y^2 \qquad (1)$$
$$\delta_y = C*y*r + D*y*x^2$$

where x, y are star standard coordinates on the GSC plate, r is the star to plate center distance, A, B, C, D are model parameters, and δ_x, δ_y the errors under investigations.

First terms in both equations are the projections on the x and y axes of pure radial formula $\delta_r = k*r^2$ which can be considered as a plate radial distortion. Second terms are included to correct plate corner zones, as plate errors have 4 axes of symmetry.

Model parameters A, B, C and D were calculated separately for 6 magnitude ranges and 32 declination zones, corresponding to the GSC observational program.

In northern hemisphere, radial term dominates. It depends evidently on zenith distance and has physically clear magnitude dependence. Vice versa, in the southern sky, both terms are significant to the same extent

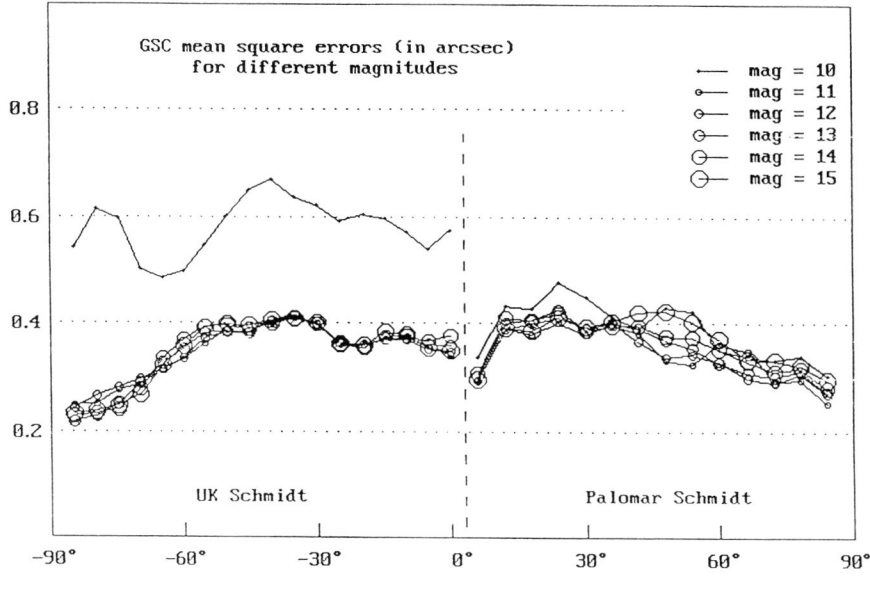

Figure 1. GSC accuracy

and their behavior is more complex. This means that GSC errors behavior is more sensitive to star brightness on the UK Schmidt plates than on the Palomar ones, as a result of the longer exposure time of the UK Schmidt plates (50–75 minutes) vs. 20 minutes for the Palomar plates. In all cases, the magnitude dependence is stronger for bright stars than for the faint ones, as it could be expected.

Systematic errors still remain in our model, but their values are significantly smaller than the source ones. According to our primary aim of investigating GSC systematic errors, we have included these results into our GSC correcting software as the third kind of correction.

3. GSC accuracy

Dispersion of the GSC–GSC' differences for the same stars from different plates after elimination of the three kinds of systematic errors gives an idea about the corrected GSC accuracy. Its mean values for different magnitudes and for different declination zones are shown on the Fig.1. As it is seen, the corrected GSC accuracy does not depend practically on star brightness for stars fainter than 10 magnitude; it depends on zone declination; it is practically the same for both telescopes. ±0.4 arcsec can be taken for the mean accuracy of the corrected GSC. It can be taken also for the upper estimation of the real GSC accuracy, which must be better because not all systematic errors are canceled by our corrections.

4. Concluding remarks

Our examination of the GSC is a very generalized one without taking into account the individual GSC plates properties and deeper investigations will give more information on the subject. Meanwhile, we hope that our simple software can be useful for those who need a really dense reference catalog, at least till appearance of the Hipparcos catalog or of a new Guide Star Catalog version.

Acknowledgements

The authors wish to thank the Conselleria de Ciencia y Educacion de la Generalitat Valenciana for financial support of this work.

References

Bucciarelli B., Lattanzi M.G., Taff L.G.: 1993, *Ap. J. Suppl.* **84**, 91.
Lasker B.M., Sturch C.R., McLean B.J., Russel J.L., Jenkner H. and Shara M.M.: 1990, *Astron.J.* **99**(6), 2019-2058.
Lopez Garcia A., Martinez Gonzalez J.M., Ortiz Gil A., Yagudin L.I.: 1994, *IAU Comission 9, WG on "Wide - field imaging", Newsletter* 6, 11-14.
Lopez Garcia A., Martinez Gonzalez J.M., Ortiz Gil A., Yagudin L.I.: 1995, *Proc. of III International Workshop on Cel. Mech. and Posic. Astrom., Cuenca, Spain, 1994*, in press.
Roeser S., Bastian U., Kuzmin A.: 1994 *Proc. of IAU xxx*, in press.
Roeser S., Morrison J.E. et al.: 1995 *Astron.J.*, in press.
Russel J.L., Lasker B.M., McLean B.J., Sturch C.R. and Jenkner H.: 1990, *Astron.J.* **99**(6), 2059-2081.
Taff L.G., Lattanzi M.G., Bucciarielli B., Gilmozzi R., McLean B.J., Jenkner H., Laidler V.G., Lasker B.M., Shara M.M. and Sturch C.R.: 1990, *Astrophis.J.* **353**, L45–L48.

APPLICATION OF GSC CATALOGUE TO MINOR PLANETS PLATE REDUCTION

A. LOPEZ GARCIA
Valencia University Obs., Dept. Astronomy and Astrophysics. Valencia, Spain.

L.I.YAGUDIN
Pulkovo Observatory, Pulkovo, St.Petersburg, Russia.

AND

M.J. MARTINEZ
Castellon University, Castellon, Spain.

Abstract. An automatic reduction process is applied to observations of minor planets obtained at Valencia Observatory, using PPM and GSC catalogues. Systematic corrections of GSC, obtained by the same authors in a previous study, are also considered. The comparison of results is done by the statistical analysis of several parameters. Preliminary conclusions about the improvement obtained using corrected GSC are included.

1. Introduction

The astrometrical properties of the GSC (Guide Star Catalog) were largely studied by different authors (Russel et al, 1990; Taff et al. 1990; Lasker et al, 1990; etc.)

In a previous study, we proposed (Lopez et al., 1994) a plate analysis of the GSC systematic errors by means of its comparison with another catalogue, the PPM, whose mean precision for astrometric reductions (0.2" for positions and 0.003"/year for proper motions) is high.

In particular, we described two kinds of errors (Lopez et al, 1995), due to the transformation into the FK5 system and to star position on GSC photographic plates. We concluded that, including corrections to these errors, the GSC was suitable to be used as a reference catalogue with a level of accuracy of about 0.4".

In this paper, we use PPM and both the standard and the corrected versions of GSC in an automatic reduction process of 267 plates with 647 exposures of minor planets observations done at Valencia Observatory from January 1985 to February 1995.

2. Results

The statistical analysis of the results obtained with standard GSC (S) and corrected GSC (C) plates vs. the PPM ones gives the following results.

1. (O–C) star residuals in RA and Dec. show similar mean values less than 1" with mean square errors (m.s.e) in the order of 0.1".
2. Due to the great differences between the northern and the southern part of the GSC, minor planet differences in R.A. and Dec. are analyzed separately for positive and negative declinations.
3. Differences GSC (C) – PPM for positive declinations are about 0.1". The improvement of GSC (C) vs. GSC (S) positions is also in the order of 0.1".
4. For negative declinations a similar result is obtained, but the differences GSC (C) – PPM remain great (about 0.2").
5. Our comparison of PPM and GSC is limited to low magnitude stars, affected of high magnitude errors. Nevertheless, we consider that results with corrected GSC are better than those obtained with the standard GSC and are similar to those got with PPM.

Acknowledgements

The authors wish to thank the Conselleria de Educacion y Ciencia of the Generalitat Valenciana for financial support of this work.

References

Lasker B.M., Sturch C.R., McLean B.J., Russel J.L., Jenkner H. and Shara M.M, 1990, *Astron.J.* **99**(6), 2019-2058.
Lopez Garcia A., Martinez Gonzalez J.M., Ortiz Gil A., Yagudin L.I., 1994, *IAU Commission 9 WG on "Wide-field imaging" Newsletter* 6, 11-14.
Lopez Garcia A., Martinez Gonzalez J.M., Ortiz Gil A., Yagudin L.I., 1995, *Proc. of III International Workshop on Cel. Mech. and Posic. Astrom., Cuenca, Spain, 1994*, in press.
Russel J.L., Lasker B.M., McLean B.J., Sturch C.R. and Jenkner H., 1990, *Astron.J.* **99**(6), 2059-2081.
Taff L.G., Lattanzi M.G., Bucciarelli B., Gilmozzi R., McLean B.J., Jenkner H., Laidler V.G., Lasker B.M., Shara M.M., Sturch C.R.: 1990 *Astrophys.J.* **353** L45-L48.

THE HIPPARCOS MISSION AND THE RE-REDUCTION OF BELGRADE ZENITH-TELESCOPE OBSERVATIONS

G. DAMLJANOVIĆ
Astronomical Observatory
Volgina 7, 11050 Belgrade, Yugoslavia
E-mail: gdamljanovic@aob.aob.bg.ac.yu

Abstract. We prepared the results (in a computer-readable form) of the Belgrade ZT observations made in the period 1949-1985 and finished their new reduction in the FK5 reference frame.

1. Introduction

At the XXI IAU General Assembly, Buenos Aires 1991, Commission 19 of the IAU, "Rotation of the Earth", formed the Working Group on Earth Rotation in the HIPPARCOS reference frame–WG ERHRF to collect the past observations and to analyse them in that reference system. Our investigations are in accordance with it. The WG ERHRF has set up a list of the best observations performed in the past. The Belgrade Observatory is on that list (Vondrák, Feissel and Essaifi, 1992) with the observations in the period 1949-1990 obtained with ZT (the visual zenith–telescope Askania–Bamberg No 77241, 110/1287 mm) by applying Talcott's method.

The Old Belgrade Latitude Programme – OP (Djurković,Ševarlić, Brkić, 1951) was observed in the period 1949-1960. The New Belgrade Latitude Programme – NP (Ševarlić and Teleki, 1960) was started in 1960 and the observations are still carried out. We used the PPM Star Catalogue (Röser & Bastian, 1991) for the re–reduction.

2. Procedure and results

The re–reduction is in accordance with MERIT standards (Melbourne et al., 1983). The new IAU(1976) coordinate system of astronomical constants, the IAU(1980) nutation model, the new dynamical reference system (JPL DE200/LE200 Ephemeris, 1984), and the FORTRAN programme for refraction (Abalakin, 1985) are used. The trigonometric stellar parallaxes (Jenkins, 1952) and the stellar radial velocities (Wilson, 1953) are used for the calculation of the apparent places of OP and NP stars.

We used the Student–Fisher criterion for eliminating the excessive instantaneous latitudes resulting from some Talcott's pairs.

The polar motion was eliminated from the material and the observations were brought in accordance with the mean pole BIH1979. After that we made determinations of the systematic errors of declinations and proper motions of Talcott's pairs and (sub)groups of OP and NP.

We used the new instrument's constants applied in the re–reduction (the angular value of the micrometer screw revolution, the angular division values and the temperature coefficients of the Talcott's levels) and the numerous systematic errors are taken into account, see (Damljanović, 1994, 1995).

3. Conclusion

The mean error of the instantaneous latitude from one Talcott's pair is less than before. From the preliminary ZT observations made 1947 the mean error was $\pm 0.''255$ (Djurković, Ševarlić, Brkić, 1951). The mean error of the OP (1949–1960) was $\pm 0.''220$ (Ševarlić and Teleki, 1960), and of the NP was $\pm 0.''272$ (1960–1965.5) and $\pm 0.''146$ (1969–1974) (Grujić et al. 1989). After our re-reduction the mean error of the OP is $\pm 0.''199$ (1949–1960) and the mean error of the NP is $\pm 0.''148$ (1960–1985).

References

Abalakin, V. K. (1985), Refraction Tables of Pulkovo Observatory (V edition), NAUKA, Leningrad, GAO AN SSSR.
Djurković, P., Ševarlić, B., Brkić, Z. (1951) *Publ. Obs. Astron. Belgrade*, **4**.
Damljanović, G.(1994) *Bull. Astron. Belgrade*, **150**, pp. 29.
Damljanović, G.(1995) *Bull. Astron. Belgrade*, **152**, pp. 71.
Grujić, R., Djokić, M., Jovanović, B.(1989) *Bull. Obs. Astron. Belgrade*, **141**, pp. 7.
Jenkins, L. F. (1952) Gen. Cat. of Trigonometric Stel. Parallaxes, Yale University Obs.
Melbourne, W., et al. (1983) Project Merit Standards, USNO Circular, **No. 167**.
Roeser, S., Bastian, U. (1991) Positions and Proper Motions (PPM) Star Catalogue, Astron. Rechen-Institut, Heidelberg.
Ševarlić, B., Teleki, G. (1960) *Bull. Obs. Astron. Belgrade*, **24**, No. 3/4, pp. 19.
Vondrák, J., Feissel, M., Essaifi, N. (1992) *Astron. Astrophys.*, **262**, pp. 329.
Wilson, R. E. (1953) Gen. Cat. of Stel. Radial Velocities, *Carnegie Institution of Washington Publ.*, Mount Wilson Obs., **601**.

INDIRECT LINKING OF THE HIPPARCOS CATALOG TO EXTRAGALACTIC REFERENCE FRAME VIA EARTH ORIENTATION PARAMETERS

J. VONDRÁK
Astronomical Institute
Boční II, 141 31 Prague 4, Czech Republic

Abstract. The indirect method of linking the Hipparcos reference frame to the frame defined by extragalactic sources is described. To this end, two independent time series of Earth orientation parameters observed by two different techniques with respect to the two reference frames are used: a) Optical astrometry observations (referred to Hipparcos stars), b) VLBI observations (referred to extragalactic objects). The parallel use of both techniques during the last decade enables to determine the orientation of the two reference frames at a fixed epoch and their mutual slow rotation with precision of at least 1mas and 1mas/year, respectively. In order not to raise confusion, the potentiality of the method is demonstrated on the example based on the star catalogues originally used at the participating observatories, not on any of the existing preliminary versions of the Hipparcos catalog.

1. Introduction

It is well known that the original reference system of the Hipparcos Catalogue has six degrees of freedom; its orientation and rotation with respect to conventional extragalactic reference system must be determined from other observations than Hipparcos project itself. The overview of the methods to be used is published by Lindegren & Kovalevsky (1995). The prevailing majority of these methods is based on more or less direct observations of the positions of the Hipparcos stars with respect to extragalactic objects.

Here we describe an indirect method, based on comparing two independent Earth Orientation Parameters (EOP) referred to the two reference systems (optical and extragalactic). The Earth provides an intermediary (rapidly rotating) reference system whose orientation, with respect to the two aforesaid systems, is monitored by two distinct methods using completely different instruments.

The new global adjustment of the EOP from optical astrometry observations since the beginning of the century is presently being prepared by a Working Group set up by IAU Commission 19 (Vondrák, 1991). Its final goal is to derive the EOP in the Hipparcos reference frame; the expected accuracy of the prepared solution is, for the last decades, at the level of 10mas per 5-day value (Vondrák et al., 1992). In the absence of the final Hipparcos Catalogue, a set of preliminary test solutions has been made based on the original local star catalogues used at the observatories, the most recent one being published by Vondrák et al. (1995). EOP are regularly monitored with respect to extragalactic objects by Very Long-Baseline Interferometry (VLBI); the results obtained by different analysis centers are combined into a single solution at the Central Bureau of the International Earth Rotation Service (IERS). The nominal precision of these observations is less than 1mas, the adopted positions of the observed extragalactic sources define the extragalactic celestial reference frame (Carter & Robertson, 1993).

2. The Earth Orientation Parameters

There are five EOP that define the position of the spin axis of the Earth, both in the terrestrial and celestial reference systems, and the phase of the spin around the axis. They are as follows:

1. The two components of the axis in the Earth x, y (polar motion),
2. The two components of the axis in space wrt standard Celestial Ephemeris Pole $\Delta\psi, \Delta\varepsilon$ (celestial pole offsets in longitude and obliquity),
3. Universal time offset from the International Atomic Time UT1–TAI.

When determining these parameters from optical astrometry observations, the diurnal part of the observed variations is substantial for determining celestial pole offsets, while longer periodic part is used to determine polar motion and UT1. The first two parameters are thus practically useless for linking the two reference frames, being sensitive only to terrestrial reference frame changes. On the contrary, the celestial pole offsets are insensitive to terrestrial reference frame changes (these changes can produce only diurnal variations in $\Delta\psi, \Delta\varepsilon$) but any change in the celestial reference frame orientation is fully reflected in these values. The most 'problematic' component is the Universal time UT1 – it is fully sensitive to both terrestrial and celestial reference systems.

3. Orientation of Optical and Extragalactic System

The relative orientation of the two reference systems and their mutual rotation can be described by six parameters. Here we use the notation introduced by Lindegren & Kovalevsky (1995) – the transformation of a column vector given in the Hipparcos system as $\vec{H} = (x_H, y_H, z_H)^T$ into the extragalactic system $\vec{E} = (x_E, y_E, z_E)^T$ at the epoch t_0 is given by the formula

$$\vec{E} = \begin{bmatrix} 1 & \varepsilon_{0z} & -\varepsilon_{0y} \\ -\varepsilon_{0z} & 1 & \varepsilon_{0x} \\ \varepsilon_{0y} & -\varepsilon_{0x} & 1 \end{bmatrix} \vec{H} \qquad (1)$$

and the time derivatives of the angles ε define the rate of mutual rotation of both systems: $\omega_x = \dot{\varepsilon}_x$, $\omega_y = \dot{\varepsilon}_y$, $\omega_z = \dot{\varepsilon}_z$.

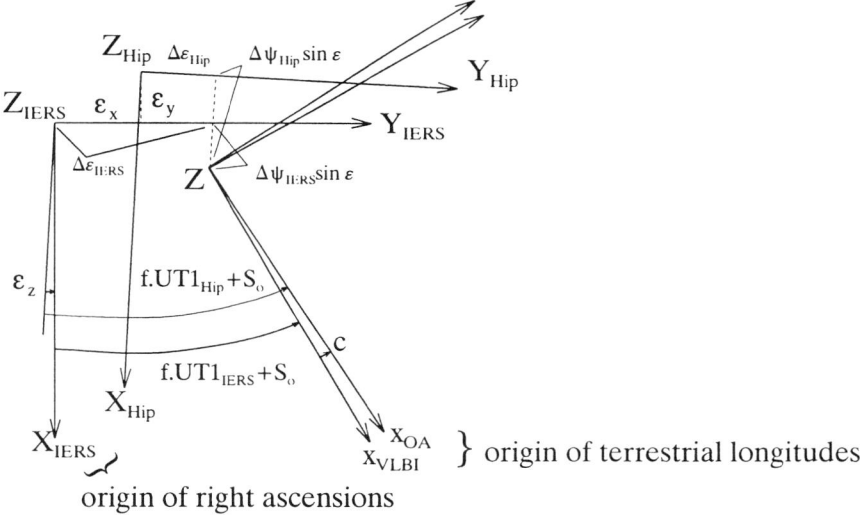

Figure 1. Relation between celestial pole offsets and the orientation of the Hipparcos and extragalactic reference frames

The relation between the EOP and the angles defining the orientation of both celestial reference systems is explained at Fig. 1. It shows the position of the axes X, Y, Z of both celestial reference frames and the position of the axis x (origin of longitudes of the terrestrial reference frames used by VLBI and optical astrometry) as seen from the north pole. Z_{Hip} and Z_{IERS} denote the positions of the north pole of Hipparcos and extragalactic reference frames, both moving with standard precession-nutation, Z is the position of the actual celestial pole. The offsets between Z and Z_{Hip}, Z_{IERS} are measured by optical astrometry and VLBI and expressed by means of celestial pole offsets $\Delta\varepsilon_{\text{Hip}}, \Delta\psi_{\text{Hip}}\sin\varepsilon$ and $\Delta\varepsilon_{\text{IERS}}, \Delta\psi_{\text{IERS}}\sin\varepsilon$. They contain not only the misalignments of the two systems of reference, but also the deviations of actual nutation from the standard model used. Provided the same model is used for optical astrometry and VLBI, the latter deviations disappear in the difference. Universal time is calculated from the measured angle between the plane of zero meridian of the terrestrial reference frame and the XZ plane of the corresponding celestial reference frame. In this case, these are the angles $\angle x_{\text{OA}} X_{\text{Hip}}$ and $\angle x_{\text{VLBI}} X_{\text{IERS}}$; they are equal to $f.\text{UT1}_{\text{Hip}} + S_0$ and $f.\text{UT1}_{\text{IERS}} + S_0$, respectively, where $f = 1.0027\ldots$ and S_0 is the standard expression for GMST of 0^h UT1, including the equation of equinoxes (Seidelmann, 1992). Since the origins of terrestrial longitudes for optical astrometry instruments (defined by their adopted longitudes) and for VLBI (defined by the adopted directions of the baselines) are not generally identical, there is no possibility of independent determination of ε_z and c. The formulas for calculating the ε values then read

$$\begin{aligned}
\varepsilon_x &= \Delta\varepsilon_{\text{IERS}} - \Delta\varepsilon_{\text{Hip}} \\
\varepsilon_y &= (\Delta\psi_{\text{Hip}} - \Delta\psi_{\text{IERS}})\sin\varepsilon \\
\varepsilon_z + c &= 15.041(\text{UT1}_{\text{Hip}} - \text{UT1}_{\text{IERS}}).
\end{aligned} \quad (2)$$

Having the two sets of EOP observed in the same time interval, we can interpolate them for the same time arguments and calculate the values ε using eq. (2) at various epochs. The time interval of common observations of both techniques being sufficiently long, the time derivatives of these values (defining the values ω) can also be obtained. In doing so, one should also take into consideration all possible correlations of the input data. Since the precision of VLBI is much higher than that of optical astrometry, we consider here only the correlations coming from our optical astrometry solution. It is a result of a rather complicated system of normal equations with about 30 thousand estimated parameters. The only significant correlations exist between the values obtained at the same 5-day interval – the adjusted values of EOP for different epochs are practically de-correlated. Consequently we treat the values calculated from eq. (2) as observed correlated quantities, taking the weight matrix from optical astrometry solution in which only certain non-diagonal elements have non-zero values.

4. Example of Linking Quasi-FK5 System, Conclusions

Before the Hipparcos catalog is available, let us demonstrate the potentiality of the proposed method on the example of a quasi-FK5 catalogue. The solution for the years 1900.0–1991.0 presented in (Vondrák et al., 1995), based on more than 3 million of observations made with 30 instruments located at 19 observatories, is referred to local catalogues partially homogenized into a system that is very close to FK5. The last decade of the solution is parallel to VLBI observations; here, we take the IERS combined solution C 02 (5-day values with epochs close to optical astrometry solution) in the years 1980.3–1991.0. In order to calculate the differences on the *rhs* of eq. (2) we interpolated the IERS values for the epochs of optical astrometry solution using cubic splines. The differences are graphically displayed in Figs. 2 and 3. The trends in all three components are obvious,

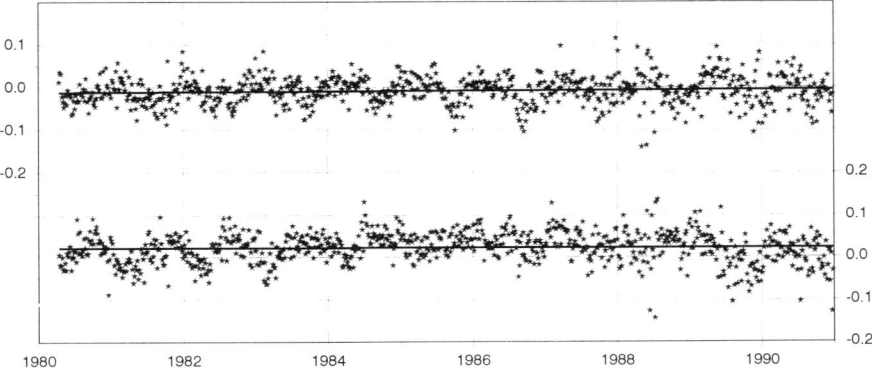

Figure 2. Differences of celestial pole offsets (in arcseconds) $\Delta\varepsilon$ (upper part) and $\Delta\psi \sin\varepsilon$ (lower part) in the sense optical astrometry minus VLBI.

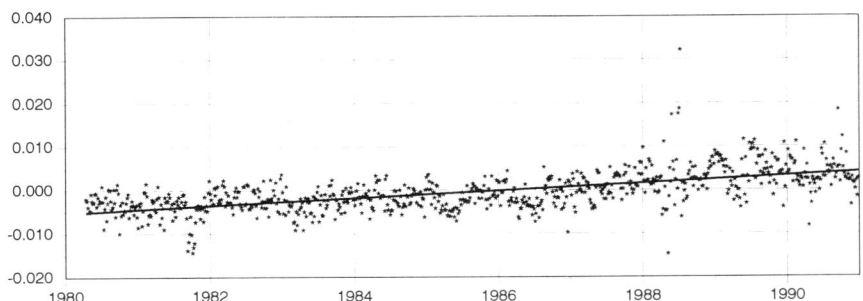

Figure 3. Differences in UT1 (optical astrometry minus VLBI) in seconds of time.

especially the one in UT1. Also remarkable is a quasi-periodic character of the differences; it seems to prompt that there might be some systematic differences in certain nutation terms (very probably in annual term) as detected by optical astrometry and VLBI.

The resulting six parameters of the mutual orientation between optical and extragalactic reference frame, their standard errors and matrix of correlation coefficients are given in Tab. 1. The values ε_0 (for the epoch

$t_0 = 1985.25$) and their standard errors (derived from the dispersion of 5-day values as seen in Figs. 2 and 3) are given in milliarcseconds, the values of ω and their standard errors in milliarcseconds per year. One can see that the standard errors in ε_{0z} and ω_z are nearly twice as big as for the other two angles. It clearly demonstrates the presence of larger systematic errors in UT1 observations than in latitude observations; the effect of higher latitudes of the observatories also plays certain role. Standard errors σ are three times bigger than one would expect from the formal precision of the global optical astrometry adjustment. The difference in terrestrial longitude systems used by optical astrometry and VLBI and their possible drift prevents us from determining the angles ε_{0z} and ω_z by this method.

TABLE 1. Example of orientation angles ε, ω, their standard errors σ (in milliarcseconds) and correlation coefficients, calculated for original star catalogs.

Orientation	σ	Correlations					
		ε_{0x}	ω_x	ε_{0y}	ω_y	$\varepsilon_{0z}+c$	ω_z
ε_{0x}	8.59	±1.11	1.000	0.063	0.012	-0.002	0.095	0.024
ω_x	-1.09	0.39	0.063	1.000	-0.002	0.017	0.023	0.090
ε_{0y}	23.00	1.11	0.012	-0.002	1.000	0.052	-0.066	0.007
ω_y	0.99	0.39	-0.002	0.017	0.052	1.000	0.006	-0.068
$\varepsilon_{0z}+c$	-19.59	1.65	0.095	0.023	-0.066	0.006	1.000	0.119
$\omega_z+\dot{c}$	9.82	0.60	0.024	0.090	0.007	-0.068	0.119	1.000

Acknowledgements

This study was made possible thanks to the grant No. 303502 awarded by the Grant Agency of the Academy of Sciences of the Czech Republic.

References

Carter, W.E., Robertson, D.S. (1993) Very-Long Baseline Interferometry applied to Geophysics, In: Mueller, I., Kołaczek, B. (eds.) *Developments in Astrometry and their Impact on Astrophysics and Geodynamics*, Kluwer, Dordrecht p. 133.
Lindegren, L., Kovalevsky, J. (1995) Linking the Hipparcos Catalogue to the extragalactic reference system, *Astron. Astrophys.*, in press.
Seidelmann, P.K. (1992) *Explanatory supplement to the Astronomical Almanac*, Univers. Science Books, Mill Valley, California.
Vondrák, J. (1991) Calculation of the new series of the Earth orientation parameters in the Hipparcos reference frame, *Bull. Astron. Inst. Czechosl.*, **42**, 283–294.
Vondrák, J., Feissel, M., Essaïfi, N. (1992) Expected accuracy of the 1900–1990 Earth orientation parameters in the Hipparcos reference frame, *Astron. Astrophys.*, **262**, 329–340.
Vondrák, J., Ron, C., Pešek, I., Čepek, A. (1995) New global solution of Earth orientation parameters from optical astrometry in 1900–1990, *Astron. Astrophys.*, **297**, 899–906.

ASTROMETRY VLBI IN SPACE (AVS)

V.I. ALTUNIN
Jet Propulsion Laboratory
4800 Oak Grove Dr., Pasadena, CA 91109 USA

V.A. ALEKSEEV
Radio Physical Research Institute
25 Bolshaya Pechorskaya St., Nigny Novgorod, 603600 Russia

E.L. AKIM
Institute of Applied Mathematics
4 Miusskaya Sq., Moscow, 125047 Russia

T.M. EUBANKS
U.S. Naval Observatory
3450 Massachusetts Ave., Washington, D.C., NW 20392-5420 USA

K.A. KINGHAM
U.S. Naval Observatory
3450 Massachusetts Ave., Washington, D.C., NW 20392-5420 USA

R.N. TREUHAFT
Jet Propulsion Laboratory
4800 Oak Grove Dr., Pasadena, CA 91109 USA

AND

K.G. SUKHANOV
Lavochkin Association
24 Leningradskaya St., Khimky-2, Moscow Region, 141400 Russia

1. MISSION GOALS

This paper describes a proposal for a new space radio astronomy mission for astrometry which uses very-long-baseline interferometry (VLBI) called Astrometry VLBI in Space (AVS). The ultimate goals of AVS are to improve the accuracy of radio astrometry measurements to the microarcsecond level in one epoch of measurements and to improve the accuracy of the transformation between the inertial radio and optical coordinate reference frames. The scientific objectives of the mission cover a few categories of astrometry tasks such as astrometry of the solar system, reference frames ties, tests of general relativity and cosmology (for references see Lowe and Treuhaft,1994, Russell et al.,1992, Eubanks et al.,1994).

2. MISSION CONCEPT

Current ground-based VLBI radio astrometry angular accuracy is primarily limited by atmospheric propagation effects and by the length of the longest attainable Earth baselines ($\sim$ 10000 km). Efforts to tie the radio and optical frames with observations of radio stars will be limited to the few-tenths of a milliarcsecond level by the unmodeled angular difference between the optical and radio centers of emission.

The above astrometric limitations can be circumvented if the radio interferometer is placed in space. It is possible to establish a unified Celestial Reference Frame and to tie an inertial Radio Reference Frame, an Optical Reference Frame, and a Geocentric-Equatorial Reference Frame with unprecedented accuracy.

A basic element of the proposed mission is a space-based radio interferometer composed of two free-flying antennas operating simultaneously with a 70m ground-based telescope at 8.4, 22.2, 32, 43 GHz. The space antennas will be located in orbit such that they will be visible to each other most of the time. A microwave (or laser) link will be established between the Space Radio Telescopes (SRTs) to provide a direct measurement of the radio interferometer baseline length, and synchronization of the SRTs' local oscillators and clocks. Along with radio interferometry equipment, each spacecraft will carry an optical beacon and optical (CCD) astrometry camera. The camera will determine the position of the optical beacon (the spacecraft with the radio telescope) relative to the optical reference stars (Figure 1).

The advantages of such a configuration are (see Alekseev, 1981,1993):

i) Such system excludes the limitations of ground-based radio astrometry due to atmospheric turbulence and refraction, Earth's motions, impossibility to view the entire sky with a single instrument;

ii) The baseline of the space-based radio interferometer (and, accordingly, its angular resolution) will not be limited by the Earth's diameter;

iii) Optical astrometry devices will provide the orientation for the radio interferometer baseline relative to the optical reference frame, thus the coordinates of the radio sources will be determined directly in the optical reference frame;

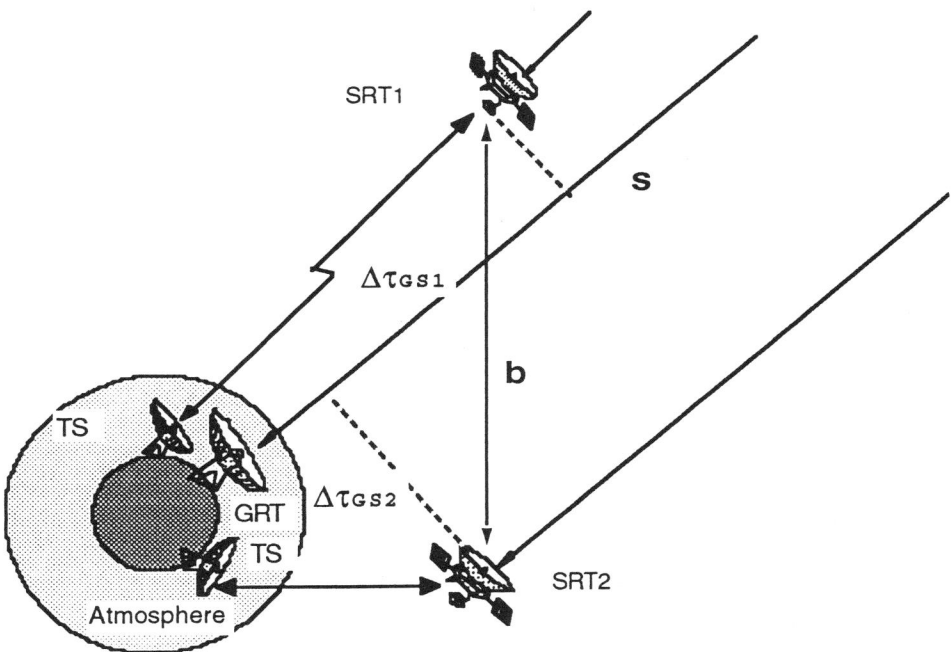

Figure 1. AVS mission configuration

iv) Direct determination of the baseline length and synchronization of the SRTs' local oscillators and clocks by a microwave link established between two spacecraft can, in principle, allow the use of a fringe phase for angular measurements, as in connected-element interferometry, instead of the group delay as used in Earth-based VLBI.

Astrometric VLBI observations by this system can provide a relative time group delay $\Delta\tau_{12}$ (main value used in VLBI astrometry measurement) between two space radio telescopes, and, accordingly, the coordinates of the radio sources as determined from the equation $c\Delta\tau_{12} = (\vec{b}\,\vec{s})$, which will not include an atmospheric impact (τ_{ATM}) and instrumental time delays in GRT equipment (τ_{ID}):

$\Delta\tau_{GS1} - \Delta\tau_{GS2} = (\tau_{SRT1} - \tau_{GRT} - \tau_{ATM} - \tau_{ID}) - (\tau_{SRT2} - \tau_{GRT} - \tau_{ATM} - \tau_{ID}) = \Delta\tau_{12}$

At the same time, simultaneous observations of a space radio interferometer combined with relatively small antennas with diameter d_{SRT} and a large ground-based radio telescope with diameter D_{GRT} will increase the signal-to-noise ratio (SNR) and, accordingly, decrease the stochastic error of radio astrometry measurements with a space-based radio interferometer:

$$\sigma(\Delta\tau_{12}) = \sqrt{\sigma^2(\Delta\tau_{GS1}) + \sigma^2(\Delta\tau_{GS2})} \sim \sqrt{2}(D_{GRT} \cdot d_{SRT})^{-1}$$

An orbiting interferometer operated at $\lambda \sim 1$ cm with a baseline between two 4-m diameter space radio telescopes $|\vec{b}| \sim 50.000$ km observing with a 70m ground-based telescope with system parameters of both space and ground telescope: Tsys= 50 K, T=300 sec, $\Delta\nu$ =128 MHz can provide angular measurements with an accuracy better than 10 and 100 microarcsec for the sources with flux density 1Jy and 0.1 Jy, respectively.

3. MISSION IMPLEMENTATION

This proposal could be implemented under the NASA category of "Midex Missions" (cost $\sim$$100 Mln). In order to meet this requirement, the mission design should be based on an existing (or feasible in near future) technology and existing supporting infrastructure. In order to keep the cost of this mission low, the space-based antennas should be small, and non deployable. If the SRTs will be launched in a geostationary orbit, the baseline for a space-based radio interferometer can be as long as 50,000-70,000 km. Two 4-m class non-deployable space antennas can be launched simultaneously by the Russian Proton or French Arian-5 boosters. The effectiveness (scientific return) of the proposed mission crucially depends on the support of large ground-based radio telescopes such as the 70m DSN antennas, the Effelsberg 100m radio telescope, the VLA phase array and the newly-developed Green Bank 100m telescope. VLBI observations are now (or will be) routinely provided by these radio telescopes at frequencies as high as 43 GHz.

The research described in this paper was partially carried out by the Jet Propulsion Laboratory, California Institute of Technology, under a contract with the National Aeronautics and Space Administration.

References

Alekseev, V.A. (1981), *Astrometry and Astrophysics (in Russian)* Vol. no. 45, pp.74–77
Alekseev, V.A. (1993), *Proceedings of the Workshop, Leningrad, USSR (published NAIC, Arecibo Observatory)* pp.102–103
Eubanks T.M., Matsakis D.N., Josties F.J. et al. (1994), *IAU Symp. S166*, in press
Lowe S.T., and Treuhaft R.N. (1994), *in VLBI Technology*, T. Sasao, S. Manabe, O. Kamea and M., Inoue, eds., pp.319–322
Russell, J.L., Jauncey, D.L., Harvey, B.R., et al. (1992), *Astron.J.*Vol. no. 103, pp.2090–2098

ANALYSIS OF THE SUN'S OBSERVATIONS WITH PRISMATIC ASTROLABE

N.V.LEISTER, P.C.R.POPPE, M.EMILIO AND F.LACLARE
Inst.Astron.Geof.USP, Caixa Postal 9638, 01065, Brazil
Observatoire de la Côte D'Azur, 06460 Saint Vallier, France

1. Introduction

A regular program of observations of the Sun with prismatic astrolabes is under way since 1974 in both, the "Abrahão de Moraes" ($\phi = -23°0.1'$) and "Caussols" ($\phi = +43°44.9'$) observatories. The zenith distance of 2968 transits of the solar limbs observed during the period from 1988 January to 1994 April, in both centers, are analysed. The primary goal of this programme is to determine corrections ΔE and ΔA to the FK5 equinox and equator, as well as to the Earth's orbital constants. Secondly, we present here possible variations of the solar radius, which are of interest because of their astrophysical significance. The results of the analysis of the observations made with these instruments depend on both the prism angles used during the observation and the latitude of the site (Leister 1989, Poppe 1994). With the use of the different reflector prisms, instead of the refracting equilateral, it was possible to observe at differents zenith distances, allowing to sweep, at the brazilian observatory, the apparent orbit of the Sun during a period of about 10 months per year, with two prisms. The instrument of the french observatory works with 11 differents zenith distances, permitting an increase of about 10 months per year in the observational period either.

2. Orientation of the frame system

The method of the data reduction is based upon the comparison between the observed zenith distance obtained by means of the transit time and the zenith distance defined by the prisms. The analysis here developed in the same way that adopted by Leister (Leister 1989) and could be made directly by mean of the condition equation:

$$\Delta z = f(\Delta E, \Delta A) + g(\Delta L, \Delta \epsilon, \Delta h, \Delta k),$$
where $f = \cos\phi \sin Z \, \Delta E + \cos Z \, \Delta A$ and
$$g = (\cos S \sin\epsilon \cos\alpha + \sin S \cos\epsilon \sec\delta) \Delta L$$
$$+ (\cos S \sin\alpha - \sin S \cos\alpha \sin\delta) \Delta\epsilon$$
$$+ 2(\sin\delta \cos S \cos^2\alpha + \sin S \sin\alpha) \Delta h$$
$$- 2(\cos S \sin\epsilon \cos\delta \cos\alpha + \sin S \cos\epsilon \cos\alpha) \Delta k.$$

S is the paralactic angle and Z is the azimuth.

The solution, together with the standard errors, calculated in the usual way from the variance-covariance matrix, is displayed in table 1.

The variance of the solution is 0.60", giving a standard error of 0.78" for an observation of unit weight. The value of latitude and clock correction, that we utilized here, was provided by the International Earth Rotation Service (IERS). The mean values of the latitude and longitude were refered to the FK5 reference system, obtained for the stars observations.

TABLE 1. General solution (arcseconds)

ΔE	$+0.10 \pm 0.04$
ΔA	$+0.07 \pm 0.07$
ΔL	-0.37 ± 0.03
$\Delta\epsilon$	$+0.63 \pm 0.07$
Δe	$+0.04 \pm 0.01$
$e\Delta\pi$	-0.37 ± 0.08
Dispersion	0.78
N	1473
Mean date	1990.7

3. Solar diameter measurements

The solar diameter follows directly from the timing of the limb transits. Here, we present the results of the diameter measurements from 1980 September to 1993, comprising 5,000 limbs observations. The advantage of prismatic astrolabes is that the observations are not affected by error in atmospheric refraction r'. An error in r' affecting the apparent zenith distance cancels out, as the radius is half the difference between the zenith distances of upper and inner solar edges. The correction R for the apparent diameter can be written:
$$R = 15 \cos\phi \sin Z (\Delta t_o - \Delta t_c) + C + kF$$
where C denote the sum of measured corrections arising from small refraction variations and geometrical effects; F is unknown plate prismatic error

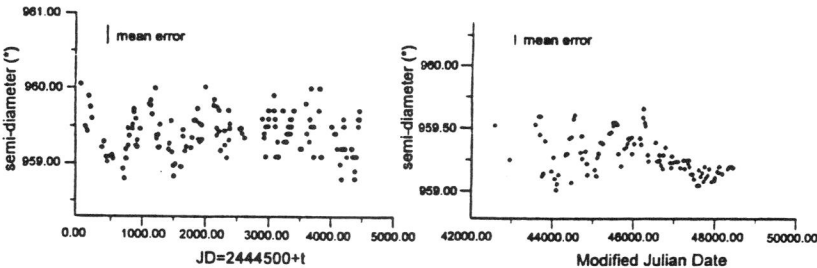

Figure 1. Monthly means of the apparent solar radius obtained at "Abrahão de Moraes" and "CERGA" observatories

and the value of the k (+1 or −1) depend on the initial plate position. The above equation was solved for the unknowns R and F at monthly intervals, with an average of 20 transits for the brazilian set and 60 transits for the french set. The results are displayed in Figures 1 and 2 for the same intervals.

TABLE 2.
The average observed solar semi-diameter

set	R(")
brazilian set	959.40 ± 0.05
french set	959.37 ± 0.02

The standard deviation of the monthly means is about 0.30" for the brazilian program. The CERGA program allows a large number of observations each day so the standard deviation of daily average is 0.15".

4. Conclusion and remarks

Concerning the determination of the equator and equinox corrections the solution obtained by least square method reveals the correction 0.10"±0.04" and 0.07"±0.07" to equinox and equator respectively, at mean epoch of observations 1990.7. The good quality of this data set shows that the FK5 do not support a significant correction. For all series of data analysed, we obtained important corrections of the obliquity of the ecliptic. In spite of the different values of the dispersion to be found among the set data of the measurements of the semi-diameters, a sinusoidal component is visible on the plots of Figure 1. A harmonic analysis reveals a significant term of period 1,000±50 days (Leister et al 1990, Laclare 1983). The amplitude of

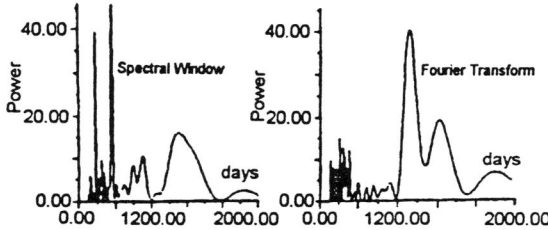

Figure 2. Fourier Transform and Spectral Window - brazilian set

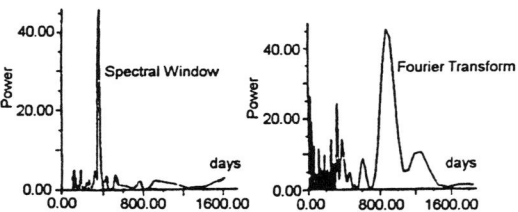

Figure 3. Fourier Transform and Spectral Window - french set

the signal in both cases is 0.25", so the signal is similar, in amplitude and phase (Figure 2 and 3). The main difference between the two time series does not lie in the thecnique but rather in the geographic location of the observatories. A second difference could be the heliographic latitudes of the apparent radius measured at the two sities; this fact could suggest a phenomenon of radial pulsation.

Acknowledgements

This research was supported by FAPESP (grants n° 92/3762-0 and n° 95/0723-2)

References

Leister, N.V. (1989) Orientação do sistema de referência. Observações do Sol com o astrolábio de Valinhos, *Ph.D.Thesis*, IAGUSP.
Poppe, P.C.R. (1994) Análise das observações do Sol com o astrolábio solar no período de 1988-1992, *Msc. Thesis*, IAGUSP.
Laclare, F. (1983) Measures du diamètre solaire à l'astrolabe, *A&A* **125**, pp. 200-203
Leister, N.V. and Benevides-Soares, P. (1990) Variations du diamétre solaire, *C.R.Acad.Sci. Paris* **311**, pp. 399-404

THREE CAMPAIGNS OF SOLAR OBSERVATIONS WITH AN ASTROLABE AT SIMEIZ

YU. B. KOLESNIK
*Institute for Astronomy of the Russian Academy of Sciences,
109017,48 Piatnitskaya St.,Moscow, Russia*

Positional observations of the Sun have become, in recent years, one of the most important contribution of astrolabes to fundamental astrometry. After pioneer observations at CERGA and Sao Paulo, both in 1974, other astrolabes have been adapted for observations of the Sun in Paris (now at Malatya, Turkey), Santiago de Chile (Chollet & Noël 1993) and San-Fernando (Sánchez et al. 1993,1995). First experimental campaign of solar observations with an astrolabe installed at Simeiz Observatory (Crimea, Ukraine) has been undertaken in 1986 (Kolesnik 1987). After some instrumental improvements, observations covering about 2.5-3 months were continued in 1987, 1990, 1991. The results are reported here.

Classical Danjon astrolabe OPL-23 equipped with an equilateral 60° transparent glass prism has been adapted for observations of the Sun. An attenuating 20 mm thick chrome coated quartz filter was mounted at the capote of the astrolabe. The angle of the equilateral prism was monitored by autocollimation with an uncertainty of about 0.06″. The classical technique of observations and reduction first developed in CERGA (Chollet & Laclare 1977) has been applied. Corrections to adopted mean longitude and latitude of an astrolabe linking observations of the Sun to a reference catalogue were derived from night-time observations of stars. Actually, all reductions have been performed in the IAU 1976,1980 system of constants with the FK5 as a reference catalogue and DE200 as an ephemeris of comparison.

The latitude of Simeiz (44° 24′ 12″) allows only 3-month campaigns of solar observations each year from May to beginning of the August. Statistical summary of 3 campaigns in 1987, 1990, 1991 is given in Table 1.

Comparison of the random and systematic accuracy of Simeiz observations in right ascension with 10 meridian instruments has been made by Kolesnik (1995). Here, using the same method, they are compared with the respective accuracies of other active astrolabes in this field, at CERGA,

TABLE 1. Number of observations (NO) and the mean standard errors of unknowns $\Delta\alpha, Y$ and Δd for the three campaigns of solar observations with the Simeiz astrolabe

Year	NO	$\sigma(\Delta\alpha)$	$\sigma(Y)$	$\sigma(\Delta d)$
1987	19	0.043^s	$0.32''$	$0.34''$
1990	31	0.039^s	$0.30''$	$0.30''$
1991	20	0.026^s	$0.19''$	$0.19''$

TABLE 2. Internal ε_i^2 and external ε_e^2 standard deviations of solar observations in right ascension made with Simeiz, CERGA, San-Fernando and Santiago astrolabes

Instrument	NO	Years	ε_e^2	ε_i^2	$\varepsilon_i^2/\varepsilon_e^2$
San-Fernando	61	91.3-92.7	0.056^s	0.048^s	0.74
Santiago	155	90.3-93.0	0.051^s	0.046^s	0.81
CERGA	687	76.3-89.0	0.041^s	0.040^s	0.97
Simeiz	69	87.5-91.6	0.042^s	0.040^s	0.89

San-Fernando and Santiago (see Table 2). Ratio $\varepsilon_i^2/\varepsilon_e^2$ gives an idea of the systematic deviation of the a series with respect to the mean instrumental system formed with 15 meridian and astrolabe series.

As follows from table 3, random accuracy of all astrolabe observations is around $0.040^s - 0.050^s$. This is comparable with errors of the best transit circles (see Kolesnik 1995). It may also be concluded that the random and systematic accuracy achieved by the Simeiz astrolabe is comparable with that provided by the best optical instruments which observe now the Sun.

Acknowledgements

The financial support from BDL and Russian Foundation for Fundamental Research is gratefully acknowleged

References

Chollet F., Laclare F., 1977, A & A 56, 207
Chollet F., Noël F., 1993, A & A 276, 655
Kolesnik Yu.B., 1987, Astron. Circ. 1509
Kolesnik Yu.B., 1995, A & A 294, 876
Sánchez M., Moreno F., Parra F., Soler M., 1993, A & A 280, 324
Sánchez M., Parra F., Soler M., Soto M, 1995, A & AS 110, 351

INDEX OF AUTHORS

Akim, E. L., 497
Alexeev, V. A., 497
Altunin, V. I., 497
Andrienko, D. A., 221
Arlot, J.-E., 1, 145, 423
Assafin, M., 419
Bange, J.-F., 447
Barkin, Yu. V., 243
Bashkirov, A. G., 327
Batllo, V., 213
Bec-Borsenberger, A., 447
Berthier, J., 423
Bougeard, M.-L., 447
Bretagnon, P., 17
Brumberg, V. A., 89, 101
Buontempo, M. E. 399
Burns, J. A., 229
Bykov, O. P., 451
Carpino, M., 203
Cattaneo, L., 193
Cavelier, C., 357
Champenois, S., 143
Chandler, J. F., 105
Chernetenko, Yu., 353
Chuichenko, O. E., 407
Colas, F., 423
Coma, J. C., 345
Damljanović, G., 489
Débarbat, S., 339
Déprit, D., 267
Descamps, P., 145, 423
Donnison, J. R., 53
Duncan, M. J., 229
Duriez, L., 117, 143
Dvorak, R., 71
Emelianov, N. V., 355
Emilio, M., 501
Erdi, B., 171
Eubanks, T. M., 497
Feissel, M., 455

Ferrandiz, J. M., 233, 243
Ferraz-Mello, S., 1, 177
Floria, L., 299
Froeschlé, C., 293
Fukushima, T., 245, 461
Gambi, J. M., 325
Garcia del Pino, M. L., 325
Getino, J., 233, 243
Ghil, M., 57
Giorgilli, A., 293
Gladman, B. J., 229
Gomes, R. S., 223
Goncharov, G. A., 435
Grebenikov, E. A., 285
Hadjidemetriou, J. D., 255
Hadjifotinou, K. G., 151
Harper, D., 151
Hestroffer, D., 437
Ivanova, T. V., 283
Ivanova, V., 251
Jedicke, R., 389
Jones, D. H. P., 431
Kaula, W. M., 57
Kharin, A. S., 407
Kingham, K. A., 497
Kinoshita, H., 61
Kirsanov, N. O., 137
Kisseleva, T. P., 427
Klačka, J., 215
Klioner, S. A., 101, 309
Knežević, Z., 203
Kolesnik, Yu. B., 407, 477, 505
Kovalevsky, J., 455
Kristensen, L. K., 443
Kuzmanoski, M., 207
Laclare, F., 501
Laques, P., 423
Lara, M., 345
Laskar, J., 75
Lecacheux, J., 423

Lega, E., 293
Leister, N. V., 501
Lemaitre, A. 165
Levison, H., 229
Lohinger , E., 71
Lopez, J. A., 199
Lopez Garcia, A., 483, 487
López Moratalla, T. J., 345
L'vov, V., 353
Marco, F. J., 199
Martinez, M. J., 199, 487
Marsden, B. G., 153
Message, P. J., 127
Miguelote, A. Y., 223
Mikkola, S., 209
Mikulskis, D. F., 53
Milani, A., 193
Mishchishina, I. I., 221
Molina, R., 249
Morando, B., 3, 423, 437
Morbidelli, A., 293
Morrison, L. V.,399
Nakai, H., 61
Newhall, X X, 29, 37
Ostro, S. J., 365
Pang, K. D., 113
Pascu, D., 373
Pitjeva, E. V., 45
Pittich, E. M., 187, 215
Poppe, P. C. R., 501
Qiao, R., 141
Ricklefs, R. L., 409
Rickman, H., 209
Ries, J. G., 409
Rocher, P., 357
Romero , P., 325
Röser, S., 481
Ruatti, Ch., 423
Ryabov, Y. A., 289
Sansaturio, M. E., 193
Scholl, H. 183
Scotti, J. V., 389

Seidelmann, P. K., 331
Shelus, P. J., 409
Shen, K. X., 141
Shevchenko, I. I., 183
Shkodrov, V., 251
Shor, V., 353
Shuygina, N. V., 469
Simon, J. L., 49
Smekhacheva, R., 353
Soffel, M. H., 303
Solovaya, N. A., 187
Souchay, J., 239
Standish Jr., E. M., 29, 37
Sukhanov, G.,497
Sveshnikov, A. M., 453
Sveshnikov, M. L., 453
Thuillot, W., 423
Toulmonde, M., 361
Tsekmejster, S., 353
Treuhaftk, R. N., 497
Trubitsina, A. A., 347
Valtonen, M. J., 209
Varadi, F., 57
Vasundhara, R., 145
Veiga, C. H., 419, 423
Vieira Martins, R., 419
Vienne, A., 143
Vigueras, A., 249
Vityazev, A. V., 327
Vokrouhlický, D., 321
Vondrak, J., 491
Vu, D. T., 423
Walter, H. G., 475
Whipple, A. L., 409
Wiant, J. R., 409
Williams, J. G., 37
Yagudin, L. I., 483, 487
Yagudina, E. I,. 469
Yau, K. K., 113
Yershov, V. N., 415
Zamorano, P., 325
Zheng, J. Q., 209